T0406381

Climate Change Management

Series editor

Walter Leal Filho, Faculty of Life Sciences, Research and Transfer Centre,
Hamburg University of Applied Sciences, Hamburg, Germany

The aim of this book series is to provide an authoritative source of information on climate change management, with an emphasis on projects, case studies and practical initiatives—all of which may help to address a problem with a global scope, but the impacts of which are mostly local. As the world actively seeks ways to cope with the effects of climate change and global warming, such as floods, droughts, rising sea levels and landscape changes, there is a vital need for reliable information and data to support the efforts pursued by local governments, NGOs and other organizations to address the problems associated with climate change. This series welcomes monographs and contributed volumes written for an academic and professional audience, as well as peer-reviewed conference proceedings. Relevant topics include but are not limited to water conservation, disaster prevention and management, and agriculture, as well as regional studies and documentation of trends. Thanks to its interdisciplinary focus, the series aims to concretely contribute to a better understanding of the state-of-the-art of climate change adaptation, and of the tools with which it can be implemented on the ground.

More information about this series at http://www.springer.com/series/8740

Paula Castro · Anabela Marisa Azul
Walter Leal Filho · Ulisses M. Azeiteiro
Editors

Climate Change-Resilient Agriculture and Agroforestry

Ecosystem Services and Sustainability

Editors
Paula Castro
CFE - Centre for Functional Ecology - Science for People & the Planet, Department of Life Sciences
University of Coimbra
Coimbra, Portugal

Anabela Marisa Azul
Center for Neuroscience and Cell Biology
University of Coimbra
Coimbra, Portugal

Walter Leal Filho
Faculty of Life Sciences
HAW Hamburg
Hamburg, Germany

Ulisses M. Azeiteiro
Department of Biology and CESAM, Centre for Environmental and Marine Studies
University of Aveiro
Aveiro, Portugal

ISSN 1610-2002 ISSN 1610-2010 (electronic)
Climate Change Management
ISBN 978-3-319-75003-3 ISBN 978-3-319-75004-0 (eBook)
https://doi.org/10.1007/978-3-319-75004-0

Library of Congress Control Number: 2018961713

This Springer imprint is published by the registered company Springer Nature Switzerland AG
The registered company address is: Gewerbestrasse 11, 6330 Cham, Switzerland

Preface

Climate change and its impacts on agriculture and on agroforestry have been observed across the world during the last 50 years. They have been threatening the stability in ecosystems and people well-being. Increasing temperatures, droughts and biotic stresses and the impacts of extreme events have continuously decreased agroforestry systems' resilience to climate change. As it is now, we need a new vision for research, management, education and learning, since they are important drivers in achieving the Sustainable Development Goals as a whole, and SDG 13 (climate action) in particular.

There is a perceived need to adapt farming and agroforestry systems so as to make them better able to handle ever-changing climate conditions, and to preserve habitats and ecosystems services. More efficient management practices and new/innovative agroforestry solutions are required and must incorporate the regional and local abiotic factors of climate, soil, water and nutrient balances as well as the biotic conditions (e.g. pests, diseases and dispersal agents). Also fostering activities related to traditional culture and improving education and learning on biodiversity and ecosystem services are decisive as they are stepping stones for sustainable development supporting ecosystem's conservation, livelihood and sustenance of populations.

It is based on the need to tackle this topic that this book has been produced. It aims at assembling wide-ranging contributions from case studies, reviews, reports on technological developments, outputs of research/studies and examples of successful projects, as well conceptual approaches, which document current knowledge, raise awareness and help the agriculture and forest sectors to adapt to climate change as it brings the theme ecosystems' services closer to education and learning, as an added value to strategic principles for healthy and valued ecosystems and sustainable human development.

The book entails contributions in a variety of areas, including ecosystem services and incentive mechanisms for environmental preservation, unlocking the social–ecological resilience of High Nature Value farmlands to future climate change, climate-smart agricultural practices (CSA) adoption by crop farmers in semi-arid regions, future climate change impacts on agriculture based on multi-model results from WCRP's CMIP5 and the urgent need for enhancing forest ecosystem resilience under the anticipated climate portfolio.

The book contains also papers addressing the issue of sustainable food systems in culturally coherent social contexts, and multifunctional urban agriculture and agroforestry for sustainable land use planning in the context of climate change. All in all, an interesting set of papers gathering information and knowledge which outline the potentials and environmental risks related to agricultural and agroforestry landscapes under a changing climate.

We hope that this publication will prove useful to all those working in the field of climate change as it relates to agriculture and agroforestry, and that it may catalyse further initiatives in this important field.

Coimbra, Portugal Paula Castro
Coimbra, Portugal Anabela Marisa Azul
Hamburg, Germany Walter Leal Filho
Aveiro, Portugal Ulisses M. Azeiteiro

Contents

Terraced Agroforestry Systems in West Anti-Atlas (Morocco): Incidence of Climate Change and Prospects for Sustainable Development

Mohamed Ziyadi, Abdallah Dahbi, Abderahmane Aitlhaj, Abdeltif El Ouahrani, Abdelhadi El Ouahidi and Hafid Achtak

Abstract The Moroccan Anti-Atlas region contains all the "ingredients" of a hostile environment including an arid climate, a highly rugged topography, a low vegetation cover due to insufficient rainfall, and inexorable soil erosion. These harsh conditions incited local peasants to adopt simple but ingenious agricultural practices that fit the prevailing rigours and ensure their livelihood survival: Terraced Agroforestry System (TAS). TAS, one of the most ancestral agricultural practices, becomes a dominant feature of the Anti-Atlas landscape. This study aims to explore the Anti-Atlas TAS as a resilient approach to counter climate change impacts and ensure a sustainable development of this region. To this end, a prospective study was conducted to survey the indigenous peasants, to assess the status of TAS, to describe its biodiversity trends, and ultimately to ensure its sustainable development. The primary results revealed that the Anti-Atlas TAS are based essentially on the Argan tree (*Argania spinosa* L.) as the predominant vegetation crown layer. Accordingly, goats represented the main integrated livestock. The related annual crops are mainly represented by local varieties of cereals and legumes. Other dryland fruit trees, such as almond, fig, olive, and date palm are also sparsely planted. Beyond their purely aesthetic and economic role, this agro-cultural heritage contributes greatly to the conservation of several local varieties and their associated fauna. Furthermore, the results allows us to identify some serious climatic and

M. Ziyadi · A. El Ouahidi
Department of Geography, Polydisciplinary Faculty of Safi,
Cadi Ayyad University, Safi, Morocco

A. Dahbi · H. Achtak (✉)
Department of Biology, Polydisciplinary Faculty of Safi,
Cadi Ayyad University, Safi, Morocco
e-mail: achtak@gmail.com

A. Aitlhaj
National Agency for Development of Oasis
and Argane Areas (ANDZOA), Agadir, Morocco

A. El Ouahrani
Department of Biology, Faculty of Sciences-Tetouan,
Abdelamlek Essaadi University, Tétouan, Morocco

P. Castro et al. (eds.), *Climate Change-Resilient Agriculture and Agroforestry*,
Climate Change Management, https://doi.org/10.1007/978-3-319-75004-0_1

social challenges faced by the persisting TAS in the Anti-Atlas region. In this regards, the regional climate change scenarios predict warmer and dryer conditions over the studied region, meanwhile the new generation of local peasants increasingly lacks interest to maintain TAS and prefers to seek new opportunities in the Souss-Massa plain valley. Consequently, this paper investigates major issues threatened social, economic, and ecological balances and provides a combination of adaptation solutions to help revive the agro-cultural heritage of TAS from the process of extinction.

Keywords Terraced agroforestry systems · Anti-Atlas · Morocco
Sustainable development · Climate change

1 Introduction

The Kingdom of Morocco is an African-Mediterranean country where agriculture sector, mostly rain-fed dependent, represents more than 15% of its GDP and employs about 4 millions of its active population (ADA 2009; RMS 2011; El Bilali et al. 2012). Despite the climatic hazards striking Morocco, including drought, flood, and storms, the key challenges facing the country today is its ability to ensure adequate food security and to shift its economy to low-carbon energy by investing in renewable alternatives. This vision was well reflected in the 22nd Conference of the Parties (COP22) held in Marrakech on November 2016, where Morocco launched the triple A initiative (Adaptation of African Agriculture) to stress the high impact of climate change on its economy tightly linked to climate variability.

Aware of this critical situation, and based on the national strategy for sustainable development, agriculture and food security remain among the key strategic sectors targeted by Morocco. Since 2008, Morocco has developed the "Morocco Green Plan" (MGP) which aims to modernize the agriculture sector to become more competitive and more integrated in the global market by creating wealth over the entire value chain, meanwhile considering human, social, and territorial development. The MGP promotes sustainable management of natural resources and identify necessary policies to support sustainable growth. Although there were specific programs of MGP which particularly targeted small scale farmers in remote areas, the Moroccan Anti-Atlas villages, where survival is largely depending on rainfed agriculture and pastoralism, are unable to fully benefit from MGP.

The Moroccan Anti-Atlas, 18,000 km^2 (Gunn 2004), is a mountainous chain with all the "ingredients" of a hostile environment including an arid climate, a highly rugged topography, a low vegetation cover, and a non-stop soil erosion. This region is by far the most vulnerable to climate change and therefore the most threatened in terms of sustainable livelihood assets in Morocco. Facing these harsh conditions, local peasants have adopted simple but effective agricultural practices that fit the prevailing rigours: Terraced Agroforestry System (TAS), which become a major component of the Anti-Atlas landscape and the main farming system.

TAS are certainly more adequate and most effective practice for soil quality and stability, for landscape preservation, and for biodiversity conservation in similar agro-ecosystems. Thought the aforementioned advantages of TAS, its overall situation seems deteriorating and its maintenance is abandoned in the Anti-Atlas region which put in danger all the potential roles played by these traditional agro-systems to maintain livelihood and survival of the local peasants. In this context, the aim of this study is to explore the human and the physical complexities of the Anti-Atlas that determine the future of TAS with specific focus on the province of *Chtouka Aït Baha*. The ultimate goal is to promote TAS as a resilient approach to counter climate change incidences and prospecting for best fit sustainable development model for this region.

2 Methodology

In order to explore the TAS of the western Anti-Atlas region, a prospective investigation was conducted in the province of *Chtouka Ait Bah*a with focus on rural communities in the mountains districts (Fig. 1). A multidimensional approach including the literature review to describe the state of the art and the evolutionary trends of these TAS, the social and the human dimensions via surveying elders of the communities and interviewing local groups. The analysis of cartographical data, and eventually the analysis of past and future climate change of this region, were also conducted.

2.1 Bibliographical Information for the State of the Art

A state of the art is reviewed based on available information and publications (reports, monograph, papers …) about the study area. A synthesis was developed in order to provide a thoughtful description of the study area including geology, geography, demography, infrastructure, land cover, climate, water resources and agriculture.

2.2 Survey and Data Analysis

Qualitative and quantitative data were collected from local population through semi-structured surveys. During two years, at least ten transects in the study area were carried out to survey individuals and within focus groups at least 100 individuals from different rural communities were questioned. We mostly targeted elderly farmers but also young people. The questionnaire was designed to explore information about the history of TAS, the farming system, the agro-diversity

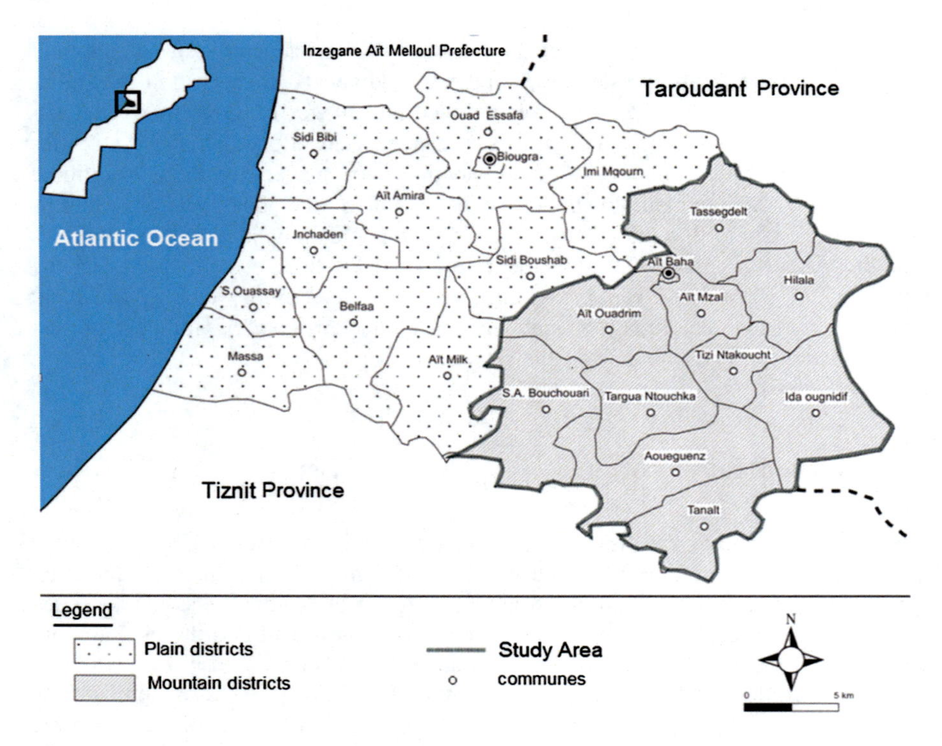

Fig. 1 Study area: the province of *Chtouka Ait Baha*

assessment, crop selection, and peasant perception of climate change, and possible adaptation solutions. These surveys allowed us to explore the trends of this traditional agroforestry practice in terraces.

2.3 Cartographic Data

The study area map and geographical information data were collected and developed using the MapInfo 12.5 software.

2.4 Analysis of Past and Future Climate Data

Past and future climate data were retrieved and analysed using DivaGIS 7.5 and SPSS 17.0. Drought incidence for Morocco was estimated using Palmer Drought Severity Index (PDSI) which is basically a proxy data based on cedar tree rings width covering a period of 1049–2001 (data source: Esper et al. 2007). The long term change in climate type over the study area was analysed using the Köppen-Geiger climate type classification change (Kottek et al. 2006). Observed and projected data covering two centuries (1901–2100) were retrieved from the World Maps of Köppen-Geiger climate classification then projected over the study area using DivaGIS software. The IPCC business as usual (BAU) scenario (A1FI) was adopted to estimate the worst climate change scenario until 2100 (IPCC 2007).

3 Results and Discussions

3.1 Geological Features of the Study Area

The Moroccan Anti-Atlas, a very old geological formation, is mountainous ranges with very steep slopes and ridges that culminate up to 2374 m above sea level in '*Jbel Lkest*'. The western part, where the famous '*Kerdous*' inlier, is dominated by the Precambrian (I) with schist, mica schist, and granite followed by the Precambrian (II) which provides a spectacular relief formed by a quartzitic cover of the '*Jbel Lkest*'. The whole is surmounted by a series of calcareous and dolomite sedimentary coverts which form the main mass of the Anti-Atlas. Epirogenic seizures followed with oscillating marine transgressions that only sporadically reach the mountain mass, causing large sedimentary gaps in its coverings series (Choubert 1963; Choubert and Faure-Muret 1972). The study area (Fig. 1) is located on this western part of Anti-Atlas (30° north latitude and −9° west longitude), fairly representative of the Moroccan Anti-Atlas. However, for a better observation of the agrarian systems, the study was focused on its Mountain districts slopes covering an area of approximately 3523 km^2.

3.2 Human Footprint and Demographic Data

Due to its abrupt and steep relief, and its hostile and uncertain climate, this region gives the sense of a harsh and inhospitable countryside. But as one go through the massif, the anthropogenic footprint appears omnipresent marking the landscapes. Indeed, the total population of these mountainous areas has been fluctuated throughout the last century. After being quite large along the three first quarters of the 20th century (Podeur 1995), the total population underwent a steady decline.

Though this observation contrasts oddly with the peasants' secular tying-up to their land, different settlements "*douars*" have lost more than the third of their population in just few decades. This continuous migration is clearly shown through contemporary censuses (1994, 2004, and 2014) (Table 1). The peasants' younger generation becoming more and more attracted to better jobs and opportunities offered elsewhere.

Table 1 Demographic evolution according to the last three consecutive censuses, in Mountain and Plain communes of *Chtouka Ait Baha* province of the Anti-Atlas

Communes	Census 1994		Census 2004		Census 2014		GR
	CTP	D	CTP	D	CTP	D	
Communes in mountain (n = 11)	63,281	148,47	57,189	179,82	47,157	201,81	**−1.63**
Communes in plain (n = 12)	176,811	217,19	240,056	344,58	323,945	482,42	**1.96**

Source Adapted from Haut Commissariat au Plan
CTP Commune total population, *D* Commune mean density, *GR* Growth rate during 20 years (1994–2014)

3.3 *Terraced Agroforestry Systems*

3.3.1 Learning from History

For many millennia, faced with abrupt relief and the scarcity of flat surfaces suitable to agriculture, many rustic techniques have been developed through peasants' generations in order to "overcome" the ungrateful environment. The slopes were completely transformed into innumerable floors, a kind of staircase of small terraced plots, conducive to a more rational use of soil and water resources. The use of these terraces goes back to a long history. According to most historians, this type of sloppy land management for soil conservation and water resources optimization would go back more than 1000 to 500 years before Christian era (Harfouche 2003). While for some others, such as Despois (1956), they were derived from the much older *'Amazigh'* civilization as the Anti-Atlas remained totally outside the area of Roman occupation.

3.3.2 Associated Management

Regularly restored and maintained, these terraced plots have not lost their authenticity. They remain perfectly arranged according to altitude contour lines and supported by walls made with dry stones of variable geological nature, locally called '*Igherman*'. This term evokes a common concept that embraces protection, resistance, and reinforcement. TAS is associated with multitude constructions which can be described as ancillary developments derived from the original topography of the land. Among these developments that can be observed while travelling across these mountains, the threshing floor locally called '*Inraren*' which testify the past cereal mass production in Anti-Atlas. Others constructions have in common the use of slope to harvest rainwater such as those many underground cisterns called '*Tanodfi*' that are found almost everywhere near the dwellings (Aziz et al. 2014). These ancillary constructions were built in the same way as the terraced fields: the peasants used the materials extracted on site and followed the same construction rules as the case of *Igherman*. Another associated structural development that can also be considered as complementary, since it is closely linked to agricultural terraces, is the famous fortified granaries named '*Igoudar*' built on the tops of the hills, where peasants store their cereal supplies that have been produced on the terraced fields. Certainly, this type of management is found elsewhere, but those of the Anti-Atlas have their own peculiarities.

3.3.3 TAS: An Adapted Food Security Management in Anti-Atlas

The slopes of the Anti-Atlas, particularly those exposed to north-west influence, possess a well-diversified natural vegetation cover dominated by the endemic *Argania spinosa* "Argan tree". This emblematic tree constitutes by far the major component of vegetation land cover under which lays not only agricultural activity in terraces but also other plants and animal species are intimately associated. It represents, along with other fruit trees, the basic natural resources for the dwellers. However, above 1500 m of altitude, and under the effect of low temperatures and high humidity, emerge other natural trees such as the green oak (*Quercus ilex* L.), more adapted to these extreme conditions (El Aboudi 1990; Peltier 1982).

To convene their needs in food, the peasants of Anti-Atlas have sought to make better use of the small land plots available by applying the concept of agroforestry. This concept consisted in associating seasonal crops such as cereals and legumes, and perennials ones including almond (*Prunus dulcis*), prickly pears (*Opuntia ficus-indica)*, and figs (*Ficus carica*). In addition to natural vegetation cover already spruce in place, the implementation of this type of agroforestry systems and the stable quantitative but also qualitative yield associated (Olivier et al. 2015) have largely contributed to the sedentariness of the local populations, which have largely focused their activities on agriculture and livestock breeding.

3.3.4 Advantages Beyond Food Security

Indeed, and beyond ensuring subsistence farming for peasants, agroforestry in terraces have some great advantages and positive repercussions at various levels:

Agronomic attributes: Soil erosion is a permanent concern for Anti-Atlas farmers. All surveyed farmers do not value their land unless it retains arable soil. The terraced fields are traditional but genius systems that effectively retain soil vitality. The presence of trees with rigorous root is an added value to limit soil erosion.
Water management: Agroforestry systems in terraces contribute not only to regulate rainwater flows (Noorka and Heslop-Harrison 2014), but also and above all prevent the siltation of small water reservoirs (Alahiane et al. 2016). At field level, TAS optimize irrigation system on numerous small flat surfaces using the dew water captured by the leaves of fruit trees and retaining by the walls, these systems retain and recycle significant amount of water resources. Down the valley, the

terraces which are closer to the water sources and stream paths benefit from a fairly regular irrigation. However those in high altitudes depend exclusively on rainfall. **Ecosystem sustainability**: Agroforestry systems in terraces play multiple eco-systemic roles which go beyond a food production. These roles includes a number of important ecological benefits such as carbon sequestration, the creation of a microclimate that dampens thermal shocks, and the preservation of biodiversity (flora & fauna). Soils of TAS retain more water and are richer in micro-organisms and other invertebrates which enhance soil fertility. A wide range of animal and plant species are tightly associated with TAS. For instance, the interstices created at the level of *Igherman* shelter a multitude of very specific flora and fauna perfectly contribute to the ecological balances of these terraces. A non-exhaustive species list of Anti-Atlas TAS includes at least 44 endemic plants species and another 27 non-endemic, 9 mammal species, including the Barbary ground squirrel (*Atlantoxerus getulus* L.), 7 bird species, and 14 reptile species (source: Centre for Information Exchange on Biodiversity in Morocco). In addition, they constitute an ideal site for the development of an important Arthropod fauna, like spiders and *Hymenoptera*, mainly ant species, which constitute key actors in the biological control of various phytophagous that threaten crops. Thus, similarly to the concept of biodiversity "hotspot" areas (Myers et al. 2000), which identify priority areas for wildlife conservation, the traditional agroforestry areas in the Anti-Atlas deserve to carry such a designation, according to their high and specific associated biodiversity and landraces (Achtak et al. 2010; Brush 1995).
Socio-economic assets: Agroforestry systems in terraces are true mountain foci which concentrate and retain an important agrobiodiversity. Indeed, in the middle of Anti-Atlas, they form by excellence a kind of reservoir/refuge for traditional crops and very old seeds, genetically adapted to local conditions. These plant genetic resources, made up of varieties selected by traditional knowledge and maintained throughout the generations, constitute a "gene bank" available in situ in these systems. Moreover, the charred remains of wood, seeds and various fruits discovered on the site of *Igîlîz* by Ruas et al. (2015), are undeniably witness of great diversity of plant species that have been able to thrive terroirs (a unique micro-climate and traditional farming practices that affect a crop's phenotype) in terraces for centuries (96 plants of which 18 are cultivated).

In this context, our present investigation, far from being exhaustive, has nevertheless enabled us to detect the existence of well-adapted plant species of great socio-economic importance for the sedentary populations. Among these, and beyond Argan tree which dominate the arboreal stratum, almond and fig represent a second important source of income for peasants, then consecutively followed by the preckly pears, and Carob tree (*Ceratonia siliqua* L.) which is scattered all over, but with low density, green oak and Atlas pistache (*Pistacia atlantica*) in the highest altitudes. Medicinal plants such as lavender (*Lavandula stoechas* L.), cistus (*Cistus villosus* and *C. salviifolius*), cytise (*Teline segonnei*), mugwort (*Artemisia herba-alba*) and thyme (*Thymus satureioides*), continue to decorate, here and there, the cultivated terraces. For obvious reasons related to local knowledge and developed through centuries by the population in these areas, the medicinal virtues of these plants are keenly sought and used by the peasant population. This have recently evolved as a good income opportunity for women through the cooperative network, taking advantage of the renewed interest for natural products. This great diversity of plants also includes a melliferous flora, representing the basic variety of honey with a very high gustatory and medicinal quality and firmly associated with the sub-species of North African bees, *Apis mellifera sahariensis*. Less and less flourishing, its production in traditional apiaries according to secular methods and commercialization still provides a quite important income for numerous local households.

3.4 Terraced Agroforestry System: Significant Regression Trend in the Anti-Atlas

Given the many advantages associated with the practice of agroforestry in terraces, one should expect a prosperous state of these systems after a constructive repercussion of local population guided by their necessity to survive in a harsh environment. Yet, paradoxically, our investigations, although over a few limited slopes of the Anti-Atlas, reflect a clear regression of this practice. This regression affects both the quality and the number of terraces exploited. The diversity of species and plant varieties is greatly affected. They are gradually regressing and are experiencing changes both in species composition and in cultivated varieties. The number of fruit trees, for instance, and in comparison with agrosystems located in the Rif Mountains (northern Morocco) where the number of varieties exceeds a hundred, remains much lower in the Anti-Atlas slopes. A similar trend is also observed in cereals and legumes, which have always been important for food and feedstuffs. Considering cereals, for example, only three species are still cultivated (barley, wheat, and maize), often with only one variety per species.

The majority of the questioned peasants (only the elderly were considered), recognized a drastic decrease in numbers of cultivated species and local varieties. They agreed on the responsibility of climatic factor, and linked such a regressive trend with inadequate water resources and drought which is increasingly thump this region. However, down the valley, along the water streams and near natural water springs, several species persist. But the few streams that feed an extensive regional hydrographic network are largely dependent on rainfall irregularities. Increasingly dry, these streams only run for a few months, from November through April, sometimes with short but destructive and violent floods. In these conditions, species and traditional varieties, which have been selected for generations, and which are sometimes peculiar to the region, gradually give a lead to new varieties which are indeed more productive and therefore economically profitable but less relevant in many aspects. Consequently, a good number of terraced plots have been totally abandoned or at least poorly maintained and are no longer the centre of interest for "young" peasants (Fig. 2).

Fig. 2 An illustration of maintained (**a**) and abandoned (**b**) TAS

In reality, it would be fairly simplistic to attribute all these reasons to climate change alone, the problem seems to be more complex. To draw a more realistic and comprehensive view of the issue, it is wise to estimate the relative share inherent in other factors. In the case of Anti-Atlas region, the socio-economic and socio-cultural factors play significant role as well. The sharp decline in agroforestry practice among the Anti-Atlas farmers is partially due to their passive behaviour and incapability to endorse new adaptive techniques.

Throughout the survey, the primary impression is a dominant sense of carelessness of farmers towards the seriousness of the situation. But, after in-depth analysis, one realizes the rational and the purely economic reasons that explain the situation of TAS in Anti-Atlas. The high production costs combined with high risks due to rainfall irregularity have often been reported as the main reason by the surveyed farmers. Wild boar (*Sus scrofa*) visits become more and more frequent causing costly damages to the crops. Increasingly, these undesirable visits of a purely wildlife, in themselves translate the beginning of ecological unbalances that are not unrelated to climate change. On the other hand, and unlike farmers in other similar regions such as in the Rif Mountains (north of Morocco), those from our prospecting area seem to be investing less and less in improving agricultural practices and maintaining agricultural terraces. According to our observations, the cultivation and the maintenance of the installations and the arboriculture are limited to the irrigated terroirs and the surroundings of the farmers' houses. The further away one gets, the degradation of terraced fields becomes evident and very pronounced. Moreover, farmers use a very detrimental solution by overexploitation of natural forest through overgrazing and medicinal plants harvesting.

The traditions and the agricultural know-how of the sedentary populations outlined an important part of the cultural heritage accumulated and transmitted through peasants' generations. It is noted that women play major role in this transmission. The emotional connection that unites farmers to their land constitutes an indispensable prerequisite for the survival and the perpetuation of this traditional agroforestry system. Yet, there is no need to highlight that this link becomes more and more eroded and generally threatens the survival of these ecosystems.

Rationally destined to take over, the younger generation is facing the dilemma of "perpetuating the agricultural heritage versus significant decline in yield and future uncertainties".

This particular population aspires to a better future and look forward to explore other prosperous opportunities. The vast valley of *Souss-Massa* attracts most workforces from various regions, including from villages of the Anti-Atlas. This attraction contributes widely to the desertion of the Anti-Atlas agroforestry ecosystem.

3.5 Drought Incidence in Morocco and Anti-Atlas Region

There are many definition of drought, but all of them usually resulted from an accumulated water deficiency over a period of time (Natsagdorj 2012; Palmer 1965; Wilhite and Glantz 1985). As a Mediterranean country, Morocco is susceptible to alternation of drought hazards. A proxy data from Cedar tree rings (*Cedrus atlantica*) show clear intermittent of dry/wet regimes covering a period of 20–25 years. Two dry patterns periods were revealed; 1860–1900 and 1925–1950 (Chbouki et al. 1995). Based on these facts, Morocco is likely to experience a dry period between 2025 and 2050. For instance, an assessment of drought hazard near Rabat based on annual rainfall anomaly analysis identified 20 drought incidences over the period 1951–1997 (Natsagdorj 2012). To illustrate the consistency of these results, we will investigate the drought incidence over Morocco using the Palmer Drought Severity Index, and the Köppen-Geiger climate classification over Anti-Atlas region.

3.5.1 Palmer Drought Severity Index (PDSI)

Palmer Drought Severity Index (PDSI) for Morocco was calculated based on Cedar tree rings width data (Esper et al. 2007, 2009), which allows capturing long-term changes in PDSI over the period 1049–2001. Since 1900, PDSI highlights drier years than wetter years (around 1960s). It shows a remarkable trend in sever (<-3) and extreme (<-5) droughts mainly since 1980s (Fig. 3).

During this period (1900–2001), PDSI ranged from −5.97 to +3.98 and expected to be −3.1 by 2050 and −4.1 by 2100. The correlation between date and PDSI is negatively significant ($r = -0.317$; $p < 0.001$). In other words, Morocco is expecting dryer future ahead.

This expected dryness was also confirmed in Anti-Atlas. Several studies investigated the impact of climate change on argan forest (Charrouf and Guillaume 2009; Kenny and De Zborowski 2007; Le Polain de Waroux and Lambini 2012; Morton and Voss 1987). The results predicted noticeable deficiency of an annual precipitation and prolonged drought periods which put extra risk toward the extinction of argan forest. Certainly, the area of argan forest in Anti-Atlas has

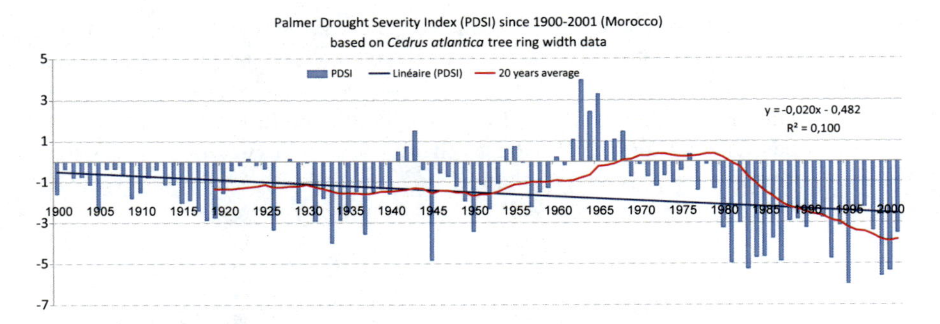

Fig. 3 PDSI covering a period of 1900–2001, based on *Cedrusatlantica* tree ring width data (*data source* Esper et al. 2007)

decreased by 44.5% between 1970 and 2007. This deforestation is consensually believed to be largely caused by the increasing frequency of drought incidences.

3.5.2 The Köppen-Geiger Climate Classification

The climate type classification data (Köppen-Geiger) make it possible to reveal a changing climate over a specific region with resolution of 0.5° latitude and longitude. We downscale this data to fit the Anti-Atlas region, specifically over Chtouka Ait-Baha province. The data allows presentation of observed (including proxy data) as well as future estimates based on IPCC scenarios (Kottek et al. 2006). Data, 1901–2100, were retrieved from World Maps of Köppen-Geiger climate classification then imported to DivaGIS software.

The result (Fig. 4) shows three maps representing observed (Ob 1901_1925) and projected (A1FI 2001_2025 and A1FI 2076_2100) data of 25 years each. The legend gathers three letters that follows Köppe-Geiger climate classification which designs the type of dominated climate. The first letter provides an indication about the main climates (here: B refers to arid climate and C to warm temperate climate), the 2nd letter designates the main precipitation regime (here: W refers to desert, S refers to steppe, and miniscule “s” refers to dry summer), and the 3rd letter designates a type of temperature regime (here: k refers to cold arid temperature, and h to hot arid temperature). The legend of Fig. 4 can be read as, BWh: arid climate with deserted environment and hot arid temperature; BSk: arid climate with steppe vegetation cover and cold arid temperature; BSh: arid climate with steppe vegetation cover and hot arid temperature; Csa: warm temperate climate with dry and hot summer; and Csb: warm temperate climate with dry and warmer summer.

From Fig. 4, we can state that the climate type over the study area (white circle) shifted from BSh in the first quarter of the 20th century to BWh in the last quarter of the 21st century. In other words, this northwards shift results mainly from the change in precipitation regime which shifted from a steppe to a desert environment. The precipitation deficiency is expected to be the major determinant to put this

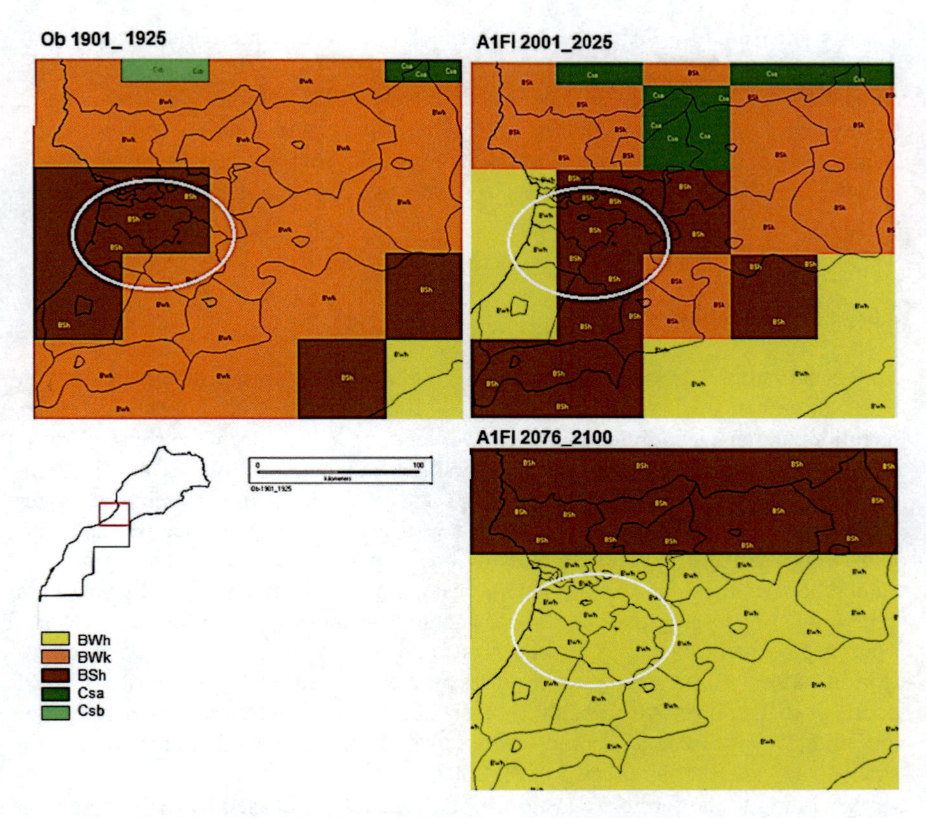

Fig. 4 Observed (Ob) and projected (A1FI) climate type shifts: Ob1901_1925, A1FI2001_2025, and A1FI2075_2100, based on the Köppen-Geiger climate classification world maps, using IPCC fossil intensive (A1FI IPCC) scenario. White circle refers to the study area, more legend embedded in the text

region under high risk of desertification. Therefore, the agroforestry systems are expected to face an arid/desert climate featured with low precipitation and hot temperature in summer.

In brief, the Anti-Atlas region is better to be prepared for more drought incidences if we are going to safeguard this famous agro-biodiversity heritage from total extinction.

3.6 *Prospect for Sustainable Development in Anti-Atlas (*Chtouka Ait Baha*)*

According to our results, prospect for sustainable development in Anti-Atlas can only be though by ensuring food security and prosperity for its communities. Food security is ensured by TAS productivity which in turn linked to climatic conditions (Brown and Funk 2008; David et al. 2008). While prosperity of its community depends on their resilient capacity. Unfortunately, these communities are living at the expense of their surrounding natural resources, and thus are economically fragile and socially vulnerable. The situation is more complex and emanates from a set of factors which can be summarized into two major but non-exclusive categories: climatic and human factors.

3.6.1 Climatic Factors

Mixed agriculture (trees/intercropping) in TAS maintains high quality soils with better microbial resilience (Lacombe et al. 2009; Rivest et al. 2013), high water retention, and enhance mineral uptake (Olivier et al. 2015). Also, the rhizosphere is conducive to the development of a diversified pedofauna (Cromack et al. 1988). Nevertheless, the ongoing warming and the ever-increasing scarcity of water resources put into risk the subsistence of well featured crop species in these systems. Our investigations show that many traditional varieties, known as "*beldi*" (or autochthon), underwent a long selection history and adaptation process by peasant's generations, have progressively or even totally disappeared over the last decades. Their replacement by mono-varietal crops and the ongoing intensive genetic erosion in these agroforestry systems represent a true threat to food security of local population (Esquinas-Alcazar 2005). This situation is unlikely to end here, rather the climatic disturbances could even have an impact in the near future on small-scale cereal crops (Lobell et al. 2008).

According to the farmers questioned, the water stress caused by the dysfunction of the climate is to a large extent no longer allowing a satisfactory yield. Trees, such as argan and almond, were always and will continue to be the backbone element of these traditional agro-systems. The microclimate generated by these trees in TAS is indeed helping the good development of intercropping. However, although trees persist today, they have experienced a significant decrease in density, which exacerbates the already detrimental effects of global warming in this region. Alarming signs have already shown up in the argan forest. During the twentieth century, its area has been reduced by half and, in some places, tree density is 66% lower than it was 50 years ago (Charrouf and Guillaume 2008, 2009; M'Hirit et al. 1998). This severe forest decline was also highlighted using gridded tree counts on aerial photographs and satellite images and was linked mainly to the increasing aridity. However, no effect of grazing by local livestock was found (Le Polain de Waroux and Lambin 2012). Therefore, the expected drought incidences found in our analysis will certainly affect a large wild fauna and flora that is dependent on argan forest agrosystems. The consequences will no longer be limited to merely agricultural shortfall rather it will affect all ecological balances and ecosystem services in a more holistic manner in this region.

3.6.2 Human Factors

Much of these factors represent a straightforward, sometimes obsolete, response to the consequences of climate change. Increasingly overwhelmed by the adverse consequences of drought incidences, farmers were continually forced to seek alternatives to mitigate such effects. Thus, some crops considered more resistant to thermal and water stresses have already substituted old crops, just few decades ago were prosperous in terrace systems. In our opinion, the example of the lentil (*Lens culinaris* L.), virtually disappeared from these systems, is most representative of this upheaval change in terraced crops. This disappearance is not due to a lack of interest for this food; rather it is an actively sought-after crop as an essential ingredient of farmers' everyday meal. Sadly, it is only available in the week markets (souks). Worse still, the local varieties become progressively more dominated by commercial ones, which is less tasty with low gastronomic quality. This impoverishment in the "reservoir" of plant species and genetic resources occurring in the Anti-Atlas agroforestry systems, have certainly led, in just few decades, to a total transformation of the local botanical landscape, with an alarming decline in species richness accompanied with a parallel erosion of traditional genetic resources.

Thus, like other traditional agro-systems (Achtak et al. 2010; Brush 1995; Hmimsa and Ater 2008), those in terraces that formed a sort of hotspot zone for cultivated plants, have gradually lost such a patrimonial characteristic. This situation has indisputably been exacerbated by a total ignorance of climate change issue among farmers and climate-terraces associated particularities. Furthermore, the total absence of any land/crop maintenance using either classical or innovative techniques (e.g.: density and distribution of cultivated plants for better management of root competition over water/mineral intake and light) is rather striking and has hardly resisted this trend of decline in agrobiodiversity.

Agroforestry systems have proved to be the best fit to adapt to climate change. For instance, the Rif agroforestry systems (northern Morocco) are sheltering panoply of traditional plant species despite noticeable but less severe climate change effects (Achtak et al. 2010; Hmimsa and Ater 2008). The study of mountain oases in northern Oman (Gulf) recorded a rich agro-biodiversity with 107 crop species belonging to 39 families, including 33 fruit species (Gebauer et al. 2007). However, the future of agroforestry terraces systems in the Anti-Atlas is strongly threatened and evolves towards extinction. Particularly, the increasingly deprived peasants resort to the deforestation of the argan forest which does little to help the situation. Furthermore, the greed for argan oil, increasingly sought-after for its culinary, cosmetic virtues, and better valued on the national and international markets in particular, which ultimately poses a real threat to this endemic and highly emblematic specie (Lybbert et al. 2011).

The most important issue and probably the most critical one is that the young generation is more and more disinterested in this kind of livelihood and look for alternatives income even outside the agricultural sector. Only the elderly peasants persevere with limited techniques, funding, as well as motivation based hope. The plots are not only less maintained, but they become less and less abundant and sometimes even left without regular farming practices. Such a social shift was inevitable and was likely to occur as a result of the growing economic needs of an increasingly demanding young generation combined with plentiful job opportunities in the high-yielding cultivated plains of the great adjacent Souss valley (El Fasskaoui 2009). The accelerating effect of climate change is certainly the main cause promoting the migration of young peasants after experiencing decades of a continuous fall in terraced agro-systems productivity, and subsequently the subsistence incomes of peasant household. In these regions, the rate of population growth, normally positive and even higher than in urban areas, remains strangely negative. The shock wave of climate disturbances is therefore no longer socially restricted and limited geographically, but its resonance even reaches the social equilibrium of large cities with a strong economy. In addition to this, food insecurity adds another social dimension, no less threatening, felt more and more in the megacities, hub of immigration.

We certainly believe it is urgent time to revive the relationship between the younger generation and their land. For instance, financial incentives may be one of the alternatives. However, such an approach should be only to trigger the rehabilitation process and cannot constitute the core resilience measure in the long term.

In this study, we intended to emphasize the Moroccan Anti-Atlas as one of the most fragile and the most destabilized region affected by global warming, and to urge all stakeholders to take action to safeguard this precious ancestral heritage. Although the Moroccan policy promotes the protection, conservation, and valorisation of ancestral agricultures in remote areas, including the Anti-Atlas, it is still far from reaching the expected results. By far, such policy requires the involvement of all actors and ensuring enough financial resources which public fund cannot afford alone. If no resilience and adaptation measures are undertaken at different levels to preserve crop diversity and to improve production for a sustainable development (Mooney et al. 2005), the situation will undoubtedly lead to the disruption of ecosystem equilibrium. The dramatic and multidimensional consequences, already on the horizon, will emerge on the social, heritage, and ecosystem levels.

If it is not possible to act on the problem upstream, as climate change is global and depends on a complexity and a conjuncture of events and macroeconomic factors, it is quite possible to propose appropriate actions further downstream. These should not be merely operations of technical supervision of farmers, distribution of selected seeds and seedlings, creation of cooperatives here and there, valorisation of local products such as argan oil, or establishment of tourist activities, but rather regional/communal/local strategies based on long term commitment and participatory approaches involving all stockholders. The aim is not only to warn and prevent, but also, and above all, to develop decision-making tools in order to better cope with a less favourable and rapidly changing climate, environment, and social realities.

4 Conclusion

Generally, the impacts of climate change deeply affect the structure and the balance of an ecosystem. In Morocco, the Anti-Atlas region is by far the most affect, since it is located at the gates of the Sahara. The regression of agroforestry systems has multiple and interdependent consequences operating at various levels. If the ecosystem component is largely unbalanced via the central ecological role played by this very particular and advantageous agricultural system, the social aspects are even more unbalanced and seem to be highly fragile by this regression.

Faced with this scenario, and in addition to a major ecological disruption, the threat of food insecurity hangs seriously in Anti-Atlas. Indeed, our finding of climate data analysis over the study area are in agreement with the 4th and 5th IPCC report predict dry periods in North Africa in the coming decades (IPCC 2007, 2014). The resilience measures undertaken remain timid, minor, less integrated, and unable to cope with the magnitude of such a metamorphosis of the landscape and the surrounding climate. If more influential measures are more urgent than ever, they must emerge from preliminary research of multidisciplinary studies carried out in these regions, which integrate different sectors and dissect all the parameters, causes, and effects relationships that govern the complex functionalities of this yet unknown landscape.

Although the threat perspectives are certainly diverse, two synergistic approaches seem to be of great interest within the perspective of safeguarding of this traditional agronomic abode. First, the use of the Payment for environmental services (PES) process can be effective in this context and could persuade and encourage the younger generation to take back these increasingly neglected systems. Nevertheless, it is clear that such a purely financial motivation, devoid of any local identity and cultural values, cannot by itself produce the expected results. A parallel support cantered on the true valorisation of local and traditional products will certainly contribute to revive these agro-systems from the process of extinction.

Acknowledgements Authors present their gratitude to Dr. Ahmed AMRI for the proof reading of this publication.

References

Achtak H, Ater M, Oukabli A, Santoni S et al (2010) Traditional agroecosystems as conservatories and incubators of cultivated plant varietal diversity: the case of fig (*Ficus carica* L.) in Morocco. BMC Plant Biol 18:10–28

ADA (2009) Green Morocco plan. Moroccan Agricultural Development Agency

Alahiane N, Elmouden A, Aitlhaj A et al (2016) Small dam reservoir siltation in the Atlas Mountains of Central Morocco: analysis of factors impacting sediment yield. Environ Earth Sci 75:1035

Aziz F, Farissi M, Khalifa J, Ouazzani N et al (2014) Les réservoirs de stockage d'eau traditionnel: caractéristiques, popularité et problèmes. Int J Innov Sci Res 11(1):83–95

Brown ME, Funk CC (2008) Food security under climate change. Science 319:580–581
Brush SB (1995) In situ conservation of landraces in centers of crop diversity. Crop Sci 35:346–354
Charrouf Z, Guillaume D (2008) Argan oil, functionnal food, and the sustainable development of the Argan forest. Nat Prod Commun 3:283–288
Charrouf Z, Guillaume D (2009) Sustainable development in northern Africa: the argan forest case. Sustainability 1:1012–1022. https://doi.org/10.3390/su1041012
Chbouki N, Stockton CW, Myers DE (1995) Spatio-temporal patterns of drought in Morocco. Int J Climatol 15:187–205
Choubert G (1963) Histoire géologique du Précambrien de l'Anti-Atlas. Tome 1, Notes et Mémoires du Service géologique du Maroc, 162, 352p, 33 fig., 81 photos, 5 cartes, 7 cartes géol. Couleurs
Choubert G, Faure-Muret A (1972) Carte géologique du Massif du Kerdous (Aït Baha, Tanalt, Anzi, Tafraout) [Document cartographique 1/200000]. Service géologique du Maroc, Rabat
Cromack K Jr, Fichter BL, Moldenke AM, Entry JA et al (1988) Interactions between soil animals and ectomycorrhizal fungal mats. Agr Ecosyst Environ 24:161–168
David BL, Marshall BB, Claudia T, Michael DM et al (2008) Prioritizing climate change adaptation needs for food security in 2030. Science 319(5863):607–610. https://doi.org/10.1126/science.1152339
Despois J (1956) La culture en terrasses en l'Afrique du Nord. Annales Économies Sociétés Civilisations, pp 42–50
El Aboudi A (1990) Typologie des arganeraies inframéditerranéennes et écophysiologie de l'arganier (*Argania spinosa* (L.) Skeels) dans le Sous (Maroc). Thèse de doctorat, Université de Grenoble I
El Bilali H, Berjan S, Driouch N, Ahouate L et al (2012) Agriculture and rural development gouvernance in Morocco. In: Conference proceeding: third international scientific symposium "Agrosym 2012", At Jahorina (East Sarajevo), Bosnia, Herzegovina, https://doi.org/10.13140/rg.2.2.35533.64485
El Fasskaoui B (2009) Fonctions, défis et enjeux de la gestion et du développement durables dans la Réserve de Biosphère de l'Arganeraie (Maroc). Études caribéennes, vol 12. http://etudescaribeennes.revues.org/document3711.html
Esper J, Frank DC, Büntgen U, Verstege A et al (2007) Long-term drought severity variations in Morocco. Geophys Res Lett 34. https://doi.org/10.1029/2007gl030844
Esper J, Frank DC, Büntgen U, Verstege A et al (2009) Morocco millennial palmer drought severity index reconstruction. IGBP PAGES/World Data Center for paleoclimatology data contribution series #2009-032. NOAA/NCDC Paleoclimatology Program, Boulder CO, USA
Esquinas-Alcazar J (2005) Protecting crop genetic diversity for food security: political, ethical and technical challenges. Nat Rev Genet 6:946–953
Gebauer J, Luedeling E, Hammer K, Nagieb M et al (2007) Mountain oases in northern Oman: An environment for evolution and in situ conservation of plant genetic resources. Genet Resour Crop Evol 54:465–481
Gunn J (2004) Encyclopedia of caves and karst science. Taylor & Francis, p 902. ISBN1579583997, 9781579583996
Harfouche R (2003) Histoire des paysages méditerranéens au cours de protohistoire et de l'antiquité: aménagements et agriculture. Thèse de Doctorat, Aix-en-Provence
Hmimsa Y, Ater M (2008) Agrodiversity in the traditional agrosystems of the Rif Mountains (North of Morocco). Biodiversity 9:78–81
IPCC (2007) Climate change 2007: synthesis report. In: Core Writing Team, Pachauri RK, Reisinger A (eds) Contribution of working groups I, II and III to the fourth assessment report of the intergovernmental panel on climate change. IPCC, Geneva, Switzerland, p 104
IPCC (2014) In: Climate change 2014: synthesis report. Core Writing Team, Pachauri RK, Meyer LA (eds) Contribution of working groups I, II and III to the fifth assessment report of the intergovernmental panel on climate change. IPCC, Geneva, Switzerland, p 151

Kenny L, De Zborowski I (2007) In Atlas de l'arganier et de l'arganeraie. IAV Hassan II, Rabat, Morocco, p 190

Kottek M, Grieser J, Beck C, Rudolf B et al (2006) World map of the Köppen-Geiger climate classification updated. Meteorol Z 15:259–263. https://doi.org/10.1127/0941-2948/2006/0130

Lacombe S, Bradley RL, Hamel C, Beaulieu C (2009) Do tree-based intercropping systems increase the diversity and stability of soil microbial communities? Agr Ecosyst Environ 131(1–2):25–31. https://doi.org/10.1016/j.agee.2008.08.010

Le Polain de Waroux Y, Lambini EF (2012) Monitoring degradation in arid and semi-arid forests and woodlands: the case of the argan woodlands (Morocco). Appl Geogr 32:777–786

Lobell DB, Burke MB, Tebaldi C, Mastrandrea MD et al (2008) Prioritizing Climate Change Adaptation Needs for Food Security in 2030. Science 319(5863):607–610. http://dx.doi.org/10.1126/science.1152339

Lybbert TJ, Aboudrare A, Chaloud D, Magnan N, Nash M (2011) Booming markets for Moroccan argan oil appear to benefit some rural households while threatening the endemic argan forest. Proc Nat Acad Sci 108(34):13963–13968. http://dx.doi.org/10.1073/pnas.1106382108

M'Hirit O, Benzyane M, Benchekroun F, El Yousfi SM et al (1998) L'Arganier. Une espèce fruitière-forestière à usages multiples. Mardaga, Sprimont, Belgium p 150

Mooney H, Cropper A, Reid W (2005) Confronting the human dilemma: how can ecosystems provide sustainable services to benefit society? Nature 434:561–562

Morton JF, Voss GL (1987) The Argan tree (*Argania sideroxylon*, Sapotaceae), a desert source of edible oil. Econ Bot 41:221–233

Myers N, Mittermeier RA, Mittermeier CG, Da Fonseca GAB et al (2000) Biodiversity hotspots for conservation priorities. Nature 403:853–858

Natsagdorj O (2012) Assessment of drought hazard: a case study in Sehoul Area, Morocco. Thesis submitted to the Faculty of Geo-information Science and Earth Observation of the University of Twente. Enschede, The Netherlands. Retrieved from http://www.itc.nl/library/papers_2012/msc/aes/otgonjargal.pdf, 30 Mar 2017

Noorka IR, Heslop-Harrison JS (2014) Agriculture and climate change in Southeast Asia and the Middle East: breeding, climate change adaptation, agronomy, and water security. In: Leal Filho W (ed) Handbook of climate change adaptation. Springer, Berlin, Heidelberg, pp 1–8. http://dx.doi.org/10.1007/978-3-642-40455-9_74-1

Olivier A, Paquette A, Cogliastro A, Rousseau AN et al (2015) Contribution de systèmes agroforestiers intercalaires à l'adaptation aux changements climatiques des agroécosystèmes. In: XIVe Congres Forestier Mondial, Durban, Afrique du Sud

Palmer WC (1965) Meteorological drought. Research paper no. 45, U.S. Department of Commerce Weather Bureau, p 58. Available online by the NOAA National Climatic Data Center at http://www.ncdc.noaa.gov/temp-and-precip/drought/docs/palmer.pdf

Peltier JP (1982) La végétation du bassin versant de l'Oued Sous (Maroc), thèse de doctorat ès sciences, université de Grenoble I, p 208

Podeur J (1995) Textes berbères des Aït Souab (Anti-Atlas, Maroc), Institut de recherches et d'études sur le monde arabe et musulman, Aix-en-Provence, p 159

RMS (2011) Doing business in Morocco. RSM International Publication, England, United Kingdom

Rivest D, Lorente M, Olivier A, Messier C (2013) Soil biochemical properties and microbial resilience in agroforestry systems: effects on wheat growth under controlled drought and flooding conditions. Sci Total Environ 463–464:51–60

Ruas MP, Ettahiri AS, Fili A, Van Staëvel JP et al (2015) Recherches archéobotaniques sur l'arganeraie médiévale dans la montagne d'Îgîlîz (Anti-Atlas, Maroc). In: Proceedings congrés international de l'arganier. Agadir. Morocco, 17–19 Dec 2015

Wilhite DA, Glantz MH (1985) Understanding the drought phenomenon: the role of definitions. Water Int 10(3):111–120

Increasing Pulse Consumption to Improve Human Health and Food Security and to Mitigate Climate Change

Beatriz Oliveira, Ana Pinto de Moura and Luís Miguel Cunha

Abstract Human food security requires both the production of sufficient quantities of high-quality protein and dietary change. This is particularly relevant given the present concurrence of rising human population, climate change and changing consumption habits. Pulses are recognized as being readily available sources of protein, complex carbohydrates, fibres, vitamins and minerals. Additionally, pulses, as well as other legumes, have the exceptional capacity of significantly increasing soil fertility, yields of companion or subsequent crops, biodiversity, environmental protection and climate change mitigation. Despite this, the use of pulses for food purposes is low in Western Europe, where pulses are mainly used for feed. This chapter reviews some of the environmental, nutritional and health benefits of pulses, and presents the main results of a campaign developed in a food service setting as an example of ways to increase the amount of pulse consumption. Results show a high acceptability of pulse consumption, whenever food services present alternatives to meat protein based on pulse protein.

Keywords Food service · Environment benefits · Health benefits
Pulses · Sustainability

B. Oliveira
Eurest Portugal, Compass Group PLC, GreenUPorto, Faculty of Sciences,
Faculty of Food Sciences and Nutrition, University of Porto, Porto, Portugal
e-mail: beatriz.oliveira@eurest.pt

A. P. de Moura (✉)
GreenUPorto & LAQV/REQUIMTE, DCeT, Open University of Portugal,
Porto, Portugal
e-mail: apmoura@uab.pt

L. M. Cunha
GreenUPorto & LAQV/REQUIMTE, DGAOT, Faculty of Sciences,
University of Porto, Campus de Vairão, Rua da Agrária, 747,
Vila do Conde 4485-646, Vairão, Portugal
e-mail: lmcunha@fc.up.pt

P. Castro et al. (eds.), *Climate Change-Resilient Agriculture and Agroforestry*,
Climate Change Management, https://doi.org/10.1007/978-3-319-75004-0_2

1 Introduction

Food is essential to our survival, yet the Food and Agriculture Organization of the United Nations (FAO) estimates that about 805 million people were undernourished in 2012–14. About 14.6 million undernourished people lived in developed countries (<5% of their overall population), whereas 790.7 million undernourished people lived in developing countries, corresponding to 13.5% of their overall population (FAO et al. 2014). On the other hand, food scarcity coexists with excessive consumption. Globally, 35% of adults aged 20 and older were overweight, with half a billion of them obese (WHO 2011). At least 2.8 million people die each year as a result of being overweight or obese. The increasing global prevalence of overweight and obesity has serious implications on health, increasing the risk of type 2 diabetes, cardiovascular disease, strokes and some cancers (WHO 2011). Furthermore, rising population levels combined with shifting dietary patterns in emerging economies will put increasing pressure on the global food supply as more food is necessary to feed the people. The United Nations predicted that the world population would reach 9.6 billion by 2050 (UN 2012), which will require either a 70% increase in food production, excluding crops used for biofuels (FAO 2009) or a more efficient food production and use of natural resources (European Commission 2014).

Additionally, over the past decades, rapid changes in diets and lifestyles have occurred with industrialisation, urbanisation, economic development and market globalisation. These drivers promote the process of diet Westernization among other cultures/countries, namely towards upper-to middle-income developing economies (Drewnowski and Popkin 1997), and even in populations which have rich and deeply rooted culinary traditions, such as Japan (Morinaka et al. 2013) or Southern Europeans countries (Varela-Moreiras et al. 2010). This dietary pattern, belonging to the common eating habits in developed countries of Western Europe and the United States of America, is characterized by a high consumption of red meat, refined grains, processed meat, dairy products, processed and artificially sweetened foods, and salt, with minimal intake of fruits, vegetables, fish, legumes, and whole grains (Cordain et al. 2005). In fact, food consumption of meat is increasing, with the largest increments occurring in developing countries. Developing countries between 1969/1971 and 2005/2007 have accounted for a large share of this increase in food consumption per capita (kcal/person/day) from 11 to 28, contrasting with the 63–80 increase for industrial countries over this period (FAO 2006). In addition, most developing countries especially China and Brazil, have largely completed the transition to animal products: meat and livestock products (eggs and dairy foods). On other hand, in both developing and industrial countries (and again notably in China), declines were observed for the consumption of pulses, roots and tubers between 1963 and 2003. There is also a great consensus in the scientific community about the importance of changing the Western diet to have a positive outcome for both people's health and the environment (Friel et al. 2009).

In this context, better promotion and broader use of pulses could be an important solution towards making food systems more sustainable and nutrition-sensitive. In order to support the spread of knowledge on pulses and trigger positive transformations in the pulses sector, FAO declared 2016 the International Year of Pulses under the slogan "International Year of Pulses: Nutritious Seeds for a Sustainable Future" (UN 2014).

Despite the various benefits and uses of pulses, the consumption of pulses, per capita, has seen a slow but steady decline in both developed and developing countries, dropping from 9.4 kg/person/year in 1961, to 7.2 kg/person/year in 2013. For Europe, the food supply of pulses, ranged between 3.4 kg/person/year in 1961 and 2.5 kg/person/year in 2013 (FAOSTAT 2017). As a result, there is a huge need to increase production and ensure that pulses are widely available for consumption throughout the world. The purpose of this paper is to evaluate nutritional, health and environmental benefits associated with the consumption of pulses; to evaluate the consumption of pulses, particularly in Europe. It also presents an example of a campaign that attempted to increase pulse consumption in a food service setting.

2 Benefits Deriving from Pulses

Pulses, a subgroup of legumes, are the crop plant members of the *Leguminosae* family that produce edible seeds (e.g. dry beans, chickpeas, lentils, dry peas), used for human and animal consumption (FAO 1994). Only legumes harvested for dry grain are classified as pulses. Likewise, pulses do not comprise legume crops that are harvested in their green stage for food, which are classified as vegetable crops (e.g. green beans, green peas). Also, that crops that are mainly used for oil extraction (e.g. soybeans, peanuts and groundnuts) and leguminous crops that are used exclusively for sowing purposes (e.g. seeds of clover and alfalfa) have been excluded.

2.1 Nutritional and Health Benefits

Although nutrient composition varies among different pulses, in general, they provide high protein and fibre content and are a significant source of vitamins and minerals (Table 1). The protein content of most pulses falls within the range of 18–32% of dry weight, twice the amount found in cereals, providing well balanced essential amino acid profiles when consumed with cereals (Boye et al. 2010). As a result, proteins from pulses are valuable supplements to cereal-based diets (Rebello et al. 2014). Moreover, pulses contain both soluble and insoluble fibres (Tosh and Yada 2010). The soluble fibre ferments, positively affecting colon health through the production of short chain fatty acids (SCFA), lowering pH, and promoting

Table 1 Nutritional values of pulses (per 100 g)

Name/scientific name	Energy (kcal)	PROT (g)	Fat (g)	CHO (g)	Fibre (g)	Fe (mg)	Mg (mg)	P (mg)	K (mg)	Ca (mg)
Adzuki beans/*Vigna angularis*	272	19.9	0.5	50.1	16.8	4.2	130	380	1,220	84
Cowpeas/*Vigna unguiculata*	311	23.5	1.6	54.1	10.6	7.6	140	410	1,170	81
Fava beans/*Vicia faba*	245	26.1	2.1	32.5	25.0	5.5	190	590	1,090	100
Butter beans/*Phaseolus lunatus*	290	19.1	1.7	52.9	19.0	5.9	190	320	1,700	85
Chikepeas/*Cicer arietinum*	320	21.3	5.4	49.6	12.2	5.5	130	310	1,000	160
Mung beans/*Vigna radiata*	279	23.9	1.1	46.3	16.3	6.0	150	360	1,250	89
Pigeon peas/*Cajanus cajan*	317	20.0	1.9	58.6	15.0	3.4	100	290	1,390	140
Peas/*Pisum sativum*	303	21.6	2.4	52.0	5.1	4.7	120	300	990	61
Lentils/*Lens culinaris*	297	24.3	1.9	48.8	10.7	3.9	110	350	940	71
Red kidney beans/*Phaseolus vulgaris*	266	22.1	1.4	44.1	15.2	6.4	150	410	1,370	100

Source McCance and Widdowson (2015)
CHO Carbohydrate; *PROT* Protein

microbiotal changes. They also help to control blood glucose level and to reduce blood LDL-cholesterol levels, which can lower the risk of heart attack and stroke (Curran 2012; Ha et al. 2014; Tosh and Yada 2010). Pulse consumption also increases gastric distention and helps to slow gastric emptying rate (Howarth et al. 2001). The insoluble fibres help with digestion and regularity: promoting the movement of material through the digestive system, thereby improving laxation, and being associated with faecal bulking through its water-holding capacity (McCrory et al. 2010; Tosh and Yada 2010).

For a food that is high in carbohydrates, pulses have a low glycaemic index (Foster-Powell et al. 2002), meaning that they do not cause a fast rise in blood sugar after eating, which is particularly important for people with diabetes. Other complex carbohydrates present in pulses are resistant and slowly digestible starch, and the raffinose family of the oligosaccharides (McCrory et al. 2010). These oligosaccharides resist digestion and absorption in the digestive system and are fermented by the colon's microflora, producing gases that give flatulence, and SCFA that support the health of the intestinal mucosa (Hoover et al. 2010; Tosh and Yada 2010).

Pulses are also nutrient dense, providing substantial amount of minerals (e.g. potassium-K, phosphorus-P, magnesium-Mg and calcium-Ca) and vitamins (C, E, folate, thiamine and niacin). They contain a variety of phytochemicals, including phenolic compounds, tannins, phenolic acids and flavonoids, acting as antioxidants and other minor constituents (saponins and phytates). These exhibit bioactivity and improve glycaemic control, protecting against hypocholesterolaemia, cancer and diabetes. The high content of fibre and protein in pulses and the low glycaemic index leads a positive association between pulse consumption and increased satiety, thus having an important role in body weight management and overweight reduction (McCrory et al. 2010). Finally, pulses are a low fat source, containing zero cholesterol and are free of gluten, hence they are also an ideal food for celiac patients.

2.2 *Environmental Benefits*

Crop production, food processing and product marketing generate greenhouse gases (GHG) that absorb infrared radiation in the atmosphere, trapping heat and warming the surface of the Earth, contributing to global climate change. In 2010, emissions from agriculture, forestry and other land use produced approximately 12 gigatonnes of CO_2 equivalent emissions ($GtCO_2eq$), accounting for 24% of the 49 $GtCO_2eq$ total GHG emissions (IPCC 2014). While CO_2 emissions are concentrated in the energy sector, the non-carbon dioxide (non-CO_2) GHG are dominated by the agricultural sector, as it was estimated to be 5.2–5.8 $GtCO_2eq$ in 2010 and comprising about 10–12% of global anthropogenic emissions (IPCC 2014; Tubiello et al. 2013). Organic and inorganic materials provided as input or output in the management of agricultural systems are typically broken down through bacterial processes, releasing significant amounts of CO_2, CH_4, and N_2O to the atmosphere.

Although CO_2 remains the major anthropogenic GHG, accounting for 76% (38 $GtCO_2eq$) of the total anthropogenic GHG emissions in 2010, the non-CO_2 GHG are more potent than CO_2 (per unit weight), trapping more heat within the atmosphere. The three major GHG associated with agriculture are carbon dioxide (CO_2), methane (CH_4) and nitrous oxide (N_2O) (EPA 2012). Each molecule of methane (CH_4) released into the atmosphere is 23 more times effective at trapping heat compared to an equivalent unit of CO_2, and 296 times for N_2O (IPCC 2001).

As pulses are relatively high in protein content, they present a low carbon footprint because they reduce farmers' dependence on synthetic N fertilizers (N inputs) responsible for generate N_2O emissions (Lemke et al. 2007; Tongwane et al. 2016). This implies that pulses indirectly reduce GHG emissions. Pulse crops have the ability to supply most of their nitrogen with the help of the symbiotic Rhizobia bacteria living in their roots (Ali and Venkatesh 2014). Bacteria fix atmospheric nitrogen (N_2) into the NH^{4+} that is largely used by pulse crops to form protein compounds, reducing the N availability for nitrification and subsequent denitrification that is associated with N_2O emission (Synder et al. 2009). The adoption of a diversified cropping system (crop rotation or inter-cropping) promotes the reduction of the carbon footprint of crop products. According to Gan et al. results, cereals grown in diversified cropping systems have a lower carbon footprint, as the same amount of yield can be achieved with less inputs like N fertilizer. Durum wheat grown after a pulse crop had a lower carbon footprint than when grown after a monoculture cereal system (Gan et al. 2011).

The unique ability to biologically fix nitrogen has also a direct and positive impact on soil biodiversity. Healthy soils contain a diversity of living organisms that range from bacteria and fungi to tiny insects, earthworms, moles and plant roots, considering their symbiotic relationships and interactions with other soil components. These diverse organisms interact with one another and with the various plants and animals in the ecosystem, forming a complex web of biological activity that contributes to a wide range of essential services to the sustainable function of all ecosystems (FAO 2002). The microbial biomass decomposes plant and animal residues and soil organic matter to release carbon dioxide and plant available nutrients (Blanchart et al. 2006). The presence of the residues of legume crops in the soil returns nitrogen to the system and maintains/increases microbial biomass, because they nourish microorganisms with carbon and high N contents that they need. Some pulses (e.g. chickpea, pigeon pea) also have the ability to increase the availability of phosphorus, which plays an important role in plant nutrition (Rose et al. 2010), naturally providing two fertilizers.

By producing their own fixed nitrogen in the soil, pulses contribute to higher yields in subsequent crop rotations, while at the same time boosting soil fertility, which in turn helps to decrease the carbon footprint of future crops. If pulse crops are used as green manure, considerable amounts of N can be supplied to the succeeding crop as pulse crops residue decomposes and contributes with N to the next crop. In dryland agriculture pulses are a part of the crop rotation, as sole crops, intercrops and relay crops are important both for improving productivity and sustainability of agriculture (Venkatesh et al. 2014).

Additionally, pulses require less water compared to other natural sources of proteins as water used to produce one kilogram of beef, mutton, chicken and lentils amounts to 13,000; 5,520; 4,325 and 1,250 L, respectively (FAO 2016). As a result, they can be cultivated in regions with limited or often erratic rainfall and they can tolerate drought stress better than other crops like wheat or canola (Cutforth et al. 2009). Comparing to other crops (oilseeds and wheat), pulses have a deep rooting systems (e.g. chickpea), which enable them to draw nutrients and moisture from deeper layers of the soil, without competing with other crops for water (Gan et al. 2009).

Finally, pulses can remain in edible condition for a long duration of time (more than one year) with their nutritional proprieties being retained, if stored in dry places (at room temperature), in airtight containers and protected against rodent, insect and micro-organism attacks (Tiwari and Singh 2012). Since pulses are shelf stable with no refrigeration, the proportion of food waste or spoilage is very low (FAO 2016). This is particularly relevant, since in developing countries, food waste arose mostly during the early and middle stages of the food chain (production, harvest, processing, storage and transportation stages), due to lack of infrastructures within the food chain, and lack of knowledge or investment in technologies (FAO 2012).

3 Pulses Consumption in Europe

Europe displays a low consumption of pulses, representing, in 2013, 9% of the world supply, with 2.6 kg/capita (Table 2). These values clearly contrast with Asia, the continent with the highest consumption of pulses and particularly with India, the single main contributor to pulse consumption in the World, as pulses are an important source of protein for their largely vegetarian population. In 2013, the domestic consumption of food pulses in Asia and India represented 54 and 30% of the world food supply of pulses, respectively, with an annual food supply of pulses of 14.9 kg/capita in India. The European situation may be explained by the changing dietary patterns and consumer preferences, where the sources of protein shift from vegetable proteins to animal proteins (dairy and meat sources). In fact, in the early stages of the Common Agricultural Policy (PAC), grain legume crops that are used exclusively for human consumption (chickpea, cowpea, groundnut, lentil, and common bean) dominated grain legume cropping in Europe with 67% of the area. This dropped to 22% by 2010 (Bues et al. 2013). The higher consumption of meat, demanding for protein-rich feed has been met by a higher production of grain legumes being used as animal feed and greatly increasing imports of soya bean that provide the necessary protein enrichment for cereal-based animal feeds (Bues et al. 2013).

Table 2 World food supply of pulses in 2013

Regions	Food supply		World food supply used (%)
	(t)	(kg/capita/year)	
World	72,195,000	7.2	100
Africa	15,840,000	11.8	22
America	10,440,000	9.1	14
Asia	38,859,000	6.6	54
Europe	6,295,000	2.6	9
Northern Europe	744,000	3.1	1
Southern Europe	1,453,000	5.2	2
Eastern Europe	3,015,000	1.8	4
Western Europe	1,083,000	1.3	1.5
Oceania	762,000	2.1	1

Source FAOSTAT (2017)

Consumption of pulses varies among the four different European regions, as there are different regional food habits and differences in production (Schneider 2002). In 2013, Southern Europeans had an average consumption of 5.2 kg/capita, Northern Europeans had 3.1 kg/capita, and Eastern and Western Europeans had 1.8 kg/capita and 1.3 kg/capita, respectively. As a result, we are confronted with North–South consumption disparity and similar lower consumption in the Eastern and Western regions (FAOSTAT 2017).

The Mediterranean Diet is responsible for the majority of the use of pulses for food in Europe, particularly for the Mediterranean European countries, namely Cyprus, Croatia, (Southern) France, Greece, Italy, Portugal and Spain. Amongst those countries, Spain maintains its position as one of the principal consumers (5.3 kg/per capita, in 2013), where pulses appear in many traditional recipes and modern creative reinterpretations from the country's top chefs, with the inclusion of chickpeas, beans and lentils (O'Broin and Tucci 2016). In fact, in the Mediterranean Diet Pyramid, pulses appear at the base, alongside fruits, vegetables, grains, olive oil, herbs, and spices (Bach-Faig et al. 2011), as the Mediterranean Diet is a plant-based pattern, where vegetables, fruits, cereals (preferably as whole grain), legumes and nuts should be consumed in high amount and frequency (Willet et al. 1995).

In the same way, the New Food Guide for the Portuguese population recommends 1–2 daily food portions of pulses (fresh and dried), corresponding to 25 g dried or 80 g cooked dried/fresh, in accordance with the established range of energy values corresponding to 2,200 kcal (Rodrigues et al. 2006). In this guide, pulses form a new and individual group. The rational for this approach was both an attempt to bring back their importance to Portuguese food habits and to substitute animal protein for vegetal protein, thus reducing saturated fat consumption (Rodrigues et al. 2006). Nevertheless, following the Portuguese Food Balance Sheet, for the quinquennium 2012–2016, a deficit of pulse availability was observed, corresponding to 0.6% of the total food availability, against the recommended 4.0% of the total food intake, as promoted by the New Food Guide recommendations (INE 2017).

4 An Example on: How to Increase the Amount of Pulse Consumption in a Food Service Setting? the "Choose Beans" Campaign

4.1 The Relevance of Catering Sector in Food Provision

In Europe, the catering sector is characterised by a high level of market concentration in the majority of the member states. In recent years, the main growth areas for EU contract caterers have been in sectors other than business and industry, even though this segment has remained dominant in terms of market value. Market development and sales growth for contract catering have been higher than average in the health and social welfare sector as well as in the education sector (Gira Foodservice 2009).

In the catering sector, sustainability related challenges arise from a number of aspects along the value chain, ranging from climate change, which affects food production, to the provision of a food service with impacts on the health, safety and well-being of consumers. Thus, the responsibilities of these businesses should include a social role in food education and raising awareness as to conscious and healthy lifestyle and food choices (Oliveira et al. 2016).

The catering company considered as object of study operates in Portugal for the corporate market and the broader public, diversifying their activities across contract catering, cafeterias and canteens, vending and specialised on-site catering activities. In 2017, the company was responsible for providing 144,000 meals per day, across 1,027 units, prepared and served by a workforce of 3,800 employees, and recording a turnover of 119 M€. The company embraces a set of initiatives that bring together environmental, social and food campaigns launched at their food service units. The success of this contribution is achieved by working with customers and consumers —who share similar concerns on these issues—in the form of events, campaigns or more long-term initiatives.

4.2 The Choose Beans Campaign

The Choose Beans campaign is integrated into a broad project that has the goal of incorporating all campaigns and practices aimed at environmental, social and nutritional improvement, where in the company has invested. The objective of this campaign was to encourage the consumption of legumes (and within these, pulses), among their final consumers. This was done by diversifying products on offer, as a way to promote healthier and more balanced meals, with smaller ecological footprints.

The Choose Beans campaign has been implemented across the corporate market segment, lasting from October 2011 to September 2017, and was implemented in 30 units.

The implementation phase consisted in the introduction of 12 new dishes on the menu with the incorporation of five pulses (fava beans, peas, beans, chickpeas and lentils) and soybeans included in the soup and main course. To raise awareness about pulses, their consumption recommendations and benefits, and to arouse curiosity and consumer interest, the company promoted workshops at these units. Graphic communication for each pulse was also developed (Fig. 1) and partakers were given a participation gift corresponding to a gauze bag with the recommended daily amount of the various dried pulses. During the introduction of new dishes, each unit counted with the presence of a nutritionist to raise consumers' awareness about pulses. This nutritionist presented various pulse samples, information leaflets about recommendations and benefits of pulse consumption, as well as the referred offer. The duration of this phase took one week.

To evaluate the efficiency of those campaigns, the consumption of pulses was analysed. These values were based on overall quantities of pulses expedited and the total number of meals produced at each of the units, during the company's management year: from November to October. Values were collected during two different times: the year before the Choose Beans campaign and the year of the campaign. The values were analysed for all Choose Beans campaign units and were compared with the yearly average values for all units that did not receive the campaign. For all the dried pulses, the amount of pulses consumed was measured in kilos and then standardized to an equivalent cooked dry/fresh measurement, by converting its dried mass, multiplying it by a 3.2 factor (Rodrigues et al. 2006).

Data was recovered for all 30 food service units adhering to the Chose Beans campaign, since 2011/12, with only two not presenting results for the year before the campaign, considering that they adhered to the campaign during the first year of hiring the company's services. Overall, during the years of intervention, a total of 3.3 million meals were served at those units, while an average of 26.1 million meals/year were served at the units with no intervention. Data from 28 units shows that for the vast majority (73%), the program was accompanied with an increase in the average consumption of pulses (Fig. 2).

The units under intervention strongly varied in their initial average consumption of pulses, going from 3.8 g/meal to 37.2 g/meal, while for the whole set of units with no intervention the average consumption of pulses has varied from 12.8 g/meal to 15.1 g/meal, between the 2011/12 and 2015/16 operating years. Over 70% (20/28) of the units under intervention yielded a positive variation in the average pulse consumption, with an increase of up to 71% of the previous year value occurring in one of the largest units, serving over 385,000 meals/year.

Despite the positive impact of the campaign, there is still a long way to go in order to achieve a relevant contribution for the recommended 80 g of cooked dry or fresh

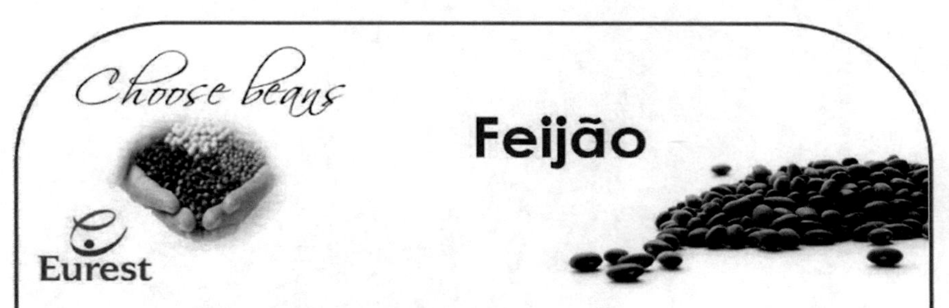

Existem mais de **300 variedades** de feijão.
As mais comuns em Portugal são:

- Feijão – branco
- Feijão - manteiga
- Feijão – encarnado
- Feijão – catarino
- Feijão – frade
- Feijão - preto

Os portugueses consomem 4.1 kg de leguminosas secas, por ano.
80% desse valor diz respeito ao consumo de feijão seco.

Benefícios
Boa fonte de fibra alimentar, tiamina (vit. B1) e niacina, ácido fólico, potássio, fósforo, ferro e magnésio.

A proteína das leguminosas, onde se inclui o feijão, é incompleta. No entanto, a qualidade pode ser aumentada se combinar o consumo de feijão com cereais, como o arroz, pão, massa, etc. Ao associar estes dois alimentos, é possível obter uma proteína de boa qualidade. Substituindo ou complementando a carne e o peixe.

Feijão branco
Usado na sopa, feijoada e cozido à portuguesa, é das leguminosas mais consumidas em Portugal.

45 a 60 minutos a cozer, após a demolha de 12 ou 6 horas, respectivamente.

	Feijão branco cozido (80g)	Feijão manteiga cozido (80g)
Calorias	73	75
Proteínas	5.3g	6.2g
Hidratos de carbono	11.7g	11g
Gordura	0.4g	0.5g
Fibra	5.4g	6g
Ácido fólico	79mg (20% DDR) *	
Ferro	1.9mg (24% DDR) *	

*DDR – Dose diária recomendada

Fig. 1 Example of the communication posters, in Portuguese, used to promote the consumption of different pulses within the Choose Beans campaign developed by Eurest (Compass Group) in Portugal at different food services. This example, promoting the consumption of beans presented different pieces of information on the most common varieties, national consumption level, recommended uses, benefits and nutritional content. Reprinted with permission from Eurest (Compass Group), Portugal

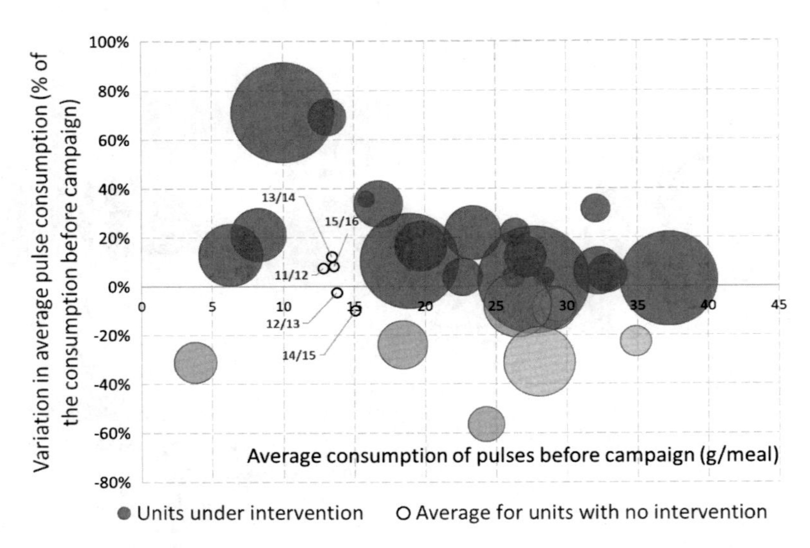

Fig. 2 Variation in the average pulse consumption during the intervention campaign Choose Beans as a function of the average consumption during the previous year, for 28 of the 30 units under intervention. Area of the circles is proportional to the number of meals served during the year of the campaign (varying from 10,947 to 453,653 meals) at each single unit. Small circumferences report average yearly variations—campaign period signalled- for all the units not undergoing intervention

raw or cooked pulses/day (Rodrigues et al. 2006). Possible actions to be taken into consideration would be the extension of the campaign over longer periods of time, or its cyclic presentation, with a week dedicated to pulses on every month or so.

5 Conclusions

Promotion of pulses for consumption and particularly as a source of protein brings many different advantages, related to the health and nutrition of individuals.

Moreover, the production of pulse crops, as well as other legume crops, promotes higher agricultural productivity levels, when used in rotation with other crops.

Due to its physiology and interaction with the surroundings, pulse crops have a major positive impact on the environment by reducing the production of GHG associated with agricultural practices and with the production of animal protein.

Taking all those factors into consideration, consumption of pulses should be taken to a higher level, increasing its level in developed countries and inverting the declining consumption in developing countries.

Acknowledgements Authors thank José B. Cunha from Oporto British School for revising English usages and grammar throughout the manuscript.

References

Ali M, Venkatesh MS (2014) Role of pulses in conservation agriculture. In: Ghosh PK, Kumar N, Venkatesh MS, Hazra KK, Nadarajan N (eds) Resource conservation technology in pulses. Scientific Publishers (India), Jodhpur, pp 75–82

Bach-Faig A, Berry EM, Lairon D et al (2011) Mediterranean diet pyramid today. Science and cultural updates. Public Health Nutr 14(12A):2274–2284

Blanchart E, Villenave C, Viallatoux A et al (2006) Long-term effect of a legume cover crop (*Mucuna pruriens var. utilis*) on the communities of soil macrofauna and nematofauna, under maize cultivation, in southern Benin. Eur J Soil Biol 42:S136–S144

Boye J, Zare F, Pletch A (2010) Pulse proteins: processing, characterization, functional properties and applications in food and feed. Food Res Int 43:414–431

Bues A, Preißel S, Reckling M et al (2013) The environmental role of protein crops in the new common agricultural policy. Directorate General for Internal Policies Agriculture, Rural Development, European Union, Brussels

Cordain L, Eaton SB, Sebastian A (2005) Origins and evolution of the Western diet: health implications for the 21st century. Am J Clin Nutr 81(2):341–354

Curran J (2012) The nutritional value and health benefits of pulses in relation to obesity, diabetes, heart disease and cancer. Br J Nutr 108(1):S1–S2

Cutforth HW, Angadi SV, McConkey BG et al (2009) Comparing plant water relations for wheat with alternative pulse and oilseed crops grown in the semiarid Canadian prairie. Can J Plant Sci 89:823–835

Drewnowski A, Popkin BM (1997) The nutrition transition: new trends in the global diet. Nutr Rev 55(2):31–43

EPA (2012) Global anthropogenic non-CO_2 greenhouse gas emissions: 1990–2030. United States Environmental Protection Agency, EPA, Washington DC

European Commission (2014) Communication from the commission to the European parliament, the council, the European economic and social committee and the committee of the regions—towards a circular economy: a zero waste programme for Europe. Brussels. http://eur-lex.europa.eu/legal-content/EN/TXT/?uri=celex:52014DC0398. Accessed 18 Nov 2017

FAO (1994) Definition and classification commodities, 4. Pulses and derived products. http://www.fao.org/es/faodef/fdef04e.htm. Accessed 18 Nov 2017

FAO (2002) Soil biodiversity—the root of sustainable agriculture. Paper prepared as a background paper for the ninth regular session of the commission on genetic resources for food and agriculture (CGRFA). FAO, Rome. http://www.fao.org/search/en/?cx=018170620143701104933%3Aqq82jsfba7w&q=Paper+prepared+as+a+background+paper+for+the+Ninth+&cof=FORID%3A9&siteurl=www.fao.org%2Fagriculture%2Fcrops%2Fthematic-sitemap%2Ftheme%2Fspi%2Fsoil-biodiversity%2Finitiatives%2Fen%2F&ref=www.google.pt%2F&ss=41j1681j2. Accessed 18 Nov 2017

FAO (2006) Livestock's long shadow: environmental issues and options. FAO, Rome

FAO (2009) How to feed the world in 2050. FAO, Rome

FAO (2012) The role of producer organizations in reducing food loss and waste. FAO, Rome

FAO (2016) Pulses—nutritious seeds for a sustainable future. FAO, Rome

FAO, Ifad, WFP (2014) The state of food insecurity in the world 2014: strengthening the enabling environment for food security and nutrition. FAO, Rome

FAOSTAT (2017) Food balance, food balance sheets, Rome. http://www.fao.org/faostat/en/#data. Accessed 18 Nov 2017

Friel S, Dangour AD, Garnett T et al (2009) Public health benefits of strategies to reduce greenhouse-gas emissions: food and agriculture. Lancet 374(9706):2016–2025

Foster-Powell K, Holt SH, Brand-Miller JC (2002) International table of glycemic index and glycemic load values. Am J Clin Nutr 76:5–56

Gan Y, Campbell CA, Liu L et al (2009) Water use and distribution profile under pulse and oilseed crops in semiarid northern high latitude areas. Agric Water Manag 96:337–348

Gan Y, Liang C, Wang X et al (2011) Lowering carbon footprint of durum wheat by diversifying cropping systems. Field Crops Res 122:199–206
Gira Foodservice (2009) The contract catering market in Europe 2006–2010: 25 countries, France
Ha V, Sievenpiper JL, de Souza RJ et al (2014) Effect of dietary pulse intake on established therapeutic lipid targets for cardiovascular risk reduction: a systematic review and meta-analysis of randomized controlled trials. Can Med Assoc J 186(8):E252–E262
Hoover R, Hughes T, Chung HJ et al (2010) Composition, molecular structure, properties, and modification of pulse starches: a review. Food Res Intl 43:399–413
Howarth NC, Saltzman E, Roberts SB (2001) Dietary fiber and weight regulation. Nutr Rev 59:129–139
INE (2017) The Portuguese food balance sheet 2012–2016. Statistics Portugal, Lisbon
IPCC (2001) Climate change 2001: the scientific basis. In: Contribution of working group I to the third assessment report of the intergovernmental panel on climate change. Cambridge University Press, Cambridge. United Kingdom, New York, USA
IPCC (2014) Climate change 2014: mitigation of climate change. Summary for policymakers, technical summary. In: Part of the working group III to the fifth assessment report of the intergovernmental panel on climate change. Cambridge University Press, Cambridge. United Kingdom, New York, USA
Lemke RL, Zhong Z, Campbell CA et al (2007) Can pulse crops play a role in mitigating greenhouse gases from north American agriculture? Agron J 99:1719–1725
McCance RA, Widdowson EM (2015) McCance and Widdowson's the composition of foods integrated dataset. https://www.gov.uk/government/publications/composition-of-foods-integrated-dataset-cofid. Accessed 18 Nov 2017
McCrory MA, Hamake BC, Lovejoy JC et al (2010) Pulse consumption, satiety, and weight management. Adv Nutr 1:17–30
Morinaka T, Wozniewicz M, Jeszka J et al (2013) Westernization of dietary patterns among young Japanese and Polish females—a comparison study. Ann Agric Environ Med 20(1):122–130
O'Broin S, Tucci LR (eds) (2016) Pulses: nutritious seeds for a sustainable future. FAO, Rome
Oliveira B, Moura AP, Cunha LM (2016) Reducing food waste in the food service sector as a way to promote public health and environmental sustainability. In: Filho WL, Azeiteiro UM, Alves FA (eds) Climate change and health improving resilience and reducing risks. Springer, Cham, pp 117–131
Rebello CJ, Greenway FL, Finley JW (2014) Whole grains and pulses: a comparison of the nutritional and health benefits. J Agric Food Chem 62(29):7029–7049
Rodrigues SS, Franchini B, Graça P et al (2006) A new food guide for the Portuguese population: development and technical considerations. J Nutr Educ Behav 38:189–195
Rose TJ, Hardiputra B, Rengel Z (2010) Wheat, canola and grain legume access to soil phosphorus fractions differs in soils with contrasting phosphorus dynamics. Plant Soil 326(1):159–170
Schneider AVC (2002) Overview of the market and consumption of pulses in Europe. Br J Nutr 88 (S3):243–250
Snyder CS, Bruulsema TW, Jensen TL et al (2009) Review of greenhouse gas emissions from crop production systems and fertilizer management effects. Agr Ecosyst Environ 133(3–4):247–266
Tiwari B, Singh N (2012) Pulse chemistry and technology. Royal Society of Chemistry, Cambridge
Tongwane M, Mdlambuzi T, Moeletsi M et al (2016) Greenhouse gas emissions from different crop production and management practices in South Africa. Environ Dev 19:23–35
Tosh SM, Yada S (2010) Dietary fibres in pulse seeds and fractions: characterization, functional attributes, and applications. Food Res Int 43:450–460
Tubiello FN, Salvatore M, Rossi S et al (2013) The FAOSTAT database of greenhouse gas emissions from agriculture Environmental. Res Lett 8(1):015009
UN (2012) World population prospects, the 2012 revision. United Nations Department of Economic and Social Affairs, Population Estimate and Projections Section, Rome

UN (2014) Resolution adopted by the general assembly on 20 December 2013. Resolution 68/231. International year of pulses, 2016. http://www.un.org/en/ga/search/view_doc.asp?symbol=A/RES/68/231&referer=/english/&Lang=E. Accessed 18 Nov 2017

Varela-Moreiras G, Ávila JM, Cuadrado C et al (2010) Evaluation of food consumption and dietary patterns in Spain by the food consumption survey: updated information. Eur J Clin Nutr 64:S37–S43

Venkatesh MS, Hazra KK, Katiyar R (2014) Nutrient acquisition and recycling through pulses. In: Ghosh PK, Kumar N, Venkatesh MS, Hazra KK, Nadarajan N (eds) Resource conservation technology in pulses. Scientific Publishers (India), Jodhpur, pp 190–198

WHO (2011) Global status report on noncommunicable diseases 2010. World Health Organization, Geneva

Willet WC, Sacks F, Trichopoulou A et al (1995) Mediterranean diet pyramid: a cultural model for healthy eating. Am J Clin Nutr 61:1402S–1406S

Beatriz Oliveira is a Nutritionist from the Faculty of Nutrition and Food Sciences of the University of Porto. She is the Quality Director at Eurest Portugal, Compass Group. She is a Ph.D. student from the Doctoral Programme on Food Consumer Sciences and Nutrition, University of Porto. Her main research topics relate to sustainable practices at the food service sector. She is a researcher at the GreenUPorto.

Ana Pinto de Moura is a Food Engineer from the Portuguese Catholic University (ESB-UCP), with a Ph.D. on Industrial Engineering Systems—with a focus on food consumption- from the INPL-Nancy, France. She is an Assistant Professor at the Universidade Aberta (Open University of Portugal), where she coordinates the distance learning (online) M.Sc. course on Food Consumption Sciences. She is a researcher at the GreenUPorto and at the Food Quality and Safety research group from the Associated Laboratory LAQV, REQUIMTE. Her main research topics relate to consumer behaviour towards food. She has specific interest in sustainable food consumption, food risk perception, attitudes towards food and nutrition, and perception of nutritional information.

Luís Miguel Cunha is a Food Engineer with a Ph.D. on Biotechnology with specialization on Food Science and Technology, from ESB-UCP. He is an Associate Professor w/tenure at the Faculty of Sciences, University of Porto. He is a researcher at GreenUPorto and at the Food Quality and Safety research group from the Associated Laboratory LAQV, REQUIMTE.

His research focuses on consumer attitudes towards food safety and quality and on consumer oriented sensory evaluation of foods applied to new product development. He has authored more than 70 articles in international refereed indexed journals (over 1000 citations) and 15 book chapters and supervised over 70 M.Sc. and Ph.D. theses in Food Science and Technology. Also, he has wide experience as leader and researcher in academic and industrial research projects. He has a particular interest on the application of sensory evaluation for new product development and on the interfaces between sensory science, consumer science, and marketing, which led to the proposal of joint M.Sc. and Ph.D. degrees on Consumer Sciences and Nutrition, between the Faculty of Sciences and the Faculty of Nutrition and Food Sciences, the first starting in 2007 and the second in 2009.

Ecosystem Services and Incentive Mechanisms for Environmental Preservation in Brazil

Andréia Faraoni Freitas Setti, Walter Leal Filho and Ulisses M. Azeiteiro

Abstract The evidence that environmental imbalances pose a serious and imminent threat to the future of mankind has prompted concrete actions in the environment. However, in Brazil, the challenge is to integrate the various regulations, public policies, new opportunities, and incentive mechanisms for forest protection and restoration. This paper discusses specific forms of action in Brazilian systems of Law, Economics and Politics that can influence environmental issues. It systemically presents economic instruments and analyzes the adoption of programs of Payment for Environmental Services to encourage voluntary practices of environmental protection. We suggest a mandatory strengthening of local power to increase the effectiveness of environmental legislation. The local sphere relates more closely with the more tangible reality and, therefore, is the closest instance of political decision-making that most directly affects people's lives. It is also where the exercise of citizenship is more fruitful and where popular participation is more intense. We assume that sustainable development is unachievable without governance because it promotes common goals through collective action and requires structural changes in the dominant institutions.

Keywords Ecosystem services · Public policy · Payment for environmental services · 2030 agenda for sustainable development

---

A. F. F. Setti (✉) · U. M. Azeiteiro
Department of Biology and CESAM Centre for Environmental and Marine Studies, University of Aveiro, 3810-193 Aveiro, Portugal
e-mail: andreiasetti@gmail.com

W. L. Filho
School of Science and the Environment, Manchester Metropolitan University, Manchester, UK

P. Castro et al. (eds.), *Climate Change-Resilient Agriculture and Agroforestry*, Climate Change Management, https://doi.org/10.1007/978-3-319-75004-0_3

1 Introduction

Both the economy and human welfare depend on ecosystem services to maintain life and productive activities. Mineral resources and energy sources fuel the economy, but we are also handed other essential elements for our survival: food, water, wood, biomass; regulation services, climate balance, flood control, disease control, air purification; recycling of nutrients, soil formation, oxygen production, as well as other benefits related to our culture, like scenic, recreational, touristic, spiritual and educational elements (Finvers 2008; Bastian et al. 2012).

Although the number of benefits is great, there is no encompassing scientific description both of the multiple relationships between different ecosystem services and of the impact of such approach on political decision-making (Cox and Searle 2009; Haines-Young and Potschin 2010).

Instead, we have been witnessing a growing degradation of ecosystems throughout the world (MEA 2005; Heinberg 2010) and an increase in the risks posed by natural and extreme events (floods, landslides, forest fires, heatwaves), which shows that there has been a decline in the resilience of ecosystems throughout the world (Bastian et al. 2012).

The current mode of production and consumption—created with the premise that the exploration of resources can increase limitlessly—is destroying biodiversity and changing the ability of our ecosystems to produce essential goods and services for life.

The environmental legislation of many countries does not include the idea of "ecosystem services," but, on the other hand, science hasn't also been able to cause a political impact, given the lack of data, standards or evaluation (Cox and Searle 2009).

The restoration of ecosystems, in turn, can be hard or expensive in the case degradation isn't irreversible, causing, nonetheless, extinctions and changes in its defining traces (Keith et al. 2013). Moreover, complex and dynamic systems, such as oceanic and atmosphere circulation, for instance, if systemically disrupted, may reach a tipping point, a point of no return (Medeiros et al. 2017).

Despite the vast increase in ecosystem services (ES) studies in recent years, it has been shown that the outputs of these assessments are not yet suitable for decision making (Martínez-Harms and Balvanera 2012; Schägner et al. 2013). The studies are focusing on key ES and yet information on many other services is scarce but essential for sound decision making. The ecosystem services often span across several administrative structures, i.e., address different policy aspects, which are often covered by different governmental ministerial or departmental units. This requires the integration of different sectors and disciplines once different value dimensions are considered, from biophysical to socio-cultural to economic (Grêt-Regamey et al. 2016).

Such challenges, including limited capacities of relevant policy units or dispersed authorities, complicate the operationalization of the ES concept. There are recommendations for a better implementation of ES into decision making spanning from the further development of policy instruments and financial mechanisms to a better integration of the regulations, public policies, and incentive mechanisms for forest protection and restoration, and a better representation of methods and results to a more interdisciplinary research (Scarlett and Boyd 2015; Grêt-Regamey et al. 2016).

Given the relevance and the urgency, the conservation of biodiversity and the maintenance of healthy ecosystems have been included in 8 of the 17 Sustainable Development Goals (SDGs), both directly, such as SDG 15 (protect, recover and promote the sustainable use of land ecosystems), and indirectly in its relations with ending poverty (SGD 1); ending hunger, achieving food security and improved nutrition and promoting sustainable agriculture (SGD 2); ensuring healthy lives and promoting well-being for all at all ages (SDG 3); promoting sustained, inclusive and sustainable economic growth (SDG 8); ensuring sustainable consumption and production patterns (SDG 12); taking urgent action to combat climate change and its impacts (SDG 13); conserving and sustainably use the oceans, seas and marine resources (SDG 14).

This paper analyzes specific ways of interfering with the Brazilian Law, Economy and Politics which may influence their treatment of ecosystem services, especially considering not only their potential of being included in regulations, policies and mechanisms dedicated to incentivizing the protection or restoration of forests, as well as the fulfillment of SDGs as from ecosystem services-oriented forms of governance.

2 Environmental Economics and the Conservation of Ecosystems

The economic literature understands ecosystem services as a mechanism that corrects a type of market failure called "negative externality," that is, the costs that circulate externally to that market, do not burden production, but falls over other parties or a population that has no relation with the activity (Nusdeo 2012).

Establishing a monetary value to environmental resources does not mean treating it like a product, but only measuring ecological benefits and damages to be applied in accountable and efficient forms of management, considering that the ecological functions of natural resources have indirect value (Slootweg et al. 2008; Chan et al. 2012).

The economic valuation of ecosystem services must influence the decision-making process regarding the importance of conserving the biodiversity, both identifying and distributing the costs and benefits of different actors, producers and consumers (Paggiola et al. 2004).

Measurement mechanisms depend on the type of environmental good or service is under scrutiny and the dimension of their contribution for (individual or collective) wellbeing. Information on availability, scarcity and willingness of people to work for its conservation must be accounted for (Bateman and Willis 1999).

The conceptualization of the idea of "externality" lies in the intersection between Economics, Ecology and Law, including also the management of public policies. The understanding of reality and its problems (as well as the promotion of solutions) in environmental law and environmental sciences is, usually, interdisciplinary.

For this paper, we considered the relations between individuals, institutions and nature, as well as the premise that, in order to be effective and efficient in terms of sustainability, policies must necessarily focus on equity (Setti et al. 2016).

The destruction of ecosystems and the unsustainable use of the services they provide generate serious environmental problems and intensify social inequalities and poverty throughout the world, affecting especially traditional communities (MEA 2005). Therefore, environmental policies need to be linked to both the economic and social spheres of the development process, altering the cost-benefit analysis of certain economic activities given their negative socioenvironmental impacts.

3 Mechanisms and Instruments to Preserve Ecosystem Services in Brazil

States use mechanisms of command and control (standards, limitations and prohibitions), as well as economic instruments (charging for the use of water, redistributing taxes on goods and services to pay for environmental services; direct payment for environmental services; concessions of forests, etc.) to correct socioenvironmental distortions and injustices produced by the market.

Brazil has been implementing economic instruments related to management and preservation of the environment in the last thirty years due to greater demographic occupation, irregular urban occupation, the expansion of the agricultural frontier and the uncontrolled use of natural resources (Born and Talocchi 2002).

One of the economic instruments that promotes the preservation of ecosystem services is charging for the use of natural resources. This can restrict use and improve quality through taxes or fees on economic activities that degrade the environment and incentives for the sustainable use of natural resources and environmental protection (Nusdeo 2012).

One of the first financial policies dedicated to environmental services in Brazil was the Tax on the Circulation of Ecological Goods and Services ("ICMS Ecológico"), created in the 1990s. The *ICMS Ecológico* is a tax that allows that part of the taxes collected by the state government is handed to municipalities that have conservation units created by the state or federal governments within their boundaries. This has been leading local leaders to understand environmental protection as something positive, given that it stimulates sustainable development, as well as ecotourism and organic food production, expanding employment and income opportunities (Born and Talocchi 2002).

Another initiative is charging for water use, which was established by the National Policy on Water Resources (Law 9.433/1997). Resources must be used to protect drainage basins, including reforestation and forest conservation (DOU 1997). Drainage basins were adopted as regional units for the planning and management of water.

Such reorganization of the system hands power back to decentralized institutions within the basin, stimulating the negotiation between various public agents, users and the organized civil society. Popular participation expands the access of people to basic urban services and infrastructure, developing the civil society and strengthening democratic mechanisms (Jacobi and Barbi 2007).

In the last ten years, not only there was great progress in environmental legislation, but also consultative and advisory councils were strengthened in several areas and in all levels (federal, state and municipal), with the active participation of representatives from social movements and NGOs (Jacobi and Barbi 2007).

A similar system was established by the National System of Conservation Units (NSCU—Law 9.985/2000) as an attempt of providing more resources to Conservation Units (CU), given the ecosystem services they provide to society. NSCU also exempts owners of Private Natural Heritage Reserves (PNHR) of Rural Land Taxes for the protected area (DOU 2000).

The country has increased the extension of its CUs, but did not obtain the necessary budget to maintain the existing CUs, allocating 20% less than what is considered the minimum necessary (Semeia 2014).

For the CUs to get financial resources from direct donations, the distribution of water and the generation of electricity, articles 34, 47 and 48 of the Law that created NSC must still be "regulated." Another source could be the so-called "legal reserve compensation," through the regularization of land ownership at CUs and implementation of the National Policy on Payment of Environmental Services (NPPES), which Bills 792/2007 and 213/2015—already introduced in Congress—address.

The Payment for Environmental Services (PES) is the voluntary payment of those who promote the conservation, recovery, expansion or the management of areas with vegetation considered to provide environmental services (Nusdeo 2013).

Although there are PES tools available, there is no national policy for PES methods currently in force. However, Bill 5.586-A/2009 on certified Reductions of Emissions from Deforestation and Forest Degradation (REDD), the Brazilian Forest Code (Law 12.561/2012) and the National Policy on Water Resources (Law 9.433/97) are all related to ecosystem services. States have more than 20 PES-related laws, decrees and bills (WWF 2014).

The goal is assuring coherence between public and private initiatives regarding the recovery and conservation of biodiversity on the national, state, regional and municipal scales, as well as promoting sectoral government policies that may produce an impact over ecosystem services, especially through policies in the following areas: Environment, Water Resources, Climate Change, Protected Areas, Technical Assistance and Rural Extension for Family Farming and Land Reform and the Forest Code.

4 Governance to Preserve Ecosystem Services

Ecosystem services-oriented forms of governance demands that intersectoral and participatory public management concepts, practices, mechanisms and tools are further deepened, promoting equity and socioenvironmental sustainability in the territory. Traditional communities contribute for that equitable model of development which both defends the environment and natural resources and promotes the solidarity economy and quality of life improvements (Setti and Azeiteiro 2016).

These communities have an ethical behavior regarding the conservation and preservation of life and the environment and, therefore, have a dialectic relationship with society, both denying and affirming its values, continuously recreating survival strategies. Unfortunately, indigenous populations, the "quilombolas" (remnant populations of fugitive slaves) and traditional populations—which are protected by the Federal Constitution of 1988—have been considered hindrances to development.

The lifestyles of traditional communities—including the way they materially and symbolically appropriate nature, their forms of knowledge, their technologies, their cultural practices and actions they take in the territory –, given the need to promote diversity in all its forms, are a counterpoint to the unsustainable lifestyles of urban/ industrial societies (Zhouri and Laschefski 2010).

Natural areas and traditional communities are being decimated in the environmental conflicts that emerge due to the unequal distribution of natural resources and the territorial disputes between groups that use the environment differently. Establishing commitments of reaching consensuses are difficult because of the clashing rationalities (modes of being, doing and thinking).

Besides deforestation, other activities such as large hydroelectricity generation projects also represent threats to the provision of environmental services and contribute to climate change (Fearnside 1989; Oyama and Nobre 2003; Betts et al. 2004).

In this scenario of inequality and conflict, traditional communities not only are excluded from development, but also have to take a greater load of the consequences of environmental degradation, fighting for autonomy and resisting the current modes of production and consumption and social organization.

In Brazil, there are 305 indigenous ethnicities and 274 languages (IBGE 2010). The protection of their territories and the traditional use of natural resources – which are fundamental to their culture and lifestyle—are part of their identity (Setti and Azeiteiro 2016). This has been identified in SDG 4, which recommends that the contributions of African descendants and indigenous populations for the development of nations are included in school curricula (UN 2017).

Goals were established to protect and assure the existence of traditional peoples' and communities' ways of creating, doing and living: preserving historical sites; carrying out mappings, inventories and studies on traditional memory, rites and celebrations; preserving linguistic diversity, as well as expressions, artistic expressions and cultural practices of the various ethnicities; promoting a culture of diversity, solidarity, equality and inclusion in the media; and strengthening peoples and cultures for climate change-related planning (UN 2017).

Environmental sustainability, in the context of traditional communities, has to do with the sustained use of natural resources and a more equitable distribution of wealth, and should be considered in environmental policies and licensing (Zhouri and Laschefski 2010).

Therefore, any public policy dedicated to the preservation of ecosystems must be strategically linked to territorial programs that allow for the inclusion of the idea of sustainable development in the territorial spheres and jurisdictions of government.

5 Integration of Public Policies, Regulations, and Incentive Mechanisms for Protecting Ecosystem Services

The integration of public policies concerning ecosystem services has to do with the hierarchy and territoriality of the relations between national, state and municipal spheres of government and their most varied combinations—given their territorial distribution, which often include more than one municipality, state or region, such as, for instance, the territorial distribution of metropolitan areas, coastal zones, areas with specific land-use conditions (Gallo and Setti 2014; IBAM 2016).

Public policies related to environmental services that includes traditional communities must intersectorally integrate the different themes reflected in government administrative structures from a systemic perspective that recognizes individuals in their contexts.

This demands participatory governance processes capable of ranking priorities based on the needs of the territory (Gallo and Setti 2014), whose institutional format promotes greater social participation in the processes of dialogue, negotiation, representation, planning and evaluation of public policies.

To meet the growing demand for environmental services and, at the same time, benefit traditional communities, the Brazilian government created policies and programs that generate intergovernmental cooperative arrangements that assure complementarity, synergy and the optimization of technical and financial resources.

The National Program for the Sustainable Development of Rural Territories (2003) considered the demands of traditional populations and indigenous peoples, although the processes of planning and implementation did not go beyond the municipal level and were restricted to small groups of beneficiaries (Bonnal 2013).

The Program for the Socioenvironmental Development of Rural Family Production (*Proambiente*) included priority groups (family farmers, artisanal fishermen, traditional populations and indigenous peoples), was based on the balance between environmental conservation and family farming and was implemented through the environmental management of rural areas, the integrated planning of productive units and the rendering of environmental services. *Proambiente* has the interesting history of having been a project created by the civil society (2000/2002) that went through a phase of transition (2003) to become a program of the federal government (Hall 2008; Neto 2008).

The National Policy for the Sustainable Development of Traditional Peoples and Communities—introduced by Decree 6.040/2007 and implemented by the National Commission for the Sustainable Development of Traditional Peoples and Communities (SDTPC)—emphasizes the recognition, the strengthening and the assurance of the territorial, social, environmental, economic and cultural rights of traditional peoples and communities, respecting and valorizing their identity, their forms of organization and their institutions.

This policy is an advance in the fight against the invisibility and the social exclusion of traditional communities by economic or land-ownership pressures or through discriminatory process in which the recognition of *quilombola* lands and the demarcation of indigenous lands, for instance, clash with the advance of agriculture, logging and mining, as well as the construction of large infrastructural projects, especially in the areas of transportation and energy, to meet the economic demands.

In the context of the creation of the markets for Payments for Environmental Services (PES), the National Water Agency created the Water Production Program (2001) to incentivize the preservation of riparian forests around water springs in private properties, assuring the increase in quantity and in quality of the water offered to the population and including technical and financial support (Wunder et al. 2009).

To assure resources for the projects, studies or any undertaking aiming at the mitigation and adaptation to climate change, the Climate Fund (Law 12.114/2009) was created to support initiatives related to PES policies, providing support to activities that stabilize the concentration of greenhouse gases and that demonstrably contribute to carbon sequestration and the provision of other environmental services, such as the recovery of degraded areas and forest restoration (Santos et al. 2012).

The Green Stipend is an environmental conservation support program created in 2011 (Law 12.512/2011) to support families in extreme poverty living in areas that have been considered priorities for conservation. The program provides a stipend for family farmers, traditional communities and people settled through land reform. The environmental services provided include the maintenance of vegetation in the property where the family resides and the sustainable use of its resources (Santos et al. 2012).

The Brazilian environmental legislation is supported by international declarations and is based on the "polluter pays" principle that charges polluters for the social costs of the productive process, an encompassing mechanism of accountability for ecological damage not exclusively related with immediate reparation (Milaré 2001). The "protector receives" principle is the rationale behind the payment of environmental services and the basis for economic incentives for the protection of areas and the preservation of their resources (Nusdeo 2012).

At the municipal scale, programs for the payment of environmental services are usually dedicated to the preservation of riparian forests and water resources. 7.5% of all municipalities in the country, mostly in the Central-West Region, already pay for environmental services (IBGE 2013).

Ecosystem services have not been discussed exclusively on the public sphere in Brazil. The private sector has also been recognizing the importance of these services for their own businesses. The degradation of ecosystems is relevant for companies that both produce and impact and depend on ecosystems and the services they provide.

The social accountability of companies is a voluntary commitment with a responsible form of management regarding their partners, employees, suppliers, consumers, community and environment that goes beyond the simple fulfillment of their legal obligations. Many organizations have been investing in social projects and taking on the responsibility for the impact their productive processes cause. Companies benefit both because it consolidates their image or brand as a modern and sustainable company and because productivity and competitiveness are linked to the quality of life of the community the institution is a part of (Garcia 2002).

On the other hand, the idea of "social function" is part of the Brazilian constitutional legal order under Art. 170 of the Federal Constitution of 1988, which establishes that entrepreneurs and administrators need to carry out their activities harmoniously, complying with their positive and negative duties and respecting the interests of the society.

The Corporate Partnership for Ecosystem Services (CPES) was created to help companies reduce their negative impacts over ecosystem services; show the value of ecosystems and of the conservation of biodiversity through business strategies that maintain such services; and attain practical results that further expand these business strategies (CEBDS 2013).

To conserve natural resources and use them sustainably, it is fundamental that policies that promote environmental integrity and social inclusion are effective and not a mere mobilization of financial resources and a creation of new markets. The growth of productive chains that cause pollution and the degradation of forests must be limited, and a mode of production and consumption based on solidarity and sustainability must be established.

Therefore, mechanisms for paying for environmental services—in combination with other regulations and tools for social control—must be accompanied by continuous processes of participatory evaluation aiming at suppressing the biodiversity and ecosystem markets, which would subject public and social interests to private, corporate ones.

6 Ecosystem Services and the Fulfilment of Sustainable Development Goals—SDGs

The idea of sustainable development emerged historically with evidence of ecological unbalance and an increase in social inequality, highlighting the unsustainability of the hegemonic mode of production and consumption. This triggered a "social cycle" in the agenda of international organizations, which globally consolidated the agenda of sustainability expressed through the SDGs.

This opens possibilities for the implementation of policies for promoting equity and environmental integrity, expanding the access to citizenry, the preservation of the environment, the solidarity economy and the quality of life (Kumar et al. 2013).

There are serious of issues related to ecosystems and biodiversity that require articulation of socioenvironmental policies between all government spheres. This strengthens shared management strategies focused on the co-accountability for the fulfilment of SDGs.

The eradication of poverty remains a priority. In terms of health, poverty shortens life almost as much as physical inactivity and much more than obesity, hypertension and the excessive consumption of alcohol. Low socioeconomic status is one of the strongest predictors of premature morbidity and mortality all over the world. Poor nutritional status is the cause of 45% of deaths among under-fives. One in every four children in the world is stunted (in developing countries the rate is one in every three); 66 million children in primary school age attend school while hungry (Stringhini et al. 2017).

Damages caused by climate change increase losses caused by the economic crisis, affecting the poorest disproportionally (TEEB 2010), which shows that social inequalities are always ecological inequalities. They determine the modes and levels of access to ecosystem goods and services, which, in turn, also determine both health and disease processes and environmental sustainability.

Other strategic areas for intervention include: (1) the conservation of ecosystems and its services by strengthening protected areas; (2) the implementation of equitable strategies for managing natural resources; (3) the prevention and control of deforestation, as well as the conservation and the sustainable management of forests; (4) the strengthening of inclusive environmental governance mechanisms; (5) the access to genetic resources and the fair distribution of the benefits of their use; (6) the identification and monitoring of exotic invasive species of plants, animals and microorganisms that affect the environment and the health of human beings; (7) the recovery and restoration of degraded ecosystems and the recovery of threatened species; (8) community-based evaluations of ecosystems based on ecosystem services and human wellbeing.

7 Conclusion

The absence of a legal framework for environmental services, the little information on methodologies for measuring and monitoring them, the absence of continued sources of financing and the unfamiliarity with the social and environmental function of traditional communities keep the national policy of environmental services from becoming a strategic action and a positive agenda.

To implement this environmental rationale, environment preservation values need to become an integral part of personal ethics, human rights and the law. The access to and the appropriation of nature cannot continue a privilege of a few, the production of knowledge must be redirected toward transdisciplinarity, and the State must be reformed to promote the participatory management of natural resources.

The actual participation of the society in the economic development process is crucial for the protection of natural resources, in the development of a political-institutional framework for articulated public policies in the fields of health, urban development, the environment, natural resources and education.

Therefore, development ranges from the protection of human rights to a deepening of democracy, to both the effective possibility of everyone participating politically and the expansion of human capacities.

In this sense, the empowerment of traditional communities through education, valorization of traditional culture and knowledge and dissemination of social technologies is achieved through the participation of the people in the management of the territory including the creation of public and multicultural articulation spaces, focused on sustainability and socialenvironmental justice.

Theory seeks to offer decision-makers alternatives that promote the conservation of biodiversity, pointing to the economic tools that make decisions in that direction viable. The challenge is to change the conservation of biodiversity into a technical, political and social issue that is vitally important for sustainable development.

Investments in new and more efficient technologies that reduce the environmental impact of industrial processes have become increasingly less viable. In order to preserve our natural capital, we need sustainable economic reforms based on the assumption that the economic crisis is an opportunity for a paradigm shift regarding our modes of production and consumption.

The use of economic tools in environmental management demands efforts in three areas: macroeconomic coherence, legal compliance and technical capacity. Such areas depend, however, on institutional management and governance capacity for sustainable development.

Given scarcity and the uncertainty involved in the relations between the economic and natural systems, we should be cautious and skeptical given the possibility of irreversible and unreplaceable loss of ecosystem services that are essential for our well-being and survival.

References

Bastian O, Haase D, Grunewald K (2012) Ecosystem properties, potentials and services—the EPPS conceptual framework and an urban application example. Ecological Indicators. Elsevier. v.21, https://doi.org/10.1016/j.ecolind.2011.03.014

Bateman IJ, Willis KG (1999) Valuing environmental preferences: theory and practice of the contingent valuation method in the US, EU, and developing countries. Oxford University Press, Oxford. ISBN-13: 9780199248919, https://doi.org/10.1093/0199248915.001.0001

Betts RA, Cox PM, Collins M, Harris PP, Huntingford C, Jones CD (2004) The role of ecosystem-atmosphere interactions in simulated Amazonian precipitation decrease and forest dieback under global climate warming. Theor Appl Climatol 78:157. Springer, New York. https://doi.org/10.1007/s00704-004-0050-y

Bonnal P (2013) Referências e considerações para o estudo e a atuação dos programas de desenvolvimento territorial (PRONAT e PTC) na perspectiva da redução da pobreza em territórios rurais. In: MIRANDA, Carlos; TIBURCIO, Breno (Org.). Políticas de Desenvolvimento Territorial e Enfrentamento da Pobreza Rural no Brasil. Série Desenvolvimento Rural Sustentável, vol 19. Brasília: IICA

Born RH, Talocchi S (2002) Proteção do capital social e ecológico por meio de compensações por serviços ambientais (CSA). Vitae Civilis, São Paulo. ISBN 9788585663926

CEBDS—Conselho Empresarial Brasileiro para o Desenvolvimento Sustentável. Biodiversidade e Serviços Ecossistêmicos: a experiência das empresas brasileiras. 2013. Available in: http://cebds.org/publicacoes/biodiversidade-e-servicos-ecossistemicos-a-experiencia-das-empresas-brasileiras/#.VZnDnPlVhBc

Chan KMA, Guerry AD, Balvanera P (2012) Where are cultural and social in ecosystem services? Framework for constructive engagement. Bioscience 62:744–756. Oxford University Press on behalf of the American Institute of Biological Sciences. https://doi.org/10.1525/bio.2012.62.8.7

Cox S, Searle B (2009) The state of ecosystem services. The Bridgespan Group, Boston. Available in: https://www.bridgespan.org/bridgespan/images/articles/executive-summary-the-state-of-ecosystem-services/The-State-of-Ecosystem-Services_1.pdf?ext=.pdf

Fearnside PM (1989) Brazil's Balbina dam: environment versus the legacy of the Pharaohs in Amazonia. Environ Manage 13. Springer, New York. https://doi.org/10.1007/BF01867675

Finvers MA (2008) Application of DPSIR for analysis of soil protection issues and an assessment of British Columbia's soil protection legislation. M. Sc. Thesis. Cranfield University, UK

Gallo E Setti AFF (2014) Território, intersetorialidade e escalas: requisitos para a efetividade dos Objetivos de Desenvolvimento Sustentável. Ciência e Saúde Coletiva 19, Rio de Janeiro. http://dx.doi.org/10.1590/1413-812320141911.08752014

Garcia BG (2002) Responsabilidade social das Empresas: a contribuição das Universidades. Editora Petrópolis, São Paulo. ISBN 85-7596-052-0

Grêt-Regamey A, Sirén E, Brunner SH, Weibel B (2016) Review of decision support tools to operationalize the ecosystem services concept. Ecosyst Serv 26:306–315. Zurich, Switzerland. https://doi.org/10.1016/j.ecoser.2016.10.012

Haines-Young R, Potschin M (2010) Proposal for a common international classification of ecosystem goods and services (CICES) for integrated environmental and economic accounting, Report to the European Environmental Agency, University of Nottingham, UK

Hall A (2008) Better RED than dead: paying the people for environmental services in Amazonia. R Soc Philos Trans R Soc 363, London. https://doi.org/10.1098/rstb.2007.0034

Heinberg R (2010) Turning the corner on growth. The Solutions Journal – Perspectives. ISSN 2154-0896. Available in: https://www.thesolutionsjournal.com/article/turning-the-corner-on-growth/

IBAM—Instituto Brasileiro de Administração Municipal (2016) Políticas públicas para cidades sustentáveis: integração intersetorial, federativa e territorial. Rio de Janeiro, IBAM, MCTI. ISBN: 978-85-88063-33-4. Available in: http://www.iicabr.iica.org.br/wp-content/uploads/2016/12/livro-Pol%C3%ADticas-Públicas-para-Cidades-MCTI.pdf

IBGE—Instituto Brasileiro de Geografia e Estatistica (2010) Os indígenas no Censo Demográfico 2010—Primeiras Considerações com Base no Quesito Cor ou Raça: In Censo Demográfico. Rio de Janeiro, IBGE, MPOG. Available in: https://www.ibge.gov.br/indigenas/indigena_censo2010.pdf

IBGE—Instituto Brasileiro de Geografia e Estatística (2013) Pesquisa de Informações Básicas Municipais. Perfil dos municípios brasileiros: 2012. Rio de Janeiro, IBGE, ISBN 978-85-240-4292-8. Available in: ftp://ftp.ibge.gov.br/Perfil_Municipios/2012/munic2012.pdf

Jacobi PR, Barbi F (2007) Democracia e participação na gestão dos recursos hídricos no Brasil. Rev Katál Florianópolis 10(2):237–244. https://doi.org/10.1590/S1414-49802007000200012

Keith DA, Rodríguez JP, Rodríguez-Clark KM, Nicholson E, Aapala K, Alonso A, Asmussen M, Bachman S, Basset A, Barrow EG, Benson JS, Bishop MJ, Bonifacio R, Brooks TM, Burgman MA, Comer P, Comín FA, Essl F, Faber-Langendoen D, Fairweather PG, Holdaway RJ, Jennings M, Kingsford RT, Lester RE, Nally RM, McCarthy MA, Moat J, Oliveira-Miranda MA, Pisanu P, Poulin B, Regan TJ, Riecken U, Spalding MD, Zambrano-Martínez S (2013) Scientific foundations for an IUCN Red List of Ecosystems. PubMed. https://doi.org/10.1371/journal.pone.0062111

Kumar P, Esen ES, Yashiro M (2013) Linking ecosystem services to strategic environmental assessment in development policies. Environ Impact Assess Rev 40:75–81. https://doi.org/10.1016/j.eiar.2013.01.002

Martínez-Harms MJ, Balvanera P (2012) Methods for mapping ecosystem service supply: a review. Int J Biodivers Sci Ecosyst Serv Manage 8:17–25. https://doi.org/10.1080/21513732.2012.663792

MEA—Millennium Ecosystem Assessment (2005) Ecosystems and human wellbeing: synthesis. Island Press, Washington, DC. ISBN 1-59726-040-1. Available in: https://www.millenniumassessment.org/documents/document.356.aspx.pdf

Medeiros ES, Caldas IL, Baptista MS, Feudel U (2017) Trapping Phenomenon attenuates the consequences of tipping points for limit cycles. Sci Rep 7:42351. https://doi.org/10.1038/srep42351

Milaré E (2001) Direito do ambiente: doutrina, prática, jurisprudência, glossário. 2a ed. São Paulo: Revista dos Tribunais. ISBN: 8520320759, 9788520320754

Neto PSF (2008) Avaliação do Proambiente - Programa de Desenvolvimento Socioambiental da Produção Familiar Rural. Ministério do Meio Ambiente. Brasília. Available in: http://www.mma.gov.br/estruturas/sds_proambiente/_arquivos/33_05122008040536.pdf

Nusdeo AMO (2012) Pagamento por serviços ambientais: sustentabilidade e disciplina jurídica. Atlas, São Paulo. ISBN 9788522470778

Nusdeo AMO (2013) Pagamento por serviços ambientais. Do debate de política ambiental à implementação Jurídica. In: Direito e mudanças climáticas: Pagamento por Serviços Ambientais, fundamentos e principais aspectos jurídicos. Lavratti, P.; Tejeiro, G. São Paulo: Instituto O Direito por um Planeta Verde. Available in: http://www.planetaverde.org/arquivos/biblioteca/arquivo_20131201182658_5649.pdf

Oyama MD, Nobre C (2003) A new climate-vegetation equilibrium state for Tropical South America. Geophys Res Lett 30. AN AGU Journal, USA. https://doi.org/10.1029/2003gl018600

Paggiola S, Von Ritter K, Bishop J (2004) Assessing the economic value of ecosystem conservation. The World Bank Environmental Department. Washington, USA. Available in: http://citeseerx.ist.psu.edu/viewdoc/download?doi=10.1.1.578.4617&rep=rep1&type=pdf

Santos P, Brito B, Maschietto F, Osório G, Monzoni M (2012) Marco Regulatório sobre Pagamento por Serviços Ambientais no Brasil. Imazon. FGV. GVces, Belém. ISBN 978-85-86212-45-1

Scarlett L, Boyd J (2015) Ecosystem services and resource management: institutional issues, challenges, and opportunities in the public sector. Ecological Economics. Elsevier, Amsterdam. https://doi.org/10.1016/j.ecolecon.2013.09.013

Schägner JP, Brander L, Maes J, Hartje V (2013) Mapping ecosystem services' values: current practice and future prospects. Ecosyst Serv 4:33–46. https://doi.org/10.1016/j.ecoser.2013.02.003

Semeia (2014) Unidades de conservação no Brasil: a contribuição do uso público para o desenvolvimento socioeconômico. Instituto Semeia, São Paulo: Semeia. Available in: http://www.terrabrasilis.org.br/ecotecadigital/images/abook/pdf/1sem2015/Passivo/UC%20Brasil.pdf

Setti AFF, Azeiteiro UM (2016) Education for sustainable development in Brazil: challenges for inclusive, differentiated and multicultural education. In: Castro P, Azeiteiro UM, Bacelar PN, Filho WL, Azul AM (ed) (Org.) Biodiversity and education for sustainable development, vol 1. 1ed.Switzerland: Springer International Publishing. https://doi.org/10.1007/978-3-319-32318-3_15

Setti AFF, Ribeiro H, Azeiteiro UM, Gallo E (2016) Governance and the promotion of sustainable and healthy territories: the experience of Bocaina, Brazil. J Integr Coast Zone Manage 16:57–69. https://doi.org/10.5894/rgci612

Slootweg R, Pieter JH, Beukering V (2008) Valuation of ecosystem services and strategic environmental assessment: lessons from influential cases. Report for the Netherlands Committee for Environmental Assessment, Utrecht. ISBN 978-90-421-2537-7. Available in: http://api.commissiemer.nl/docs/mer/diversen/valuation.pdf

Stringhini S, Carmeli C, Jokela M, Avendaño M, Muennig P, Guida F, Ricceri F, d'Errico A, Barros H, Bochud M, Chadeau-Hyam M, Clavel-Chapelon F, Costa G, Delpierre C, Fraga S, Goldberg M, Giles GG, Krogh V, Kelly-Irving M, Layte R, Lasserre AM, Marmot MG, Preisig M, Shipley MJ, Vollenweider P, Zins M, Kawachi I, Steptoe A, Mackenbach JP, Vineis P, Kivimäki M (2017) Socioeconomic status and the 25 × 25 risk factors as determinants of premature mortality: a multicohort study and meta-analysis of 1.7 million men and women. The Lancet 389, n. 10075. https://doi.org/10.1016/s0140-6736(16)32380-7

TEEB—The Economics of Ecosystems and Biodiversity (2010) A quick guide: the economics of ecosystems and biodiversity for local and regional policy. Printed by Progress Press, Malta. ISBN 978-3-9812410-2-7. Available in: http://www.teebweb.org/media/2010/09/TEEB_D2_Local_Policy-Makers_Report-Eng.pdf

UN (2017) Report of the special Rapporteur on the issue of human rights obligations relating to the enjoyment of a safe, clean, healthy and sustainable environment. Human Rights Council. Available in: https://documents-dds-ny.un.org/doc/UNDOC/GEN/G17/009/97/PDF/G1700997.pdf?OpenElement

Wunder S, Börner J, Tito MR, Pereira L (2009) Pagamentos por serviços ambientais: perspectivas para a Amazônia Legal. 2a ed. Brasília: Ministério do Meio Ambiente. ISBN 978-85-7738-114-2. Available in: http://www.mma.gov.br/estruturas/168/_publicacao/168_publicacao17062009123349.pdf

WWF-Brasil (2014) Diretrizes para a Política Nacional de Pagamento por Serviços Ambientais. Iniciativa Diretrizes PNPSA. Brasília. Available in: http://d3nehc6yl9qzo4.cloudfront.net/downloads/diretrizes_pnpsa__final.pdf

Zhouri A, Laschefski K (2010) Desenvolvimento e conflitos ambientais: um novo campo de investigaçao. In: Zourhi AE, Lachefski K (eds) (Org.) Desenvolvimento e conflitos ambientais, UFMG, Belo Horizonte, pp 11–33. ISBN: 9788570417749

Legislation

DOU (1997). Lei n. 9433, Institui a Política Nacional de Recursos Hídricos, cria o Sistema Nacional de Gerenciamento de Recursos Hídricos, regulamenta o inciso XIX do art. 21 da Constituição Federal, e altera o art. 1º da Lei nº 8.001, de 13 de março de 1990, que modificou a Lei nº 7.990, de 28 de dezembro de 1989. Brasília. http://www.planalto.gov.br/ccivil_03/leis/L9433.htm. Accessed 10 April 2017

DOU (2000). Lei n. 9.985, Regulamenta o art. 225, § 1º, incisos I, II, III e VII da Constituição Federal, institui o Sistema Nacional de Unidades de Conservação da Natureza e dá outras providências. Brasília. http://www.planalto.gov.br/ccivil_03/leis/L9985.htm. Accessed 10 April 2017

DOU (2011). Lei n. 12.512, Institui o Programa de Apoio à Conservação Ambiental e o Programa de Fomento às Atividades Produtivas Rurais; altera as Leis nºs 10.696, de 2 de julho de 2003, 10.836, de 9 de janeiro de 2004, e 11.326, de 24 de julho de 2006. Brasília. http://www.planalto.gov.br/ccivil_03/_ato2011-2014/2011/lei/l12512.htm. Accessed 10 April 2017

DOU (2012), Lei n. 12.561, Dispõe sobre a proteção da vegetação nativa; altera as Leis n°s 6.938, de 31 de agosto de 1981, 9.393, de 19 de dezembro de 1996, e 11.428, de 22 de dezembro de 2006; revoga as Leis n°s 4.771, de 15 de setembro de 1965, e 7.754, de 14 de abril de 1989, e a Medida Provisória nº 2.166-67, de 24 de agosto de 2001; e dá outras providências. Brasília. http://www.planalto.gov.br/ccivil_03/_ato2011-2014/2012/lei/l12651.htm. Accessed 10 April 2017

DOU (2009). Lei n. 12.114, Cria o Fundo Nacional sobre Mudança do Clima, altera os arts. 6º e 50 da Lei nº 9.478, de 6 de agosto de 1997, e dá outras providências. Brasília. http://www.planalto.gov.br/ccivil_03/_ato2007-2010/2009/lei/l12114.htm. Accessed 10 April 2017

DOU (2009). Lei n. 12.187, Institui a Política Nacional sobre Mudança do Clima - PNMC e dá outras providências. Brasília. http://www.planalto.gov.br/ccivil_03/_ato2007-2010/2009/lei/l12187.htm. Accessed 10 April 2017

Are We Missing the Big Picture? Unlocking the Social-Ecological Resilience of High Nature Value Farmlands to Future Climate Change

A. Lomba, A. Buchadas, João P. Honrado and F. Moreira

Abstract Agriculture is a dominant form of management and a major driver of global change in the Anthropocene. Whilst farmland intensification and expansion came at high costs for the global natural capital, the role of low intensity, traditionally managed High Nature Value farmlands (HNVf) for achieving food security while contributing to the conservation of biodiversity and the delivery of ecosystem services has been acknowledged. Yet, most research on HNVf has focused the impacts of land-use change, potentially overlooking the challenges that climate change may pose to their future persistence. In this chapter, we present an overview of the impacts of climate change on HNVf, based on the analysis of current scientific evidence. Overall, while an increasing number of studies was observed, there is a lack of data-driven assessments at the global scale, reflecting an inability to forecast the impacts of climate change on low-intensity farming systems at this scale. High variation was observed regarding the impacts addressed and the topics tackled in space and time in the research records scrutinized. A paradigm shift towards integrated approaches considering both climate and land-use change impacts in the social-ecological systems underlying low-intensity, traditionally managed HNVf is discussed and future perspectives and research needs are outlined.

Keywords Agro-biodiversity · Climate change · Ecosystem Services (ES)
Environmental Sustainability · High Nature Value farmlands (HNVf)
Knowledge gaps · Social-ecological systems (SES)

A. Lomba (✉) · A. Buchadas · J. P. Honrado · F. Moreira
CIBIO/InBIO, Centro de Investigação em Biodiversidade e Recursos Genéticos da Universidade do Porto, Campus Agrário de Vairão, R. Padre Armando Quintas, nº 7, 4485-661 Vairão, Portugal
e-mail: angelalomba@fc.up.pt

P. Castro et al. (eds.), *Climate Change-Resilient Agriculture and Agroforestry*,
Climate Change Management, https://doi.org/10.1007/978-3-319-75004-0_4

1 Introduction

1.1 A Global View of Traditional Farming and Farmlands

Agriculture is among the dominant uses of land (currently ca. 40% of Earth's terrestrial surface) and one of the key drivers of global change of the Anthropocene (Foley et al. 2011; Rockström et al. 2017). Over the 20th century, population growth and increasing demand for food placed an unprecedented demand on agriculture, with expansion and intensification of agricultural practices eroding world's natural capital (Rockström et al. 2017). While recent studies claim the need to roughly double agricultural production to keep pace with projected demands for population growth, achieving food security and environmental sustainability became a hot topic on both scientific and policy agendas (Cumming et al. 2014; Fischer et al. 2017; Godfray et al. 2010).

The role of low-intensity, often traditionally managed, High Nature Value farmlands (HNVf) to achieve global food security, while contributing to environmental sustainability has been increasingly acknowledged (Fischer et al. 2017; Rockström et al. 2017). As all farmlands, low-intensity farming systems are social-ecological systems, resulting from the intertwined relation between Man and nature through centuries (Benayas et al. 2008; Fischer et al. 2012; Plieninger and Bieling 2012). Overall, the cultural and natural heritage of such farmlands relies on the maintenance of specific farming practices that have evolved with, and are adapted to local climatic, topographic and environmental conditions (Altieri 2004; Fischer et al. 2012; Lomba et al. 2014). Understanding the complexity of these social-ecological systems is thus of utmost relevance to effectively anticipate the impacts of social-ecological change and assure their persistence in the future. Mainstream conservation strategies highlight the contribution of low-intensity farmlands to environmental sustainability, including maintenance of biodiversity and ecosystem services (Fischer et al. 2017; Garibaldi et al. 2017; Rockström et al. 2017). In such context, the pivotal role of extensively managed farmlands towards the delivery of a wide range of ecosystem services beyond biodiversity e.g. carbon sequestration, aesthetic landscapes has been claimed (Power 2010; Swinton et al. 2007). While most low-intensity agricultural landscapes and related agroecosystems have coped with anthropogenic pressures through time, increasing current social-ecological change may undermine the inherent resilience of agroecosystems, promoting the widespread loss of biodiversity and ecosystem services, thus jeopardizing their future sustainability (Venter et al. 2016). Fostering the potential contribution of traditional, extensively managed farmlands to meet the societal demands of food security while assuring environmental sustainability is currently a major challenge (Garibaldi et al. 2017).

While recent research targeted the effects of land-use change e.g. abandonment or intensification, and rural development policies on High Nature Value farmlands (e.g. Queiroz et al. 2014; Ribeiro et al. 2014), few studies have tackled the potential impacts of climate change. As HNVf systems are adapted to local climatic conditions, they are particularly prone to climate change, which is expected to affect their resilience to cope with anthropogenic environmental disturbance (Fischer et al. 2008; Foley et al. 2011; Tscharntke et al. 2012). Thus, understanding the effects of climate-change on farmlands under low-intensity farming systems is an essential task to foster the persistence of their inherently high social and ecological heritage at the global scale.

Our aim in this chapter is to present an overview of the current scientific evidence on climate change impacts on HNVf, and evaluate the potential risks of such change for these farming systems. Such overview builds on a comprehensive review of published research in which the effects of climate change in High Nature Value farming systems were assessed. Overall, using predefined keywords, scientific publications were screened in major international databases (including ISI Web of Science and Scopus), and resulting records analysed through descriptive statistics. The resulting dataset was first characterized for its spatial and thematic scope, and temporal evolution of the research performed. Additionally, the dataset was analysed to highlight patterns regarding: (1) the type of data used by study region; (2) the topic studied by study region; (3) reported impacts by study region; and, (4) reported impacts by topic addressed. Results are discussed for implications, namely in the context of future research perspectives and needs.

1.2 What Are High Nature Value Farmlands and Why Are They Important?

In the European Union (EU), the pivotal role of low-intensity farmlands for the maintenance of natural capital and protection of the countryside converged into the 'High Nature Value farmlands' (HNVf) concept (Beaufoy et al. 1994; Lomba et al. 2014). The relevance of HNVf for nature conservation and rural development goals is embedded within EU's agricultural and environmental policies (Jongman 2013; Ribeiro et al. 2014; Strohbach et al. 2015), namely the Common Agricultural Policy and the Common Biodiversity strategy to halt biodiversity loss by 2020 (target 3; O'Rourke et al. 2016).

HNVf are agriculture-dominated landscapes whose high natural and/ or conservation value rely on the maintenance of specific, usually low-intensity farming systems (Beaufoy et al. 1994; Halada et al. 2011; Lomba et al. 2014). Overall, HNV farming systems are characterized by low levels of agro-chemical inputs, mechanization and livestock levels, and by rotational uses of the land (Lomba et al. 2014; Oppermann et al. 2012; Plieninger and Bieling 2013). HNVf owe their ecological value to high proportions of natural and semi-natural vegetation e.g. pastures and meadows—HNVf type 1 (Andersen et al. 2004; Oppermann et al. 2012), and/or to the presence of small-scale elements in the agricultural landscapes e.g. tree lines, hedgerows, and field margins—HNVf type 2 (Andersen et al. 2004; Lomba et al. 2014). Other farmlands, often under more intensive farming practices, are also considered as HNVf due to their role on the maintenance and survival of populations

Fig. 1 High Nature Value farmlands in the North Western of Portugal: livestock-based systems (left), and mixed farming systems (right)

of species with unfavourable conservation status, known to be largely dependent on habitat or landscape conditions maintained by agricultural management—HNVf type 3 (Andersen et al. 2004; Moreira et al. 2005). HNVf farming systems include livestock-based systems (e.g. the *montados* in southern Portugal; or mountain livestock-based systems e.g. in northern Iberia, Fig. 1), arable-based systems (e.g. dryland, non-irrigated, systems in Italy and Greece), permanent crop oriented systems (e.g. olives, fruit and vines, historically important in the Mediterranean) and several types of mixed farming systems (e.g. in northern Portugal; Fig. 1).

As multifunctional farmlands, HNVf provide a wide variety of habitats, conditions and resources, enabling the coexistence of wildlife alongside farming activities, thus enhancing agro-biodiversity (Lomba et al. 2014). Their contribution for the delivery of a wide range of ecosystem services at several scales has also been highlighted (Bernués et al. 2016; O'Rourke et al. 2016; Plieninger and Bieling 2013). These include provisioning (e.g. high-quality food, fibre and maintenance of genetic resources), cultural (e.g. recreation and agro- and ecotourism, maintenance of cultural heritage, scenic landscapes) and regulating services (e.g. climate regulation, soil erosion prevention, pollination), sustained by key ecological supporting functions (e.g. primary production, nutrient cycling, soil formation) (Keenleyside et al. 2014; Oppermann et al. 2012).

Recent estimates reveal that over 30% of all agricultural land in the EU correspond to High Nature Value farmlands (O'Rourke et al. 2016). While being a European concept, the HNVf concept may be applicable to other social-ecological and geographical contexts, where the maintenance of high levels of biodiversity and the delivery of ecosystem services are related to the persistence of specific farming systems (Bennett et al. 2016; Plieninger and Bieling 2013). Yet, HNVf have been declining due to rural depopulation, agricultural abandonment and afforestation in marginal farming areas, and intensification in the most productive areas (Bernués et al. 2014; Oppermann et al. 2012; Plieninger et al. 2006, 2014; Strohbach et al. 2015).

1.3 Resilience of Traditional, Low-Intensity Farmlands to Environmental Change

As complex social-ecological systems, HNVf landscapes reflect a long history of persistence of region-specific low-input farming systems (Dorresteijn et al. 2015). HNVf are inherently rich in biodiversity and known to contribute to the provision of ecosystem services, thus contributing to EU countryside environmental sustainability and resilience (Bernués et al. 2016; O'Rourke et al. 2016; Plieninger and Bieling 2013). Beyond acknowledged biophysical properties e.g. heterogeneous landscapes, which reflect the diversity of uses and management of land parcels and linear elements (Bignal and McCracken 2000; Plieninger et al. 2006), the natural capital supported by HNVf relies on social-ecological interactions between rural communities and ecosystems (Dorresteijn et al. 2015; Fischer et al. 2012). Traditional farming practices co-evolved with the environment in a synergistic relationship in which farmers shaped rural landscapes and ecosystems granted people with essential goods and services, thus assuring and motivating a sustainable use of natural resources (Altieri 2004; Dorresteijn et al. 2015; Fischer et al. 2012; Oppermann et al. 2012). Moreover, detailed agro-ecological knowledge held by rural communities supported a sustainable appropriation of ecosystems while assuring local food self-sufficiency (e.g. Altieri 2004). Traditional knowledge, the adaptive nature of traditional farming systems and their inherent multifunctionality are the backbone of social-ecological resilience and sustainability of HNV farmlands (e.g. Oppermann et al. 2012). Yet, recent global social-ecological changes, namely rapid land-use and climate changes, may hinder the ability of HNVf and other traditionally managed farming systems to adapt and persist in the future, thus threatening their inherent natural and social capital.

Land-use change, including agricultural abandonment in marginal areas (Terres et al. 2015) or intensification in the most productive areas (Bernués et al. 2014; Plieninger et al. 2014; Strohbach et al. 2015), is among the major threats to HNVf persistence. Abandonment is as a major driver of change of farmlands, especially those managed under traditional, low-input farming systems (Queiroz et al. 2014). Farmland abandonment stems from three types of drivers (Benayas et al. 2007): ecological or biophysical (e.g. elevation, fertility, soil depth), socio-economic (market incentives, migration and rural depopulation, technology, often linked with agrarian policies), and adaptation and management (reflecting the adaptation of farming systems to local conditions). Whilst land-use change may affect farmlands with high nature value, climate change is also likely to impact low-intensity farming systems (Downing 1993; Rockström et al. 2017).

1.4 Potential Impacts of Climate Change on Traditional Low-Intensity Rural Landscapes

Climate change may pose a serious challenge to the resilience of traditional farmlands and thus to farmers' persistence in these territories (Altieri et al. 2015; Vignola et al. 2015). As climate change impacts are expected to widely vary geographically (Porter et al. 2014), traditional smallholder farmers in marginal or risk-prone environments are considered disproportionately vulnerable to climate change. Such vulnerability stems from many stressors, namely the location of farms, various socio-economic, demographic and policy trends, coupled with a limited capacity to adapt or screen alternative livelihoods (Altieri and Koohafkan 2008; Morton 2007; Vignola et al. 2015). In fact, traditional techniques and traditional ecological knowledge might be insufficient to cope with different climatic conditions, so humanity may face the loss farmers' historical adaptation to the environment (Oppermann et al. 2012).

Climate drivers as warming or extreme temperature trends, drying or extreme precipitation trends combined with carbon dioxide fertilization will synergistically act to create increased yield variability (Porter et al. 2014). In the specific case of Mediterranean HNVf, but not exclusively, water scarcity is expected to impact HNV farming systems known to be more dependent on rainfall patterns or on water available for irrigation. As projected climate change move beyond historical conditions, the adaptive capacity of traditional farmers may be simultaneously affected by a reduced reliance in traditional knowledge and by policies and regulations leading to the substitution of traditional livelihoods through changes of farming practices, as well as the loss of transmission of traditional knowledge (Porter et al. 2014; Salick and Ross 2009). Among the foreseeable impacts of climate change on low-intensity farming systems are: (i) changes in productivity (e.g. through reduction of expected yields (Morton 2007; Rounsevell et al. 2006); (ii) expansion of lowland crops and intensive farming systems to higher elevations; and, (iii) erosion of agro-biodiversity and degradation of essential ecosystem functions. Ultimately, the degradation of ecosystem functions is expected to affect the delivery of essential ecosystem services such as pest regulation or pollination services (Dale and Polasky 2007; Pedrono et al. 2016; Salick and Ross 2009). The depletion of supporting functions and regulating services (e.g. water retention and nutrient cycling) may further enhance the adverse impacts of climate change on the low productivity of traditionally managed farmlands (Bommarco et al. 2013; Pedrono et al. 2016).

Overall, putative negative impacts of climate change may strengthen recent trends for rural exodus and thus the potential collapse of the social-ecological systems underlying the nature value of low-intensity farmlands (Honrado et al. 2016). Synergic interactions between climate and land-use change, including farmers' management decisions (e.g. agricultural abandonment), policy decisions and market related stressors may further impact agricultural production and the delivery of societally relevant services from extensively managed farmlands (Bommarco et al. 2013; Pedrono et al. 2016; Power 2010; Rounsevell et al. 2006).

2 Methods

2.1 Literature Search

A literature review aiming to scrutinize previously published research tackling the impacts of climate change on traditional, low-intensity, High Nature Value farming systems was performed. Classifying farming systems according to the intensity of agricultural practices and resulting management is challenging, as it depends on several factors (namely biophysical conditions, climate, …). Besides operational challenges, classifying farming systems as low-intensity or traditional, raises conceptual challenges. Here, we considered farming systems characterized by low-intensity practices, reflected by indicators such as rotation of crops or small-scale traditional management of holdings and rainfed farms. The selection of keywords was based on a review of previous keywords from reference papers (e.g. Heller and Zavaleta 2009, Plieninger et al. 2014), and following a participatory approach with a team of researchers specialised in the research topic. Such literature search was implemented in the ISI Web of Science (ISI WOS) and Scopus databases for all years up to 2016, by using the following combination of keywords: "extensi* agricult*" OR "extensi* farm*" OR "traditional agricult*" OR "traditional farm*" OR "high nature value" OR "low-input farm*" OR "low-input agricult*" OR "farm* low-input" OR "agricult* low-input" OR "cultural landscape*" OR "farm* system*" AND "climat* chang*" OR "global warm*" OR "chang* climat*". Potential keywords were first iteratively tested on ISI WOS. Keywords selected were those most-commonly and unequivocal used in the literature focusing the targeted research topics. ISI WOS and Scopus were chosen as ISI WOS was designed with the intention of satisfying users in citation analysis; and Scopus has a more expanded spectrum of journals than ISI WOS, though having the citation analyses more limited to recent articles when compared to ISI WOS (Falagas et al. 2008). For the sake of analysing patterns in published research on the impact of climate change in low-intensity farming systems, the use of both databases and the selected keywords allowed a coherent picture of the principal patterns in the research done so far, though not being an exhaustive review on the subject. We retrieved 2171 records, from which duplicate records were withdrawn, thus resulting in 1624 potential records. Two criteria were used to ascertain the eligibility of records to be included in the final dataset and analysis: (1) focus on low-intensity farming systems and underlying biodiversity, ecosystems or social-ecological systems; and, (2) focus on the impacts of climate change (post industrial revolution) in low-intensity farming systems. Challenges related with farming systems classification were identified during the process of inclusion of records. As strategy to overcome them, we used a participatory approach between reviewers to agree on the farming systems to be included or excluded based on the characteristics underlying legible low intensity farming systems. From these, 100 records were considered legible for further analysis, described in detail below.

2.2 Analysis of Data Gathered from Literature Review

Records were classified according to the following criteria: type of research (concept or data driven); type of data used (records classified as not applicable, whenever no data was used, qualitative data or quantitative data); and, geographic/study region (classified according to the continent where the research was developed). Further, screened publications were categorized according to their research focus, by assigning them to the following classes: 'Production', 'Biodiversity', 'Ecosystems regulation' or 'Cultural'. Records were classified as 'Production' whenever the topic concerned yields and food security, including farmers' perceptions on the impacts of climate change on farm productivity. 'Biodiversity' targeted the impacts of climate change on agrobiodiversity. Within the category 'Ecosystems regulation', records focusing on regulation processes, such as water or nutrient cycle, or other processes contributing to the maintenance of physico-chemical components of the farming system were considered. Finally, records classified as 'Cultural' included those tackling recreation, tourism or aesthetics issues. Finally, records were classified according to the reported impacts of climate change as follows: (1) potential positive impacts (hereafter 'positive'), (2) potential negative impacts (hereafter 'negative'), (3) mixed positive and negative impacts (hereafter 'mixed'), and (4) no clear statement on impact type (hereafter 'not clear') when the characterization of impacts was not clearly articulated. The rationale supporting the selection of the metrics used for the analysis of data retrieved from the scientific literature is provided in detail on Table 1.

To scrutinize the resulting data, a set of descriptive statistics were implemented in Microsoft Excel© (2016). Results are presented as stacked plots of the number of records per year to express the temporal evolution of records' data type and targeted topics. Additionally, radar graphs were constructed for each of the considered classifications of data to show the percentage of records as follows: (1) the type of data used by study region; (2) the research topic by study region; (3) the reported impact by study region; and, (4) the reported impact by research topic. Regarding the last categorization, i.e. reported impact by research topic, records that focus on more than one research topic were excluded from the analyses to avoid double counting. Overall, the resulting set of bibliographic references retrieved and analysed, presented a wide variation in what concerns the type of approach, methods, scale, and study design, and therefore the selected variables are considered to cover the most important variability among them, allowing for a synoptic view of the published research on the impacts of climate change on low-intensity farming systems.

Table 1 Rationale supporting the adopted framework implemented for the analysis and characterization of the research literature screened

Questions	Options	Rationale
Type of research	- Data driven - Concept driven	Conceptual research express records devoted to theoretical hypothesis and discussion on the topic; data driven express empirical approaches testing the impacts of climate change in the targeted farming systems. Overall, analysing the type of research provides an overview of what have been the main trends of the scientific field.
Type of data	- Qualitative data - Quantitative data - Not applicable	Classification on either qualitative or quantitative data reflect how the components of the socio-ecological system are being captured. Qualitative data refers to e.g. data obtained through surveys and interviews and mostly related to the capture of social impacts (Garibaldi et al. 2017). Quantitative data e.g. yields, reflect data quantified. Not applicable, classification used in the case of concept driven records, where no data is used. By analysing the type of data used in the published scientific records scrutinized, we aimed to understand the availability of empirical research on the impacts of climate change on the social-ecological systems underlying the targeted farming systems.
Study topic	- Production - Biodiversity - Ecosystems regulation - Cultural	As multifunctional landscapes, low-intensity farming systems contribute to the production of food and other goods, while contributing to the support of biodiversity and the delivery of a wide range of ecosystem services (regulation services e.g. climate, and cultural services). By analysing the topic of the research in each record, we aimed to understand how integrative has been the research on the impacts of climate change in such social-ecological systems.
Study region	- Europe - Asia - South America - North America - Africa - Australia - Global	The distribution of small-scale, low-intensity farmlands varies across continents (e.g. Queiroz et al. 2014). By analysing the study region of the scientific records analysed, we expect to achieve insights on the potential relation between the spatial distribution and the respective thematic scope, reported impacts and types of data used.
Reported impact	- Positive - Negative - Mixed - No clear statement on the type of impact	As climate change is expected to affect distinctly farming systems across geographical regions (Porter et al. 2014), this metric aimed to qualify the reported impacts of climate change on low-intensity farming systems and relate them with both the thematic scope and study regions of the scrutinized scientific records.

3 Results and Discussion

3.1 Research on Impacts of Climate Change in Traditional, Low-Intensity Farmlands: When, Where and How?

A sharp increase in the research published on climate change and traditional, low-intensity farmlands occurred over the last five years (e.g., 4 records in 2011 to 13 records in 2016; Fig. 2a) with the first record found in 1989. The published records retrieved by the bibliometric search referred mostly to data driven research (78%). Still, 22% of records tackled the topic from a conceptual perspective only. Among data driven records, 78% applied quantitative data, 17% qualitative data, and 9% both qualitative and quantitative data. Overall, the temporal evolution of the types of data used since 2012, depict a trend for increasing number of assessments build partially on 'qualitative data' (Fig. 2a). This may reflect a growing awareness of the relevance of gathering and analysing this type of data (e.g. resulting from surveys and interviews) to complement and eventually improve researchers' ability to capture and understand the impacts of climate change on the social-ecological components of farming systems (Garibaldi et al. 2017).

Regarding the geographic context, Africa and Europe were found to be the continents where most records were originated (both with 28% of records), followed by Asia (17%), Oceania (16%), and South America (6%). Only 5% of analysed records resulted from Global assessments. The prevalence of research tackling the effects of climate change in Africa, Europe, Asia, converges with previous research highlighting the higher occurrence of (small-scale), low-intensity, traditional farming in those continents (e.g. Queiroz et al. 2014). Interestingly, the same does not apply to Oceania, where farming systems have been mainly described as large-scale farms, with high average farm size (Lowder et al. 2016). Yet, the observed prevalence of Australian records seem to reflect specific rainfed

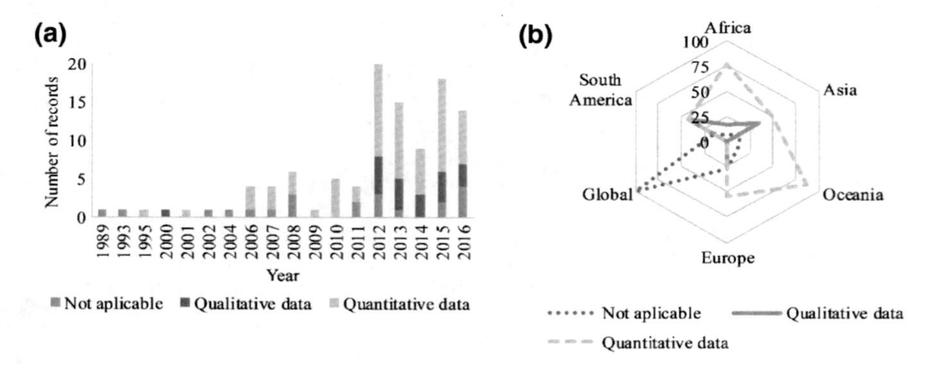

Fig. 2 **a** Number of records identified per year and according the types of data considered. "Not applicable" reflects concept-driven records, whilst 'Qualitative' and 'Quantitative' data are associated to data driven records. **b** Radar graph reflecting the percentage of records retrieved from each continent and by each type of data considered. Some records reported multiple types of data

rotational farming systems (e.g. Asseng and Pannell 2013). The gap observed for North America (expressed as the lack of records) may be due to the prevalence of large scale farmlands (>50 ha), which are known to account for ca. 94% of all farmed area in the USA and for 72% of holdings in Canada (FAO 2014). Additionally, it may also relate to differences on the specific thresholds considered to discriminate farmlands under low-intensity management (traditional or small-scale farmlands) across continents, which may hinder comparability across social-ecological contexts (FAO 2014). Interestingly, patterns observed in the USA converge with claims that Europeans tend to focus more on the conservation of rural landscapes, unlike North Americans that value 'wilderness' and non-agricultural landscapes (Batáry et al. 2015; Dobbs and Pretty 2004). Remarkably, conservation programs in the USA integrate a specific component of agricultural-related legislation since 1930s (Lichtenberg 2014), even though targeting a small proportion of all farmed area ($\sim$3%; Barbarika 2011; FAO 2014).

When linking records based on the types of data and geographic distribution, we found that 'qualitative data' prevails in South America (43%), Asia (35%), Africa (17%) and Europe (17%) (Fig. 2b). Yet, at the global scale, impacts of climate change in low-intensity farming systems are mainly discussed within concept driven manuscripts (corresponding to the 'not applicable' category of type of data; with 100% of the records; Fig. 2b). Conversely, quantitative data appears to be mostly applied in research developed in Oceania (88%), Africa (77%), Europe (55%) and Asia (50%), targeting distinct scales of analysis. The application of 'qualitative data' in Asia, Africa and South America may relate with an interest in smallholders' perspective on the impacts and potential adaptive responses, since it has been acknowledged that such farming systems may strive under climate change due to higher socioeconomic vulnerability (Morton 2007). Finally, the lack of data-driven assessments at the global scale, seems to be related to operational limitations (e.g. lack of suitable/compatible data across the globe) to address the impacts of climate change in low-intensity, traditional farming systems at such scale of analysis. Such limitations may hamper the ability to anticipate how climate change will impact such farmlands either on production, either on the wide delivery of ecosystem services (namely support for biodiversity, regulation,…).

3.2 Which Impacts of Climate Change on Traditional High Nature Value Farmlands Have Been Under Researchers' Scrutiny?

Overall, the analysed records covered mostly impacts of climate change on farmland production (70%), followed by impacts on biodiversity and ecosystem regulation (each attaining 14% of the records), and finally on cultural aspects (2%). Interestingly, the impacts of climate change on the cultural aspects of traditional farming systems seem to have been the last to get researchers' attention, with the

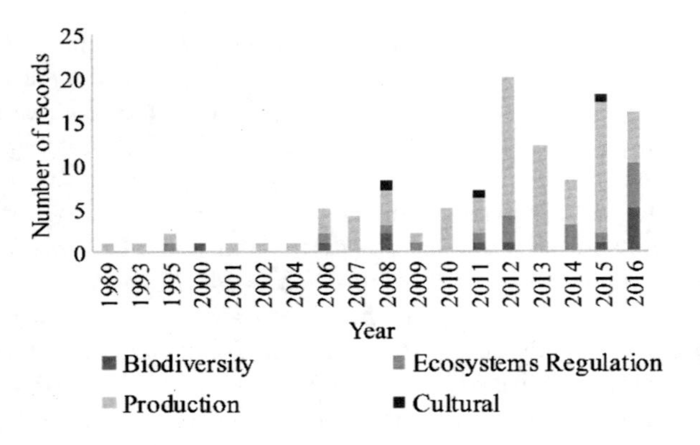

Fig. 3 Number of records through years across research topics. Note that some records report to multiple categories, and thus the number of papers across categories sums up more than the total number of records

first studies dated from 2008 (Fig. 3). A deeper analysis on the topics under scrutiny across geographic regions highlight that research performed in the African continent focus mainly agricultural yields (86%), with few studies targeting either biodiversity or ecosystems regulation (7%, each). Similar trends were found also for the Asian continent (production with 73% of the records, ecosystem regulation with 14%, and biodiversity with 9% of the records), but here one record targeted the impacts of climate change on the cultural capital of such farmlands (corresponding to 5% of the records). Contrastingly, European research seems to focus the impacts of climate change on production (56%) and agro-biodiversity (28%), with fewer efforts invested on understanding how such change will reflect on regulation (14%) and cultural services (3%) in HNV farming systems. Records reported for Oceania tackled mainly production (74%), followed by ecosystems regulation (21%) and biodiversity (5%). As for South America, records targeted mainly production (60%), followed by ecosystems regulation (20%), biodiversity and cultural (each with 10%). Likewise, global records focus mostly on production (71%), with only one record tackling biodiversity and ecosystems regulation (14% each). Overall, an emphasis of research on yields converges with an increasing societal demand for food production and security (Fischer et al. 2017; Rockström et al. 2017), with this being particularly relevant in the context of subsistence farming (see for e.g. Morton 2007). In developing countries, a high percentage of the population rely on small-scale, traditional farming systems for subsistence, even though they are particularly vulnerable to social-ecological change, e.g. climate change (Morton 2007). As so, the ability of such farming systems to provide food security is of the utmost importance, and assuring their persistence in the future is still a challenge to tackle.

In Europe, research priorities seem to derive from a concern of maintaining farming systems, by preventing their abandonment or intensification in the future. The higher focus on biodiversity in European records may be due to the convergence of agrobiodiversity within Rural Development Policies, through recognition of the importance of High Nature Value farmlands (Lomba et al. 2014). Interestingly, only 3 out of 28 European records retrieved by our search mention

HNVf in the main text, even though no clear reference to HNV is done when describing the targeted farming systems. This is consistent with the conceptual and methodological challenges that the implementation of the HNV concept faces (for a review see Lomba et al. 2014), especially in what concerns data availability, resolution (spatial and temporal) and quality (Pôças et al. 2014), and the lack of a common framework or approach (Lomba et al. 2014).

Topics such as ecosystem regulation and, more critically, cultural impacts are less covered when impacts of climate change are being scrutinized. Overall, this trend relates to an inability to forecast the potential impacts of climate change on the complex social-ecological systems underlying traditional farming systems, and thus to anticipate threats to their future persistence e.g. policy and market-driven changes, land-use change. Such trends are, nevertheless, symptomatic of the lack of (short or long-term) assessments targeting the value of traditional farming systems in what concern their ability to provide ecosystem services, and support to species and habitats of high nature and/or conservation value.

Most records reported or predicted negative impacts from climate change (69%), followed by mixed (26%), with only 3% reporting positive impacts on farming systems. This converges with previous research, in which negative impacts on food production are reported as consequence of expected climate change (Porter et al. 2014). Climate change is expected to impact food production differently across geographical regions, with negative impacts expected to affect mainly low-latitude countries, while mixed impacts are expected to high latitude countries (Porter et al. 2014). From the revised records, the highest percentage of negative impacts by continent were reported for Asia (88%) and Global assessments (80%), followed by Africa (68%), South America (67%), Oceania (63%), and Europe (61%) (Fig. 4).

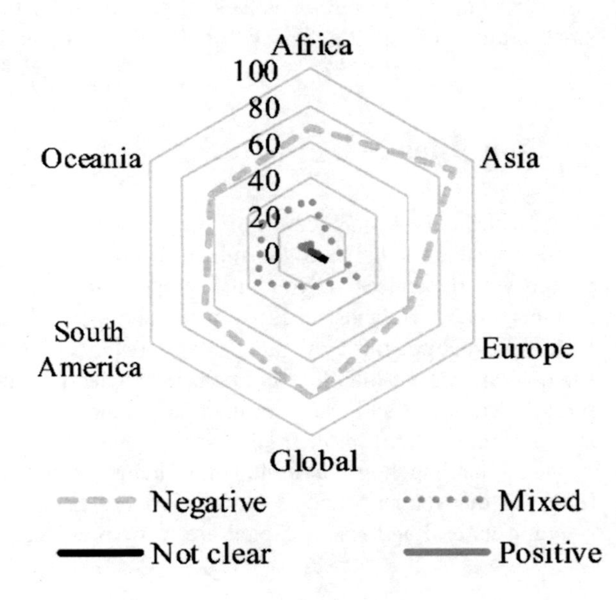

Fig. 4 Radar graph reflecting the percentage of records by study region and by impact type categories

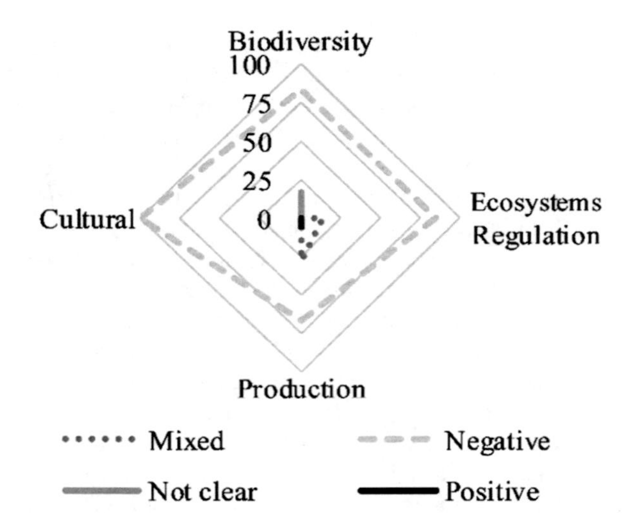

Fig. 5 Radar graph reflecting the percentage of records by topic and by impact type categories. Records that focus on more than one topic were excluded from the analysis to avoid double counting

While mixed effects were reported across all continents, the highest percentage were reported in South America (33%), Oceania (31%), Africa and Europe (both with 29%), followed by Global (20%) and Asia (12%). Positive effects were only reported in Africa (6%), Europe (4%) and Oceania (4%).

Finally, records focused only on biodiversity reported mostly potentially negative impacts (83.3%). The same was observed for records focused only on ecosystems regulation (85.7%). Records focused only on cultural topics reported only potential negative impacts. Records on production reported mostly potential negative impacts (66.2%), followed by mixed impacts (27.9%) (Fig. 5). The observed impact variability or lack of it may, nevertheless, relate more with the small number of records than with the realm of potential impacts.

4 Conclusions

Worldwide, there's an increasing awareness of the pivotal role of low-intensity farming systems and resulting farmlands for food security and for the wide provision of ecosystem services, beyond the support to biodiversity. As social-ecological systems, such farmlands result from the complex relations and dynamics between Man and nature through time. Yet, social-ecological changes, e.g. lack of socio-economic profitability and consequent land abandonment or intensification, rural development and/or environmental policies, are threatening the persistence of such farming systems at the global scale.

Understanding how environmental change may impact such farmlands in the future so that their resilience may be fostered is of utmost relevance if their outstanding natural and social capital are to be preserved. Whilst most of the debate has focused on the impacts of land-use change (e.g. abandonment) on low-intensity farmlands, climate change is also foreseen to significantly impact such farmlands in the future. Here, we tackled this challenge by scrutinizing the existent scientific knowledge, by reviewing previous research relating climate change to low-intensity, traditionally managed farming systems.

Overall, our results depict substantial differences among social-ecological contexts in what concerns the reported potential impacts (positive, negative or mixed) of climate change on low-intensity managed farmlands. Records focusing on production originate from areas characterized by subsistence farming, while impacts on ecosystem functions (regulation, cultural or biodiversity) are scrutinized in distinct contexts where these farming systems are currently facing abandonment. Specifically, there is a lack of data-driven assessments at the global scale. Still, forecasting the impacts of climate change on low-intensity, traditionally managed farmlands at the global scale is essential if they are to be anticipated (and mitigated) at this scale.

Africa and Europe were found to be the continents where most records were originated from. Most records focused largely on farmland production, followed by impacts on biodiversity and ecosystem regulation and at a lesser extent on cultural aspects. Interestingly, a focus on food production was found in developing regions (e.g. Africa) where subsistence farming prevails, while higher focus on biodiversity was observed on developed regions (e.g. Europe) probably due to environmental goals within rural development policies.

Our results point to a general inability to forecast the potential impacts of climate change on the complex social-ecological systems underlying traditional farming systems. Nevertheless, most records reported or predicted negative impacts from climate change, followed by mixed, with very few reporting positive impacts on farming systems.

In a nutshell, more research focusing the impacts of climate change on low-intensity, traditionally managed farmlands is needed if a more holistic understanding of food security and human well-being is to be reached. Furthermore, as the resilience of such farming systems rely on their multifunctionality, integrative approaches are needed to fully understand the impacts of climate change in such systems. Such approaches entail considering the potential impacts of both climate and land-use change on biodiversity and the provision of ecosystem services.

While an increasing number of studies analyzing the impacts of climate change on low-intensity, traditionally managed farming systems was observed, there is still room for improvement in what concerns the conceptual and analytical approaches developed and implemented. In fact, there is a pressing need to develop modelling tools and frameworks able to model and forecast dynamics of coupled social-ecological systems, as well as interactions and feedbacks between them. Investing efforts in the improvement of climatic projections, including alternative scenarios of development, at multiple decision scales of analysis, is also among the research needs essential to foster current knowledge. Ultimately, integrated approaches in which the social-ecological systems functioning and dynamics of such farmlands are considered in relation to both climate and land-use change

would provide the backbone of a paradigm shift in current knowledge, supporting adaptive management and monitoring of low-intensity, traditionally managed farmlands under current environmental change.

Acknowledgements AL and FM were supported by the Portuguese Science and Technology Foundation (FCT) through Post-Doctoral Grant SFRH/BPD/80747/2011 and contract IF/01053/2015, respectively. This research is a result of project FARSYD-2011–2016—POCI-01-0145-FEDER-016664, supported by Norte Portugal Regional Operational Programme (NORTE 2020), under the PORTUGAL 2020 Partnership Agreement, through the European Regional Development Fund (ERDF), and by national funds through FCT—Portuguese Science Foundation (PTDC/AAG-EC/5007/2014).

References

Altieri MA (2004) Linking ecologists and traditional farmers in the search for sustainable agriculture. Front Ecol Environ 2(1):35–42. https://doi.org/10.1890/1540-9295(2004)002%5b0035:leatfi%5d2.0.co;2

Altieri MA, Koohafkan P (2008) Enduring farms: climate change, smallholders and traditional farming communities, Environment and Development Series 6. Third World Network, Penang

Altieri MA, Nicholls CI, Henao A, Lana MA (2015) Agroecology and the design of climate change-resilient farming systems. Agron Sustain Dev 35(3):869–890. https://doi.org/10.1007/s13593-015-0285-2

Andersen E, Baldock D, Bennett H, Beaufoy G, Bignal E, Bouwer F, Elbersen B, Eiden G, Giodeschalk F, Jones G, McCracken D, Nieuwenhuizen W, Eupen Mv, Hennekes S, Zervas G (2004) Developing a high nature value farming area indicator. Internal Report. European Environment Agency, Copenhagen

Asseng S, Pannell DJ (2013) Adapting dryland agriculture to climate change: farming implications and research and development needs in Western Australia. Climatic Change 118(2):167–181. https://doi.org/10.1007/s10584-012-0623-1

Barbarika A (2011) Conservation reserve program annual summary and enrollment statistics. United States Department of Agriculture, F.S.A., Washington DC

Batáry P, Dicks LV, Kleijn D, Sutherland WJ (2015) The role of agri-environment schemes in conservation and environmental management. Conserv Biol 29:1006–1016. https://doi.org/10.1111/cobi.12536

Beaufoy G, Baldock D, Clarke J (1994) The nature of farming: low intensity farming systems in Nine European Countries. IEEP, London. ISBN 1-873906-01-3

Benayas JMR, Martins A, Nicolau JM, Schulz JJ (2007) Abandonment of agricultural land: an overview of drivers and consequences. CAB Rev: Perspect Agric Vet Sci Nutr Nat Resour 2 (057). https://doi.org/10.1079/pavsnnr20072057

Benayas JMR, Bullock JM, Newton AC (2008) Creating woodland islets to reconcile ecological restoration, conservation, and agricultural land use. Front Ecol Environ 6(6):329–336. https://doi.org/10.1890/070057

Bennett EM, Solan M, Biggs R., McPhearson T, Norström AV, Olsson P, Pereira L, Peterson GD, Raudsepp-Hearne C, Biermann F, Carpenter SR, Ellis EC, Hichert T, Galaz V, Lahsen M, Milkoreit M, Martin López B, Nicholas KA, Preiser R, Vince G, Vervoort JM, Xu J (2016) Bright spots: seeds of a good Anthropocene. Front Ecol Environ 14:441–448. https://doi.org/10.1002/fee.1309

Bernués A, Rodríguez-Ortega T, Ripoll-Bosch R, Alfnes F (2014) Socio-cultural and economic valuation of ecosystem services provided by Mediterranean Mountain Agroecosystems. PLoS ONE 9(7):e102479. https://doi.org/10.1371/journal.pone.0102479

Bernués A, Tello-García E, Rodríguez-Ortega T, Ripoll-Bosch R, Casasús I (2016) Agricultural practices, ecosystem services and sustainability in high nature value farmland: unraveling the perceptions of farmers and nonfarmers. Land Use Policy 59:130–142. http://dx.doi.org/10.1016/j.landusepol.2016.08.033

Bignal EM, McCracken DI (2000) The nature conservation value of European traditional farming systems. Environ Rev 8(3):149–171. https://doi.org/10.1139/a00-009

Bommarco R, Kleijn D, Potts SG (2013) Ecological intensification: harnessing ecosystem services for food security. Trends Ecol Evol 28(4):230–238. http://dx.doi.org/10.1016/j.tree.2012.10.012

Cumming GS, Buerkert A, Hoffmann EM, Schlecht E, von Cramon-Taubadel S, Tscharntke T (2014) Implications of agricultural transitions and urbanization for ecosystem services. Nature 515(7525):50–57. https://doi.org/10.1038/nature13945

Dale VH, Polasky S (2007) Measures of the effects of agricultural practices on ecosystem services. Ecol Econ 64(2):286–296. http://dx.doi.org/10.1016/j.ecolecon.2007.05.009

Dobbs TL, Pretty JN (2004) Agri-environmental stewardship schemes and "multifunctionality". Rev Agric Econ 26(2):220–237. https://doi.org/10.1111/j.1467-9353.2004.00172.x

Dorresteijn I, Loos J, Hanspach J, Fischer J (2015) Socioecological drivers facilitating biodiversity conservation in traditional farming landscapes. Ecosyst Health Sustain 1(9):1–9. https://doi.org/10.1890/ehs15-0021.1

Downing TE (1993) The effects of climate change on agriculture and food security. Renew Energy 3(4–5):491–497. https://doi.org/10.1016/0960-1481(93)90115-w

Falagas ME, Pitsouni EI, Malietzis GA, Pappas G (2008) Comparison of PubMed, scopus, Web of science, and Google scholar: strengths and weaknesses. FASEB J 22(2):338–342. http://dx.doi.org/10.1096/fj.07-9492LSF

FAO (2014) The state of food and agriculture: innovation in family farming. Food and Agriculture Organization of the United Nations, Rome

Fischer J, Brosi B, Daily GC, Ehrlich PR, Goldman R, Goldstein J, Lindenmayer DB, Manning AD, Mooney HA, Pejchar L, Ranganathan J, Tallis H (2008) Should agricultural policies encourage land sparing or wildlife-friendly farming? Front Ecol Environ 6(7):380–385. https://doi.org/10.1890/070019

Fischer J, Hartel T, Kuemmerle T (2012) Conservation policy in traditional farming landscapes. Conserv Lett 5(3):167–175. https://doi.org/10.1111/j.1755-263x.2012.00227.x

Fischer J, Abson DJ, Bergsten A, French Collier N, Dorresteijn I, Hanspach J, Hylander K, Schultner J, Senbeta F (2017) Reframing the food–biodiversity challenge. Trends Ecol Evol 32(5):335–345. doi:http://dx.doi.org/10.1016/j.tree.2017.02.009

Foley JA, Ramankutty N, Brauman KA, Cassidy ES, Gerber JS, Johnston M, Mueller ND, O/'Connell C, Ray DK, West PC, Balzer C, Bennett EM, Carpenter SR, Hill J, Monfreda C, Polasky S, Rockstrom J, Sheehan J, Siebert S, Tilman D, Zaks DPM (2011) Solutions for a cultivated planet. Nature 478(7369):337–342. http://www.nature.com/nature/journal/v478/n7369/abs/nature10452.html#supplementary-information

Garibaldi LA, Gemmill-Herren B, D'Annolfo R, Graeub BE, Cunningham SA, Breeze TD (2017) Farming approaches for greater biodiversity, livelihoods, and food security. Trends Ecol Evol 32(1):68–80. https://doi.org/10.1016/j.tree.2016.10.001

Godfray HCJ, Beddington JR, Crute IR, Haddad L, Lawrence D, Muir JF, Pretty J, Robinson S, Thomas SM, Toulmin C (2010) Food security: the challenge of feeding 9 billion people. Science 327(5967):812–818. https://doi.org/10.1126/science.1185383

Halada L, Evans D, Romão C, Petersen J-E (2011) Which habitats of European importance depend on agricultural practices? Biodivers Conserv 20(11):2365–2378. https://doi.org/10.1007/s10531-011-9989-z

Heller NE, Zavaleta ES (2009) Biodiversity management in the face of climate change: a review of 22 years of recommendations. Biol Conserv 142:14–32. https://doi.org/10.1016/j.biocon.2008.10.006

Honrado JP, Lomba A, Alves P, Aguiar C, Monteiro-Henriques T, Cerqueira Y, Monteiro P, Barreto Caldas F (2016) Conservation management of EU priority habitats after collapse of traditional pastoralism: navigating socioecological transitions in Mountain Rangeland. Rural Sociol 82:101–128. https://doi.org/10.1111/ruso.12111

Jongman RHG (2013) Biodiversity observation from local to global. Ecol Ind 33(0):1–4. http://dx.doi.org/10.1016/j.ecolind.2013.03.012

Keenleyside C, Beaufoy G, Tucker G, Jones G (2014) High nature value farming throughout EU-27 and its financial support under the CAP. Report Prepared for DG Environment, Contract No ENV B.1/ETU/2012/0035, Institute for European Environmental Policy, London

Lichtenberg E (2014) Conservation, the Farm Bill, and U.S. agri-environmental policy. Choices 29 (3):1–6

Lomba A, Guerra C, Alonso J, Honrado JP, Jongman R, McCracken D (2014) Mapping and monitoring High Nature Value farmlands: challenges in European landscapes. J Environ Manage 143(0):140–150. http://dx.doi.org/10.1016/j.jenvman.2014.04.029

Lowder SK, Skoet J, Raney T (2016) The number, size, and distribution of farms, smallholder farms, and family farms worldwide. World Dev 87:16–29. http://doi.org/10.1016/j.worlddev.2015.10.041

Moreira F, Pinto MJ, Henriques I, Marques T (2005) The importance of low-intensity farming systems for fauna, flora and habitats protected under the european "Birds" and "Habitats" directives: is agriculture essential for preserving biodiversity in the Mediterranean Region? In: Burk AR (ed) Trends in biodiversity research. Nova Science Publishers, New York, pp 117–145

Morton JF (2007) The impact of climate change on smallholder and subsistence agriculture. Proc Natl Acad Sci 104(50):19680–19685. https://doi.org/10.1073/pnas.0701855104

Oppermann R, Beaufoy G, Jones G (eds) (2012). High nature value farming in Europe. 35 European countries: experiences and perspectives. Verlag Regionalkultur, Ubstadt-Weiher

O'Rourke E, Charbonneau M, Poinsot Y (2016) High nature value mountain farming systems in Europe: case studies from the Atlantic Pyrenees, France and the Kerry Uplands, Ireland. J Rural Stud 46:47–59. http://dx.doi.org/10.1016/j.jrurstud.2016.05.010

Pedrono M, Locatelli B, Ezzine-de-Blas D, Pesche D, Morand S, Binot A (2016) Impact of climate change on ecosystem services. In Torquebiau E (ed) Climate change and agriculture worldwide. Springer Netherlands, Dordrecht, pp 251–261. https://doi.org/10.1007/978-94-017-7462-8_19

Plieninger T, Bieling C (2012) Resilience and the cultural landscape: understanding and managing change in human-shaped environments. Cambridge University Press, Cambridge

Plieninger T, Bieling C (2013) Resilience-based perspectives to guiding high-nature-value farmland through socioeconomic change. Ecol Soc 18(4). https://doi.org/10.5751/es-05877-180420

Plieninger T, Höchtl F, Spek T (2006) Traditional land-use and nature conservation in European rural landscapes. Environ Sci Policy 9(4):317–321. https://doi.org/10.1016/j.envsci.2006.03.001

Plieninger T, Hui C, Gaertner M, Huntsinger L (2014) The impact of land abandonment on species richness and abundance in the mediterranean basin: a meta-analysis. PLoS ONE 9(5):e98355. https://doi.org/10.1371/journal.pone.0098355

Pôças I, Gonçalves J, Marcos B, Alonso J, Castro P, Honrado JP (2014) Evaluating the fitness for use of spatial data sets to promote quality in ecological assessment and monitoring. Int J Geogr Inf Sci 28(11):2356–2371. https://doi.org/10.1080/13658816.2014.924627

Porter JR, Xie L, Challinor AJ, Cochrane K, Howden M, Iqbal MM, Lobell DB, Travasso MI (2014) Chapter 7. Food security and food production systems. Climate change 2014: impacts, adaptation and vulnerability, Working Group II Contribution to the IPCC 5th Assessment Report, Geneva

Power AG (2010) Ecosystem services and agriculture: tradeoffs and synergies. Philos Trans R Soc B: Biol Sci 365(1554):2959–2971. https://doi.org/10.1098/rstb.2010.0143

Queiroz C, Beilin R, Folke C, Lindborg R (2014) Farmland abandonment: threat or opportunity for biodiversity conservation? A global review. Front Ecol Environ 12(5):288–296. https://doi.org/10.1890/120348

Ribeiro PF, Santos JL, Bugalho MN, Santana J, Reino L, Beja P, Moreira F (2014) Modelling farming system dynamics in high nature value farmland under policy change. Agric Ecosyst Environ 183:138–144. https://doi.org/10.1016/j.agee.2013.11.002

Rockström J, Williams J, Daily G, Noble A, Matthews N, Gordon L, Wetterstrand H, DeClerck F, Shah M, Steduto P, de Fraiture C, Hatibu N, Unver O, Bird J, Sibanda L, Smith J (2017) Sustainable intensification of agriculture for human prosperity and global sustainability. Ambio 46(1):4–17. https://doi.org/10.1007/s13280-016-0793-6

Rounsevell MDA, Berry PM, Harrison PA (2006) Future environmental change impacts on rural land use and biodiversity: a synthesis of the ACCELERATES project. Environ Sci Policy 9 (2):93–100. doi:https://doi.org/10.1016/j.envsci.2005.11.001

Salick J, Ross N (2009) Traditional peoples and climate change. Glob Environ Change 19(2):137–139. https://doi.org/10.1016/j.gloenvcha.2009.01.004

Strohbach MW, Kohler ML, Dauber J, Klimek S (2015) High nature value farming: from indication to conservation. Ecol Indic 57(0):557–563. http://dx.doi.org/10.1016/j.ecolind.2015.05.021

Swinton SM, Lupi F, Robertson GP, Hamilton SK (2007) Ecosystem services and agriculture: cultivating agricultural ecosystems for diverse benefits. Ecol Econ 64(2):245–252. http://dx.doi.org/10.1016/j.ecolecon.2007.09.020

Terres J-M, Scacchiafichi LN, Wania A, Ambar M, Anguiano E, Buckwell A, Coppola A, Gocht A, Källström HN, Pointereau P, Strijker D, Visek L, Vranken L, Zobena A (2015) Farmland abandonment in Europe: identification of drivers and indicators, and development of a composite indicator of risk. Land Use Policy 49:20–34. http://dx.doi.org/10.1016/j.landusepol.2015.06.009

Tscharntke T, Clough Y, Wanger TC, Jackson L, Motzke I, Perfecto I, Vandermeer J, Whitbread A (2012) Global food security, biodiversity conservation and the future of agricultural intensification. Biol Conserv 151(1):53–59. doi:http://dx.doi.org/10.1016/j.biocon.2012.01.068

Venter O, Sanderson EW, Magrach A, Allan JR, Beher J, Jones KR, Possingham HP, Laurance WF, Wood P, Fekete BM, Levy MA, Watson JEM (2016) Sixteen years of change in the global terrestrial human footprint and implications for biodiversity conservation. Nat Commun 7. https://doi.org/10.1038/ncomms12558

Vignola R, Harvey CA, Bautista-Solis P, Avelino J, Rapidel B, Donatti C, Martinez R (2015) Ecosystem-based adaptation for smallholder farmers: definitions, opportunities and constraints. Agric Ecosyst Environ 211:126–132. https://doi.org/10.1016/j.agee.2015.05.013

A. Lomba holds a Ph.D. degree in Biology since 2011 (University of Porto) and is currently a Post-Doc Researcher in CIBIO/InBIO. Currently, A. Lomba research focus the development of methods and tools for understanding and intervening in the complex interactions and dynamics of the socio-ecological dimensions underlying high nature value farmlands and farming systems. Such knowledge will then allow anticipating how environmental and policy-driven changes may impact the nature value of farmlands (biodiversity and ecosystems services provisioning) and to integrate ecosystem services and resilience-based frameworks into rural development policies, as tools to enhance sustainability and ecosystem services resilience in High Nature Value farmlands under scenarios of social-ecological change.

A. Buchadas holds a master degree in Ecology, Environment and Territory, from the Faculty of Sciences of the University of Porto (2016) and is currently a research fellow in FARSYD project, which tackles the relation between farming systems and biodiversity and ecosystem services in high nature value farmlands, whilst assessing tradeoffs between driving forces behind land-use change, and the safeguard of the natural and social capital underlying such farmlands. Current research interests include: (i) the study of ecological systems, their interaction with the human dimension and our ability to accommodate ecological and social interests; and, (ii) the application of modelling technics to understand, analyze and forecast (socio) ecological systems.

João P. Honrado is a biologist with a Ph.D. in Biology (vegetation science) by the University of Porto, where he is Assistant Professor of ecology, biodiversity and natural resources. He also leads the Predicting and Managing Ecological Change research group at InBIO. His research focuses on the drivers, processes and consequences of ecosystem and landscape change. He applies a socio-ecological research framework to address the impacts of the many processes driving the dynamics of socio-ecological systems, from climate change and biological invasions to land use change and wildfires. His research is aimed to improve conservation strategies, spatial planning and land use systems under global change, linking ecosystem and landscape processes to the services and societal benefits they sustain.

F. Moreira holds a Ph.D. degree in Biology (University of Lisbon, 1996). He is the Principal investigator of the InBIO research group "Biodiversity in Agricultural and Forest Ecosystems" and the holder of the REN Invited Research Chair in Biodiversity. Current research interests include 3 major topics: (i) the links between farmland and forest management and how they are reflected on biodiversity and the provisioning of ecosystem services; (ii) the impacts of wildfire on plants and landscapes, as well as the effectiveness of different post-fire ecosystem restoration approaches; (iii) biodiversity impacts of anthropogenic linear infrastructures.

Profitability of Cassava Based Farms Adopting Climate Smart Agriculture (CSA) Practices in Delta State, Nigeria

Anthony O. Onoja, Joshua Agbomedarho, Ibisime Etela and Eunice N. Ajie

Abstract This study assessed the profitability of Climate Smart Agriculture (CSA) practices adopted by cassava based farmers in Delta State, Nigeria. It also ascertained the differences in profits between the CSAs and then evaluated the determinants of their profits. Primary data were collected from 120 farmers with the aid of a structured questionnaire and a Focused Group Discussion approach. The data were analyzed using descriptive statistics, Gross Margin, OLS regression models and Friedman test statistic. Findings indicated that the CSAs practiced by the farmers were crop rotation (91.8%), bush fallowing (80.0%), use of organic manure (73.3%), agro-forestry (40%) and multiple cropping (55%). The mean gross margin estimates for multiple cropping, monocropping and agro-forestry were $1684, $146.07 and $700 respectively. There was a significant variation in the profitability of the three farming systems. Extension service, CSA adoption, farming experience and sex were the major determinants of profitability in the cassava farms operating the CSAs. Based on the findings, it was recommended that agricultural extension services be provided to boost productivity of CSA practices in the crop farms; while policies that will encourage CSA adoption (especially agro-forestry and multiple cropping) should be put in place. Gender mainstreaming in the CSA adoption should also be encouraged by relevant authorities.

Keywords Climate smart agriculture · Farm productivity · Resilience against climate change · Food security · Agribusiness

A. O. Onoja (✉) · J. Agbomedarho
Department of Agricultural Economics and Extension,
University of Port Harcourt, Port Harcourt 500102, Nigeria
e-mail: tonyojonimi@gmail.com

I. Etela
Department of Animal Science, University of Port Harcourt,
Port Harcourt 500102, Nigeria

E. N. Ajie
Department of Agricultural Sciences, Ignatius Ajuru University of Education,
Port Harcourt, Nigeria

P. Castro et al. (eds.), *Climate Change-Resilient Agriculture and Agroforestry*,
Climate Change Management, https://doi.org/10.1007/978-3-319-75004-0_5

1 Introduction

A recent report of The Food and Agriculture Organization, FAO (2017) noted that climate change's impact on global food security will affect not just food supply but also food quality, food access and utilization as well as the stability of food security. The report added that adoption of sustainable land, water, fisheries and forestry management practices by smallholders will be crucial to efforts at adapting to climate change, eradicating global poverty and ending hunger. FAO noted that it would be equally important to work on mitigating the effects on agriculture, as this will open up pathways for agricultural development that can increase food production in ways that release fewer GHG emissions per unit of food. Without efforts to adapt to and mitigate climate change, FAO warned, food insecurity would likely increase substantially.

FAO (2009a, b) ranked Nigeria as the global largest producer of cassava with an annual output of about 45 million metric tonnes. In addition to this her cassava transformation is the most advanced in Africa (Egesi et al. 2006). In Delta State of Nigeria, agriculture is the mainstay of the economy. Cassava and other arable crops are largely grown for local consumption and commercial purpose in Delta State (Delta 2011). However, it is worrisome to note that Nigeria is under the threats of climate change and variability which may negatively impact on the growth of agriculture and food security in the country. The Nigerian Meteorological Agency (NIMET 2013) noted that there was increasingly late onset of the rainy season in recent years such that by 1971–2000 a vast portion of the country had experienced late onset of the rains. The report also noted that the cessation of the rainy season in the country transformed from being generally "normal" between 1941 and 1970 to "early cessation" during the 1971–2000 period. Thus, the period of the rainy season in the country has been reduced since 1941 when the onset and cessation were generally normal to 1971 when signals of late onset and early cessation of the rainy season set in. Since then, the length of the rainy season had remained shrinking while annual total rainfall remained about the same, thereby giving rise to high impact rainfall, resulting in flash floods. The report also indicated that temperatures across the country showed an increasing trend from mid 20th century to date. The mean temperature anomaly shows clearly the prevalence of warming in the country. Temperatures have increased from 0.2–0.5 °C in the high ground areas of Jos, Yelwa and Ilorin in the north and Shaki, Iseyin and Ondo in the Southwest to 0.9–1.9 °C over the rest parts of the country.

Cassava farming in Nigeria and indeed, in Delta State had been largely rainfed making the farms more vulnerable to the effects of climate change. The challenges posed by climate variations on crop production, has been well documented. The Intergovernmental Panel on Climate Change, IPCC (2007) reported that climate change would likely lead to a major spatial shift and extension of crop lands as it would create a favorable or restricted environment for crop growth across different regions; besides, frequency of heat stress, drought and flood exert negative effects on crop production. Niger Delta region of Nigeria in which Delta State is located is characterized by many lowlands, with most areas less than six metres above sea level and highly vulnerable to rainfall variability. Flooding thus is a common climate extreme event. It is also a major threat to the surrounding farms with the presence of numerous creeks, rivers and water bodies. There are documented evidence of climate change impacts in various parts of Nigeria including Delta State (Odjugo 2010). However, it is not certain how the effects of climate change and the adaptation efforts being practiced by the farmers had affected the profitability of cassava farms in Delta State of Nigeria.

Current research efforts are pointing to CSA adoption as a way of coping with climate change while mitigating the causes and reducing poverty through increase in farm productivity. Agriculture is considered to be "climate-smart" when it contributes to increasing mitigation, food security and adaptation sustainably. This new concept now dominates current discussions in agricultural development due to its capacity to amalgamate the agenda of agricultural development and climate change communities under one brand (Neufeldt et al. 2013) especially the UN Sustainable Development Goals and the Malabo declaration of the African Union. Adopting Climate-Smart Agriculture practices can reduce the risks facing smallholder farmers and mitigate the effects of extreme weather events on farms. Some notable CSAs involved enhancing crop production through minimum tillage systems/conservation agriculture, evergreen agriculture, shifting of planting dates, agro-forestry systems (to mention but a few) as Climate Smart Agriculture solutions (Landscapes 2014). To a large extent, the effect of climate change on cassava which is a major crop grown in Delta State is not known; and at such we do not know if the crop yields or profits from the farms are affected by these climatic variations. Prior to this time, no research, to the best knowledge of these researchers, has been carried out on this subject in this region. It is against this backdrop that this study was designed to assess the productivity of climate smart agricultural practices adopted by cassava based farmers in Delta State of Nigeria.

A study of this nature could give an insight into ways of reducing poverty status of farmers in the face of climate change. Lessons could also be learned on what determine profitability in farming when climate smart agricultural practices are being adopted by crop farmers. Studies focusing on how CSAs adoption could affect farmers' profitability are not very common even as the concept of CSA is becoming widely accepted as a measure for mitigating climate change and adapting to climate change while ensuring poverty reduction. Hence the findings of this study will provide a lot of policy impetus for simultaneously addressing issues of climate change mitigation, adaptation and farmer's poverty reduction drive.

1.1 Objectives of the Study

The broad objective of this study is to assess the profitability of CSA practices adopted by cassava based farmers in Delta State of Nigeria and understand the factors which influence the farm profitabilities under this practice.

The study specifically:

(i) identified the major climate smart agricultural (CSA) practices adopted by cassava farmers to adapt to climate change effects in the area;
(ii) determined the profitability of cassava farms under three major CSAs regimes, and
(iii) ascertained the determinants of profitability in cassava farms adopting CSA farming systems.

1.2 Hypotheses of the Study

Two null hypotheses were formulated to guide the study. These were specified as follow:

Ho_1: There is no significant variation in profitability of farms practicing the CSAs in the study.
$Ho_{2:}$ Types of CSA adopted by the farmers have no significant effects on farm profitability in the area.

1.3 Conceptual Frameworks

Climate risks facing cropping, livestock and fisheries are expected to increase in coming decades, particularly in low-income countries characterized by weaker adaptive capacity. Impacts on agriculture threaten both food security and agriculture's pivotal role in rural livelihoods and broad-based development (CCAFS and UNFAO 2014).

Climate Smart Agriculture (CSA) refers to *agriculture that increases sustainable productivity, resilience (adaptation), reduces or removes greenhouse gases (mitigation), and enhances achievement of national food security and development goals* (FAO 2010). This is a triple objective that is intended to be achieved by employing climate smart techniques. In some parts of the world, CSA has been able to boost crop productivity and reduce loss in agricultural productivity (The World Bank 2011). CSA does not imply an expensive, outrageous practice but uses the available techniques already existing in ways that would positively affect the output of agricultural products. World Bank (2011) warned that climate change was already changing the face of farming. Increases in temperature, rainfall patterns' variability, more extreme droughts and floods, and the shifting distribution of pests and diseases are all signs that could be attributed in part to the increase in emissions of greenhouse gases resulting from human activities. Given that these factors exert some impact on food production, adapting to climate change via climate smart agriculture (CSA) must be the first priority for the agriculture sector (World Bank 2011).

Nevertheless, the future is still bright. There are many practices under 'climate-smart agriculture' which can increase food production, aid farmers to build resilience against climate change and reduce emissions of greenhouse gases (FAO 2010). According to FAO (2010) some CSAs being adopted by farmers in the developing countries are farming system which could be described as conservation agriculture. This encompasses agronomic practices such as using minimum tillage methods for land preparations, use of improved seed and stock varieties, planting of cover crops, crop rotation, changing of planting dates, mulching, among other practices. According to Rosenzweig and Livewrman (1992) and Jagtap (2007) it is important to note that in many parts of the tropics as in Nigria, traditional practices such as terracing, agro-forestry, multiple cropping or crop diversification act as measures to cushion the effect of the changing climate on crop production, leads to increased output level and also conserve the soil fertility. These are all various forms of CSAs.

2 Research Methods

2.1 Study Area

The study was conducted in Delta State, Nigeria. Located between latitude 5° 30′N and longitude 6° 00′E, Delta State is endowed with a total land area of 16,842 km^2 (6503 mi^2) (Federal Republic of Nigeria 2007). Delta State is an oil producing state one of the rich agricultural producing states of Nigeria, situated in the region known as the South-South geo-political zone with a population of 4,098,291 (National Bureau of Statistics 2006a, b). Agriculture, Forestry and Fishing are prominent sources of livelihood in the state. The food crops produced in the state include cassava, fruits, rice, yam, vegetables, mangoes, pawpaw, pineapples, tomatoes banana and pepper. Delta is a major exporter of petroleum, rubber, timber, and palm oil alongside palm kernels via the Niger Delta ports (Delta 2011). Nigeria's economy is dominated by revenue from the oil sector. However, contribution of agriculture to overall GDP had been decreasing over the years especially after the discovery of oil in the country. For instance, it was observed that agriculture share of GDP went from 64% in 1960 to 46% in 2010. This, according to Udah et al. (2015) was as a result of decimal performance of its subsectors. With the exception of crop sub sector, they noted, livestock share of agricultural GDP declined from 24% in 1980 to 6% in 2010; forestry from 4 to 1% and fishery from 11 to 3% respectively.

2.2 Sampling Plan and Data Collection Method

The study sample comprised of 120 cassava farmers randomly selected from a list of 4500 cassava farmers registered with the Delta State Agricultural Development Project (DSADP) spread across four local government areas that were randomly selected from 25 local government areas (LGAs) in the state. These LGAs included Ughelli, Ndokwa-North, Sapele and Patani LGAs. In each of these LGAs, two communities were randomly selected from which 15 farmers each were selected. This gives a total sample of 120 farmers (i.e. $2 \times 15 \times 4 = 120$).

Data was collected from the primary source through the use of a structured questionnaire. 120 copies of the questionnaire were administered to the selected farmers. The questionnaire asked questions pertaining to the socio-economic attributes of the farmers and their demography such as age, sex, household size, monthly income proxied by monthly household expenditure, types of CSA adopted, educational attainment level, location, production inputs used on the farm as well as their unit prices and quantities used. Focused Group Discussions were also conducted with selected farmers and leaders of farm associations in addition to personal interviews conducted on the respondents. The focused group discussion elicited responses on types of CSAs farmers adopted and reasons they chose to adopt them. It also gave insight as to the extent of adoption of CSAs in the area and challenges of adopting the CSAs. It was necessary to conduct the FGD in order to bring about participation of farmers in the design of the study as well as to cover some issues not raised in the structured questionnaire. The FGD involved 13 farmers who were invited to discuss the required responses to the interview of the study. The field work was conducted by the researcher in 2015 while questionnaire were administered with the help of 4 trained enumerators and one translator.

2.3 Data Analysis Technique

Objective (i) was attained with the aid of descriptive statistics such as frequencies, percentages and mean. Gross Margin analysis was used to determine the profitability of the farms under three selected CSAs (Monocropping, Agro-forestry and Multiple Cropping or multiple cropping). The crops were selected based on their high frequency of adoption during the year in review (2015). OLS Multiple regression analysis models (using three functional forms, linear, semi-log and double log models) were used to analyze the effect of socioeconomic variables and adoption of climate smart practices on the profitability of the farmers in under various CSA regimes in the study. The differentials in Gross Margin of the CSAs (hypothesis 1) was tested using Friedman test following Conover (1999).

2.3.1 Model Specification

(i) *Gross Margin Analysis*

The formula for Gross Margin of cassava farms is given as follows:

$$GM = TVP - TVC \quad (1)$$

where,

GM Gross Margin in Naira (and USD)
TVP Total Value of Production in Naira (and in USD) and
TVC Total Variable Cost in Naira (and in USD).

(ii) *OLS Multiple Regression Models*

The implicit form of the multiple regression model used to analyze the determinants of profit under various CSA regimes in the study is specified as follows:

$$Y = f(X_1, X_2, X_3, X_4, X_5, X_6, \mu) \quad (2)$$

The explicit forms of the regression models used for the analysis are specified as follows:

Linear function

$$Y = \beta_0 + \beta_1 X_1 + \beta_2 X_2 + \beta_3 X_3 + \beta_4 X_4 + \beta_5 X_5 + \beta_6 X_6 + \beta_7 X_7 + \beta_8 X_8 + \mu \quad (3)$$

Double-log model

$$\ln Y = \beta_0 + \beta_1 \ln X_1 + \beta_2 \ln X_2 + \beta_3 \ln X_3 + \beta_4 \ln X_4 + \beta_5 \ln X_5 + \beta_6 \ln X_6 + \mu \quad (4)$$

Semi-log model

$$\ln Y = \beta_0 + \beta_1 X_1 + \beta_2 X_2 + \beta_3 X_3 + \beta_4 X_4 + \beta_5 X_5 + \beta_6 X_6 + \mu \quad (5)$$

Y amount of revenue realized from climate smart agricultural practices (in Naira or USD);
β_0 Intercept of the model;
$\beta_1 - \beta_8$ slope coefficient of the respective variables;
X_1 age of farmer's (in years);
X_2 extension service access (0 = "Yes", 1 = "No");
X_3 CSA type practiced (mono cropping = 1, agro forestry = 2, mixed cropping = 3);

X_4 location of the farm (distance of farm from the nearest market in kilometres);

X_5 farming experience (years);

X_6 sex (0 = male, 1 = female); and μ = Stochastic error term (assumed to have zero mean and constant variables).

(iii) *Friedman test*

Friedman test is a nonparametric test of equality of means used for analyzing randomized complete block designs (Conover 1999). According to Conover, the test is an extension of the sign test when there may be more than two treatments. The Friedman test is built on the assumption that there are k experimental treatments ($k \geq 2$).

3 Results and Discussion

The following socio-economic attributes of the farmers were noted and discussed as follows.

Table 1 shows the marital status of the respondent in the study area. The analysis reveals that (5.0%) of the respondents were single, (65.0%) were married, (21.7%) were widowed, (3.3%) were divorced, (5.0%) were separated. This implies that more married people engaged in the business of cassava production in the study area.

Table 2 shows that (43.3%) of the respondents were female and (56.7%) were male who engaged actively in cassava production. More males engaged in cassava production than female.

Table 3 shows that 1.7% of the respondent were between the age range of 1–5, 23.3% were of the age range 31–40, 61.7% were between 41 and 50, and 13.3% were between 51–60 years. This shows that more farmers fell between the ages of 41–50 years and were engaged more in cassava production as a means of livelihood to sustain their families.

Table 4 shows that 3.3% of the farmers had a household size of 1, 38.3% had a household size of 4, (25.0%) had a size of 5, 20.0%) had a household size of 6,

Table 1 Distribution of farmers according to marital status

	Frequency	Percent (%)
Single	3	5.0
Married	39	65.0
Widowed	13	21.7
Divorced	2	3.3
Separated	3	5.0
Total	60	100.0

Source Field data, 2015

Table 2 Distribution of farmers according to sex

Gender	Frequency	Percent (%)
Male	26	43.3
Female	34	56.7
Total	60	100.0

Source Field data, 2015

Table 3 Distribution of farmers according to age

Age group (years)	Frequency	Percent (%)
21–30	1	1.7
31–40	14	23.3
41–50	37	61.7
51–60	8	13.3
Total	60	100.0

Source Field data, 2015

while (1.7%) had 9 members as family size. This shows that majority of the farmers had a family size of about 4 individuals.

Table 5 shows that 6.7% of the respondents never had formal education, 45.0% only had primary education, 43.3% of the respondents had secondary education and only a few (5.0%) had tertiary education. This indicates that a greater percentage of the respondents were literate and this would afford them the opportunity to understand and adopt modern farm practices thereby enhancing productivity and profitability.

Table 6 displays the years of farming experience of the respondents, the analysis shows that 11.7% had an experience of 1–5 years, 63.3% had experience of 6–10 years, 18.3% had 11–15 years of experience, while 6.7% of the farmers had 16–20 years of farming experience. Thus it can be stated that there were high number of experienced farmers here.

The findings from Table 7 shows that (3.3%) of the respondents have a farm size below 1 ha, 91.7% produce on a farm size of 1–5 ha, 5.0% have farm size ranging from 6 to 10 ha. This shows that a greater proportion of the cassava farmers in the study area operate small and medium scale farms.

Table 4 Distribution of farmers according to household size

Household size	Frequency	Percent (%)
1	2	3.3
2	2	3.3
3	3	5.0
4	23	38.3
5	15	25.0
6	12	20.0
7	2	3.3
9	1	1.7
Total	60	100.0

Source Field data, 2015

Table 5 Distribution of farmers according to education qualification

Level of education	Frequency	Percent (%)
Non formal	4	6.7
Primary	27	45.0
Secondary	26	43.3
Tertiary	3	5.0
Total	60	100.0

Source Field data, 2015

Table 6 Distribution of farmers according to years of farming

Farming expeience (years)	Frequency	Percent (%)
1–5	7	11.7
6–10	38	63.3
11–15	11	18.3
16–20	4	6.7
Total	60	100.0

Source Field data 2015

Table 7 Distribution of farmers according to farm size

Farm size (ha)	Frequency	Percent (%)
<1	2	3.3
1–5	55	91.7
6–10	3	5.0
Total	60	100.0

Source Field data, 2015

The frequencies of selected types of CSAs adopted in the study during the period in focus are presented in Table 8. Table 8 shows that majority of the farmers (55%) were adopting multiple cropping, growing cassava alongside maize, yam, and plantain, while those who practiced agro-forestry CSA constituted 40% of the entire farmers. A very small proportion (5.0%) of the farmers were engaged in multiple cropping as their CSA. The use of multiple cropping as preferred by the cassava farmers in the findings above indicated that the farmers were really very prepared to tackle the risks of climate change sustainably. This is because with multiple cropping, the farmers could build resilience against risks of crop failure by depending on more crops in a planting season.

Table 8 Distribution of farmers according to CSA adopted

CSA adopted	Frequency	Percent (%)
Multiple cropping	66	55.0
Monocropping	6	5.0
Agroforestry	48	40.0
Total	120	100.0

Source Field data, 2015

Moreover, in multiple cropping some crops that replenish lost nutrients as is the usual case with leguminous crop rotation or interplanting perform poorer in regaining loss nutrients under climate change could be incorporated. The farmers also agreed that they practiced some other types of CSAs such as crop rotation (91.8%), bush fallowing (80.0%), use of organic manuring (73.3%), planting of improved varieties of planting materials (65%) and early planting (91.7%). However, for the year in review, the farmers agreed to adopting only monocropping, agro-forestry and multiple cropping, that is why these CSAs were chosen.

3.1 Profitability Differentials in Three CSAs Adopted by Cassava Farms in the Study

The estimates of mean profits (Farm Gross Margins) in the three dominant CSAs found in the study are presented in Table 9. The results indicated that multiple cropping and agro-forestry were more profitable with estimated gross margins of $1684 (₦320,000.00) and $700 per hectare ₦199.00:$1. (₦133,000.00). The higher profit recorded from the multiple cropping indicates that diversification of crops could help leverage risks against variability in climate or rainfall. The result of agro-forestry showing a higher profit than monocropping also attest to the reliability of agro-forestry as a useful CSA in boosting farm profitability. The use of two different crops or more hedges risk of crop failure and low yield. It is therefore not surprising to see that the two CSAs, crop diversification and agro-forestry were more profitable than monocropping.

Results of the Friedman test gave a Chi-square statistic of 8.273 at 2 degrees of freedom and was statistically significant at 5% (p = 0.016). With this result, we reject the null hypothesis (Ho_1) which held that there was no significant variation in

Table 9 Tests of hypothesis on variation in mean gross margins of the three systems of farming

Type of CSA		Multiple cropping	Monocropping	Agro-forestry
N (120)		66	6	48
	TVP	₦ 930,000	₦ 356,000	₦ 580,000
	Less TVC	₦ 610,000	₦ 326,800	₦ 447,000
Normal parameters	Mean gross margins (GM value in USD[a])	₦ 320,000.00 ($1684)	₦ 29,200.00 ($146.07)	₦ 133,000.00 ($700)
	Std. deviation	177,400.00	105,900.00	60,230.00
Friedmann chi test statistic = 8.273				
Asymp. sig. (2-tailed) = 0.016 Degrees of freedom (df) = 2				

Source Field Data, 2015
Test distribution is normal
[a]Exchange rate of 190.024 Nigerian Naira to one US Dollar

profitability of farms practicing the CSAs in the study. The Friedman test gave the following ranks to the CSAs: Multiple cropping, 2.50; Monocropping, 2.42 and Agro-forestry, 1.08. The ranks are in order of magnitudes, implying that higher ranks indicates higher profitability. The foregoing result enabled us to conclude that the gross margins recorded were significantly different in the three CSAs with multiple cropping having the highest GM, followed by agro-forestry and monocropping respectively.

3.2 *Effects of Socioeconomic Variables and Adoption of CSA Practices on Cassava Farms' Profitability*

Results in Table 10 gives the estimated coefficients and other parameters of the OLS regression applied to ascertain the effects of type CSA adopted alongside socio-economic factors' effects on the gross margins of cassava farms. Out of the three functional forms estimated we found that the semi log model is the best fit for analysis because it has the lowest Akaike info criterion of 1.028. The R^2 (0.63), implies that 63% of variation in profitability in the model was explained by the variations in the values of the independent variables of the model. With the VIF of 1.883, it shows that there is no problem of severe multicollinearity in our model. Most of the estimated slope coefficients also returned signs that were in sync with theoretical expectations.

Four variables, extension service, CSA type adopted, farming experience and sex of the farmer exerted significance effects on the estimated profitability of cassava production in the model. Extension service had a slope coefficient estimate of 0.06 and was significant at 1% of probability, showing that an increase in services rendered by extension agents by 1% increased the profitability of cassava based farmers in the study by 0.6%. CSA type adopted, with a positive coefficient of 0.100 is significant at 5% probability, indicating that a switch to another CSA practiced by the farmers increased their revenue by 0.10%. This result gave us the confidence to reject the second null hypothesis of the study which stated that types of CSA adopted by the farmers have no significant effect on the farm profitability in the study. CSA adoption therefore can be said to exert significant effect on the farm profitability levels. This is in line with FAO (2010) and Jagtap (2007) submissions which both indicated that CSAs can enhance mitigation, adaptation of farms as well as lead to increased farm productivity. Farming experiences with a negative coefficient of −0.02 is significant at 5% probability level and this indicates that revenue decreased with increased years of farming experience. Sex with a coefficient of 0.463 is significant at 1% probability implying that gender was a significant determinant of profitability of cassava farms under various CSA regimes in the study.

Table 10 Regression analysis parameters' estimates to show the effect of CSA adoption and socio-economic characteristics of farmers on gross margin of three CSAs

Dependent variable = revenue	Linear model			Semi log model			Double Log Model		
Explanatory variables	Coeff.	t-stat.	Prob.	Coeff.	t-stat	Prob.	Coef.	t-Stat.	Prob.
Age	681.83	0.42	0.68	0.01	0.56	0.58	0.06	0.20	0.84
Extension service	18,322.82	5.59***	0.00	0.06	3.516***	0.00	0.35	3.15***	0.00
CSA type adopted	13,995.59	1.24	0.22	0.10	1.942**	0.06	0.19	1.86*	0.07
Farming experience	−2885.12	−1.74*	0.09	−0.02	−2.121**	0.04	−0.18	−1.62	0.11
Location	2874.12	0.83	0.41	0.01	0.73	0.47	0.01	0.11	0.91
Sex	102,557.70	3.16***	0.00	0.46	2.989***	0.00	0.71	3.06***	0.00
Intercept	6969.65	0.08	0.94	11.30	25.75	0.00	11.85	10.76***	0.00
R-squared	0. 602			0.638			0.627		
Adj. R squared	0.57			0.59			0.58		
F-statistic	13.09			23.55			12.49		
Prob(F-statistic)	0.00			0.00			0.00		
Akaike info criterion	5.52			**1.03**			1.06		
VIF	1.88			1.88			1.88		

Source Computed from Field Survey, 2015

***Figures are significant at 1%; statistical level; **figures are significant at 5%; statistical level; *figures are significant at 10%; statistical level

4 Conclusion

This study was able to assess the profitability of CSA practices implemented by cassava based farmers in Delta State of Nigeria. At the same time, the differentials in profits across CSAs and their farm profitability determinants were duly ascertained using descriptive methods, budgetary analysis (Gross Margin), Friedman statistics and OLS econometric approach. The study then compared three dominant CSAs adopted by the farmers which included monocropping, multiple cropping and agroforestry out of other CSAs practiced. Multiple cropping gave better profit, compared to the other two CSAs, agroforestry and monocropping. It is not surprising as monocropping lacked the kind of risk absorptive capacity of multiple cropping and that of agro-forestry. It was found that the variation in the profitability of the three CSAs were significant. The implications of this is that multiple cropping could be a better option if a farmer seeks profit or plans to undertake commercial production of cassava. Even, in terms of sustainability, it would be advisable to adopt agro-forestry and multiple cropping since they are equally profitable and have additional benefits of improving biodiversity and reducing risks of crop failure associated with climate change.

It was found that extension service, CSA adoption, farming experience and gender were the major determinants of profitability in the cassava farms in the study. Based on the foregoing findings, the following policy recommendations were made: (1) CSA adoption should be encouraged among arable crop farmers as a way of improving profitability in the farms, mitigating climate change and improving farm productivity; (2) farmers seeking higher profits in their CSAs and seeking to hedge risks associated with monocropping should adopt multiple cropping as it boosts commercialization; (3) policies to encourage CSA adoption must consider mainstreaming gender in the system; and (4) authorities including governments and stakeholders in climate change should provide agricultural extension service to farmers to enable them adopt CSA practices productively.

References

CCAFS and UNFAO (2014) Questions & answers: knowledge on climate-smart agriculture. United Nations Food and Agriculture Organisation (UNFAO), Rome

Conover (1999) Practical nonparametric statistics, 3rd edn. Wiley, pp 367–373. Available at http://www.itl.nist.gov/div898/software/dataplot/refman1/auxillar/friedman.htm. Accessed on 12 Nov 2015

Delta (2011) Encyclopedia Britannica. In: Encyclopedia Britannica ultimate reference suite. Encyclopedia Britannica, Chicago

Egesi C, Mbanasor E, Ogbe F, Okogbenin E, Fregene M (2006) Development of cassava varieties with high value root quality through induced mutations and marker-aided breeding. NRCRI Umudike annual report 2006, pp 2–6

Federal Republic of Nigeria (2007) Official gazette 94:24

FAO (2009a) Food security and agricultural mitigation in developing countries: options for capturing synergies. Rome, pp 17–18

FAO (2009b) Harvesting agriculture's multiple benefits: mitigation, adaptation, development and food security. Rome. Retrieved on 23 Feb 2017 from http://www.fao.org/3/a-i6881e.pdf

Food and Agriculture Organization, FAO (2010) Climate-smart agriculture: policies, practices and financing for food security, adaptation and mitigation. http://tinyurl.com/65nfr7k

FAO (2017) The future of food and agriculture-trends and challenges. FAO, Rome

Intergovernmental Panel on Climate Change, IPCC (2007) Climate change 2007: impacts, adaptation, and vulnerability. In: Contribution of working group II to the third assessment report of the Intergovernmental Panel on Climate Change. Cambridge, UK

Jagtap S (2007) Managing vulnerability to extreme weather and climate events: Implications for agriculture and food security in Africa. In: Proceedings of the international conference on climate change and economic sustainability held at Nnamdi Azikiwe University, Enugu, Nigeria, pp 12–14

Landsapes (2014) Climate smart alliance in Africa. Retrieved from http://www.landscapes.org/glf-2014/africa-climate-smart-agriculture-alliance-unique-partnership-systemic-approach-food-insecurity-climate-change-africa/

National Bureau of Statistics (2006) National population census. Federal Government of Nigeria, Abuja

Neufeldt H, Jahn M, Campbell BM, Beddington JR, DeClerck F, De Pinto A, Zougmoré R (2013) Beyond climate-smart agriculture: toward safe operating spaces for global food systems. Agric Food Secur 2(12):10–1186

National Bureau of Statistics (2006b) Population census. Federal Republic of Nigeria, Abuja

Nigerian Meterological Agency (NIMET) (2012) Nigerian climate review. NIMET, Abuja. Retrieved on 20 Feb 2015 from http://www.nimet.gov.ng/sites/default/files/publications/2012%20NIGERIAN%20CLIMATE%20REVIEW%20EDIT.docx

NIMET (2013) Seasonal rainfall prediction (SRP). NIMET, Abuja

Odjugo PAO (2010) General overview of climate change impacts in Nigeria. J Hum Eco 29(1):47–55

Rosenzweig C, Livewrman D (1992) Predicted effects of climate change on agriculture: a comparison of temperate and tropical regions. In: Majumdar SK (ed) Global climate change: implications, challenges, and mitigation measures. The Pennsylvania Academy of Science, PA, pp 342–361

Udah SC, Nwachukwu IN, Nwosu AC, Mbanasor JA, Akpan SB (2015) Analysis of contribution of various agricultural subsectors to growth in Nigeria agricultural sector. Int J Agric For Fish 3(3):80–86

World Bank (2011) Climate-smart agriculture: a call to action. Retrieved from http://tinyurl.com/6kkvnv4

Dr. Anthony O. Onoja (PhD) is a climate change expert, a research consultant, Senior Lecturer and Head, Department of Agricultural Economics & Extension, University of Port-Harcourt, Nigeria. He holds a PhD in Agricultural Economics (Resource and Environmental Economics). He is the President, Agricultural Policy Research Network (APRNet, http://www.aprnetworkng.org/1/). He had won scholarships, research grants and fellowships (at national and international levels). He is well published and travelled as a researcher. He has consulted for GTZ, Forum for Agricultural Research in Africa (FARA), Global Development Network, World Bank and won an award for *outstanding contribution to research and development* given by University of Port Harcourt. He is a managing editor to two reputed journals belongs to numerous professional organizations including APRNet, International Association of Agricultural Economics (IAAE), African Econometric Society (AES), African Association of Agricultural Economists (AAAE) and Nigerian Association of Agricultural Economists (NAAE). E-Mail: tonyojonimi@gmail.com

Mr. Joshua Agbomedarho is a graduate of the Department of Agricultural Economics and Extension, University of Port Harcourt, Nigeria. He holds a *B. Agriculture* degree of the same institution and is currently a member of the National Youth Service Corps (NYSC), Nigeria. His research interest includes: farm productivity, climate change adaptation and farm systems analysis for policy making.

Dr. Ibisime Etela is a Registered Animal Scientist (*RAS*) and a founding member of the Regional Centre of Expertise Port Harcourt (RCE Port Harcourt): acknowledged in June 2015 by the United Nations University—Institute for the Advanced Study of Sustainability (UNU-IAS), Tokyo, Japan. An Associate Professor at the University of Port Harcourt, Nigeria, D. Etela has published 34 journal articles, one book chapter, over 20 conference papers, and has attended different national and international agricultural and agribusiness forums, seminars and workshops. He holds a doctorate degree in Animal Science from the University of Benin, Nigeria, as well as master's and bachelor's degrees from the Rivers State University of Science and Technology, Nigeria. He is a member of the Nigerian Institute of Animal Science (NIAS), Animal Science Association of Nigeria (ASAN), and Nigerian Society for Animal Production (NSAP).

Dr. Eunice N. Ajie (PhD) is a Lecturer and researcher at the Department of Agricultural Sciences, Ignatius Ajuru University of Education, Port Harcourt, Nigeria. She holds a PhD degree in Agricultural Economics (Agricultural Marketing option) of the University of Nigeria, Nsukka (2014). Her research interest includes agricultural marketing, agricultural policy analysis, climate change impacts on agriculture and gender issues. She is a member of many professional societies and has a many peer reviewed research papers in reputed journals and chapters in edited book of readings to her credit.

Climate-Smart Agricultural Practices (CSA) Adoption by Crop Farmers in Semi-arid Regions of West and East Africa: Evidence from Nigeria and Ethiopia

Anthony O. Onoja, Amanuel Z. Abraha, Atkilt Girma and Anthonia I. Achike

Abstract The study was designed to scientifically identify two analogous African sites in semi-arid regions experiencing climate change so as to share their common experiences and then document CSA practices adopted in these regions. It identified analogous sites in Nigeria and Ethiopia for the purpose of studying their climate change adaptation experiences; assessed the socio-economic attributes of crop farmers in the semi-arid regions of these countries under stress and risk of climate change; ascertained the perception of crop farmers on climate change risks in the areas and then described the CSAs adopted in the two analogous sites. Identification of sites were done using GIS tool called CCAFs. Then 120 crop farmers each were randomly selected from the two countries (240 farmers) in a stratified manner. Primary data were collected with the aid of Focus Group Discussion method, a set of structured questionnaire and interview schedule after validating the questionnaire. Data collected were analyzed using descriptive statistics and ranking techniques; analysis of variance and t test. It was found that the socioeconomic attributes of farmers in Ethiopia and Nigerian farms varied especially with respect to food assess, types of crops cultivated, household size, education and extension contacts even though major crops in the regions were similar (sorghum, maize, millet and sesame). The two countries had similarities in the adoption of CSAs with

A. O. Onoja (✉)
Department of Agricultural Economics and Extension, University of Port Harcourt, Port Harcourt, Nigeria
e-mail: tonyojonimi@gmail.com

A. Z. Abraha · A. Girma
Institute of Climate and Society, Mekelle University, Mekelle, Ethiopia

A. I. Achike
Department of Agricultural Economics, University of Nigeria, Nsukka, Nigeria

A. Z. Abraha · A. Girma
Department of Land Resources Management and Environmental Protection, Mekelle University, Mekelle, Ethiopia

P. Castro et al. (eds.), *Climate Change-Resilient Agriculture and Agroforestry*,
Climate Change Management, https://doi.org/10.1007/978-3-319-75004-0_6

the most common CSAs being crop rotation, agro-forestry, adoption of water management techniques, terracing/bunding and contour cropping. In Nigerian farms, while changing of planting dates (76%), diversification of crops (71%) and planting of high resistant varieties (82%) were common CSAs adopted by the farmers, Ethiopian farmers did not adopt these on a high scale. There was no difference in rate of adoption of CSAs in the two countries. It was recommended that farmers should be assisted to build capacities in applying more reliable CSAs such as use of drought tolerant varieties of seeds, improved water management techniques, and to have better access to early warning information on climate; irrigation facilities and finance.

Keywords Agroforestry · Climate smart agriculture · Climate change
Farming systems · Multiple cropping

1 Introduction

There is an increasing trend of negative climate change impact in the Sub-Sahara Africa (SSA). The impact is usually worse due to deep rooted poverty and lack of infrastructure. Farmers in this region have the least ability to cope (Inter Governmental Panel on Climate Change, IPCC 2001). It appears that the most feasible policy option remains on how to adapt effectively to climate change effects as mitigation alone cannot achieve much. Adaptation cost in Africa is enormous and thus requires more efficient way of managing it. For instance, the United Nations Environment Programme (UNEP) Adaptation Gap Report (UNECA 2014b) projected an adaptation cost for SSA to be between $14 billion and $15 billion a year. It is equally predicted to reach $70 billion by 2045 if no additional mitigation action is taken. This looming cost of adaptation amidst lingering hunger, food insecurity and poverty exacerbated by climate change effects in Africa and other developing countries could have informed FAO (2010a, b) to emphasize that "agriculture in developing countries must undergo a significant transformation in order to meet the related challenges of food security and climate change". This could explain why there is an increasing emphasis on how to apply Climate Smart Agriculture (CSA) as a measure for sustainably adapting to climate change's negative effects in agriculture. CSA describes agricultural practices, approaches and systems that sustainably and reliably increase food production and the ability of farmers to earn a living, while protecting or restoring the environment (FAO 2010a, b). CSA aims to build the food and nutrition security of the rural poor so that farm families have access to enough nutritious food at all times, even in the face of a changing climate. According to FAO (2010a, b), CSA practices enable farming communities to: sustainably and reliably increase agricultural productivity and incomes; adapt and build resilience to extreme weather events and a changing climate; and where appropriate, contribute to reducing greenhouse gas emissions and concentrations. FAO (2013) noted that CSA shares objectives and principles with sustainable intensification of crop production. Sustainable crop production intensification (SCPI) can be summed up in the words “save and grow”. Sustainable intensification, FAO further expounded, means a productive agriculture that conserves and enhances natural resources. It uses an ecosystem approach that draws on nature’s contribution to crop growth—soil organic matter, water flow regulation, pollination and natural predation of pests—and applies appropriate external inputs at the right time, in the right amount to improved crop varieties that are resilient to climate change and use nutrients, water and external inputs more efficiently. A CSA approach adds a more forward looking dimension, more concern about future potential changes and the need to be prepared for them.

In Nigeria as well as other West African (WA) countries, observed temperatures have been increasing faster than the projected global warming. The increase varied between 0.2 and 0.8 °C since the end of the 1970s (Sarr 2012). This trend is stronger for minimum rather than maximum temperatures (ECOWAS-SWAC/OECD/CILSS 2008). There is large consensus that in WA one of the major climate change impacts will be on rainfall, making it more variable and less reliable. This will affect the onset and length of growing season, particularly in semi-arid areas where yields from rain-fed agriculture could be reduced by up to 20–50% by 2050 (Sarr et al. 2007). Greater climate variability which incorporates the later onset, higher temperatures and increased potential evapotranspiration will make farming systems more highly vulnerable to climate change. Climate change will significantly affect food production and requires immediate and ongoing adaptation (Gornall et al. 2010). East Africa too is not free from the threats of climate change especially on food security. According to You and Ringer (2011) extreme hydrological variability and seasonality have reportedly constrained Ethiopia’s past economic development by negatively affecting crop production—chiefly through droughts—and by destroying roads and other infrastructure due to flooding in some areas. They noted that as climate change unfolds, average climatic variables will shift, and weather variability will intensify, exposing Ethiopian agriculture to higher levels of risk and jeopardizing economic growth, food security, and poverty reduction.

A wide range of adaptive actions can be implemented to reduce or overcome some negative effects of climate change on agriculture. Ethiopia, another Sub-Sahara African in East Africa, could have some lessons based on similarities it has with Nigerian semi-arid zones and same with Nigeria too from Ethiopia. According to You and Ringer (2011), two factors critical to assuring food security, whether at the local or the global level, are increasing crop productivity and increasing access to sustainable water supplies. These factors are also vital to the economic success of agriculture, which is particularly important in Ethiopia given that the sector accounts for about 41% of the country’s gross domestic product (GDP), produces 80% of its exports, employs 80% of the labour force, and is a major source of income and subsistence for the nation’s poor.

Adaptation to climate change, according to IPCC (2001) denotes adjustment in natural or human systems in response to actual or expected climatic stimuli or their effects, which moderates harm or exploits beneficial opportunities. Some shared adaptation methods in agriculture include, but not restricted to the use of new crop varieties and livestock species that are better suited to drier conditions, irrigation, crop diversification, adoption of mixed crop and livestock farming systems, and changing planting dates (Deressa et al. 2008). Even though some studies such as those cited in Deressa et al. (2008) documented some climate change adaptation practices in some specific locations or regions, there is an ample research gap for documentation of Climate Smart Agricultural (CSA) practices across countries with similar agro-ecologies in Sub-Sahara Africa. The choice of Ethiopian semi-arid region as the study area is instructive. The economic development of Ethiopia is dependent on the performance of the agriculture sector, and the contribution of this sector depends on how the natural resources are managed. Unfortunately in Ethiopia, the quality and the quantity of natural resources are degrading, worsening food insecurity in the country, due to a multitude of factors (The Environment for Development (EfD), 2009), including climate change and variability. Temesgen et al. (2008) study in different regions of Ethiopia found that Tigray region is among the most vulnerable regions in the country because of higher frequencies of drought and floods, lower access to technologies, fewer institutions dealing with climate related hazards, and lack of infrastructures. It stressed that vulnerability to climatic change is highly correlated with poverty and living status of farmers determines their vulnerability to and adaptation with climatic changes. In more specific terms not much is known about how the two most populous countries in Africa (Ethiopia and Northern Nigeria), already enmeshed in aridity and having similar agro-ecologies in some regions have been coping with climate change. Such lessons can help farmers in semi-arid zones of Africa upscale or adopt better and more efficient Climate Smart Agricultural technologies that will boost food security in the semi-arid zones of Africa based on Nigerian and Ethiopian experiences. Results from such studies based on comparative climate scenario modeling (between Nigerian semi-arid zone and arid zones of Ethiopia) with a view of adopting best crop growing technologies to adapt to climate change has great potentials for providing useful data in addressing food security issues especially in arid and semi-arid regions of Africa facing threats of climate change and food insecurity; hence the need for this study.

(i) Conceptual Issues

This study benefits from the concepts of farm risk theory and utility function model as well as human crises theory (a product of socio-natural interaction) as documented in Onoja and Achike (2014). The farm risk theory got more empirical backings from the works of Koundori et al. (2004) who demonstrated that perception of risks by farmers affect their level of adoption of technologies. Thus this investigation is framed to enhance understanding of how societies, especially agrarian societies may adapt to climate change, and is also informed by political ecology, especially in the ideal of integrating environmental and societal processes in a balanced manner following Walker (2005).

(ii) Adaptation strategies adopted by Crop Farmers in Sub-Sahara Africa to Cope with Negative Effects of Climate Change and Variability

In the face of increasing climate change impacts on water resources available for agriculture in Africa, there have been several adaptation strategies adopted and also recommended by several authorities concerned in order to build resilience and improve productivity of crops in Africa. UNECA (2014a) for instance recommended that implementation of the following adaptation strategies can help Africa optimize the use of its available water for agriculture and other purposes. These include, but not limited to: augmenting supply by building new reservoirs and/or expanding use of groundwater where feasible; expanding large-scale irrigation; promotion of efficient use of water resources through drip irrigation, water recycling, and reuse; improvement of supply management, for example, by using ground and surface water conjunctively and improving reservoir and reservoir-system management; improvement of demand management through promoting conservation methods; and promotion of contingency planning for droughts and floods. Spielman (2013) noted that although cultivars and fertilizers were by no means passé, attention could be turning to technologies that are less discrete or harder to identify, observe, or define. Examples of such technologies include practices as varied as integrated soil fertility management, agro-ecology, agroforestry, systems of crop intensification, integrated pest management, minimum tillage systems, and conservation agriculture. recommendations on planting dates, tillage practices, plant spacing, irrigation timing, or residue disposal were, for many farmers, controversial and counterintuitive after generations of collective experience. Charles and Rashid (2007) indicated that the use of irrigation, planting early maturing and drought resistant crop varieties and soil and water conservation practices were the most important strategies used by the communities in Southern Africa to cope with climate change effects. Empirical studies by Deressa (n.d.) observed that use of different crop varieties was the most commonly used method in Ethiopia whereas use of irrigation was the least adaptation practiced among the major adaptation methods identified in the Nile Basin of Ethiopia. Kalibba and Rabele (2009) found that the most common measure adopted by the farmers is crop rotation. Others were fallowing, construction of waterways, vegetable cover and contour farming. Sandbag construction and interplanting was the least common among the respondents. However, all respondents adopted at least one soil conservation measure in their wheat fields as a way of coping with effects of climate change. Onyeneke and Nwajiuba (2010) found that diversification of crops planted, soil conservation measures, changing planting dates, planting of trees, irrigation and rainwater harvesting were the most common forms of adaptive strategies to the harsh effects of climate change by the crop farmers in South East Nigeria. They further noted that 40% of the farmers admitted not adopting any adaptation practice at all.

1.1 *Objectives of the Study*

The broad objective of this study is to document and compare the Climate-Smart Agricultural practices (CSA) adopted by crop farmers in West and East African semi-arid regions. The specific objectives of this research, therefore is to, among other things:

(I) identify analogous sites in Nigeria and Ethiopia i.e. finding of the future climate of (analogues) site of Mekelle in Nigeria for the purpose of studying their climate change adaptation experiences;
(II) assess the socio-economic attributes of crop farmers in the semi-arid regions of Ethiopia and Nigeria facing threats of climate change;
(III) ascertain the perception of crop farmers on climate change risks in the areas
(IV) identify and describe the indigenous and modern Climate Smart Agricultural strategies/technologies applied to adapt to climate variability and change in Ethiopian and Nigerian farms analogous to each other; and then
(V) analyze the policy implications of the findings.

The findings and outcomes of this research is intended to enhance understanding of local political economy factors and the mechanisms behind adoption of climate smart agriculture in the SSA especially in West Africa and Eastern Africa. The project's outcomes will also promote knowledge support for enhanced adoption of CSA especially in semi-arid regions of SSA. It will equally provide information to African research institutions and scholars seeking to better understand the proven CSA technologies that can help increase agricultural productivity and their drivers for the purpose of improving the economic wellbeing of the farmers and boosting food security in the region. This study will also generate baseline data and information to support evidence-based CSA policy, programme design and performance monitoring in the context of the Comprehensive Africa's Agricultural Development Programme CAADP 10-year Results Framework (CRF) to guide and accelerate implementation of CAADP at the country level. These are both in sync with African Union Commission (AUC) and the AU NEPAD Planning and Coordinating Agency (NPCA) framework aimed at Sustaining CAADP Momentum.

2 Research Methods

Study area:

The study area includes Nigerian Sahel Savanna region (particularly Potiskum and Katsina) and Ethiopian semi-arid northern region called Tigray Province (see Figs. 1 and 2). Potiskum is situated at latitude 11°42′ and longitude 11°02′ while Katsina lies on latitude 13°01′ North and 07°41′ East. The Nigerian Sahelian zone is marginal in terms of rainfall. According to Ati and Iguisi (n.d.) data shows that rainfall of this zone has been on the decline particularly since the mid-1960s. They

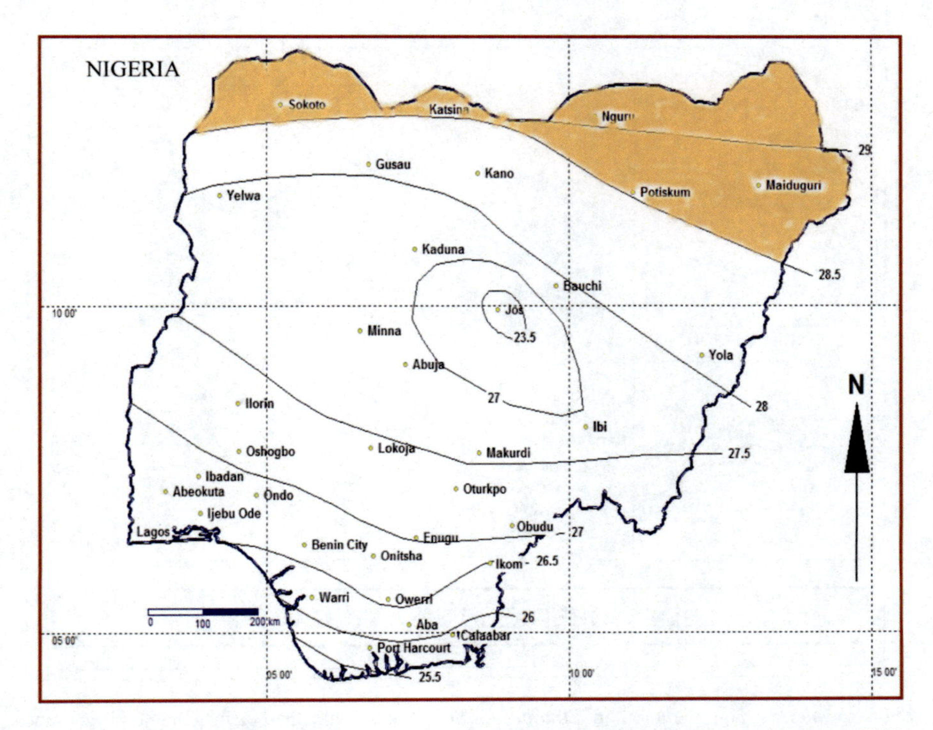

Fig. 1 The Nigerian Sahel Savanna region of Nigeria in shaded area. *Source* Nwanya (2013) shaded by the authors

observed that rainfall data for 50 years for four stations (Samaru, Potiskum, Sokoto and Katsina) indicated a decrease in annual rainfall in the zone from the mid-1960s up to the mid-1990s. Recent trends, however, show increase in annual rainfall from the mid-1990s. Rainfall in this zone, they added, is highly variable and the onset of the rain is erratic with rainfall intensity getting very high between the months of July and August. In the Sahel Savannah zone of Nigeria, rainfall start dates are 18 May to 24 June with cessation dates of 5–23 October; length of growing season days is 109–156 while seasonal rainfall amount ranges from 421 to 884 mm (NIMET 2013).

Agriculture dominates the Ethiopian economy, accounting for 80% of national employment, 41% of gross domestic product (GDP) and 33% of total exports or 70% of merchandise exports (Diao et al. 2007). More than 80% of these agricultural output and value-added (amounting to more than a quarter and a third of national output and value-added, respectively) is generated by subsistence farming. More interestingly, subsistence livestock production accounts for close to 40% of agricultural output and a third of value-added.

Data collection

With the aid of CCAFs (Climate Change Agriculture and Food Security) tools of CGIAR and following Arango and Jones (2014) two analogous sites in Nigeria and

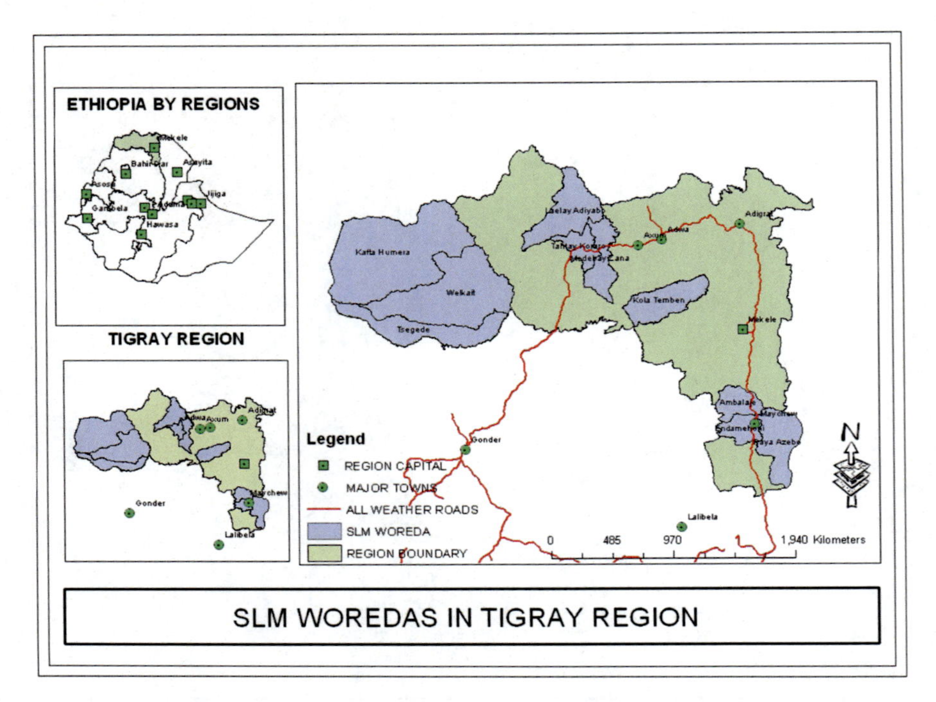

Fig. 2 Map of Tigray region in Ethiopia. *Source* Sustainable land management programme, Ethiopia (no date)

Ethiopia were selected. The selection of the analogues sites was based on BACKWARD method: The BACKWARD method answers the question "Where can I find the future climate of, for example, Sheraro site of Ethiopia in Nigeria today" (Arango and Jones 2014). The method is called the CCAFs Dissimilarity measure.

The relevance of an appropriate analogies goes beyond the number of similarities between a chosen past situation and a likely future, to the significance of the similarities. Ethiopia is known for its generally high rainfall variability. The well-known droughts of the 1970s and 80s prevailing over the whole Sahel including large parts of Ethiopia serve as an especially useful baseline. Nigeria is also known to have a large population of her farmers residing and producing crops in the Sahelan region of the country.

Sampling Method

Having detected and confirmed the two analogous sites by backward approach applying the Hallegiate model, (see Figs. 3 and 4 and Appendix 1, 2 and 3) the researchers decided to work with the four major sites located. These include, for the Nigerian sahelan Agro-climatic zone, Katsina and Potiskum; and for Ethiopia, the sites include Mekelle and Sheraro in Tigray region respectively. In each country 120 farmers from 6 communities (3 communities in Potiskum and Katsina

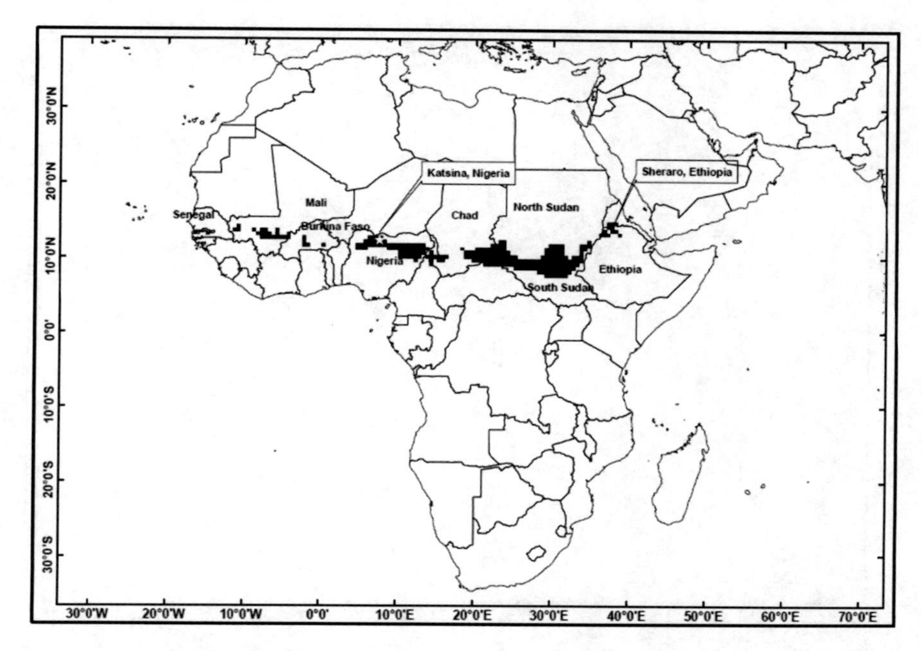

Fig. 3 Katsina backward CCAFS result (Note it indicates that Katsina in Nigeria has a similar agro-climatic region with Sheraro based on 30 years forecast for the Ethiopian location. The density of the shades in the identified spots confirmed the significance of their similarities. The shaded areas are the similar areas)

each for Nigeria) and 3 communities each for Sheraro and Mekelle respectively for Ethiopian semi-arid zone. These, gave a total of 220 farmers in the entire study.

The study would have benefitted more from use of very robust data in its analysis especially use of time series and panel data that could have included more countries in the two regions for the study, but this was not possible owing to inability of the researcher to assess such data, a situation aggravated by poor record keeping in the areas of the study and limited fund. However, these do not, in anyway invalidate the veracity and generalizability of the findings of this research as efforts were made by the researcher to adopt very unbiased approach to arrive at highly representative samples of farmers in the agro-climatic zones studied. The use of CCAFS tool to capture the most analogous sites is a very reliable way of objectively capturing similar sites to obtain data that will address the issues of concern to the two countries and zones in SSA. This approach is yet to be seen in studies involving analysis of CSAs adoption. By and large, the most appropriate methods for analyzing the data for the types of objectives pursued by the study were applied and the findings reported in a way that policy makers can understand.

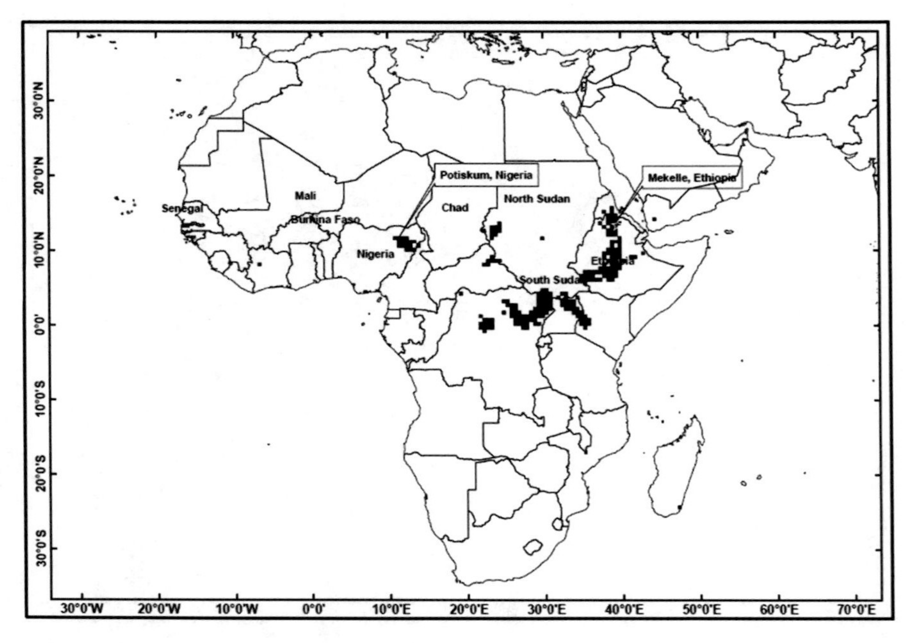

Fig. 4 Mekelle backward CCAFS result (Note that the Figure indicates that Mekelle in Ethiopia and Potiskum have similar agro-climatic conditions based on 30 years forecast. The density of the shades in the identified spots confirmed the significance of their similarities. The shaded areas are the similar areas)

Socio-economic survey

A set of structured questionnaire was used to obtain information about the adaptation technologies (CSAs) used, their perception of climate change threats and the socio-economic attributes of the crop farmers in the areas. This was augmented with focus group discussion (FGD) with key farming decision making units in the area. The set of structured questionnaire used was validated by two experts from Institute of Climate and Society, Mekelle University, Ethiopia. Four FGDs, each consisting 20 participants, 10 males and 10 females, drawn from different communities, were held for climate related risk identification and characterization, identification and prioritization of adaptive mechanisms as well as of CSAs applied. Following Mengistu (2011) we adopted tools such as hazard identification and characterization, hazard behaviour story telling (time-line), hazard ranking matrix, experiential stories telling on indigenous technologies and knowledge to acquire information on farmers' perception on climate change trends, existing hazards and their severity. The different CSAs used as coping strategies by the community were also identified and analyzed for their effectiveness. Effectiveness was rated as very satisfactory, satisfactory and not satisfactory and the rating number converted to% to assess satisfaction level.

Data Analysis

The information generated were recorded using worksheets prepared for each category of discussion while data collected on each parameter was expressed as percent of respondents. The questionnaire items collected too from interview schedules conducted on the respondents were analyzed and presented with the aid of simple descriptive statistics including percentages, means, ranking techniques and tabular analysis. Two tests of hypotheses were conducted: first, a test of mean differences in severity of climate change risks perceived by farmers in the two countries was performed using Analysis of Variance (ANOVA); while the test of significance difference in the rate of adoption of the identified CSAs in the two countries were done using t Test.

3 Results and Discussion

3.1 Identification of the Analogues Sites

The study detected and confirmed two analogous sites by using the Climate analogues technique, particularly, the backward approach applying the Hallegiate model, (see Figs. 3 and 4 and Appendix 1, 2 and 3) the researchers decided to work with the four major sites located. These include, for the Nigerian sahelan Agro-climatic zone, Katsina and Potiskum; and for Ethiopia, the sites include Mekelle and Sheraro in Tigray region respectively.

The *climate analogues* method is a technique that compares locations based on similarity in precipitation and temperature but it is also flexible to incorporate other input variables such as soils, infrastructure, social and economic conditions (Arango and Jones 2014). According to CCAFS (2014) the technique is a novel way of supporting modeled policy recommendations with on-the-ground empirical testing. Analogues, they noted, refer to sites or years that experience conditions with statistical similarity, primarily in terms of current or future climate, but they can also include additional factors such as soils, crops, and socioeconomic characteristics. This helps link top-down global models with targeted field trials or visits. In essence, the approach locates a site whose climate today is similar to the given future of a place of interest (i.e. where can we find today the future climate of Nairobi, Kenya?), or vice-versa. Strategies and technologies for adapting to climate change in particular locations should ideally be grounded in knowledge of the future climatic conditions in those locations. Estimates hold that 70% of future climates already exist somewhere in the world. That is where the analogues approach comes in. Using one or more global climate models, the analogues tool developed by CCAFS takes climate and rainfall predictions for a particular site and searches for places with similar conditions at present. With the knowledge of what they may face in future, farmers, researchers and policy makers can determine their adaptation options based on real—as opposed to crystal ball-gazing—models, noted

CCAFS. The tool is also used to track historical data to learn how communities have adapted—or have failed to adapt—to climate change over time. Importantly, it helps capture the real world capacity of farmers to adapt, which is too often not taken into account in catastrophic climate models. It was suggested that users of the tool—(available online at analogues.ciat.cgiar.org/climate) are at liberty to include variables such as crops, soils and socio-economic indicators in their searches. The Backward Method is one of the possible analogues which attempts to provide an answer to the question: "Where can I find the future climate of my site today?" With such information, Nigerian farmers, who will be where the Tigray region sites in Ethiopia were as at 2014 would therefore be able to learn what adaptation methods they will use in 30 years time from 2014 based on current experiences of the semi-arid Ethiopian sites located here.

3.1.1 Socio-economic Attributes of Crop Farmers in the Semi-arid Regions of Ethiopia and Nigeria Facing Threats of Climate Change

Results in Table 1a, b indicate the various socio-economic settings and attributes of the crop farmers in the study. From the result it could be noted that mean monthly household expenditures on food varied among the two countries. While Ethiopian

Table 1 **a** Estimates of selected socio-economic attributes of crop farmers in the study. **b** Estimated frequencies of selected socio-economic attributes of crop farmers in the study

	Tigray (Semi-arid) region Ethiopia	Nigerian Sahelian region
	Mean estimates	Mean estimates
Household food expenses per month in local currency and USD	662.5 Birr ($31.57)	10,072.95 Naira ($50.36)
Farm size	2.2	2.7
Household size	5.1	8
Age	39.1	43
Farming experience in years	18.9	12
Years of formal education	2.1	7
Frequency of agricultural extension visits	3.1	2

	Location	Ethiopian semi-arid region		Nigerian semi-arid (Sahel) region	
Item	Attributes	Observed frequency	Percentage	Observed frequency	Percentage
System of farming	Sole cropping	9	7.5	13	10.8
	Mixed cropping	6	5	22	18.3
	Agroforestry/ Alley farming	12	10	11	9.2
	Mixed farming	93	77.5	74	61.7
	Total	120	100	120	100.0

(continued)

Table 1 (continued)

	Location	Ethiopian semi-arid region		Nigerian semi-arid (Sahel) region	
Item	Attributes	Observed frequency	Percentage	Observed frequency	Percentage
Sex	Female	15	12.5	31	25.8
	Male	105	87.5	89	74.2
	Total	120	100	120	100
Marital status	Single	3	2.5	8	6.7
	Divorced	10	8.3	4	3.3
	Widowed	3	2.5	5	4.2
	Separated	6	5	3	2.5
	Married	98	81.7	100	83.3
	Total	120	100	120	100
Major occupation	Farming	15	12.5	55	45.8
	Trading	15	12.5	13	10.8
	Livestock Herding	51	42.5	43	35.8
	Artisan/Labourer	15	12.5	5	4.2
	Others	24	20	4	3.3
	Total	120	100	120	100
Types of crops grown	Teff	65	54	0	0.0
	Wheat	46	38	7	5.8
	Barley	33	28	0	0.0
	Sorghum	100	83	117	97.5
	Rice	0	0	0	0.0
	Oats	0	0	0	0.0
	Millet	88	73	112	93.3
	Maize	68	57	111	92.5
	Sesame	56	47	56	46.7
	Others	34	28	23	19.2

Source Field Data 2015

households spent an estimated monthly mean figure of $31.57 on food, their counterparts in Nigeria Sahel Savanna region spent an average of $50.36. The discrepancy may not be unrelated to the economic conditions of the two countries. This may not be surprising when one notes that Nigeria is the biggest economy in Africa with an estimated GDP of about $520 billion, a far cry from Ethiopian's GDP ($50 billion). Moreover, Ethiopia is noted to have, since 1996, been hit by climate change induced disasters about 15 times (FAO 2010a, b). This could have residual impact on food security status of the country till date thus explaining the apparent lower food security status than Nigeria recorded in this survey. FAO (1996) corroborated this in its report which stated that apart from being chronically food insecure, and food deficit, the highland areas of Ethiopia (including Tigray) have been hit by transitory food insecurity and that most of the food aid needs for 1997 were basically related to poverty and structural food insecurity, which in a number of cases has been compounded by unfavourable weather conditions. The FAO (1996) mission estimated that some 1.9 million people required some 186,000 tons of food aid until the next harvest.

Mean farm sizes recorded in the areas were 2.2 and 2.7 ha for Ethiopian and Nigerian farms respectively. Larger farm sizes have been reported to increase the likelihood of adapting to climate change. A recent study by Gebrehiwot and van der Veen (2013) found that a unit increase in farm size resulted in a 3.6% increase in the probability of using crop diversification, a 4.4% increase in the probability of conserving soil, and a 7.4% increase in the probability of planting trees to adapt to climate change.

They equally cited Amsalu and Graaff (2007)'s report who also found that farmers with large farm holdings were more likely to invest in soil conservation measures in the Ethiopian highlands.

In terms of household size, Household size Ethiopian farm households appeared to have less number of household members (mean = 5.1) than their Nigerian counterparts who recorded mean household sizes of 8. Household size has a mixed impact on farmers' adoption of agricultural technologies. While, on one hand larger family size is expected to enable farmers to take up labour intensive adaptation measures (Anley et al. 2007) on the other hand, Deressa et al. (2009) found that increasing household size had no significant effect on increasing the probability of adaptation. Hence we cannot conclude that the differences in mean household sizes in the two regions of our study could increase their potentials for adopting CSAs.

Maddison (2006) and Onoja (2014) noted that some factors which influence climate change adaptation strategies included age and gender of farmers, although they were completely beyond the control of policy makers. Onoja (2014)'s study in Nigeria indicated that extension contact, gender of the head of household, with climate variables (temperature and rainfall levels) determined choice of adaptation strategies in the country. We found that mean age of farmers in Ethiopia were 39.1 while Nigerian farmers were on the average 43 years however, Ethiopian farmers recorded higher mean farming experience (18.9) than their Nigerian counterparts (12 years). In terms of education Ethiopian farmers spent less years on formal education (mean = 2 years) compared to their Nigerian counterparts who spent on the average 7 years on formal education. In terms of access to agricultural extension services, Ethiopian farmers had higher mean frequency of access (3) compared to their Nigerian counterparts who had mean frequency of 2. In both countries males dominated farm ownerships with Ethiopian male headed households recording 87.5% and Nigerian household head's gender, 74.2% (Table 1b). In both countries majority of the farmers were married (81.7% for Ethiopia and 83.3% married in Nigeria). The most common system of farming in the two regions surveyed is mixed farming (77.5% for Ethiopia and 61.7% for Nigeria). While many Nigerian farmers in the Sahel zone (approximately 46%) were full-time farmers, their counterparts in Ethiopia were involved in so many off-farm activities. Teff (54%);

Wheat (38%) and Barley (28%) were mainly crops associated with Ethiopian farmers but the two countries commonly grow millet, sorghum, maize and sesame in large quantities (see Table 1a for details).

3.1.2 Perception of Crop Farmers on Climate Change Risks Affecting Farm Crops Production

In Table 2 the levels of severity of climate change related farm risks are itemized.

The rankings were done in decreasing importance (e.g. mean rank, 1, is more severe than 2 and so on). From the estimates it would be summarized that in Ethiopia, the most severe 5 climate risks perceived by farmers were drought (mean severity ranking = 2.55); market shocks (loss of Revenue and poor price) (mean severity ranking = 3.21); and late onset of rains (mean severity ranking = 3.21). In

Table 2 Results of estimated rankings of climate change and variability risks perceived by farmers in the two regions of the study

S. N.	Location	Tigray region (Semi-arid) Ethiopia		Sahel region of Nigeria	
	Type of risk perceived	Mean ranks of severity	Rank of severity	Mean ranks of severity	Rank of severity
1	Drought	2.55	1	1.72	1
2	Flood	6.46	11	14.28	15
3	Crop and animal diseases	4.04	4	8.09	7
4	Market shocks (loss of revenue and poor price)	3.21	2	6.09	5
5	Late arrival of rains	3.21	2	4.23	3
6	Drying up of stream and small rivers that flow all year round	6.00	9	6.73	6
7	Gradual disappearance of flood recession cropping in riverine areas	8.00	16	11.82	12
8	Increased temperature	4.78	7	2.89	2
9	Increased pests incidence including weeds	7.26	12	8.22	9
10	Salt water sweeping into land making it difficult for crops' growth	10.77	17	17.2	17
11	Erosion increase	4.07	5	8.88	10
12	Leaching/loss of soil nutrients	7.38	13	10.33	11
13	Environmental pollution	7.52	14	12.32	14
14	Reduced yield of crops harvested	5.60	8	4.32	4
15	Fire outbreak	4.57	6	8.19	8
16	Loss of some indigenous crop species	7.71	15	12.23	13
17	Hail storms and other strong winds	6.21	10	15.65	16

Source Field Survey 2015

Table 3 Results of F-test (Two-sample for variances) in severity indices perceived by the two groups of farmers (Ethiopian and Nigerian Farmers)

	Mean ranks of severity Ethiopia	Mean ranks of severity Nigeria
Mean	5.843	9.011
Variance	4.630	20.077
Observations	17	17
Df	16	16
F	0.231	
P(F $\leq$ f) one-tail	0.003	
F critical one-tail	0.429	

Source Field Survey 2015

Nigeria, the most severe climate change related risks in order of importance are as follows: drought (mean severity ranking = 1.72); increased temperature (mean severity ranking = 2.89); late onset of rains (mean severity ranking = 4.23). The above findings are in tandem with Onoja (2014)'s findings in Sahel Savannah region of Nigeria and other research elsewhere in Africa (e.g. James et al. 2013).

It was also found from our FGD that in Ethiopia 98% of the farmers admitted were aware of climate change impact while in Nigeria 78% agreed they were aware of climate change impacts. This is in consonance with earlier empirical works of Acquah-de Graft and Onumah (2011); Deressa et al. (2008) and Fosu-Mensah et al. (2010) who noted that most of the farmers in sub-Sahara Africa are aware of the impact of climate change, especially changes in temperature and precipitation.

The findings would not be conclusive without comparing the perceived severity levels which we infer from the total mean scores of both groups of farmers' mean ranking of the felt risks.

Results in Table 3 indicated an F-ratio estimate of 0.231 ($p < 0.01$) thus enabling us to reject the first null hypothesis of the study which held that there was no significant difference in estimated pooled ranked means of severity of climate change risks perceived by farmers in the two countries. Since we rank severity in ascending order, the estimated pooled mean rank of the severity of climate change related risks recorded by Ethiopia (5.84) which is lower than Nigerian estimate of 9.01 therefore implies that Ethiopian farmers had higher perception of the severity of climate related risks posed in the study at 1% level of significance. This is not surprising, given the level of exposure of Ethiopian farms to climate related hazards earlier discussed in this study.

3.1.3 Climate Smart Agricultural Strategies/Technologies Adopted to Cope with Perceived Risks of Climate Variability and Change in Ethiopian and Nigerian Analogous Farm Sites

Crop Rotation was the most common CSA adopted in both Ethiopia and Nigerian sites to cope with effects of climate change and variability. All the farmers (100%) agreed to have adopted this CSA. Other common CSAs applied in both countries include Crop residue incorporation (100% for Ethiopian farms and 75% for Nigerian farms); agroforestry systems mixed with crops (Ethiopia, 84% and Nigeria 95%); planting of windbreaks and shelter belts is equally common in both Ethiopian (89%) and Nigerian farms (100%). Ethiopia had 41% affirmed responses in favour of Integrated Pest Management while in Nigeria only 13% adopted this. The use of exclosures is more common with Ethiopian farms where 94% of them adopted this. Contour ridging/planting is another common CSA practiced in Ethiopia (68%) and Nigeria (96%). In Ethiopian farms other common CSAs adopted included planting of windbreaks and shelter belts (89%); rainwater harvesting (54%); use of earth catchment construction (87%); contour ridging/planting (57%); water management practices (87%), soil improving agroforestry (90%) as well as terracing/bunding (88%). In Nigeria, the following were the most common CSAs adopted apart from the ones earlier mentioned: cover cropping (63%); intercropping with legumes (73%); minimum tillage (55%); earth catchment construction (64%); water management practices (97%), planting of vegetative strips (55%) and terracing/bunding construction (75%) (Table 4).

The Focus Group Discussion result indicated that in Nigerian farms, changing of planting dates (76%) and planting of high resistant varieties (82%) were also common CSAs adopted by the farmers in Sahel Savannah region to cope with climate change and variability risks. On the other hand this is not so common in Ethiopian farms as only 45 and 48% agreed to the use of these two CSAs. For instance, Kalibba and Rabele (2009) found that the most common measure adopted by the African farmers they studied was crop rotation. The findings are in tandem with some earlier researchers' who noted that African farmers have used different CSAs in the past and present to adapt to the risks of climate change and variability in their environment e.g. Charles and Rashid (2007), Deressa et al (2008) and Spielman (2013).

Before concluding the research, a comparison of farmers' adoption rate of the listed adaptation strategies for the farmers in the two countries was done and the result of this analysis are presented in Table 5.

From the table, it could be seen that the t-value estimated was 0.228 ($p > 0.10$) and was not significant even at 10% level of significance. We therefore accept the null hypothesis which stated that the rates of adoption of the identified CSAs in the two countries were not significantly different from zero. In other words we are concluding that farmers in semi-arid Ethiopian farms and their counterparts in Semi-arid region (Sahel Savanna) of Nigeria do not significantly differ in their rates of adoption of CSAs examined.

Table 4 Estimated frequencies of climate smart agricultural strategies/technologies adopted to cope with perceived risks of climate variability and change in Ethiopian and Nigerian analogous farm sites

S. N.	Type of CSA adopted by farmers	Observed frequency	Percentage	Observed frequency	Percentage
1	Crop rotation	120	100	120	100
2	Agroforestry systems mixed with crops	101	84	114	95
3	Integrated pest management	41	34	15	13
4	Exclosures	113	94	26	22
5	Fire protection of vegetation	82	68	26	22
6	Cover cropping	42	35	76	63
7	Use of mulch and compost	18	15	21	18
8	Crop residue incorporation	120	100	90	75
9	Manuring	54	45	48	40
10	Intercropping with legumes	55	46	88	73
11	Minimum Tillage	29	24	66	55
12	Planting of windbreaks and shelter belts	107	89	120	100
13	Rainwater harvesting	65	54	36	30
14	Earth catchment construction	104	87	77	64
15	Tied ridges	12	10	56	47
16	Contour ridging/planting	68	57	115	96
17	Formal irrigation systems	12	10	46	38
18	Water management	104	87	116	97
19	Soil improving agroforestry	108	90	35	29
20	Vegetative strips	41	34	66	55
21	Terracing/bunding	106	88	90	75
	Mean rate of adoption of CSA in %		59.57		57.42

Source Field Survey 2015

Table 5 Results of t-test (Two-sample assuming unequal variances) used to test rate of adoption of CSAs hypothesis (HO_2)

	Mean rate of CSA adoption in Ethiopia	Mean rate of CSA adoption in Nigeria
Mean	59.57	57.42
Variance	986.96	875.97
Observations	21	21
Hypothesized mean difference	0	
Df	40	
t Stat	0.228	
$P(T \leq t)$ one-tail	0.410	
t critical one-tail	1.684	
$P(T \leq t)$ two-tail	0.821	
t critical two-tail	2.021	

Source Field Survey 2015

4 Conclusion

The study had used climate analogues approach to scientifically identify two analogous African sites in semi-arid regions experiencing climate change in order to help farmers, stake holders and scholars share knowledge and common experiences from documentation of CSA practices adopted in these regions which is home to most populous two African countries. Specifically the study identified analogous sites in Nigeria and Ethiopia for the purpose of studying their climate change adaptation experiences; assessed the socio-economic attributes of crop farmers in the semi-arid regions of these countries under stress and risk of climate change (this is a rare kind of study as scientific comparisons of CSAs in these selected areas are not existing); ascertained the perception of crop farmers on climate change risks in the areas and then described the CSAs adopted in the two analogous sites. The new aspect of this research is that no one (to the best knowledge of these researchers) has ever compared the CSAs of Ethiopian semi-arid regions with Nigerian Sahel (semi-arid regions) to inform farmers in Nigeria about possible future adaptation strategies they will need to adopt if they must cope with risks of climate change that will face the Nigerian sites in the next 30 years from 2014 based on the climate analogues forecasts. The study found that the socioeconomic attributes of farmers in Ethiopia and Nigerian farms varied to some extent, especially with respect to food security index, education (on which Ethiopian farmers lagged behind), types of crops cultivated, household size and extension contacts even though it was found that major crops in the regions were similar (sorghum, maize, millet and sesame). It was found that, in Ethiopia, the most severe climate risks perceived by farmers were drought; market shocks/loss of revenue/poor price; late onset of rains; crop and animal diseases and erosion/leaching; while in Nigeria, they included drought; increased temperature; reduced yield of crop and market shocks/ loss of revenue/poor price). The two countries had similarities in the adoption pattern of CSAs with the most common CSAs being crop rotation, agro-forestry, adoption of water management techniques, terracing/bunding and contour cropping. In Nigerian farms, while changing of planting dates (76%), diversification of crops (71%) and planting of high resistant varieties (82%) were common CSAs adopted by the farmers, Ethiopian farmers did not adopt these on a high scale. The researcher found no difference in rate of adoption of CSAs in the two countries. Based on the research outcomes the study recommends that farmers should be assisted to build capacities in applying more reliable CSAs such as use of drought tolerant varieties of seeds, improved water management techniques, and to have better access to early warning information on climate; irrigation facilities and finance by relevant institutions (NGOs, governments and farmer associations) so that they can build resilience and improve their farm productivities. Finally success stories of crops and CSAs such as cultivation of the ff, oats, peas and livestock grazing in enclosures should provide a learning point for Nigerian farmers in their bid to adapt to climate change risk by 2030.

Acknowledgements The researchers are very grateful to TRECC Africa (An Intra ACP Training) programme based in Stellenbosch University, for the fund and opportunity provided to conduct this study via its Ph.D. research exchange scholarship programme on Transdisciplinary Knowledge in climate change studies for the 2014/2015 batch awarded to the Principal Investigator, Anthony Ojonimi Onoja as a Ph.D. researcher at the Department of Agricultural Economics, University of Nigeria, Nsukka. We are also grateful to the university of the Principal Investigator, University of Port Harcourt, Nigeria, for releasing him to do the research and training. Thanks too, the Institute of Climate and Society, Mekelle University for providing some logistic support for the success of this research.

Appendix 1 (Arc-GIS Output Showing the CCAFS Result for Katsina Non)

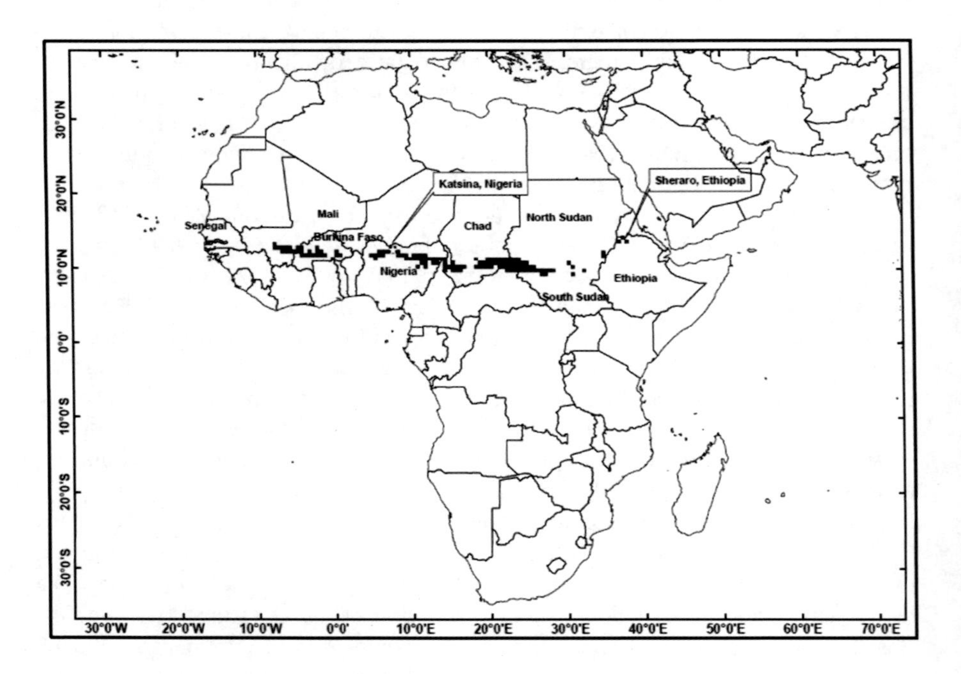

Appendix 2 (Arc-GIS Output Showing the CCAFS Result for Mekelle Non)

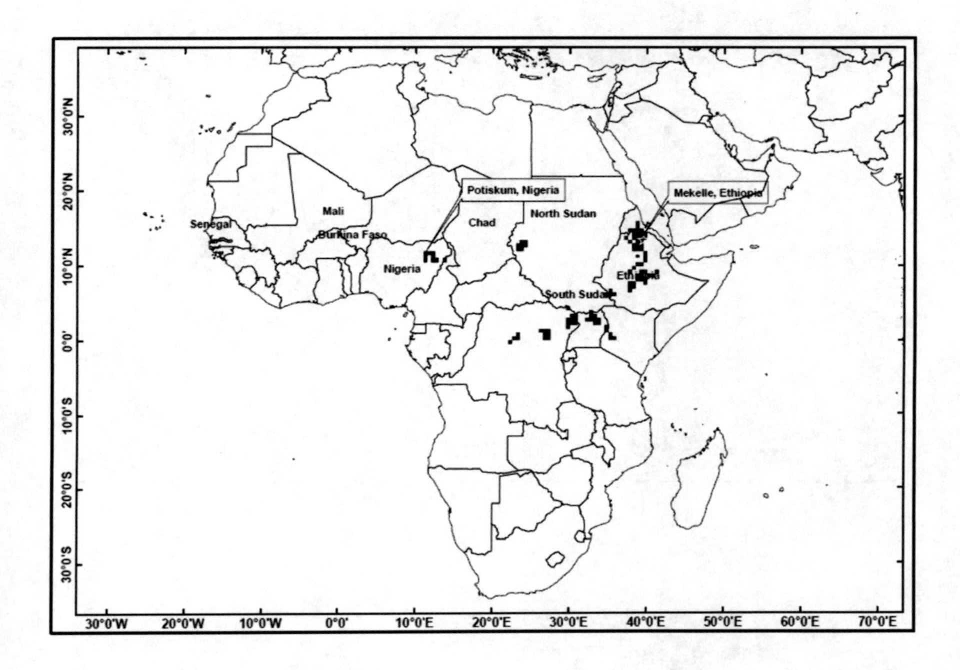

Analogues Results

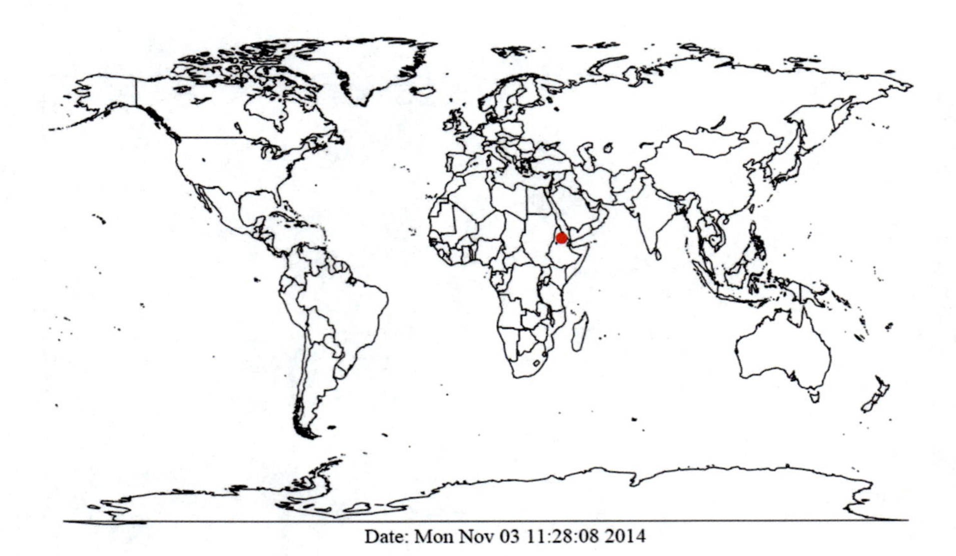

Date: Mon Nov 03 11:28:08 2014

Provided by

CIAT
Centro Internacional de Agricultura Tropical
International Center for Tropical Agriculture
Consultative Group on International Agricultural Research

References

Acquah-de Graft H, Onumah E (2011) Farmers' perceptions and adaptations to climate change: an estimation of willingness to pay. Agris 3(4):31–39

Amsalu A, de Graaff J (2007). Determinants of adoption and continued use of stone terraces for soil and water conservation in an Ethiopian highland watershed. Ecol Econs 61:294–302

Anley Y, Bogale A, Haile-Gabriel A (2007) Adoption decision and use intensity of soil and water conservation measures by smallholder subsistence farmers in Dedo district, Western Ethiopia. Land Degrad Dev 18:289–302

Arango D, Jones E (2014) The concept of climate scenarios and analogues. Retrieved on 14 July 2014 from http://ccafs.cgiar.org/regions/east-africa

Ati OF, Iguisi EO (n.d) Are we experiencing drier conditions in the Sudano-Sahelian zone of Nigeria? Being abstract a paper presented at the international conference on climate change and economic sustainability, Department of Geography, Meteorology and Environmental Management, Nnamdi Azikiwe University, Awka, Anambra State. Accessed 5th Nov 2014 from: http://www.abu.edu.ng/publications/2009-06-25-174641_5376.doc

Charles N, Rashid H (2007) Micro-level analysis of farmers' adaptation to climate change in Southern Africa. IFPRI discussion paper 00714, Washington DC, USA

Climate Change Agriculture and Food Security, CCAFS (2014) Climate analogues. Retrieved 08 Aug 2014 from http://www.ccafs-analogues.org/tool/

Deressa T (n.d.) Analysis of perception and adaptation to climate change in the Nile basin of Ethiopia. A research project conducted under the title "Food and water security under global change: developing adaptive capacity with a focus on rural Africa." Retrieved on 19th Oct 2011 from http://www.africametrics.org/documents/conference08/day1/session1/deressa.pdf

Deressa T, Hassan RM, Alemu T, Yesuf M, Ringler C (2008) Analyzing the determinants of farmers' choice of adaptation methods and perceptions of climate change in the Nile Basin of Ethiopia. International Food Policy Research Institute, IFPRI discussion paper 00798

Deressa TT, Hassan RM, Ringler C, Alemu T, Yesuf M (2009) Determinants of farmers' choice of adaptation methods to climate change in the Nile basin of Ethiopia. Glob Environ Change 19:248–255

Diao X, Fekadu B, Haggblade S, Taffesse AS, Wamisho K, Yu B (2007) Agricultural growth linkages in Ethiopia: estimates using fixed and flexible price models. International Food Policy Research Institute, IFPRI discussion paper no. 00695

ECOWAS-SWAC/OECD/CILSS (2008) Climate and climate change. The Atlas on regional integration in West Africa. Environment Series. January 2008. http://www.atlas-westafrica.org

Food and Agricultural Organization, FAO (1996) FAO global information and early warning system, World Food Programme (WFP) special report. FAO/WFP Crop and Food Supply Assessment Mission to Ethiopia; 16 December

FAO (2010a) Climate-smart" agriculture policies, practices and financing for food security, adaptation and mitigation. Food and Agriculture Organization of the United Nations, Rome

FAO (2010b) The state of food insecurity in the world. Addressing food insecurity in protracted crises. Food and Agricultural Organization of the United Nations (FAO), Rome

FAO (2013) Climate smart agriculture sourcebook. FAO of the United Nations, Rome

Fosu-Mensah B, Vlek P, Manschadi M (2010) Farmers' perceptions and adaptations to climate change: a case study of Sekyedumase District in Ghana. A contributed paper presented at World Food Systems Conference in Tropentag, Zurich: 14th–16 Sept 2010

Gebrehiwot T, van der Veen A (2013) Farm level adaptation to climate change: the case of farmer's in the Ethiopian highlands. Environ Manage 52:29–44. https://doi.org/10.1007/s00267-013-0039-3

Gornall J, Betts R, Burke E, Clark R, Camp J, Willett K, Wiltshire A (2010) Implications of climate change for agricultural productivity in the early twenty-first century. Philos Trans R Soc B 365(1554):2973–2989

Inter-governmental Panel on Climate Change, IPCC (2001) Climate change: impacts, adaptation and vulnerability. In: McCarthy JJ, Canziani OF, Leary NA, Dokken DJ, White KS (eds) A contribution of the working group II to the third assessment report of the intergovernmental panel on climate change. Cambridge University Press, Cambridge

James SJ, Zibanani K, Francis NO (2013) Farmers' perceptions and adaptations to climate change in Sub-Sahara Africa: a synthesis of empirical studies and implications for public policy in African agriculture. J Agric Sci 5(4) (Published by Canadian Center of Science and Education)

Kaliba ARM, Rabele T (2009) Impact of adopting soil conservation practices on wheat yield in Lesotho. Research paper 42, Aquaculture/Fisheries Centre, University of Arkansas at Pine Bluff, 1200 North University, Pine Bluff AR, USA. 493–608

Koundori P, Nauges C, Tzouvelekas V (2004) Technology adoption under uncertainty: theory and application to irrigation technology. Am J Agr Econ 88(3):657–670

Maddison D (2006) The perception of and adaptation to climate change in Africa. CEEPA. Discussion Paper No. 10. Centre for Environmental Economics and Policy in Africa. University of Pretoria, Pretoria, South Africa

Mengistu DK (2011) Agricultural sciences farmers' perception and knowledge of climate change and their coping strategies to the related hazards: case study from Adiha, central Tigray, Ethiopia. Agric Sci 2 (2):138–145. https://doi.org/10.4236/as.2011.22020. Retrieved 4 Dec 2014 from http://www.scirp.org/journal/AS/

Nwanya SC (2013) Study on perspectives of energy production systems and climate change risks in Nigeria. In: Singh BR (ed) Climate change—realities, impacts over ice cap, sea level and risks. InTech, https://doi.org/10.5772/55225. Available from: https://www.intechopen.com/books/climate-change-realities-impacts-over-ice-cap-sea-level-and-risks/study-on-perspectives-of-energy-production-systems-and-climate-change-risks-in-nigeria

Onoja AO (2014) Effects of climate on arable crop farmers' productivity, food security and adaptation strategies in Nigeria. A Ph.D Thesis submitted to the Department of Agricultural Economics, University of Nigeria, Nsukka. Available at https://oer.unn.edu.ng/.../effects-of-climate-on-arable-crop-farmers-productivity-food

Onoja AO, Achike AI (2014) Effects of climate change perception and socioeconomic factors on arable crop farmers' decision to adapt to climate change risks in Nigerian Agro-ecological Regions. In 17th ICABR conference book of abstracts of research findings presented at the international consortium on applied bioeconomy research (ICABR), 17–20 June 2014 at Safari Park Hotel, Nairobi, Kenya

Onyeneke R, Nwajiuba C (2010) Socio-economic effects of crop farmers' adaptation measures to climate change in the Southeast Rainforest Zone of Nigeria. In: Nmadu JN, Ojo MA, Mohammed US, Baba KM, Ibrahim FD, Yisa ES (eds) Commercial agriculture, banking reform and economic downturn: setting a new agenda for agricultural development in Nigeria. Proceedings of 11th annual national conference of national association of agricultural economists (NAAE) held at Federal University of Technology, Gidan Kwano, Minna, 30th Nov–3rd Dec, pp 369–376

Sarr B (2012) Present and future climate change in the semi-arid region of West Africa: a crucial input for practical adaptation in agriculture. Atmos Sci Let. Wiley Online Library. Retrieved 23 June 2014 from http://www.agrhymet.ne/portailCC/images/pdf/asl_368_Rev_EV.pdf https://doi.org/10.1002/asl.368

Spielman DJ (2013) Evaluating the "New Agronomy". In: IFPRI 2013 global food policy. IFPRI, Washington DC, p 44

Sustainable Land Management Programme, Ethiopia (no date) Geospatial information. SLM Woredas in Tigray Region. Retrieved 16 Aug 2017 from https://sites.google.com/site/slmethiopia/slm-woredas-in-tigray-region

Temesgen D, Hassan R, Ringler C (2008) Measuring Ethiopian farmers' vulnerability to climate change across regional states. IFPRI discussion paper no. 806, USA

United Nations Economic Commission for Africa, UNECA (2014a) Concept note. Ninth African Development Forum: innovative financing for Africa's Transformation, Marrakech, Morocco, 12–16 October

United Nations Economic Commission for Africa, UNECA (2014b) Keeping climate impacts at bay: a 6-point strategy for climate- resilient economies in Africa. UNECA, Addis Ababa

Walker PA (2005) Political ecology: where is the ecology? Prog Hum Geogr 29:73–82

You GJY, Ringler C (2011) How can African agriculture adapt to climate change? Insights from Ethiopia and South Africa, climate change impacts in Ethiopia: hydro-economic modeling projections. Accessible from http://www.ifpri.org/sites/default/files/publications/rb15_19.pdf

Anthony O. Onoja (Ph.D.) is currently a researcher/consultant, a Senior Lecturer of Agricultural Economics and Head, Department of Agricultural Economics and Extension, University of Port Harcourt, Choba, Port Harcourt, Nigeria. He is also the current President, Agricultural Policy Research Network (APRNet http://aprnetworkng.org/). He obtained his Ph.D. in Agricultural Economics (Majoring in Resource and Environmental Economics) from the University of Nigeria, Nsukka. A managing editor to two reputable journals, Dr. Onoja is widely published as a researcher and had successfully carried out funded research projects from reputable organizations such as Association of African Universities (AAU), Tertiary Education Fund (TET Fund) and Forum for Agricultural Research in Africa (FARA). He is currently working on funded research works from Global Development Network and AGRIFOSE2030/University of Sweden's CIDA funded projects. He is a member of numerous professional organizations including the International Association of Agricultural Economists (IAAE),African Association of Agricultural Economists (AAAE) and Nigerian Association of Agricultural Economists (NAAE) and Agricultural Policy Research Network (APRNet).

Amanuel Zenebe Abraha (Ph.D.) is an Associate Professor in Geography, Director, Institute of Climate and Society (ICS). He is also a lecturer at the Department of Land Resources Management & Environmental Protection (LaRMEP), Mekelle University, Mekelle, Ethiopia. He is well published as a researcher and an editor to some reputed journals.

Atkilt Girma is a Lecturer in GIS, remote sensing, soil survey, land use planning at Mekelle University, Mekelle, Ethiopia. He is also a Ph.D. researcher at ITC university of Twente, Netherlands. He is extensively published as a researcher.

Anthonia Ifeyinwa Achike (Ph.D.) is a Professor of Agricultural Economics at the University of Nigeria, Nsukka. She holds a doctorate (Ph.D.) in Agricultural and natural Resources Economics, and specialized in Agricultural Finance and Project Analysis. Professor Achike has many publications to her credit. She is a member of many local and international professional bodies as well as Associate Fellow of the African Institute for Applied Economics. Professor Achike was the Head of Department of Agricultural Economics, University of Nigeria, Nsukka (2005–2007), and also a global faculty member of the trade policy training in Africa (TRAPCA). She has won many competitive local and international research grants and has successfully completed them. At present, she is Consultant for a rice value-chain and aflasafe marketing studies in Nigeria as well as for an institutionally affiliated agribusiness project of the University of Nigeria, Nsukka, both of which are still on-going.

Challenges, Futures and Possibilities of Land Use in Rural Areas of Cela Municipality: Risks, Climate Change Impacts, Adaptation and Links to Sustainability

Bernardo Castro

Abstract The management of community rural land in Angola is aggravated by recurrent drought cycles intensifying the risk factors that can compromise economic, social and environmental sustainability. This work seeks to emphasize the role of adaptation of community rural families in an environment marked by extreme weather events that defy local economies. Data were collected through fieldwork relying upon qualitative methods (semi-structured interviews and focus groups). In addition, during field research, participant observation was used to collect detailed information through informal conversations with local people and observations of everyday life. Different strategies for coping/adapting to environment changes were also registered. The key informants were primarily comprised of traditional leaders in the village—men, women of different ages—who generally engage in local adaptive measures. Regional and local farming practices, families' migrations and the conflicts over land use were studied (and assessed existing small-scale dynamics centred on sustainability and resilience objectives). The results indicated a significant local adaptation deficit, lack of integration of climate change adaptation strategies in public instruments decision and a systemic and integrated approach to the territory management.

Keywords Adaptation · Climate change · Risk and vulnerability
Land use · Rural communities · Sustainability

B. Castro (✉)
ONG Rede Terra, Luanda, Angola
e-mail: bernardocastro9@gmail.com

P. Castro et al. (eds.), *Climate Change-Resilient Agriculture and Agroforestry*,
Climate Change Management, https://doi.org/10.1007/978-3-319-75004-0_7

1 Introduction

The risks of climate change, whose causes have been attributed to anthropic pressure, represents a challenge for rural families in Angola. These families base the expression of their identities on their relationship with the land. Some of these rural families, who are traditionally sedentary, find themselves forced to migrate to other areas, leaving behind their memories and history. Exploitation and consumption of natural resources, bereft of the principles of precaution and sustainability, coupled with increased demand for land, in a continent that is estimated to have lost 65% of its agricultural area since 1950, have led to the loss of character of some spaces and in turn caused deprivation and localised conflict. The sense of uncertainty is exacerbated by the deterioration of soil quality in many rural communities (forcing internal migratory flow). The centuries-old adaptations to climatic events are now insufficient.

The inconsistency of the rainy season, with particular reference to the central regions of Angola, along with the lack of ability to respond to its impact, exposes vulnerable families to risk, with few options to deal with extreme situations. The risks of natural disasters due to drought or flooding, repeated and intense in many areas, have grown in recent years, contributing to the depletion of resources and disrupting small local economies.

Angola is a country whose economy depends, to a great extent, on climatic factors. The country is experiencing deforestation (1600 km^2 of forests due to the burning and felling of trees for the production of paper, coal and wood), hydric stress, erosion of soil and its loss of quality.

Angola is a country affected by poverty and inequality, all of which increase the vulnerability of the affected communities and, at the same time, demonstrate how complex it is to adapt and build resilience, given the lack of knowledge of the impacts of climate change (Castro et al. 2018).

The National Plan for Adaptation to Climate Change (PNACC) points to the changes in rainfall patterns in Angola, and in the temperature. Surface temperatures in Angola increased by between 0.2 and 1 °C between 1970 and 2004 in the coastal and northern regions and between 1 and 2 °C in the central and eastern regions. It is predicted that, in the next 100 years, there will be an increase of between 3 and 4% in the temperature of the surface of eastern Angola, with a slightly lower increase in the coastal and northern regions (Lotz-Sisitka and Urquhart 2014).

Current projections indicate that warming, in relation to average annual temperature, in accordance with expectations, will, in large areas of Africa, exceed 2 °C by the final two decades of this century. This is based on the premise that 50% of its population live in areas prone to drought and also face problems related to inequality and poverty, making it difficult to implement successful adaptation strategies.

Implementing strategies to adapt to climate change into public policy is a significant challenge to risk management and sustainable soil use. Adaptation is seen as, beyond mitigation, an essential instrument in the development process.

Therefore, the incorporation of adaptation strategies into public policy is not only an obligation under The United Nations Convention on Climate Change but also a matter of equality and justice. The ability to adapt is determined by many factors, including social, material and environmental conditions and governance schemes, such as, for instance, the ability of a political system to implement adequate, large-scale public policies (Setti et al. 2016).

This paper seeks to emphasise the role of sociocultural, historical, educational and socio-environmental elements in order to provide solutions for the adaptation of rural community families when dealing with an environment marked by extreme weather conditions.

2 Background

Climate Change is a global and multi-dimensional (demographic, meteorological, environmental and socio-environmental, economic, institutional, socio-political, educational and technological factors) problem, namely in Africa (Collier et al. 2008). The response to climate change is fundamental achieving the Sustainable Development Goals (SDGs) as defined in Transforming Our World—the 2030 Agenda for Sustainable Development, namely SDG Goal 13—*Take urgent action to combat climate change and its impacts*.

Africa is a continent vulnerable to the impacts of climate change, due to extreme events (both droughts and floods), infectious diseases, population growth, urbanisation, agricultural based economy, natural disasters and low capacity to adapt to climate change (Boko et al. 2007; Patt et al. 2010; Mueller et al. 2011). African countries face this global climate change challenge which threaten the political, socio-economic (Shackleton et al. 2015) and environmental, ecological and biodiversity dimensions (conservation biology) (Midgley and Bond 2015).

Agriculture, forestry, agroforestry, biodiversity and ecosystem services are climate sensitive dimensions (Midgley and Bond 2015; Souza et al. 2015), together with land availability. Inadequate land use, tenure policies and lack of improved agricultural technology, dimensions can also contribute to the impacts of a changing climate (Cain 2013; Carranza and Treakle 2014).

3 Land Management, the Integration of the Risk of Climate Change in the Angolan Normative Judicial Context, and the Action Plan for Climate Change

According to the Constitutional Law of 1975 and of the current Constitution of The Republic of Angola, the State is the original Landowner. The first land law of independent Angola is Law no. 21-C/92 of August 28th (from 1992) and the first normative instrument to establish the basis of national policy for the concession of land rights. However, the right to own property has only been recognised in 2004 with the initiation of Law no. 9/04 of November 9th. The nature of land occupation and use of land are not governed by instruments of land planning as per Law 3/04 of June 25th. The concession of land, under Law no. 9/04 of November 9th is subject to the principle of seniority and is done without the approval of the joint executive decree that establishes the price of land.

The Angolan Constitution protects and respects the property and land of rural communities according number 2 to articles 15 and 37. This protection is carried out by the Land Law, article 37, line e. In the article 37, incise 1, 2, 3 of the Land Law states: «The occupation, possession and the right to use land is granted to rural communities who inhabit and exploit them usefully and in accordance with norms. The aforementioned recognition of rights, is done in a document issued by the appropriate authority in accordance with the regulations of this diploma. Communal rural land, while integrated into common law, cannot be the object of concession». In article 3, incise 2, line b) of Presidential Decree no. 216/11 from 2011, the State guarantees access to and use of land and also recognises legal rights and the management of communal land with a view to promoting social and economic justice in the field. The legislator, by enshrining article 87, also protects and values the historical and cultural heritage in communal rural communities that represent the history and memories of the communities. The recognition of communal lands, defined as collective native spaces subject to property law, dates back to the colonial period, starting in 1920. However communal rural laws remain without the recognition mentioned in Law no. 9/04 of November 9th. According to Angolan legal ordinance, communal rural lands are not subject to concession, as shown in Fig. 1, which classifies land according to its legal regime and purpose.

Angola has the plan for the strategic management of risk and disaster; the National Plan of Action and Adaptation, which is based on policies and national and international commitments to the prevention and reduction of risk of natural disasters, along with sustainable development, which aims to reinforce the ability to adapt to climate change on a national level. Some of the measures included in this plan are the creation of a National Solidarity and Assistance Fund and the creation and development of an Integrated System of Information Management to assess risk and disaster, the strengthening of systems to monitor and evaluate threats, and the strengthening of land planning, including risk management criteria with an emphasis on adaptation. However, The Angolan Constitution makes no reference, in article 39, to the risks associated with climate change or measures related to adaptation, limiting itself to environmental protection, the maintenance of ecological equilibrium, the rational use of resources and the respect for the rights of future generations.

Adaptation to climate change is approached in isolation from the rest of the development policy instruments through the National Plan of Action and Adaptation. In Angola, there is not sufficient information on the critical points in the agricultural systems, nor is there a reliable database of the evaluation, impacts and projections of climate change. However, according to the Angolan National Plan of

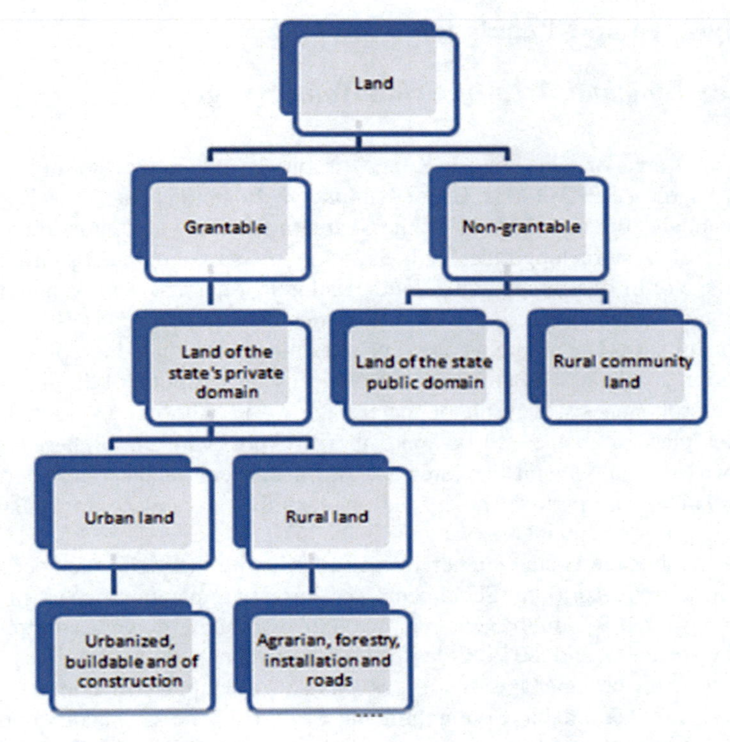

Fig. 1 Classification of Angolan land

Action for Adaptation, the agro-climatic analyses show a decrease in the duration of the rainy season in some locations, with frequent episodes of prolonged droughts during the rainy season.

Family agriculture sector is the most susceptible to the effects of climate change. The analysis of air temperature data in Luanda reveals a rate of increase of 0.2 °C per decade, resulting in a total cumulative increase of 1.9 °C between 1911 and 2005, and higher increases in the cold season, according to the Angolan National Plan of Action.

The adaptation scenarios in Angola require facing the challenges of the institutionalisation of climate change as an intrinsic factor in development adopting the climatic perspective in the production of public policies, and the scarcity of reliable data as to the forecasts and the lack of real identification of the capacity or local vulnerability are a reality yet to be overcome.

4 Study Area and Population

4.1 The Municipal Area of Cela (Waku-Kungu)

The municipal area of Cela (Waku-Kungu) is one of 12 in the South Kwanza province, Angola (Fig. 2). It is situated in the hydrographic basin of Kwanza between parallels 10°45′ and 11°40′ of the Southern latitude and the meridians 14° 45′ and 11° of Eastern longitude, with 5525 km^2 of surface area. Its limits are Kibala to the North, Kassongue to the South, Bailundo (Huambo) and Andulo (Bié) to the East and Ebo and Seles to the West. The municipal population is estimated at 255,000 inhabitants. The municipal area includes the intermediate high plains with a deforestation accelerated by anthropic factors, caused by rupture of the sills and a period of morphogenesis, a result of increased erosion. The rainy season in the region takes place from the end of September to April, with the highest rainfall typically occurring in November, March and April. Between January and February, there is a relatively dry period. The dry season (cacimbo) takes place from May and the first fortnight of September.

Cela has an average annual temperature of 21 °C, with the highest temperatures occurring in March, April and September. The lowest temperature occurs in June (Diniz and Aguiar 1998). In the Cela region, relative air humidity varies throughout the year between 70 and 80%. Cela features a savannah with primitive forest bushes, known in the language of Cela as 'kikala', a natural area dominated by typical species of Terminalia, Combretum and Erythrina. The savannah wood and the dispersed areas of open forest have Isoberlinia, Brachystegia and Julbernardia as their main species, according to Goes (2016) and mentioned in Diniz and Aguiar (1998).

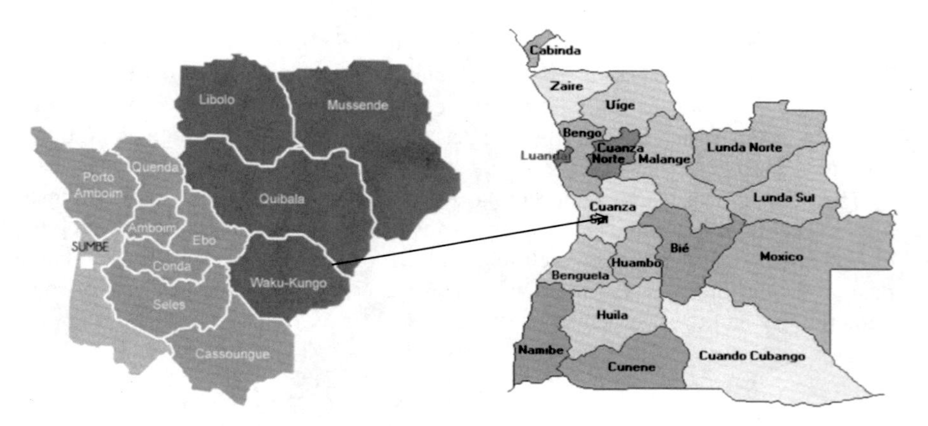

Fig. 2 Municipal area of Cela (Waku-Kungu) in the South Kwanza province

4.2 Population and Socio-economic Situation

The population is Bantu and belongs to the Ambundu ethno-linguistic group. The traditional occupation and source of family income essentially comes from agriculture. Hunting and fishing are seasonal and organised, in general, by the traditional authorities. The population lives in self-built clay houses covered with grass or in some cases zinc sheets. It is organised politically by Mbanzas (under the jurisdiction of the traditional powers) and its ownership and management are determined according to custom, enjoying the right to use and benefit from land in accordance with the Angolan Constitution and founding legislation. The land is, as a rule, the right of men. When sons reach more or less 16 years old, the age of consent, the parents attribute or reserve a part of their land, in preparation for married life, thereby land is inextricably linked to identity and wealth. Parents do not give land to their daughters, as they depend on the wealth of their future husbands. Only in cases of divorce or where a husband's death occurs, but also in the expectation of a future marriage, wives may have access to their parents' land. The wife, in this situation, will lose claim to succession rights from the first relationship.

The majority of the population is poor, to the extent that houses are without electricity or access to sanitation or tap water. The great majority consume water from dew (wells built by families). The Kwanza-Sul province, where the Cela municipality is located, the vast majority of households are not connected to the public water system. There are a total of 39,438 people with access to water by water tanks; 2583 by fountains; 66,965 by deep water holes; 17,976 by protected dew; 47,491 by unprotected dew; 2, 284 have access by rainwater and 90,569 have access to water by potholes or rivers (Censo 2014), with the highest proportion of households having an agricultural activity as their main occupation.

The levels of educational attainment are very low, and it is estimated that there is one doctor for every 22,000 inhabitants. Malaria is the primary cause of death in the province, which has the highest level of orphans under the age of 16 (Censo 2014). The solid waste is left in the open air or deposited near houses. In these communities, there is a significant lack of information, for example there are no newspapers or local radio stations. Given the lack of electricity, it is no surprise that some families use small generators and thereby can watch television news. The communities are not served by local infrastructures for the support of agriculture, there are no public transport networks, and as a result motorcycles and bicycles are the preferred means of transport in and around these communities.

With independence in 1975, the Angolan people returned to their original lands, but without land law recognition, as the first formal land law was not passed until 1992 (Law no. 12-C/92), with the introduction of a multi-party system. The communities of the area do not have licenses for the lands based on existing legislation. These communities face conflict over lands and the processes of land management. There is no information about the number of licenses that have been granted for different rights and corresponding areas.

5 Methods

In this study, informed by a qualitative methodology, there is a strong emphasis on individual and group interviews. This approach was used with individual interviews, taking into account the complexity and sensitivity of some of the issues related the confidential treatment on the part of the local authorities. With a view of creating opportunities for dialogue, participation, learning and articulation, the group interviews served to identify concepts and to capture beliefs, perceptions, expectations, motivations and traditional community needs.

Having defined and constructed the study model, three communities from the administrative sector of Kachongo to the south of the municipal area of Cela, and three to the North were selected. The rationale for selecting these communities is based on the desire to capture different dynamics, perceptions and needs, firstly in the rural communities of the South, adjacent to the central municipal administration, which since the colonial administration, has experienced some attention in terms of local investment and the establishment of health and education infrastructures. According to the rural communities to the North of the municipal area, as it is historically more remotely situated from governing centres and without investments or the establishment of basic infrastructures.

From the initiation of this process up to field work, previous/various visits to each community and a selection of relevant informants were made in order to form groups that would be interviewed as well as agenda rectifications and meeting points. Therefore, considering that it is appropriate for this study, eight relevant informants were selected for each group, making up five groups for each of the three communities. The preparatory visits showed that the topics presented were of the greatest interest to the communities and were therefore approached as openly as possible. The selected topics were: the use of geo-historical identities; sustainability in the context of climate change; and risks, vulnerabilities and adaptation.

The group interviews were semi-structured, guided by the selected topics. The interviews were applied over the course of a year from October 2015 to October 2016. The interviews were carried out in the community spaces, the Cela municipal cinema, the government headquarters and in residences of traditional authorities lasting a maximum duration of 1 h and 30 min. Proximity, informality and relaxation were points taken into account when choosing the space for interviewing, also taking into consideration the social-political framework marked by some control of information.

Prior to conducting the interviews, some testing sessions were organised for community rights activists so that during the application of the questionnaires, they would help to create a relationship of proximity and symmetrical communication through the use of the local language and occasional clarifications on the objectives of the study. Every interview was recorded, this being fundamental for the careful analysis of the most frequently used words, expressions, their meanings, consistency and pertinence as well as for the identification of relevant ideas.

In addition to the groups with relevant informants, the research was also supported by guided conversations with institutional groups. One of the groups was formed by five people from the provincial government and another one with representatives of the traditional power of authorities from different communities in the municipality, a total of 174 members of the Traditional Authorities Association. A total of 68 interviews were also held with representatives of the traditional power, as traditionally these representatives cannot reveal in public all of the community's secrets. Therefore, there were in total, 30 group interviews and 68 individual interviews, involving a total of 414 people.

Both preparatory interviews with the institutional groups were as follows: Policy and legal instruments on land management; Local resources and vulnerabilities; Strategies or resources on adaptation and prevention or monitoring climate risk; Database or information on land management; and Forms of information and participation in policies and perception on climate change.

6 Results and Discussion

The results indicated that of the 30 groups interviewed and asked as to whether the lands they occupy are either in the State's private domain or in the customary domain, only 2 groups believed that they are within the State's lands, while 28 groups, representing 93%, of the sample, claimed that they are on the formally protected lands of their ancestors, defined by the Property Constitution, as belonging to the families in rural communities and managed according to custom.

For those families who are integrated in the customary domain, we can say that the land represents, as can be interpreted in the following expressions and which summarize some of the statements given by the groups interviewed, that: (a) land as a historical repository of collective memories; (b) the land as a sign of the person's identity and dignity; (c) the land as a legacy between the past and future. Below, are some passages from the interviews with the groups that reflect this reality.

> «*Before the Colonial administration and the existence of the Angolan State these lands were always ours. We do not know what rights you are talking about but before laws were made, these lands were already inheritance and a shelter from our ancestors.*
>
> *During colonization our elderly died here and we were compulsively moved to the territories of other tribes.*
>
> *Those children were not born here. We returned after the national independence as our soul is here. This is where war found us and many of us died. Everything was destroyed but even living under a tree is considered our home land.*
>
> *Our houses are not our protection but the spirits of our ancestors that have left these lands. To sell these lands is to sell the elderly that rest in peace on them, our traditions and our history*" (Member of the Kissanga-Kungo group interviewed, August 2016).

> «*It is said by politicians that the land belongs to the one that works on it. Here, the land's purpose is not only to produce food. It is part of us and of our history. One that has no land, has no history*» (Taken from the Mbanza Mussende Group, Traditional Communities, November 2015).

The customary domain territory has its own characteristics in which values are formed by a set of guidelines and meanings according to its social, spatial and historical conditions.

When the groups were questioned on whether the development of identity relationship, which is often mistaken with the land, was always like this, 100% replied affirmatively. However, 18 groups state that the situation is changing:

> «*Today our children do not value the lands that were left to us by our ancestors. For them, the land's history and identity is meaningless. While we visit the lands because they speak to us and we speak to them, even without gripping a hoe, the children value more the economic side of the lands*» (Taken from the Group, Traditional Communities, 2016).

Justifying why the rupture of identity arises between the land and the new generation, 29 groups, which represents 97% of the groups, believe it is a recent situation, estimating that this may have occurred around 36 years ago. From the synthesis of responses of these 29 groups, many of the children were not born here and were not put through teachings of the "Jangos" (houses where elders share experiences and knowledge between different generations), in addition to the fact that the government does not appreciate nor teach in schools the historical and cultural meaning of the lands.

In response to the question of whether the communities own land deeds, it has been seen that both families that integrate the Customary domain (97%) and those who claim to be on private State domain lands (3%) do not own any land deeds. The generalised feeling is one of insecurity towards the issue of possession. The following passage is from the oldest member of the group, who is 68 years old, and expresses the feeling shared in all of the groups that were interviewed:

> «*Here we are always worried. Today, we are in one place but we never know where we will be tomorrow (…)*» (Taken from the group, Traditional Communities, 2015).

This statement agrees with the provisions of the Presidential Decree No 216/11 August, 8th (from 2011) according to which the majority of the population does not have a safe access for using the land.

There are normative references to the appreciation and preservation of the historical-cultural identities as noted in article 87 of the Constitution of the Republic of Angola as well as formal rural community land protection. However, there are no regulations embodying the land values as a collective memory repository or the identities of the families in rural communities. The country does not have a specific regulation on communal lands nor does a national land policy exist in Angola. Legislation on rural regulations and on rural development has not yet been published (to the date of this study).

In the Municipality of Cela, there is a high deficiency of information on the local governance instruments available, for example, The Municipal Integrated

Programme. Only 2 out of 30 groups confirm that they had heard of the Climate Change. The remaining 28 groups claim to never had access to the document nor were they ever given any information or had the local governance instruments made available to them. The municipal governments are in a process of adaptation which could have a successful outcome as they rely on local assessments and integration of local investments, policies and regulatory framework. This will be the case for political systems whose administration is appreciated by decentralisation and non-establishment which is not the situation in Angola. The most relevant expressions from the interviews on this topic from the groups was:

> *«They do not explain sustainability. We don't know what that is. But according to what you have explained today, we are aware that the lands belonged to our ancestors and we must leave them to our children and grandchildren. But they are not the same anymore»* (Taken from the group, Traditional Communities, 2016).

Families who are integrated in the rural communal lands do not possess the resources which allow them to work with concepts associated to sustainable development as an instrument of control, which would be enable them to manage, in a sustainable form, their natural resources. There is great lack of supervision and public policies surveillance in natural resources management in Angola and the systems of education needs to be rebuilt based on ethical valorisation towards solidarity, equity and intra and inter-generational justice.

Sustainable land management in Angola is a matter of concern not only because of the high risks associated with conflict that emerges from it but also because it places many families in vulnerable situations. Families living in suburban areas, and due to their living conditions, are exposed to risk by disorderly construction on areas that are at risk of flooding and susceptible to land mass movements.

The impacts of climate change will exacerbate environmental risks such as droughts, floods, extreme weather events which could contribute to food shortage, infrastructure damages as well as natural resource degradation, which are the source of livelihood (OCDE 2011).

The communities in the area of this study report the existence of drought and migratory cycles which are more intense and recurrent, or the extinction of some animal species. An elderly lady, Hepo, from the group of the northern community of the Cela municipality stated that:

> *«Here, we call one the "Ixoué". It has horns but its feet is more adapted to areas that have less water. That whole moist area was its habitat, but everything has changed. None of these animals exist now because the landscape was modified. Even the fish that we used to fish in those lakes and lagoons are scarce today».*

> *«We had no problems in fishing catfish or other types of fish. There was plenty of food. This has changed since the Agrarian Complex of Cela was implemented, beginning to spray the lands, by low-powered aircrafts, in order to kill the weeds. Many fertilizers and poisons were let go in the air, later on flowing into our rivers and lakes which borders with Nzúnzua. If we as humans die because of these poisons, it is obvious that animals do as well. In these last years we have been struggling with the rain season. Nowadays, we have no certainty on rain seasons dates as in the early days»* (Taken from the group, Traditional Communities, 2015).

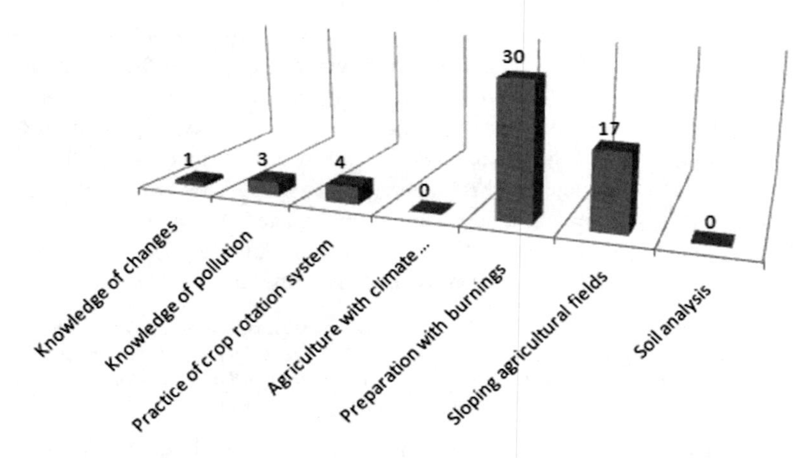

Fig. 3 What the communities know and do about the variables inscribed in the figure

Information on climate changes in Angola is scarce. However, in reference to the initial statement made by the Angolan government, Lotz-Sisitka and Urquhart (2014) point out a well proven tendency that the rainfall pattern in the capital of Angola, since 1941 has been irregular. Between 1941 and 1964 the tendency was to increase, between 1964 and 1978 a rapid decrease, and the years after 1978 a tendency to increase again.

Figure 3 illustrates the different perceptions, knowledge and practices of climate change, pollution, the practice of crop rotation system, agriculture and climate change, and preparation of burnt lands, sloping plots and soil analysis.

As to the perceptions and knowledge of climate changes, only one of the 30 groups assert to having information on this matter in question, through an informal channel. On the pollution issue, only three groups out of the 30 allege having a domain in this matter. As for the practice of using a crop rotation, there is knowledge about its importance in four groups whereas no groups practiced a guided agriculture by climate information.

In every community represented in the 30 groups, agriculture is practiced without any prior soil analysis, but they did alleged to having used agrochemicals and pesticides on land areas with a considerable slope, which resulted in chemical fertilisers being drawn either by irrigation or by rainfall to the underground waters and streams contaminating both water and soil. In the Kwanza-Sul province of which the Cela municipality is part of, out of 273,094 families, 224,740 consume untreated water in the rural areas (Censo 2014). The following excerpts were taken from the interviews held with the different groups in the communities:

«My son, we know that the world will come to an end but to say that we are causing problems provoking rain deficiency and for our lands not to produce, this we do not know».

«The problem is that the land has its own life and God made it that way. Now the lands are tired and everything is dying, not even worms come up very often».

«In the early days, we had crops here and there. Today everything is tight and we're always in the same place. The soil has lost its warmth so we must buy fertilizers to try and get something out of it».

«Here, where we are, we used to harvest a lot of corn but today if we don't plant beans and use fertilizers nothing will come up and hunger will occur». «They got hold of a great part of the moist lands and we went back to the mountains as in colonial times. No one helps us and they don't talk about climate changes» (Taken from the group, Traditional Communities, 2015).

In the Cela municipality, there are no sensitising and educational programmes on nutrition safety in context to climate changes, nor are there any tools of orientation for agricultural practices. However, interventions are held in terms of settlement and the use of the soils/lands by process of occasional allotments which imposed a segmented-spatial approach to exposing the communities to uncertainties and conflicts of different land usage.

The communities state that they live in a situation of uncertainty and insecurity. When they were questioned on whether what they harvest is more or less than what they harvested 30 years ago and why this is the case, the responses of the 30 groups was varied. A total of 24 groups (80%) of the groups claim that they harvested more in the old days; four groups (13%) claim that they harvest more today, whereas two groups (7%) of the groups claim to harvest more or less the same as they did 30 years ago.

Having more land, rainfall, in more or less forseen time with normal intensity, on more or less unfertilised soil, were some of the reasons given by the 24 groups. For the second group, today, they are able to harvest more as they use fertilisers and tractors which they can rent. As for the last group, they were not able to present any reasons. As to insecurities and uncertainties the 30 groups state:

«Rain does not fall during its rightful season but when it does, it either destroys the crops due to its heavyness, or enriched with strong winds and hail. Everything is damaged».

«We haven't got any support or anyone to protect us. We may have to leave at any time. We have lived through some droughts and if it gets worse we don't know who will be able to help us with food».

«Last year was terribly cold.» (Taken from the group, Traditional Communities (2016).

According to the IPCC (2015), the related climate risks intensify other environmental *stress* vectors, affecting the lives of the poor by the impacting on the livelihood means of these people, crop reductions and/or house destruction.

Angola is in a region of Africa in which many nations have economies linked with climate-sensitive livelihoods and are therefore considered to be highly vulnerable to climate change issues. In the southern region of the continent, 1.4 million acres of forest area are lost per year, whereas between 1999 and 2008 the area of African territory used for agriculture purposes grew 30.7%, with pasture area having grown 8.5% (Lesolle 2012).

Vulnerability is defined as a "tendency and predisposition to be adversely affected, their determinants being sensibility or susceptibility to danger and lack of capacity to resist or adapt" (IPCC 2014b). In the areas under study, the observed vulnerability is related to a lack of response in situations of extreme events, many due to considerable indications of poverty, lack of information and participation in training and decision-making processes, absence of basic infrastructures to support local initiatives, uncertainties and insecurities in terms of land ownership, a lack of diversification of livelihoods as well as lack of dialogue and cohesion on community issues.

> *«Nowadays, we chiefs are not respected. Neither the communal nor the municipal administrators respect us. Everything is State. In the villages when we ask for a task to be done in order to protect the crops and the community, few fulfill it. They think that tradition is a spell. This is why there are so many illnesses and deaths. The churches that came with the colonists continue to think like the colonists. Here, my son, we're really confused»* (Taken from the Kachongono's Traditional Authority Group, August 2016).

In the current context, ruptures in the traditional social hierarchy and local economic dynamics can be exacerbated due to the impacts of climate changes, especially in the most vulnerable rural communities, which are agriculturally dependent, under permanent state of uncertainty of land tenure, and are faced with recurrent droughts. For example, the central-southern region of Angola has been affected by recurrent drought and flood cycles since 2008. At the beginning of 2016, between 755 and 930 people from the southern region were affected by a drought, with the province of Cunene being the most affected area (PDNA 2016). Angola is located in a region which has a combination of prolonged climatic and spatial anthropogenic actions, especially hydrologic cycles have been disturbed, and underground water reserves are threatened by excessive exploitation and pollution (Lesolle 2012). Changing traditional practices and settlings can encourage a capacity of local adaptation.

In Angola, and in the perspective of this study, it is not the vulnerability of the physical system itself, whose pillars of sustainability may have been altered over time, which is the most relevant question, but that of political and social-economic vulnerability allied to the socio-environmental dimensions. The development template is paradoxical, meaning it is based on a dual paradigmatic structure between rural and urban; the traditional and modern showing strong human differences and territorial dysfunctions.

Policies defined by a lack of a strategic environmental evaluation as well as insufficient development of accountability systems have led to unsustainable practices of resource exploitation and land transformation (UNEP/GEO5 2012). Conflicting laws, values and interests undermine the capacity to develop institutional mechanisms that promote collaboration in management and response to common challenges such as droughts (UNEP/GEO5 2012). On the contrary, there are policies based on planning structures that treat the environment as a set of independent resources rather than an aggregated system which obviously weakens environmental management even more (UNEP/GEO5 2012).

The studied communities live in houses covered by grass or in some cases zinc sheets, communities in which 97% do not have access to electricity. The consequences of this fact are expressed by a representative of the traditional authority power:

> *«We have lived in these conditions since colonial times. Each one makes his adobes and builds. The problem is poverty. We live on field products that do not yield anything today. We do not have money to have town houses. Here the children have to live in the cities to continue studying and do not want to come back because there is no life here. Neither television nor newspapers nor money to buy radio and listen to news. At night everything goes dark because there is no electricity»* (Testimony of the Kilembo Communities Focal Group, 2016).

In Southern Africa, around 80% of the people who do not have electricity access live in rural areas (AFDB 2016). Essentially, in the South Kwanza province, out of 273,092 households, approximately 178,761 use oil lamps for lighting and cooking; 4639 use wood for cooking and lighting in the rural areas (97% do not have access to electricity) (Censo 2014).

In southern Africa, energy consumption per capita is the lowest of all continents estimating about 181 kWh per year except for South Africa (AFDB 2016). Deprivation of access to certain goods and services such as electricity and basic infrastructures in rural communities occur mainly due to a lack of decentralisation and unsettlement of public services insufficient development of public goods management. Further to this, dialogues within different types of knowledge (e.g., local, ecological knowledge) witnessing an increase of the State's presence in the areas, essentially in more distant rural and impoverished areas in Angola. According to Lesolle (2012), the climate changes represent a serious threat to energy safety in all of Development Community for Southern Africa (SADC).

The communities live basically on firewood for cooking and lighting or protection against the cold. However, nowadays families have to travel long distances to get firewood. In the municipality, there is no butane gas production. The butane gas used comes from other municipalities.

> *«In those days, everything was forest with many trees, medicinal plants and wild fruit. Everything has been destroyed today. Crops were grown in the mountains and inert figures were exploited. The mountains are bare and when it rains the water flows with heavy pressure and is destroying houses. It has already killed many people this year. We have to travel more than 7 km to fetch firewood. It is impossible to do anything without firewood here. It is worthless buying butane gas here. We just sit around the fire and talk all evening. Firewood means everything to us; it is not only for cooking. Charcoal is expensive and no one sits by the stove»* (Taken from the group, Kissanga-Kungu, May 2016).

Figure 4 shows the consumption rate on charcoal, butane gas and firewood in the communities.

On the topic of fuel, two issues provoke interest. The first one was related to the fact that communities in the 30 groups associated firewood usage as a cultural element, in other words, the use of firewood is not only an economic purpose. When firewood is used to build houses (Jangos) or in bonfires, it is a social factor. Sitting around bonfires, the elderly pass on knowledge and past experiences to the younger ones. In a Jango (a traditional house where the elderly of the community gather and sit to solve problems or share experiences and knowledge) according to them «it isn't a Jango without a bonfire» (Taken from the group, Traditional Communities, 2016). The second issue has to do with disassociating tree destruction for various purposes from the environmental element.

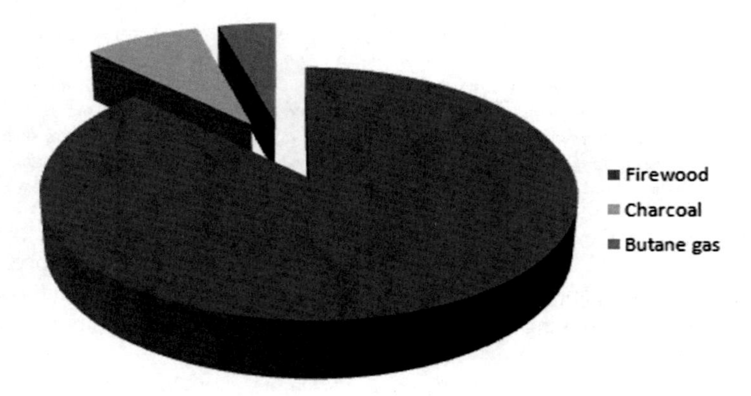

Fig. 4 Use of fossil fuels in communities

> *«Rain is God's work. It has nothing to do with the trees that are destroyed. Yes, trees help make shadows. What you are talking about, the carbon that the trees keep, straight down to the soil, we have never heard of»* (Taken from the group, Kissanga-Kungu, May 2016).

Across Africa, aridity/desertification is expected to increase significantly, mainly as a result of increasing temperatures, thereby increasing the rate of evapotranspiration of the plants, who are not able to compensate this as there is fluctuating rainfalls (Schaeffer et al. 2013). In southern Africa, the indications of aridity increases are up to 30% (Schaeffer et al. 2013) and the changes in precipitation reduce agricultural production, deteriorate food safety, increase drought incidents, and also increase the likelihood of conflicts due to water shortages (Chérif et al. 2017). It is therefore urgent that there is local adaptation to climate changes, which is worrying in the Municipality of Cela, as in of adaptation, there are low levels of perception on climate changes, a lack of information on local hydrology, and an identification inability and lack of risk monitoring (Chérif et al. 2017) as well as limited technical knowledge.

Adaptation is understood as an "adjustment platform within the human, social-political and natural systems as a response to stimulations in present or future effects which may limit the negative impacts and explore the positive aspects" (OCDE 2011).

Angola has a National Adaptation Action Plan (PANA) with priorities in the adaptation for the water, erosion and biodiversity sectors. One of the main points in the water sector is to improve the levels of information on hydrology. Most cultivated soils (highlands) have a weak structure with a low level of fertility, making them susceptible to erosion. In Africa, the traditional water management and associated watering techniques have frequently been ignored in the irrigation statistics as well as in the agricultural planning (Serrano 2013). In the communities under study, the families stated that they only used rainwater for immediate consumption.

«Here we use water from dew; we haven't stored much water when it rains for the drought periods» (Taken from the group, Kimbuata e Kaxike).

«There is hunger here when it does not rain. We live on products coming from the fields. In those days, the elderly had other types of skills, for example, making baskets, mortars, pestles, hoes, mattings and many other craft objects which they would trade for other products in different regions that haven't been affected by droughts. Today, one or another make these craft objects but no one buys them. We had big insect plagues which would come with the birds and destroy everything we planted. More than 50 years ago we had other forms of survival because Nature would give us a lot. Today, everything is destroyed» (Taken from the group, Kilembo, April, 2016).

Other priorities are related to soil/land in the sense of understanding the human and natural factors which contribute to the risk of soil erosion in many parts of the country.

Biodiversity is also a priority listed by PANA whose objective is to, amongst, collect information on actual levels of biodiversity and its conservation status. However, in the 30 groups from the communities under study, it was alleged that they were unaware of the existence of this instrument and had no information or training on climate changes adaptation and biodiversity conservation. Referring to the issues of alternative sources and when the rain ruins their crops, the statements were as follows:

«Corn cannot be watered with buckets of dew. When there is hunger or a drought in the area, we migrate to other areas that could help feed us. We have always worked on the land counting on the rain. In the old days, we always knew when it would rain, have heavy winds or hail, but today, the rain just surprises us and it comes when we least expect it or when we do expect it, it doesn't. We do not have a food bank as back up in the communities».

«When rain starts to be scarce, the traditional authorities perform a ritual so as to have rain. We have people here that take the rain from us. So when it doesn't rain, some people are accused. We have few alternatives when there's no rain. The ritual for rain calling doesn't work as well as it used to. In fact, truth be told, we didn't use to have hunger problems because we hunted, fished or picked mushrooms, wild fruit and roots. Nowadays, the forests have been destroyed and we just don't do anything. We used to live without soap because we washed ourselves with a root that made foam and it would clean our clothes. Without money, we have few alternatives. The shops do not accept the idea of exchanging products coming from our land for clothes or other food as in the colonial time. Everything is money» (Taken from the Group, Traditional Communities (2015).

For the agricultural, fishing and food production sectors, the IPCC (2014a) itemises the development of various types of crops adapted to new carbon dioxide levels (CO_2), temperature and rain (e.g. resistant to drought periods); to implement more efficient systems of watering; reinforce the ability of risk management of existing crops; compensate landowners economically due to alterations on soil use; improve financial support to low scale crops; and expand agricultural markets; and to make available advanced technological responses (more secure storage systems and preserving foods). As for the health, poverty, inequality and conflict sectors, the IPCC (2014a) indicates the following aspects for adaptation; improve drinkable water access and basic sanitation; expand vaccination and maternal and child health support services; increase the capacity to respond to natural disasters; establish efficient systems of health awareness; improve education, food supplies, healthcare services, electricity, secure housing and structural support access as well as its qualities; develop programmes to reduce gender inequalities and other forms of exclusion; increase the population's access to local resources and its control; reduce disaster risks; provide social security networks and insurance programmes; diversify and provide means of income and subsistence; and improve access to technology and decision-making forums on its development and use. These measures are brought into question as they reflect, on a large scale, the observed local vulnerabilities.

The form of preserving seeds has also changed. In the old days, the seeds were smoked in order to protect them from insects or they were deposited in well-protected barns. Even so, according to local statements, they still lost out. Nowadays, the corn or beans are deposited in well closed 200-litre containers with corks. In the communities' point of view, it is the most efficient form of protecting their products. When sowing the seeds, some communities would paint them with some plant leaves (e.g. paint the corn) and ashes so as for the partridges and other insects not destroy them.

As for the strong winds, the communities say they had no control on stopping them, especially as they are unpredictable, just like hail. But, occasionally a fence around a plantation is put up on some plots of small farmland.

In the different groups, as can be read in the passages below, it is stated that some traditional practices which had been abandoned, are starting to be restored.

> *«Here, we cultivate mostly corn and beans but nowadays we are restoring the cassava and yams plantation as well as animal breeding, for example; baby goats, sheep and chickens because time has changed. If it does not rain in the months of October, November and December, there will be hunger and many will move to others areas»* (Taken form the Kachongono Group, April de 2016).

> *«We have few solutions for too much rain or no rain at all. We are taking advantage of the plots that are up high but when it doesn't rain, it's a disaster. So, for those that have more than one plantation in different places, they take advantage to plant before the rain falls in more humid areas. In Cela's language, these areas are called "itaka". We can see that the corn seeds grow rapidly and ripen in a short period of time, but it is not practical here»* (Taken from the Futa Group, August, 2016).

> *«We used to catch a lot of fish which we would lay out in the sun and then preserve them for a long time. This would help us because, when there is hunger, we go other neighbourhoods to exchange the fish for cassava or corn. We picked kalomba or other types of insects from the trees and we would dry them, preserving them in bags, helping us a lot during hunger times, but the spaces are all being fenced in. We don't know what to do today»* (Taken from the Kissanga-Kungu Group, March, 2016).

Lesolle (2012) sets adaptation in a policy framework that takes into account contents of decision-making in the rural environment in which improvement in commerce and investments can increase market access for a small-scale agriculture. However, in terms of the Cela municipality, training and decision-making processes depend on the national instruments.

The cost of local adaptation is also a sensitive issue given the limited capacity for local resource mobilisation in a context where, given that much of Africa lacks infrastructures to adapt itself to climate change, most of its adaptation consists of capital investments for adaptation (Schaeffer et al. 2013).

According to the IPCC (2014a) the effectiveness of adaptation is conditioned to the investment made on the communities' abilities to adapt and on affected systems if populations were previously involved in the choosing and implementing adaptation responses, which is not the case of the communities in Cela. In Angola, there are no sectoral adaptation guides which make the aims of each sector from PANA feasible.

From this study's perspective, the scenarios for adaptation should be structured based on four priority aspects which the Hyogo Action 2005/2015 presents: Political Dimension—which means the guarantee of Disaster Risk Reduction (RRD) as a national and local priority supported by strong institutions. This is to say that in the context of this priority, the creation of national RRD platforms should be created and implemented among other actions; incorporation of RRD in development and planning policies as well as the introduction of legislation for RRD; Scientific Dimension—identified by actions such as identification, evaluation, risk monitoring and optimising systems of pre-warning; Social Dimension—which means the creation of social justice values, education, safety and resilience culture and sharing knowledge and finally, Vulnerability Dimension—which means a focus on the reduction of underlining risk factors.

In Cela's case, the measures for adaptation should have in consideration not only the local vulnerabilities but also the country's vulnerabilities in which it is confronted with:

(a) Vulnerability and resilience in local government and insufficient capacity and efficiency of the institutions, especially in disaster risk situations; insufficiencies in multidisciplinary land management and participation in decision-making processes; the deficient access of public services and goods to the populations and the inability to deal with the risks in a context of uncertainty;
(b) Socioeconomic vulnerability which means a very high level of impoverishment of the families and rural spaces; socio-spatial segregation and agrarian conflicts which impose occasional migrations due to increasing lack of spaces and natural resources;
(c) Technological vulnerabilities—insufficient of identification instruments, information and climate risk monitoring associated with climate changes;
(d) Axiology and educational vulnerability, which is identified by the low levels of schooling, knowledge and information on climate changes, adaptation and resilience. A lack of dialogues enhancing knowledge between experiences and traditional knowledge as well as knowledge of current civilization is another factor to take into consideration, ensuring that adaptation objectives mutually reinforce each other for sustainability;
(e) Normative vulnerability is marked by a lack of instruments and disrespect to laws on agrarian management;

(f) Communal organization vulnerability which is revealed in fragmentation values on agrarian and cultural conflicts; lack of local dialogues and the inexistent community funds for solidarity; abandonment of practices of diversification of means of local sustainability.

7 Conclusions

This study makes it evident that the climate change vulnerability in Cela, Angola, is related to a lack of response in situations of extreme events, due to considerable indications of poverty, lack of information and participation in training and decision-making processes, as well as the absence of basic infrastructures to support local initiatives.

The local fragmentation of ideals and solidarity systems has largely resulted in the inability of traditional adaptation responses in the context of climate change due to a lack of integration and articulations between the instruments of territorial planning and the uncertainties and insecurities in terms of land ownership. In this way, Angola needs an institutional and normative definition of land use and the effectiveness and implementation of a national adaptation plan on climate change.

The environmental and climate governance must be addressed, especially in terms of strategic and vigilante decision-making referring to the risks of climate changes guaranteeing the local institutions' participation and of the civil society to decision-making and inter-sectoral coordination.

References

AFDB (2016) Relatório Anual 2015 do Grupo do Banco Africano de Desenvolvimento. 364p, Abidjan, ISSN 1737-8990. Acedido, 14 de Novembro de 2016. http://www.afdb.org/fileadmin/uploads/afdb/Documents/Publications/AfDB_Annual_Report_2015_PORTUGUESE.pdf

Boko M, Niang I, Nyong A, Vogel C, Githeko A, Medany R, Osman-Elasha B, Tabo R, Yanda P (2007) Africa. In Impacts, adaptation and vulnerability contribution of working group ii. In: Parry M, Canziani O, Palutikof J, Van der Linden P, Hanson C (eds) Fourth assessment report of the intergovernmental panel on climate change. Cambridge University Press, Cambridge, pp 433–467

Cain A (2013) Angola: land resources and conflict. In: Unruh J, Williams RC (eds) Land and post-conflict peacebuilding. Earthscan, London. Available at https://environmentalpeacebuilding.org/assets/Documents/LibraryItem_000_Doc_155.pdf

Carranza F, Treakle J (2014) Land, territorial development and family farming in angola a holistic approach to community-based natural resource governance: the cases of Bie, Huambo, and Huila Provinces. Food and Agriculture Organisation of the United Nations Rome, 2014 Land and Water Division Working Paper 9 available at http://www.fao.org/3/a-mk753e.pdf

Castro B, Leal Filho W, Caetano F, Azeiteiro UM (2018) Climate change and integrated coastal management: risk perception and vulnerability in the Luanda Municipality (Angola). Leal F (ed) Climate change impacts and adaptation strategies for coastal communities. Climate change management series published by Springer. ISBN 978-3-319-70702-0 http://www.springer.com/gp/book/9783319707020

Censo (2014) Resultado Definitivos do Recenseamento Geral da População e da Habitação de Angola 2014—Instituto Nacional de Estatística, Gabinete Central do censo, Luanda-Angola. http://aiangola.com/wp-content/uploads/2016/03/Apresentaçao-Resultados-Definitivos-Censo-2014-V122203201619h28IMPRESS%C3%83O.pdf

Chérif S, Leal Filho W, Azeiteiro UM (2017) The role of farmers' perceptions in coping with climate change in Sub-saharan Africa Int J Global Warming 12(3/4):483–498. Inderscience Publishers. http://dx.doi.org/10.1504/IJGW.2017.10005907

Collier P, Conway G, Venables T (2008) Climate change and Africa. Oxford Rev Econ Policy 24 (2):337–353. https://doi.org/10.1093/oxrep/grn019

Diniz AC, Aguiar BFQ (1998) Zonagem agro-ecológica de Angola. Estudo cobrindo 200,000 km^2 do território. ICP, Fundação Portugal Africa, Fundo EFTA para o Desenvolvimento. Lisboa

Goes JMV (2016) Caracterização de um sistema sujeito de desfloresta e conversão do uso do solo na região da Quibala, Angola—62p, Dissertação de Mestrado, Instituito Superior de Agronomia, Universidade de Lisboa

IPCC (2014a) Intergovernmental panel on climate change. Climate change 2014: synthesis report. Contribution of working groups I, II and III to the fifth assessment report of the intergovernmental panel on climate change. IPCC, Geneva. Available at https://www.ipcc.ch/pdf/assessment-report/ar5/syr/AR5_SYR_FINAL_SPM.pdf

IPCC (2014b) Climate change 2014: impacts, adaptation, and vulnerability. Contribution of working group II to the fifth assessment report of the intergovernmental panel on climate change. Cambridge University Press, Cambridge, United Kingdom and New York

Lesolle D (2012) Documento de política sobre as alterações climáticas da SADC: Avaliação das opções de política para os Estados membros da SADC. 56p, SADC policy paper on climate change available at http://www.sadc.int/files/3613/6724/7855/SADC_Policy_Paper_Climate_Change_PT_1.pdf

Lotz-Sisitka H, Urquhart P (2014) Associação Regional das Universidades da África Austral (SARUA)—Climate Change Counts. Fortalecendo as Contribuições das Universidades para o Desenvolvimento com o Clima na África Austral—Relatório Nacional de Angola. ISBN: 978-0-9922355-5-0. Available at http://www.sarua.org/files/SARUA%20Vol2No1%20Relat%C3rio%20Nacional%20de%20Angola.pdf

Midgley GF, Bond WJ (2015) Future of African terrestrial biodiversity and ecosystems under anthropogenic climate change. Nat Clim Change 5:823–829. https://doi.org/10.1038/nclimate2753

Müller C, Cramera W, Hare WL, Lotze-Campen H (2011) Climate change risks for African agriculture. PNAS 108(11):4313–4315. https://doi.org/10.1073/pnas.1015078108

OCDE (2011) Integração da Adaptação às Alterações Climáticas na Cooperação para o Desenvolvimento: Guia para o Desenvolvimento de Políticas. http://dx.doi.org/10.1787/9789264110618-pt

Patt AG, Tadross M, Nussbaumer P, Asante K, Metzger M, Rafael J, Goujon A, Brundrit G (2010) Estimating least-developed countries' vulnerability to climate-related extreme events over the next 50 years. Proc Natl Acad Sci 107:1333–1337. https://doi.org/10.1073/pnas.0910253107

PDNA (2016) Seminário Nacional de Avaliação das Necessidades Pós-Desastre (PDNA) e Quadro de Recuperação Resiliente, Luanda-Angola. Available at http://www.ao.undp.org/

Schaeffer M et al (2013) Africa adaptation gap technical report: climate-change impacts, adaptation challenges and costs for Africa. In: Baarsch F, Adams S, Kelly de Bruin K, Marez L, Freitas S, Hof A, Hare B (eds) Michiel Schaeffer, climate analytics, Germany. University, Sweden. 58p, http://climateanalytics.org/files/schaeffer_et_al__2013__africao__s_a_daptation_gap_technical_report.pdf

Serrano V (2013) Pequenos regadios em Angola. O caso específico dos regadios tradicionais: panorama nos finais dos anos 60. Oportunidades e desafios. Documento preparado para a reunião da ASSESCA (Associação das Instituições de Ensino Superior Agrário)—Huambo, 17–19 Abril de 2013—Publicado na revista Outlook on Agriculture, vol 20, n. 3, 175–181, 1991, em co-autoria com Richard Carter (Prof em Gestão e Abastecimento de Água, Silsoe College, Inglaterra)

Setti AFF, Ribeiro H, Gallo E, Alves F, Azeiteiro UM (2016) Climate change and health: governance mechanisms in traditional communities of Mosaico Bocaina/Brazil. In: Leal Filho W, Azeiteiro UM, Alves F (eds) (Org.) Climate change and health: improving resilience and reducing risks, 1st edn. Springer International Publishing, Berlin, pp 329–351. https://doi.org/10.1007/978-3-319-24660-4_19

Shackleton S, Ziervogel G, Sallu S, Gill T, Tschakert P (2015) Why is socially-just climate change adaptation in sub-Saharan Africa so challenging? A review of barriers identified from empirical cases. Wiley Interdisc Rev Clim Change 6:321–344

Souza KKE, Harvey B, Leone M, Murali KS, Ford JD (2015) Vulnerability to climate change in three hot spots in Africa and Asia: key issues for policy-relevant adaptation and resilience-building research. Reg Environ Change 15:747–753

UNEP/GEO5 (2012) Crescente urbanização, globalização e fraca governação são fortes ameaças para o ambiente: A terra e a água enfrentam uma crescente pressão, mas políticas e parcerias inspiradoras demonstram que o progresso é possível. Síntese para África. Delegação Regional do PNUMA em África. In: http://www.unep.org/geo/pdfs/geo5/RS_Africa_pr.pdf (acedido em Outubro de 2016)

Future Climate Change Impacts on Malta's Agriculture, Based on Multi-model Results from WCRP's CMIP5

Charles Galdies and Kimberly Vella

Abstract Based on the World Climate Research Program's (WCRP) predicted changes in the magnitude and distribution of regional precipitation and temperature, this study assesses the future viability of agriculture in the Maltese islands, which are situated in the central Mediterranean region considered by many as a climate change hotspot. The analysis uses the latest results from an ensemble of 11 Coupled Model Intercomparison Project phase 5 (CMIP5) models addressing IPCC's four Representative Concentration Pathways (RCPs) for the years 2050 and 2070 as provided by WorldClim database. Using statistical, empirical crop- and livestock-modeling techniques, this unique study shows that future climate change is likely to negatively affect Malta's natural freshwater supplies, livestock and crop survival. As a consequence, the distribution of the already stressed local arable land will change, modifying production patterns and economics. The analysis of multi-model predictions provided a more robust evaluation of the likely impacts of physical and bioclimatic factors that are of relevance to local agriculture. Irrespective of which RCP scenario is considered, we find that the expected losses in productivity and food quality will be significant.

Keywords Malta · Agriculture · Climate change impacts · CMIP5
Evapotranspiration · Thermal-humidity index · Model clustering
Heat stress

1 Introduction

It is now widely accepted that increased atmospheric concentrations of greenhouse gases (GHGs) are enhancing the earth's natural greenhouse effect and accelerating global warming (IPCC 2014). The impacts of global warming on important sectors such as food security (Dawson et al. 2016), agriculture (Galdies et al. 2016) and

C. Galdies (✉) · K. Vella
Environmental Management and Planning Division, Institute of Earth Systems,
University of Malta, Msida Msd 2050, Malta
e-mail: charles.galdies@um.edu.mt

P. Castro et al. (eds.), *Climate Change-Resilient Agriculture and Agroforestry*,
Climate Change Management, https://doi.org/10.1007/978-3-319-75004-0_8

water resources (Arnell and Lloyd-Hughes 2014) have become worldwide concerns. Apart from the mean global temperatures increasing, it is expected that a higher incidence of heat waves may have even more far reaching effects on crop activity (Feller 2016).

Precise and credible climate projections are therefore required to both predict such future climate change impacts and to design appropriate climate adaptation action. The strong desire for such information has motivated continued improvement of sophisticated Earth System Models (ESMs) and statistical downscaling (McSweeney et al. 2014), since these are deemed as critical to capture the complex nonlinear dynamical interconnections, long-range associations, and feedback mechanisms that modulate the climate system. The latest generation of ESMs—the Coupled Model Intercomparison Project Phase 5 multi-model dataset (CMIP5; Taylor et al. 2012), reflects the most recent effort made by numerous climate modelling groups around the world to provide climate model outputs for the latest IPCC AR5. CMIP5 has been organized by the Working Group on Coupled Modelling (WGCM) on behalf of WMO's World Climate Research Program (WCRP), and several petabytes of data have been generated. Compared to its predecessor CMIP3, CMIP5 models have finer resolution processes, incorporate additional physics and have well integrated ESM components together with the inclusions of carbon cycle feedback, aerosol, atmospheric chemistry, biogeochemistry and dynamic vegetation components. According to Teng et al. (2012), the spatial resolution of CMIP5 models is better than CMIP3 models, with the former having horizontal resolution finer than 1.3°, while in the case of CMIP3 models only a single model fell within this group.

The provision of CMIP5 multi-model data has greatly simplified research on climate impact and adaptation and formally established the use of ensemble of ESMs in such studies (Carter et al. 2007). Moreover, availability of multi-model ensembles provides us with a good assessment of model uncertainty and variability among Atmosphere-Ocean Global Climate Model (AOGCM) projections, of which the ensemble mean shows better forecasting skill than individual AOGCMs (Tebaldi and Knutti 2007). Before the availability of such archives, researchers had to rely on modelling groups for the provision of specific model-derived variables, for downscaling the output to a spatial and temporal resolution appropriate for their impact studies, and then use this derived product to drive their impacts assessments (Downing et al. 2001). This modeling and data processing requires considerable expertise and computational resources and as a result such activities were quite infrequent. As an alternative, many climate change impacts modelers and decision-makers tended to rely on simple methods such as perturbations to historical meteorology sometimes based on mean AOGCM projections (e.g. Gleick 1987), and later using only a few projections (such as Hayhoe et al. 2004). The limited climate scenarios generated by a few AOGCMs represented a narrow range of future climate projections and thus limited the exploration of future impacts and adaptation studies.

However, in spite of the significant advances made in the standardization and delivery of simulations of the future climate, there still remains a gap between such projections and their meaningful use by decision-makers and stakeholders (Briley et al. 2015). There is an urgent requirement to resolve this divide especially in small island states since numerous studies show that climate change is already playing an important role in adversely affecting their food security (Baldacchino and Galdies 2015). The need to have effective adaptation measures aimed at increasing the resilience on climate change has become a necessity to ensure food security by small islands states.

The development of effective, long-term adaptation policies is thus of crucial importance in a small island state like Malta. Malta's agriculture policy is geared towards maintaining productivity against exceedingly strong competition in import crops, which has already increased marginalization of small scale farmers and farmers growing food crops in an already fragile environment. A quick literature search on the topic of food security in the Mediterranean islands points to the scarcity of such studies.

1.1 Aim of This Study

A comprehensive evaluation of Malta's climate extremes as projected by CMIP5 multi-model dataset has never been carried out in Malta. In this study we try to fill this gap by conducting a local thorough analysis of the CMIP5 models' projections of the future climate, demonstrating that it is precisely in the ability to synthesise complex information that the value of the CMIP5 multi-model dataset as a tool for specific decision-making can be found. Recently, local researchers have focused their efforts to study future climate change based on single regional model simulations (Global Warming Focus 2016); however, the analysis of this single model output was restricted to a physical understanding of the future climate and never directly extrapolated to local action within any particular sector, including agriculture. Therefore multi-model projections of local precipitation and its related extremes, together with temperature fluctuations and bioclimatic factors in Malta have yet to be investigated. Thus, the main aim of this study is to utilize CMIP5 multi-model dataset to analyse in detail the available long-term climate projections and to reflect on the relevant future risks to the Maltese agriculture. For this study we use high resolution projections from 11 different CMIP5 AOGCMs (Taylor et al. 2012). The use of downscaled climate projections by this study is important to anticipate climate change impacts at the local level.

2 Methodology

2.1 Study Area and Data Collection

The Maltese archipelago lies approximately at the centre of the Mediterranean Sea between 36° 00′ 00″ and 35° 48′ 00″ north-south latitude and 14° 35′ 00″ and 14° 10′ 30″ east-west latitude (Fig. 1). Its agricultural sector is considered to be a vital resource to the country's economy and social well-being since it is seen as crucial for the preservation of Malta's cultural and physical landscape, with a population 431,333 people in an area of just 316 km^2 (The World Bank 2018). The climate of the Maltese islands shows an annual mean temperature of 18.6 °C, a mean maximum of 22.3 °C and mean minimum of 14.9 °C (Galdies 2011). Such a temperature regime, together with an annual total precipitation of 553 mm make the islands relatively hot and dry.

The Maltese farming sector is being increasingly faced by several constraints arising from climate change (Government of Malta 2010). Local agricultural practices are mainly defined by the small size of the land parcels owned or farmed by individual farmers, where the number of holdings in 2013 amounted to 12,466

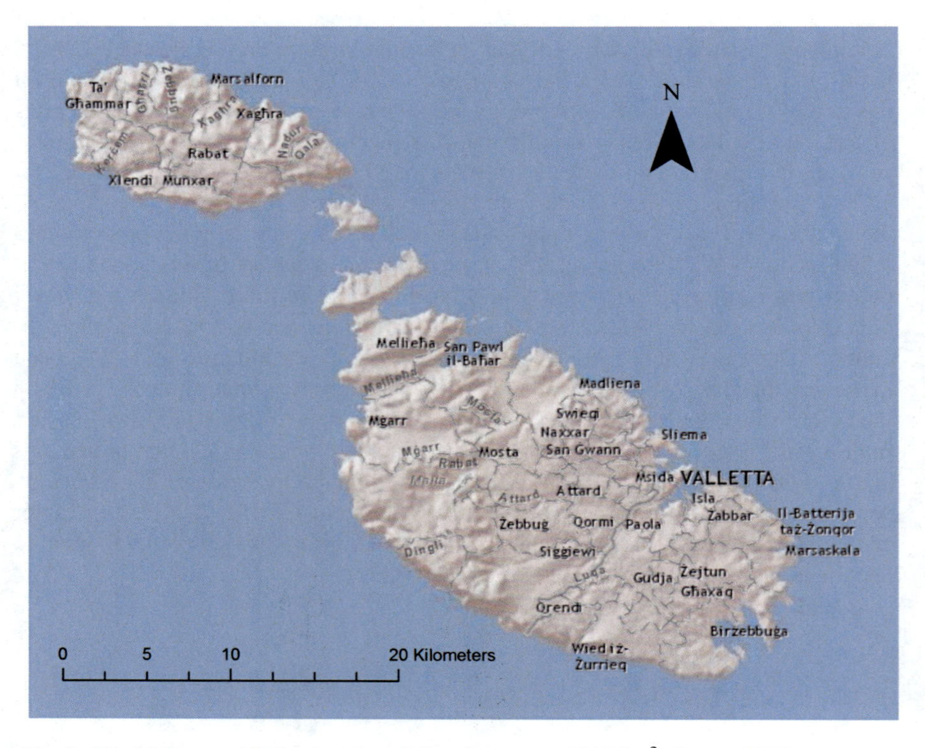

Fig. 1 The Maltese archipelago with a total surface area of 316 km^2 (*Source* ESRI)

working a total of 11,689 hectares of Utilised Agricultural Area (UAA) (National Statistical Office 2016). A total of 75.6% of these holdings have a UAA less than 1.0 hectare each; only 2.4% are considered large, each having a minimum of five hectares of UAA.

The cultivation of forage crops, covering 5290 hectares (45.3%) is predominant, followed by kitchen market produce (16%), kitchen gardens (12%), potatoes and vineyards (both at 6%). Fallow land accounted to 8% of the total. Notable increases between 2010 and 2013 were observed for vineyards (+11.2%), market gardens (+14.9%), kitchen gardens (+29.8%). However, the remaining cultivation experienced a decrease, namely other areas[1] (−22.9%), permanent crops (−8.8%), fallow land (−4.8%), forage (−4.7%) and potatoes (−1.7%). The negative values represent economically important crops that are either exported, such as potatoes and fruit, or used domestically to sustain this sector, such as forage for local cattle.

In terms of sectoral employability, a total of 19,066 persons were active farmers during 2013. The majority (70.3%) work less than 25% of 1 AWU (Annual Working Units). On the other hand 1372 farmers worked 1 AWU, these being engaged full-time in the sector.

2.2 *Multi-model Climate Projections Under Multiple Climate Scenarios*

Climate change projections were derived from WorldClim website (original data retrieved from CMIP Phase 5, http://cmippcmdi.llnl.gov/cmip5). A total of 11, downscaled and bias-corrected output of AOGCMs (Table 1) at 30-arc-second resolution (Hijmans et al. 2005) for the four main greenhouse gas (GHG) emission scenarios were available by WorldClim ver. 1.4 (http://worldclim.org/). These scenarios are based on Representative Concentration Pathways (RCPs) namely, RCP 2.6, 4.5, 6.0 and 8.5 (IPCC 2014; van Vuuren et al. 2012). RCP 8.5 is a high-end emissions scenario where, by 2100, anthropogenic forcing reaches 8. 5 W m^{-2} and atmospheric CO_2-equivalent concentrations are $\sim$1370 ppm, while RCP 2.6 reflects a peak forcing of 3.1 W m^{-2} ($\sim$490 ppm CO_2-equivalent) before it returns to 2.6 W m^{-2} by 2100. In order to reach RCP 2.6, ambitious GHG emissions reductions would be required over time. Multi-model (n = 11) projections for the Maltese islands were extracted and analysed separately for two temporal frameworks representing the twenty-first century: 2050 (averaged over the period 2041–2060) and 2070 (averaged over the period 2061–2080).

[1]This sector includes area occupied by greenhouses, flowers and ornamental plants grown in the open.

Table 1 List of AOGCMs models used by this study

Climate model	Code	Source
BCC-CSM1-1	BC	Beijing Climate Center, China Meteorological Administration
CCSM4	CC	National Center for Atmosphere Research (NCAR)
GISS-E2-R	GS	NASA Goddard Institute for Space Studies (NASA GISS)
HadGEM2-AO	HD	Met Office Hadley Centre (UK MetOffice)
HadGEM2-ES	HE	Met Office Hadley Centre (UK MetOffice)
IPSL-CM5A-LR	IP	Institute Pierre-Simon Laplace (France)
MIROC5	MC	Japan Agency for Marine-Earth Science and Technology, Atmosphere and Ocean Research Institute (The University of Tokyo), and National Institute for Environmental Studies
MRI-CGCM3	MG	Meteorological Research Institute
MIROC-ESM-CHEM	MI	Japan Agency for Marine-Earth Science and Technology, Atmosphere and Ocean Research Institute (The University of Tokyo), and National Institute for Environmental Studies
MIROC-ESM	MR	Japan Agency for Marine-Earth Science and Technology, Atmosphere and Ocean Research Institute (The University of Tokyo), and National Institute for Environmental Studies
NorESM1-M	NO	Norwegian-Climate Centre

2.3 Multi-model Analysis

Multi-model analysis was carried out to quantify the projected variability of the monthly and annual mean anomaly of ambient minimum temperature (Tn), maximum temperature (Tx) and precipitation (Pr) under all RCP scenarios for the Maltese Islands and their anomaly from the current climate norm WMO reference period of 1961–1990; (Galdies 2011). Thus time evolution of changes in the simulated indices spans the period 1961–2070. This analysis was done by (1) statistically analyzing the spread of the modelled output projections with time, and (2) multivariate analysis using hierarchical clustering was carried out to identify statistically correlated projections into distinct clusters (Knutti et al. 2013; Ahmadalipour et al. 2015). The resulting agglomerate clustering provided a measure of the similarity of multi-model behavior. Climate models with statistically similar monthly projections were clustered together, whereas models obtaining uncorrelated mean monthly values were grouped into different clusters (Tan et al. 2006; Norušis 2011). The distinctiveness of the main clusters resulting from the 11 models was statistically tested and confirmed using an independent sample t-test at 0.05 criterion (Kim 2015).

2.4 Climate Change Impact

Twelve bioclimatic variables were selected as determinants of the likely future climate projection on local agriculture (Table 2). Bioclimate variables describe seasonal conditions and climate extremes and therefore are considered to be more directly relevant to agriculture than would be monthly climate variables (Watling et al. 2012; Galdies 2015), and therefore assessments constructed from bioclimatic variables should perform better in our case. The De Martonne Aridity Index (AI_M) (De Martonne 1926) was calculated to summarise the degree of the current and projected aridity (Maliva and Missimer 2012). The algorithm used for AI_M is appropriate for the climate conditions of the Maltese islands; and ranges as follows: Arid 0–10; Semi-arid 10–20; Moderately arid 20–24; Semi-humid 24–28; Humid 28–35; Very humid 35–55; Extremely humid >55.

The current and projected local Reference Evapotranspiration (ETo) were calculated using meteorological data and the FAO Penman-Monteith equation according to FAO standards (ETo calculator ver. 3; http://www.fao.org/land-water/databases-and-software/eto-calculator/en/). Input variables included ambient maximum temperature (code 101), minimum temperature (code 103), mean relative humidity (code 202), vapour pressure (code 221), wind speed (code 303), daily sunshine duration (code 401), and solar radiation (code 421; source: Scerri 1982).

Table 2 Twelve bioclimatic variables used in this study to model present and future impact on local agriculture for 2050 and 2070 (*Sources* Galdies 2011, WorldClim ver 1.4)

Variable	Source
Annual mean temperature (BIO1)	This work
Max temperature of warmest month (BIO5)	This work
Min temperature of coldest month (BIO6)	This work
Temperature annual range (BIO7)	This work
Mean temperature of wettest quarter (BIO8)	This work
Mean temperature of driest quarter (BIO9)	This work
Annual precipitation (BIO12)	This work
Precipitation of wettest month (BIO13)	This work
Precipitation of driest month (BIO14)	This work
Precipitation of wettest quarter (BIO16)	This work
De Mortonne aridity index (AI_M)	Maliva and Missimer (2012)
Evapotranspiration (ETo)	FAO (2009)
Temperature-humidity index (THI)	Lajinian et al. (1997), Schoen (2005), Mader et al. (2006), Yousef (1985)

Table 3 Different THI equations with different weightings of temperature and humidity used to describe the combined effects of air temperature and humidity associated with the current and projected levels of thermal stress on livestock

Code	Algorithm	Source
THI_1	$(T_F - (0.55 - (0.55*(RH/100)))*(T_F - 58))$ *T_F is the mean air temperature in degrees Fahrenheit; RH is the relative humidity in %*	Lajinian et al. (1997)
THI_2	$T_F - 0.9971e^{0.02086*\ TF}\ (1 - e^{0.0445*(Td\ -\ 57.2)})$ *T_F is the mean air temperature in degrees Fahrenheit; RH is the relative humidity in %; Td is the dew point temperature*	Schoen (2005)
THI_3	$(0.8*T_C) + (RH*0.001*(T_C - 14.4) + 46.4)$ *T_C is the ambient air temperature in degrees Celsius; RH is the relative humidity in %*	Mader et al. (2006)
THI_4	$T_C + (0.36*Td_C) + 41.2$ *T_C is the ambient air temperature in degrees Celsius; Td is the dewpoint temperature*	Yousef (1985)

A selection of different Temperature-Humidity Index (THI) algorithms with different weightings of temperature and humidity (Bohmanova et al. 2007) were used (Table 3) to describe the combined effects of ambient temperature and humidity associated with the baseline (1951–2010) and projected (2050 and 2070) levels of thermal discomfort on livestock. THI, which is a thermal discomfort index for cattle, ranges as follows: <68: no discomfort; $68 \leq THI \leq 72$: mild discomfort; $72 \leq THI \leq 75$: discomfort; $75 \leq THI \leq 79$: alert; $79 \leq THI \leq 84$: danger; $THI \geq 84$: emergency (Cervigni et al. 2013).

3 Results and Discussion

3.1 Projected Climate Trends and Multi-model Variability

Figure 2a–f shows examples of the three projected climatic variables for the least and worst RCP for 2050 and 2070, compared to the Malta's current standard climate (1961–1990). The spread of the multi-model output is evident and the nature of the profile trajectories are determined by their respective RCP; for RCP 2.6, a strong increase is seen at first in both Tn and Tx, after which they more or less stabilize by 2070.

We note that the various projections show multi-model variability in the expected changes. For RCP 2.6 the projected Tx varies from +0.6 °C (MG) to +2.4 °C (MI) in 2050, and from +0.82 °C (GS) to +2.6 °C (MI) in 2070; for Tn from +0.4 °C (MG) to +2.2 °C (MI) in 2050, and from +0.7 °C (GS) to +2.4 °C (MI) in 2070; for Pr changes from −132 mm (IP) to +25.5 mm (MG) in 2050, to −109.6 mm (IP) to +25.2 mm (HD) in 2070. A similar degree of multi-model variability is shown under RCP 8.5, with higher overall temperatures

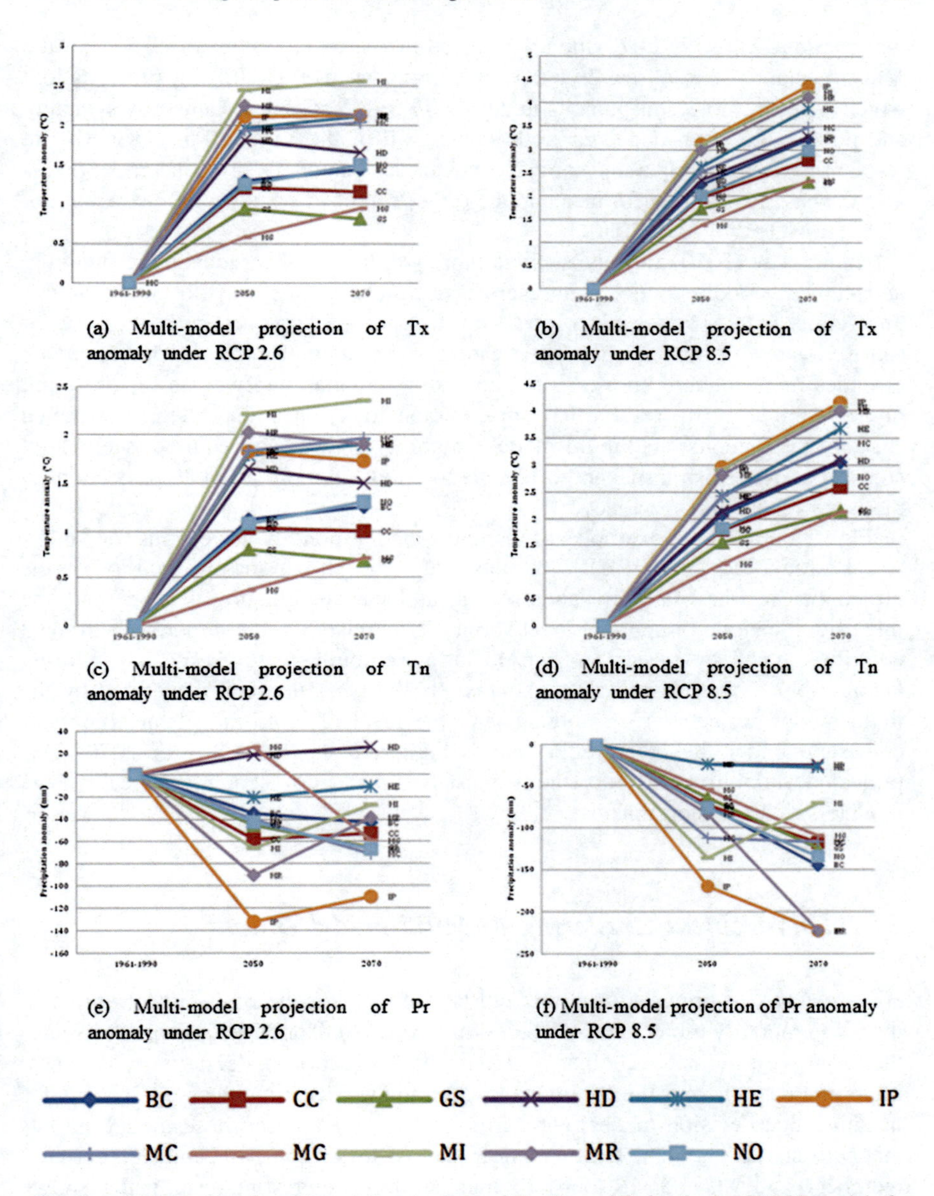

(a) Multi-model projection of Tx anomaly under RCP 2.6

(b) Multi-model projection of Tx anomaly under RCP 8.5

(c) Multi-model projection of Tn anomaly under RCP 2.6

(d) Multi-model projection of Tn anomaly under RCP 8.5

(e) Multi-model projection of Pr anomaly under RCP 2.6

(f) Multi-model projection of Pr anomaly under RCP 8.5

Fig. 2 a–f. Examples of multi-model projections of Tx, Tn and Pr by the 11 CMIP5 models for 2050 and 2070 under RCPs 2.6 and 8.5. The Y-scale has not been normalised in order to clearly portray the labeled profiles separately

that continue to rise in 2070 due to projected increased emissions of GHG. For the Maltese islands, this climatic change will bring an overall drop in precipitation, which is much more conspicuous under the RCP 8.5 scenario. Generally speaking and putting the inter-model variability aside, all of the 11 CMIP5 AOGCMs are responding in the same manner by 2050 as far as Tn and Tx are concerned; this is also the case for their performance in the projection of Pr with the only exception of HD and MG (RCP 2.6; 2050).

Knutti et al. (2010) attribute such variability to limited theoretical understanding, and choice of both model parameters and structure. One way to constrain this uncertainty is to consider multi-model results, and our approach provides a sort of sensitivity test to models' parameterization and structural choices. This approach provides more reliable information than a single model, and we can therefore put higher confidence on results that are common to an ensemble. A more detailed study of the demonstrated multi-model variability is therefore required (see below) so as to assist decision makers to be able to choose a more accurate projection on the basis of this inconsistency.

Figure 3a–c shows examples of the multi-model monthly projections for Tn, Tx and Pr for 2050 under RCP 2.6, illustrating the strong inter-variability of the projections for the Maltese islands. As far as local precipitation is concerned, the most significant and important local variability is related to the important autumnal contribution of freshwater to the local agricultural sector, especially during October-November (Galdies 2011). This variability is also being captured by the multi-model projections for autumn, and the level of multi-model uncertainty is consistent under the other RCPs, but is highest under RCP 8.5 for 2070. This projected variability and uncertainty must be taken into account by local impact modelers and decision-makers.

3.2 *Hierarchical Clustering of Multi-model Output*

Hierarchical clustering led to a two-cluster solution for the multi-model output of the mean monthly projections of Tn, Tx and Pr for 2050 and 2070. Table 4 displays the mean projections for 2050 and 2070 for the two set of model clusters. The most important result here is the commonality seen in the cluster members for Tn and Tx, meaning that decision makers and farm managers can select those model outputs that consistently show the least variation, such as for example BC and CC (cluster 1 members) and MC and MR (cluster 2 members). An interesting result is that shown by the conflicting Hadley Centre models (HD and HE) which statistically lie within separate clusters depending on the RCP type. This occurrence is presently being flagged since in its The Third, Fourth, Fifth and Sixth National Communication to UNFCCC (Government of Malta 2014), the Maltese Government bases its analysis and climate projections given by the Hadley CMIP5 models, without any reference to its variability and performance under the various RCPs. For 2070, HE seems to be consistently grouped within cluster 2, and so this problem seems to be resolvable

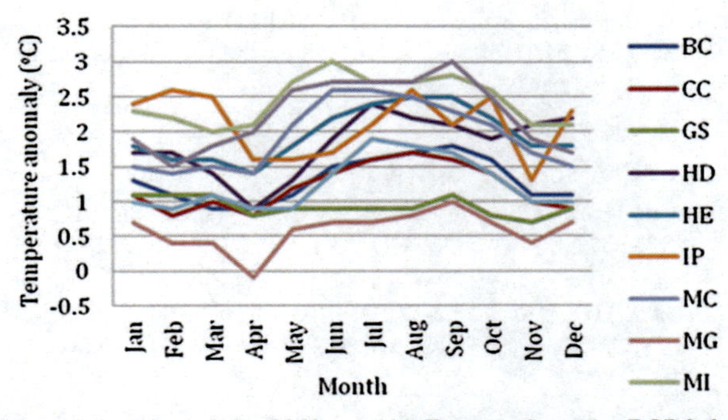

(a) Multi-model projection for 2050 monthly Tn anomaly under RCP 2.6

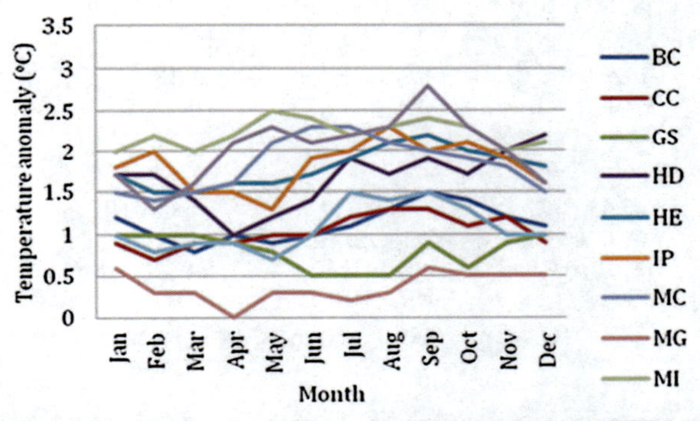

(a) Multi-model projection for 2050 monthly Tx anomaly under RCP 2.6

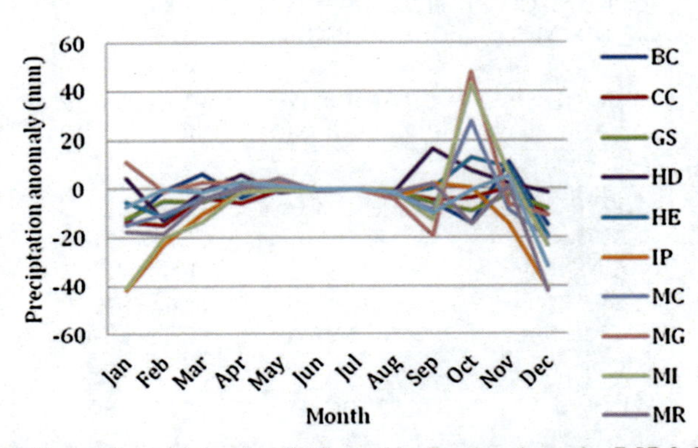

(a) Multi-model projection for 2050 monthly Pr anomaly under RCP 2.6

Fig. 3 **a–c**. Examples of multi-model projections for 2050 of the monthly Tn, Tx and Pr anomaly under RCP 2.5

Table 4 Details of the mean projections and model cluster members (p-value < 0.05 level of significance) for 2050 and 2070 for the Tn, Tx, and Pr and their respective cluster members under the four RCPs

Tn, yearly anomaly: 2050				
	RCP 2.6	RCP 4.5	RCP 6.0	RCP 8.5
Cluster 1 (members)	**+1.06** (BC, CC, GS, MG, NO)	**+1.55** (BC, CC, GS, HD, MG, NO)	**+2.08** (BC, HE, IP, MC, MI, MR)	**+1.95** (BC, CC, GS, HD, MG, NO)
Cluster 2 (members)	**+2.09** (HD, HE, IP, MC, MI, MR)	**+2.27** (HE, IP, MC, MI, MR)	**+1.36** (CC, GS, HD, MG, NO)	**+2.87** (HE, IP, MC, MI, MR)
Tx, yearly anomaly: 2050				
Cluster 1 (members)	**+0.90** (BC, CC, GS, MG, NO)	**+1.43** (BC, CC, GS, HD, HE, MG, NO)	**+1.89** (BC, HE, IP, MC, MI, MR)	**+1.76** (BC, CC, GS, HD, MG, NO).
Cluster 2 (members)	**+1.89** (HD, HE, IP, MC, MI, MR)	**+2.14** (IP, MC, MI, MR)	**+1.18** (CC, GS, HD, MG, NO)	**+2.71** (HE, IP, MC, MI, MR)
Pr, yearly anomaly: 2050				
Cluster 1 (members)	**−49.65** (BC, CC, GS, HD, HE, IP, MC, MR, NO)	**−72.47** (BC, CC, GS, HD, HE, IP, MC, MR, NO)	**−58.81** (BC, CC, GS, HD, HE, MC, MG, NO)	**−64.86** (BC, CC, GS, HD, HE, MC, MG, MR, NO)
Cluster 2 (members)	**−20.4** (MG, MI)	**−99.3** (MG, MI)	**−118.23** (IP, MI, MR)	**−153.0** (IP, MI)
Tn, yearly anomaly: 2070				
	RCP 2.6	RCP 4.5	RCP 6.0	RCP 8.5
Cluster 1 (members)	**+1.26** (BC, CC, GS, HD, MG, NO)	**+1.72** (BC, CC, GS, MG, NO)	**+1.84** (BC, CC, GS, HD, MG, NO)	**+2.08** (BC, CC, GS, HD, MC, MG, NO)
Cluster 2 (members)	**+2.20** (HE, IP, MC, MI, MR)	**+2.77** (HD, HE, IP, MC, MI, MR)	**+2.83** (HE, IP, MC, MI, MR)	**+4.15** (HE, IP, MI, MR)
Tx, yearly anomaly: 2070				
Cluster 1 (members)	**+1.17** (BC, CC, GS, HD, IP, MG, NO)	**+1.53** (BC, CC, GS, MG, NO)	**+1.67** (BC, CC, GS, HD, MG, NO)	**+2.64** (BC, CC, GS, HD, MG, NO)
Cluster 2 (members)	**+2.04** (HE, MC, MI, MR)	**+2.67** (HD, HE, IP, MC, MI, MR)	**+2.64** (HE, IP, MC, MI, MR)	**+3.86** (HE, IP, MC, MI, MR)
Pr, yearly anomaly: 2070				
Cluster 1 (members)	**−71.19** (BC, GS, IP, NO)	**−14.73** (BC, HD)	**−34.27** (BC, HD, HE, MG)	**−149.11** (BC, CC, GS, IP, MC, MG, MR, NO)
Cluster 2 (members)	**−33.49** (CC, HD, HE, MC, MG, MI, MR)	**−99.9** (CC, GS, HE, IP, MC, MI, MG, MR, NO)	**−105.06** (CC, GS, IP, MC, MI, MR, NO)	**−41.33** (HD, HE, MI)

for the 2070 projections. Model clustering for Pr seems to be less problematic; here BC (cluster 1 member) and MI (cluster 2 member) can be considered as representative CMIP5 models that encompass the overall variability identified for Pr.

3.3 Projected Aridity

Unlike droughts, aridity is a long-term climatic phenomenon. A good number of aridity indices are available; however they do not have much relevance from a water management perspective. However, such indices can be significant to track the effects of climate change on local water resources. The projected AI_m trend for the Maltese islands shows a drastic decrease under all four RCPs for both 2050 and 2070 (Fig. 4a). By 2070, the increased temperature will lead to a higher stress on freshwater resources, especially by agriculture. The rate of evapotranspiration is expected to increase in view of the increased positive trend of BIO8 for 2050 and 2070, compared to present levels (Fig. 4b). This is because aridity is affected by the temperature and the annual timing of precipitation in areas with a sufficiently high temperature for plant growth, since less water is lost directly by evapotranspiration during cold seasons (Walton 1969).

3.4 Projected Evapotranspiration

The reference evapotranspiration (ET_o) referred to in this study is used to determine the actual water use rate for various crops, and therefore is crucial for effective irrigation management. This is because the actual crop water use can be determined by multiplying ET_o with crop-specific coefficients (K_c) in accordance with the

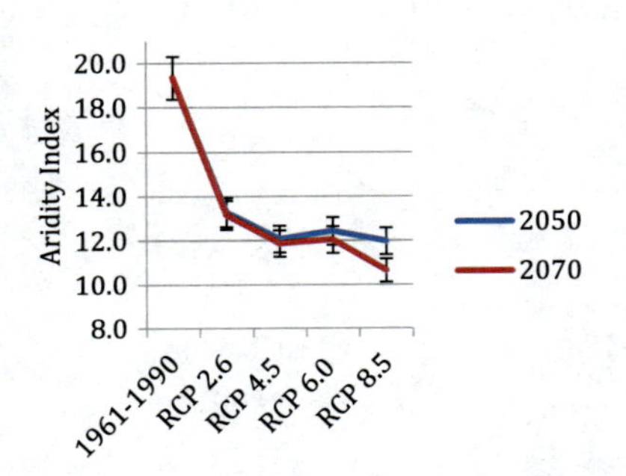

(a) Current and projected De Martonne Aridity Index (AI_M).

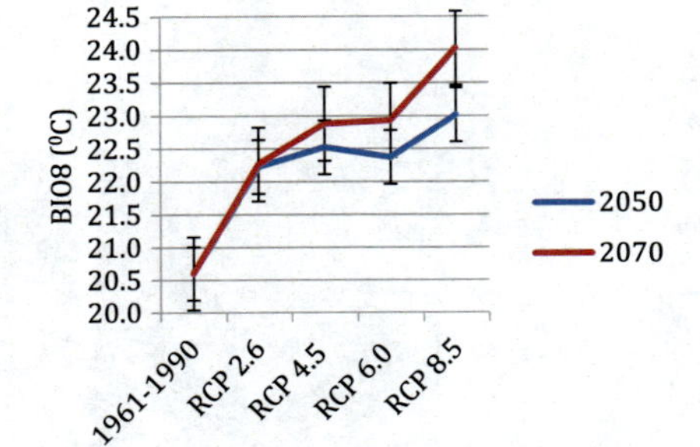

(b) The trend of the mean temperature of the wettest quarter (BIO8).

Fig. 4 **a** Increased aridity trend on the basis of CMIP5 model projections compared to present conditions, and **b** current and future trend of the mean temperature of the wettest quarter, both under the four RCPs for 2050 and 2070

crop's various growth stages. No published monthly evapotranspiration profiles exist for Malta, making the present study even more relevant. Here we preferred to calculate a local monthly average rather than taking a long-term average so as to better understand the current and future seasonal variation. At the same time, the daily ET_o value may deviate from these monthly climatological values, especially during extreme weather conditions (Galdies et al. 2016). Ideally, a measured or locally estimated ET_o value should be used. The present values may however be feasible for design or comparative purposes, and certainly for comparative climate change impacts on freshwater resources and irrigation rates.

Present results show that ET_o is low early in the year and gradually increases through spring and into summer, then again decreases gradually toward the end of the year (Fig. 4). The 1961–1990 ET_o ranges from 1.2 mm day^{-1} in winter to 4.9 mm day^{-1} in July. This range is slightly higher for the latest climate reference period of 1981–2000 (January: 1.3 mm day^{-1}; June–July: 5.1 mm day^{-1}). It is important to note that the local average climatological Relative Humidity for January is as high as 80%. The current range of ET_o compares well with that measured independently for the nearby Italian island of Pantelleria of 1.7–5.0 mm day^{-1} (Liuzzo et al. 2016).

Figure 5 also shows how ET_o is projected to increase by 2050 and 2070 under the worst case scenario of RCP 8.5. In this example, the average values derived from cluster 2 CMIP5 models have been considered in order to increase the accuracy of the calculated projected $ET_{o.}$ These results point towards a substantial increase of the ET_o for all months, especially from May till December, attributed to projected warmer monthly Tn and Tx (see Table 4).

A detailed analysis of monthly ET_o trends (Fig. 6a–d) for 2050 under the four RCPs indicate an increasing anomaly from the least to the worst pathway for both model clusters, but especially for cluster 2. This can be explained by the projected higher Tn and Tx projected by this cluster (see Table 4) which ultimately leads to higher evapotranspiration rates. Results also show that the projected scenario for 2070 under the four RCPs is much worse than for 2050 in view of the expected

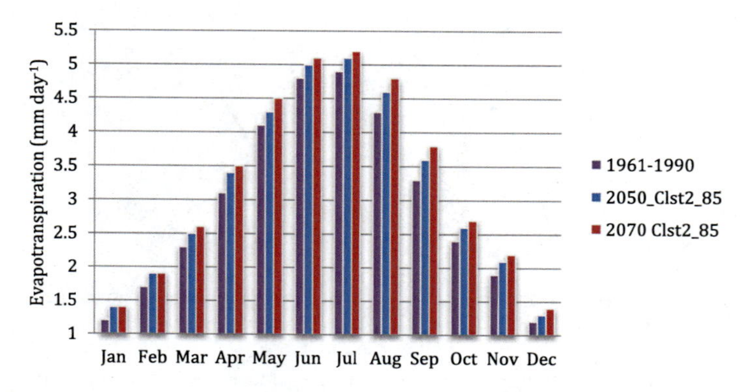

Fig. 5 Current and projected monthly average for ET_o for the Maltese islands on the basis of the average CMIP5 multi-model output of cluster 2 members

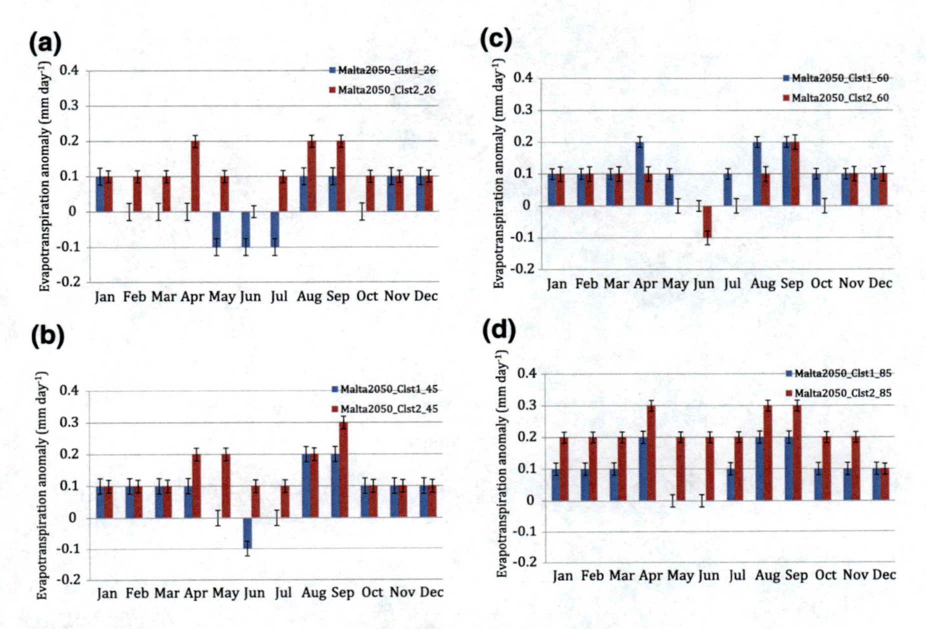

Fig. 6 a–d. Projected ET_o trends for 2050 under the four RCPs

increased levels of GHG concentrations in the atmosphere by then. For instance, results show that in October 2070, the projected ET_o by cluster 2 under RCP 6.0 will be +0.6 mm day^{-1}, which is equivalent to an *increased loss* of 6 m^3 of water per hectare per day (FAO 2017).

3.5 *Projected Heat Stress*

Figures 7 and 8 show the current and projected THI for the local livestock. Current climatic conditions are already subjecting local livestock to a significant level of 'discomfort'(June–October) and 'alertness'(July–September) as per threshold levels published by Cervigni et al (2013). These stressful periods to local livestock are due to either current elevated levels of ambient temperature and humidity, or both occurring at the same time. A cumulative effect of heat stress on cattle has also been studied (Mader et al. 2006), which comes into effect when nighttime temperature stays above 21 °C. The increased nighttime temperature does not allow the animal to shed the heat from the day before and three days in a row of high temperatures can be dangerous for ruminants. This is often the local case, especially during the summer period. To counteract such heat stress, local livestock farmers resort to technical solutions to reduce such discomfort (Galdies et al. 2016), especially during summer alertness.

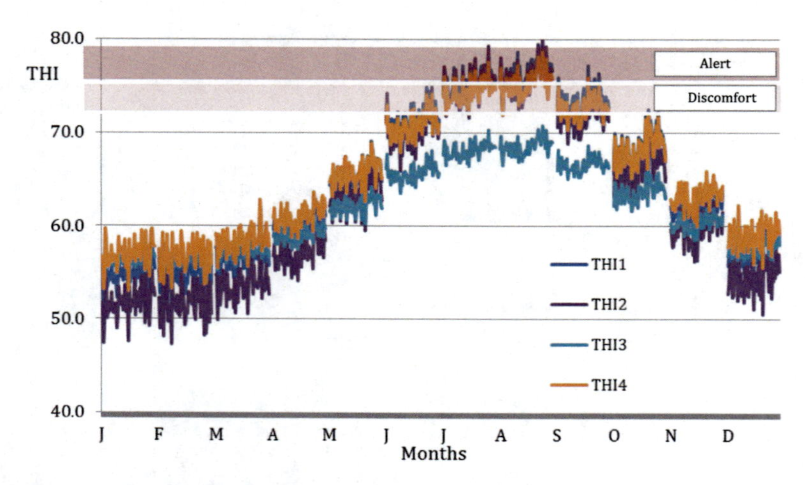

Fig. 7 THI1-4 levels for the period 1951–2010 based on various THI algorithms (see Table 4)

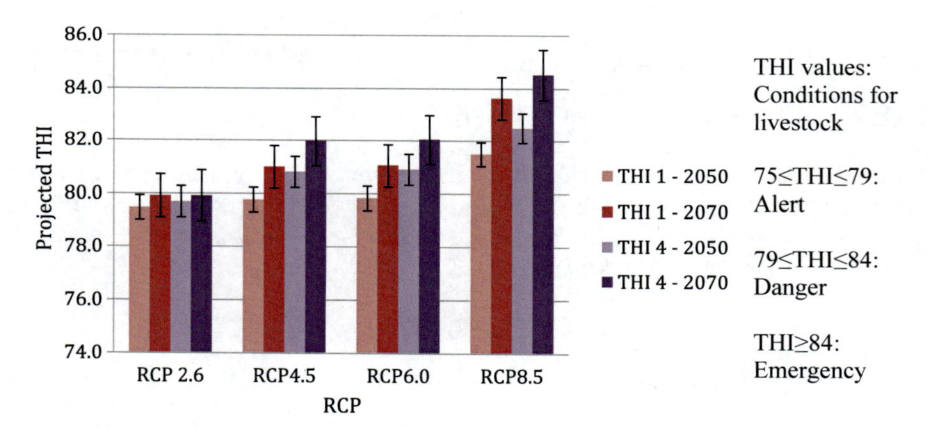

Fig. 8 Projected THI 1 and 4 levels for 2050 and 2070, under the four RCPs

It is important to note that the THI does not incorporate the effects of solar radiation or wind speed and therefore is a cruder index than more advanced ones that do include these inputs (Galdies et al. 2016). However, the use of the THI should pose no significant local uncertainties since local livestock are housed within large, well-ventilated covered sheds.

Malta's projected THI for 2050 and 2070 under the four RCPs is shown in Fig. 8. Only the temporal trends of THI1 and THI4 have been included for clarity. Results show that in contrast to RCP 2.6, the increased THI to 'dangerous' levels for both 2050 and 2070 under the remaining RCPs is statistically evident. In 2070, both THI1 and THI4 project an 'emergency' level under RCP 8.5.

3.6 Projected Water Requirements

This study shows that the future water requirements by a number of locally important crops (such as potatoes, beans, onion, strawberry, fodder, marrow etc.,) are projected to increase in view of the increased ET_o and ETc for the years to come. Results specifically show an overall decrease in rainfall during the wettest quarter (BIO16) for 2050 and 2070, which when coupled to increasing evapotranspiration trends, will naturally lead to an increased demand for irrigation for important cash crops, especially potatoes. Given that Malta has a semi-arid climate with an aridity index expected to fall by 50% by 2070, the archipelago is expected to encounter a considerable water deficit especially during the growth period of important crop types, and this deficit is set to become exacerbated at least until 2100.

For example, projected shifts in BIO5 (Max Temperature of Warmest Month) and BIO6 (Min Temperature of Coldest Month) suggest that Maltese winters (summers) will become increasingly milder (hotter) as the century progresses. By 2070, the percentage change in BIO6is expected to be very similar to that shown by BIO5 both by approximately 3 °C from present values, with a slightly higher rate shown by BIO5. It is interesting to note that the positive trend in the annual mean minimum temperature for the period 1951–2010 has been so far higher than that for the annual mean maximum temperature for the same period (Galdies 2011).

This study shows that Malta's projected temperature and rainfall are expected to significantly change and become more extreme by the end of 21st century. Higher temperatures and less and variable rainfall tend to increase water demand per unit of irrigated area. Now, most of Malta's main arable land ($\sim$7604 ha, or 68%) is entirely reliant on rainfall, while the rest ($\sim$3653 ha, or 32%) is currently estimated to use an unsustainable amount of 28,176,000 m^3 yr^{-1} (NSO 2012). The projected decrease in the future precipitation (Fig. 2e–f) will negatively impact this sector in the near future, making it even less competitive than it is now. Unfortunately, no estimates of water consumption by indoor livestock units are locally available.

According to our projected bioclimatic and evapotranspiration estimations for 2070 under a realistic RCP 6.0 (intermediate GHG emissions), Malta's arable land would need at least an additional 6 m^3 of water ha^{-1} day^{-1} to make up for the expected increased water loss. A 2010 Census of Agriculture revealed that the total volume of water used for irrigation in the Maltese islands is 28,176,000 m^3 yr^{-1} applied to at least 3653 ha of irrigated arable land (equivalent to 7713 m^3 ha^{-1} yr^{-1}). By 2070, we estimate that this amount has to be augmented by an additional 2190 m^3 ha^{-1} yr^{-1}, equivalent to 7.9 million m^3 of freshwater per annum. The already existent scarcity of surface water supply through reservoirs and ground water is likely to spatially limit the future potential for irrigation, which has critical implications for future crop production.

4 Conclusions

This study analyses and interprets the results generated by the latest suite of AOGCMs. It provides important information useful to researchers, policy-makers, managers and stakeholders whose task is to understand and adapt to climate change and its impacts in Malta. It is hoped that this original and thorough assessment will improve our understanding of plausible local climate futures as far as Malta's agriculture sector is concerned. It is important that the new scenarios highlighted by this study are communicated to local agronomists and policy makers.

In our opinion, the way forward for Malta's agriculture is to align itself to a future reality of increased aridity. For example, policy makers must start a plan that addresses a resolute, long-term and adaptive water management by (1) introducing a gradual water pricing, (2) demanding a highly efficient (at least 80%) irrigation method/s, such as by precision irrigation, (3) focusing on the cultivation of high value crops, and the (4) utilization of treated wastewater.

By translating the significance of CMIP5 multi-model results into tangible and measurable scenarios, we hope that national authorities will be in a better position to understand the importance of modulating climate-driven risks and further refine the adaptive capacity and resilience of Malta's agricultural sector.

References

Ahmadalipour A, Rana A, Moradkhani H, and Sharma A (2015) Multi-criteria evaluation of CMIP5 GCMs for climate change impact analysis. Theor Appl Climatol pp 1–44. https://doi.org/10.1007/s00704-015-1695-4. Accessed 28 Apr 2017

Arnell NW, Lloyd-Hughes B (2014) The global-scale impacts of climate change on water resources and flooding under new climate and socio-economic scenarios. Clim Change 122(1): 127–140. https://doi.org/10.1007/s10584-013-0948-4

Baldacchino G, Galdies C (2015) Global environmental change: economic and labour market implications for small island territories. Xjenza Online 3:81–85

Bohmanova J, Misztal I, Cole JB (2007) Temperature-humidity indices as indicators of milk production losses due to heat stress. J Dairy Sci 90:1947–1956. https://doi.org/10.3168/jds.2006-513

Briley L, Browna D, Kalafatis SE (2015) Overcoming barriers during the co-production of climate information for decision-making. Clim Risk Manage 9:41–49

Carter TR et al (2007) New Assessment methods and the characterisation of future conditions. climate change 2007: impacts, adaptation and vulnerability. In: Parry ML, Canziani OF, Palutik of JP, van der Linden PJ, Hanson CE (eds) Contribution of working group II to the fourth assessment report of the intergovernmental panel on climate change. Cambridge University Press, Cambridge, UK

Cervigni R, Valentini R, Sartini M (eds) (2013) Toward climate-resilient development in nigeria. international bank for reconstruction and development. The World Bank. ISBN (electronic): 978-0-8213-9924-8

Dawson TP, Perryman AH, Osborne TM (2016) Modelling impacts of climate change on global food security. Clim Change 134:429. https://doi.org/10.1007/s10584-014-1277-y

De Martonne E (1926) Une nouvelle function climatologique: L'indice d'aridité. La Meteorologie, pp 449–458

Downing TE, Nishioka S, Parikh KS, Parmesan C, Schneider SH, Toth F, Yohe G (2001) Methods and tools. In McCarthy JJ et al (eds) Climate Change 2001: impacts, adaptation and vulnerability, Cambridge University Press, 105–143 http://digitalcommons.unl.edu/animalscifacpub/608. Accessed 28 Apr 2017.

FAO (2009) ETo calculator version 3.1, issued in January 2009. Land and water digital media series No 36. Available via http://www.fao.org/land-water/databases-and-software/eto-calculator/en/. Accessed 28 Apr 2017

FAO (2017) Chapter 1—Introduction to evapotranspiration.http://www.fao.org/docrep/X0490E/x0490e04.htm. Accessed 28 Apr 2017

Feller U (2016) Drought stress and carbon assimilation in a warming climate: reversible and irreversible impacts. J Plant Physiol 20(203):84–94. https://doi.org/10.1016/j.jplph.2016.04.002

Galdies C (2011) The climate of malta: statistics, trends and analysis, 1951–2010. National Statistics Office, Malta. ISBN 9789995729196

Galdies C (2015) Potential future climatic conditions on tourists: a case study focusing on malta and venice. Xjenza Online 3:6–25

Galdies C, Said A, Camilleri L, Caruana M (2016) Climate change trends in Malta and related beliefs, concerns and attitudes toward adaptation among Gozitan farmers. Eur J Agron 74:18–28

Gleick PH (1987) Regional hydrologic consequences of increases in atmospheric CO_2 and other trace gases. Clim Change 10:137–160. https://doi.org/10.1007/BF00140252

Global Warming Focus (2016) Climate research; new climate research study findings have been reported from University of Malta (An analysis of teleconnections in the Mediterranean region using RegCM4), Atlanta, 124

Government of Malta (2010) National climate change adaptation strategy. Climate change committee for adaptation, Malta. Consultation Report, 143 pp

Government of Malta (2014) The third, fourth, fifth and sixth national communication of malta under the United Nations framework convention on climate change, 195 pp

Hayhoe K et al (2004) Emissions pathways, climate change, and impacts on California. In: Proceedings of the National Academy of Sciences of the USA, 101: 12 422–12 427. https://doi.org/10.1073/pnas.0404500101

Hijmans RJ, Cameron SE, Parra JL, Jones PG, Jarvis A (2005) Very high resolution interpolated climate surfaces for global land areas. Int J Climatol 25:1965–1978

IPCC (2014) Climate Change 2014: Impacts, Adaptation, and Vulnerability. http://www.ipcc.ch/pdf/assessment-report/ar5/wg2/WGIIAR5-Chap7_FINAL.pdf

Kim TK (2015) T-test as a parametric statistic. Korean J Anesthesiol 68(6):540–546. https://doi.org/10.4097/kjae.2015.68.6.540

Knutti R, Furrer R, Tebaldi C, Cermak J, Meehl GA (2010) Challenges in combining projections from multiple climate models. J. Climate 23:2739–2758. https://doi.org/10.1175/2009JCLI3361.1

Knutti R, Masson D, Gettelman A (2013) Climate model genealogy: generation CMIP5 and how we got there. Geophys Res Lett 40(6):1194–1199. https://doi.org/10.1002/grl.50256

Lajinian S, Hudson S, Applewhite L, Feldman J, Minkoff HL (1997) An association between the heat-humidity index and preterm labor and delivery: a preliminary analysis. Am J Public Health 87:1205–1207

Liuzzo L, Viola F, Noto LV (2016) Wind speed and temperature trends impacts on reference evapotranspiration in Southern Italy. Theor Appl Climatol 123:43–62. https://doi.org/10.1007/s00704-014-1342-5

Mader TL, Davis MS, Brown-Brandl (2006) Environmental factors influencing heat stress in feedlot cattle. Faculty Papers and Publications in Animal Science. Paper 608.

Maliva R, Missimer T (2012) Arid lands water evaluation and management, environmental science and engineering. https://doi.org/10.1007/978-3-642-29104-3_2. Springer-Verlag, Berlin Heidelberg

Maslin M, Austin P (2012) Climate models at their limit? Nature 486:183–184. https://doi.org/10.1038/486183a

McSweeney CF, Jones RG, Lee RW, Rowell DP (2014) Selecting CMIP5 GCMs for downscaling over multiple regions. Clim Dyn 44:3237–3260. https://doi.org/10.1007/s00382-014-2418-8

National Statistics Office (2012) Census of agriculture 2010: Results-news release. at:http://www.nso.gov.mt/statdoc/document_file.aspx?id=3215. Accessed 28 Apr 2017

National Statistics Office (2016) Agriculture and fisheries 2014. National Statistics Office, Malta, p 152
Norušis MJ (2011) Cluster Analysis in: IBM SPSS statistics 19 advanced statistical procedures companion (Chapter 17). Prentice Hall, pp 375–404. Retrieved from http://www.norusis.com/pdf/SPC_v19.pdf
Scerri E (1982) The radiation climate of Malta. Sol Energy 31(1):129–133
Schoen C (2005) A new empirical model of the temperature–humidity index. J Appl Meteorol 44:1413–1420
Tan PN, Steinbach M, Kumar V (2006) Introduction to data mining, vol 1. Pearson Addison Wesley, Boston. ISBN-13: 978-0321321367
Taylor KE, Stouffer RJ, and Meehl GA (2012) An overview of CMIP5 and the experiment design. American Meteorology Society April 2012: 485–498. http://dx.doi.org/10.1175/BAMS-D-11-00094.1
Tebaldi C, Knutti R (2007) The use of the multi-model ensemble in probabilistic climate projections. Philos Trans Roy Soc Lond A365:2053–2075. https://doi.org/10.1098/rsta.2007.2076
Teng J, Chiew FH, Vaze J (2012) Will CMIP5 GCMs reduce or increase uncertainty in future runoff projections? American Geophysical Union—Fall Meeting, 3–7 December 2012, San Francisco, USA
The World Bank (2018) Data. https://data.worldbank.org/country/malta. Last accessed 14 November 2018
van Vuuren DP, Edmonds J, Kainuma M et al (2012) Clim Change 109:5. https://doi.org/10.1007/s10584-011-0148-z
Walton K (1969) The arid zone. Aldine Publishing Co, Chicago, IL
Watling JI, Romanach SS, Bucklin DN, Speroterra C, Brandt LA, Pearlstine LG, Mazzotti FJ (2012) Do bioclimate variables improve performance of climate envelope models? Ecol Model 246:79–85
Yousef MK (1985) Stress physiology in livestock. CRC Press, Boca Raton, FL

Charles Galdies Dr Charles Galdies is a senior lecturer with the Division of Environmental Management and Planning within the Institute of Earth Systems. He has received his Ph.D. in Remote Sensing and GIS from Durham University (UK) in 2005. He studied ways to improve small-scale weather and ocean forecasting in the central Mediterranean region using novel remote sensing observations of the ocean and atmosphere. Dr Galdies previously served as Chief Meteorological Officer of the Malta Meteorological Office from 2007 to 2011, and Deputy Executive Director of the International Ocean Institute Headquarters. He also acted as the Permanent Representative of the Government of Malta with the World Meteorological Organisation. Dr Galdies' expertise focuses on weather and climate, the application of remote sensing for coastal, benthic and terrestrial ecological mapping, as well as environmental data processing and analysis. He has provided consultancy to the Food and Agriculture Organisation, the European Commission, the International Union for the Conservation of Nature (IUCN) and to private companies and non-governmental organization working in the field of environmental management, ecology and policy formulation.

Kimberly Vella Ms Kimberly Vella read for the Bachelors of Science in Earth Systems from the Institute of Earth Systems at the University of Malta in 2016. As partial fulfillment of the degree she has focused her research study on assessing the future climate of the Maltese Islands by analysing the output of CMIP5 models. Recently, she has completed a certification in Aeronautical Meteorological observing from the UK Meteorological Office. Vella is currently serving as an Assistant Meteorological Officer at Malta's Meteorological Office. She aims to further broaden the area of knowledge in the ones most interested in, in particular climate change research and meteorology field areas.

The Urgent Need for Enhancing Forest Ecosystem Resilience Under the Anticipated Climate Portfolio Over Kerala Under RCP 4.5 and Its Possible Implications on Forests

Praveen Dhanya and Andimuthu Ramachandran

Abstract Regional changes in climate have been observed in many parts of the world posing significant risk to all kinds of ecosystem and livelihood especially in climate sensitive sectors. This study was carried out to understand the plausible future changes that may occur to the biologically rich forested areas of kerala, using a regional climate model (RCMs)-RegCM4 by downscaling HadGEM-ES global climate model outputs at 25 km resolution. The downscaled data obtained from the RCMs were used to project the day time and night time temperature of Kerala under Representative Concentration Pathway (RCP) 4.5. The weather variables viz., maximum temperature, minimum temperature were extracted and projected for three time slices namely 2010–2040, 2040–2070 and 2070–2100 based on the reference period 1971–2000. The maximum and minimum temperature is projected to rise 2.79 and 2.59 °C at the end of 21st century in the forested areas of Kerala. The rise in day time warming was seen to be between 1.6 and 2 °C during mid-century and ranges 2.1 and 2.6 °C in the end of 21st century. The likely rise in night time warming was seen to be between 1.5 and 1.9 °C during mid century and 2.1 °C and 2.5 °C in the end of 21st century. Periyar Tiger Reserves, Silent Valley, Wayanad areas are projected to experience the severe warming in future in the range of 2.7 °C. The projected night time warming was in the range of 0.8 °C in the western coastal districts and 1.04 °C in the hilly areas during the near century (2010–2040) period. Comparatively higher levels of warming was observed in the Wayanad Palakkad and Malappuram districts. The possible impacts may threaten the forest biodiversity. As the simulation results indicates significant warming under even under mid emission trajectory RCP 4.5, further enhanced research is required to understand how different predominant endemic species behave under

P. Dhanya (✉) · A. Ramachandran
Centre for Climate Change and Adaptation Research, College of Engineering, Guindy Campus, Anna University, Sardar Patel Road, Chennai 600025, India
e-mail: dhanyapraveen.cc@gmail.com

P. Castro et al. (eds.), *Climate Change-Resilient Agriculture and Agroforestry*, Climate Change Management, https://doi.org/10.1007/978-3-319-75004-0_9

the drastic or slow alterations in the climate and growing conditions in future. This study also attempted to evaluate the existing adaptive conservation plans in the forest sector in the state and the need for strengthening its resilience.

Keywords Climate change · Climate variability · Forest · Forest ecosystem projections · RegCM · CMIP5

1 Introduction

The scientific reports have indicated that climate change is likely to pose definite challenges on most of the economic sectors in India that are driven by climate such as water resources, agriculture and allied services, biodiversity and forests (IPCC 2013a, b). The accelerated warming of 0.78 °C for the last one decade can be mainly attributed to the human induced greenhouse gas emissions over the past and is anticipated to contribute more in the future (IPCC 2012). Globally, forests cover 4 billion hectares (ha) of land, or 30% of the earth's land surface (FAO 2005b). Based on a range of vegetation modeling studies, IPCC reports (IPCC 2007a, b, c) suggests potential forest dieback towards the end of this century and beyond, especially in tropics, boreal and mountain areas (IPCC 2007a, b).

Climate change is widely viewed as the single greatest threat to both humans and natural systems in the next century. Accordingly, an increasing proportion of ecological, environmental, and conservation research is aimed at understanding how climate shapes our biosphere, and how changes in climate will affect the distribution and persistence of biodiversity on the planet in the future (Chaturvedi et al. 2012a, b; IPCC 2013a, b). Climate change introduces considerable uncertainty into forest management planning and outcomes, potentially undermining efforts at achieving sustainable practices (Gopalakrishnan et al. 2010). The future projections are considered as the preliminary step forward in any climate change impact assessment (Chaturvedi et al. and Kumar et al. 2011).

Rao et al. (2009) have stated that there was an increase in maximum temperature over Kerala by 0.64 °C during the period of 49 years, commencing from 1956 to 2004 while the increase in minimum temperature was 0.23 °C. Overall increase in annual average temperature over Kerala was 0.44 °C. He also stated that the increase in annual mean temperature, maximum temperature and minimum temperature were 0.49, 0.76 and 0.22 °C over a period of 103 years (1901–2003) respectively. The increase in annual temperature over the country was 0.54 °C

Kerala's dense populations, higher dependence on monsoon large coast line and limited land resources make the state one of the most vulnerable to climate change. State has the third highest population density in India. (Kerala: 819, India: 324). Drought like situations creates acute food and livelihood insecurity which leads to import of food grains and poverty. Land hunger in the state for housing and lively hood leads to encroachment of forests and low lying wetlands. The State has rich biodiversity and tropical rain forests and is spread in 13 agro-ecological zones under the humid tropics. Biodiversity refers to all species of plants, animals and micro-organisms existing and interacting within an ecosystem (Vandermeer and Perfecto 1997). Kerala forest fall in two biogeographic provinces, viz Western Ghats and the Western Coast, and are rich in biodiversity and vital for environmental protection and considered to be a repository of rare and endangered flora and fauna (McCarty et al. 2009). However, Climate change/variability and deforestation over a period of years have changed the typical agro-ecosystems moreover, many of the rural poor in Kerala are dependent on the forests and wetlands for their livelihood (Pounds et al. 2006).

Modeling future climate change provides a basis for effective visualization of probable changes in climate and helps in further impact assessments in various sectors including forests. It provides a futuristic picture of the climate scenario for the next few decades so that appropriate policies and programmers can be put in place to cut down potential damages. There are very few simulation studies reported based on the IPCC Assessment Report5 Representative Concentration Pathway (RCP) for India (Ravindranath et al. 2006, 2008).

Due to their finer resolutions, the regional climate models outputs enables a better and reliable outcomes for impact studies. The main endeavor of this study is to project the potential climate change of Kerala using the latest IPCCs RCP 4.5 emission trajectories and its implications on forests areas which is rich in its biodiversity. As forests have a major role to play in the ecosystem sustainability of Kerala, the scope of this paper includes the simulations of potential of future climate change, quantitative and spatial analysis of the existing change and understanding the existing adaptive management strategies for forest conservations.

2 Materials and Methods

2.1 Study Area

Kerala is located in the southern tip of Indian subcontinent and extends from 8°15′N and 12°50′N latitude and 74°50′E and 77°30′E longitude. Its is also "Gateway of monsoon" to the country as it is the entry point of monsoon to the Indian subcontinent. It is also one of the wettest places in the humid tropics. The temperature in Kerala normally ranges from 28° to 32 °C (82° to 90 °F) on the plains but drops to about 20 °C (68 °F) in the highlands. The mean monthly Climatology of Kerala is been given in the Figs. 1, 2 and 3.

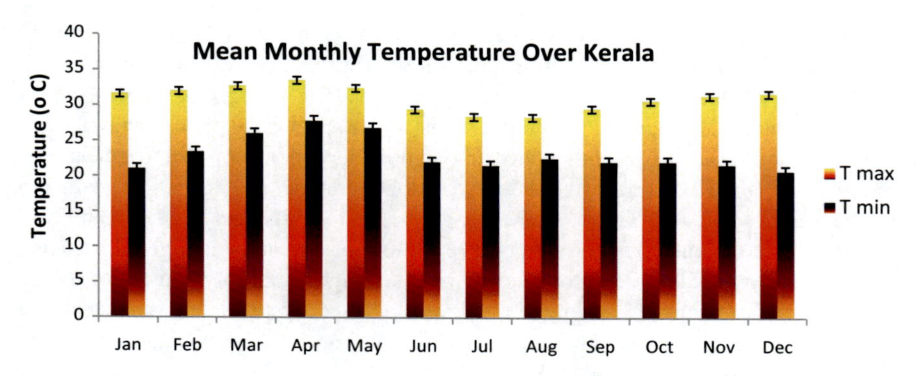

Fig. 1 Mean monthly temperature of Kerala for the period 1901 to 2002 (*Source* IMD)

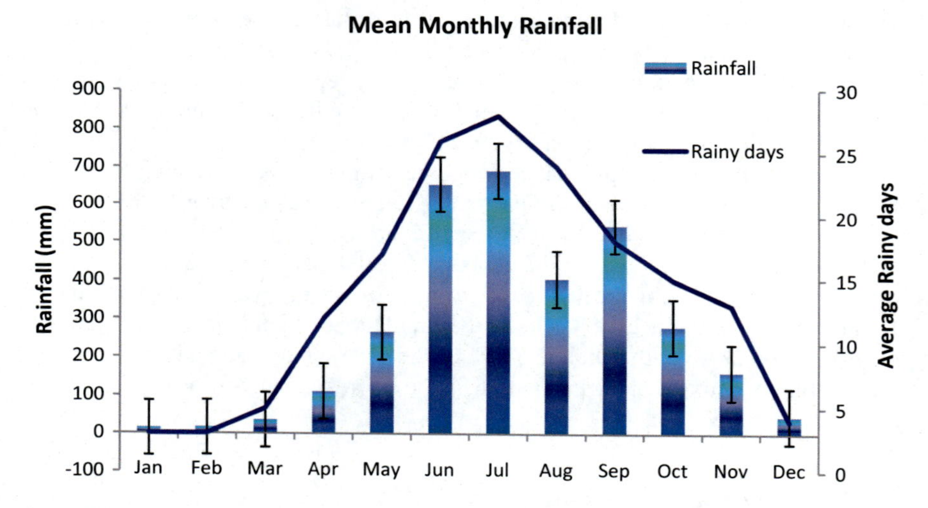

Fig. 2 Mean monthly rainfall of Kerala for the period 1901–2002 (*Source* IMD)

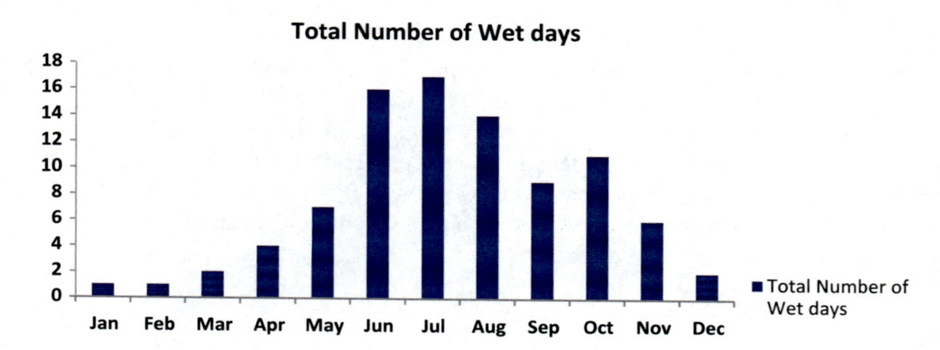

Fig. 3 Average total number of wet days in Kerala for the period 1901–2002

Owing to its diversity in geographical features, the climatic condition in Kerala is also diverse. It can be divided into 4 seasons-Winter, Summer, South-West Monsoon and North-East Monsoon The annual rainfall of Kerala is about 2.7 times the national average, receiving about 3000 mm as against 1150 mm of the national average. The State is relatively rich in rainfall endowment; with an annual precipitation of around 2600 mm. Ninety percent of this precipitation is during the two monsoons, that is from June to August (south west) and October to November (north east). About 60% of annual rainfall is received during southwest monsoon period and about 30% during northeast monsoon. From December to March there is very little rainfall, but the occasional rainfall during this period is a very critical requirement for cultivation as we still depend upon rainfall for raising many of the crops. The State is bestowed with 44 rivers and a number of backwaters, streams, canals and other inland water bodies and has rich biodiversity. The tropical rainforests spread in 13 agro-ecological zones. The major group under the soils of Kerala is laterite (http://www.kau.edu/pop/agro-ecologicalzonesofkerala.htm).

Kerala has three floristic hot spots-Agastyamala, Anamala and Silent Valley. Western Ghats in Kerala have 4500 species of flowering plants. Western Ghats have 145 species of mammals (of which 14 are endemic to Western Ghats), 169 species of fresh water species, 93 species of amphibians (of which 40 are endemic). There are 486 species of birds (of which 16 are endemic to Western Ghats). There is also innumerable micro flora and fauna. The State has two biosphere reserves—Nilgiri and Agastyamala (GoK 2008). The state has three out of twenty-five wetlands of international importance included in the Ramsar list viz., Ashtamudi, Sasthamkotta and Vembanad-Kolland (Figs. 4, 5, 6).

2.2 *Forest Cover in Kerala*

Much of the forest cover of Kerala is spread over the Western Ghats. The Western Ghats represents one of the world's 18 hot spots of biodiversity and is considered to be a repository of endemic, rare and endangered flora and fauna. The percentage of forest cover in Kerala is 28.90 which is higher than the national coverage of 19.50. About 51% of the total forest cover is in the southern districts and the remaining 49% is in the central and northern regions (Tables 1 and 2). Idukki and Pathanamthitta districts have the largest area under forest cover. The classification of forest types are mentioned in Figs. 7 and 8.

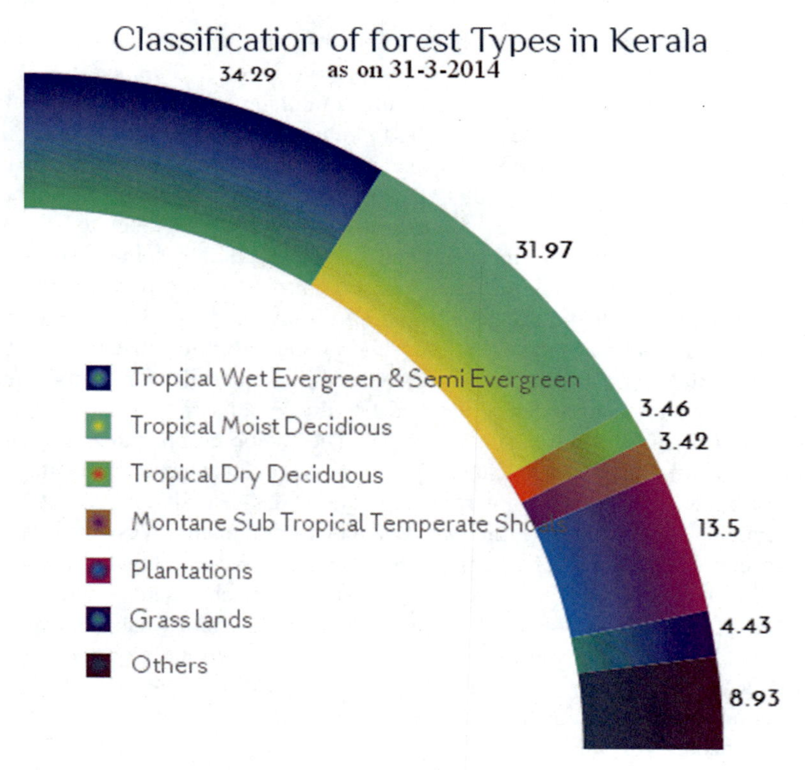

Fig. 4 Projected night time temperature changes over forest areas spanning three continuous time slices (near-2010–2040 mid-2041–2070, and end century-2071–2100)

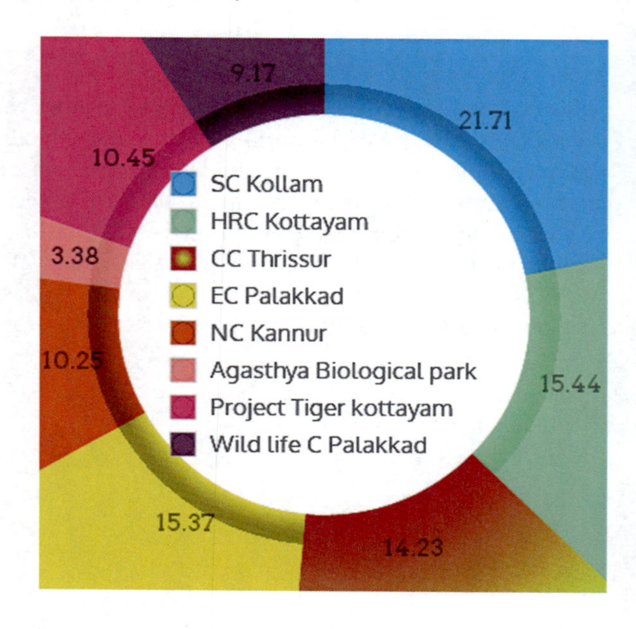

Fig. 5 Potential impacts of climate change on forest ecosystems

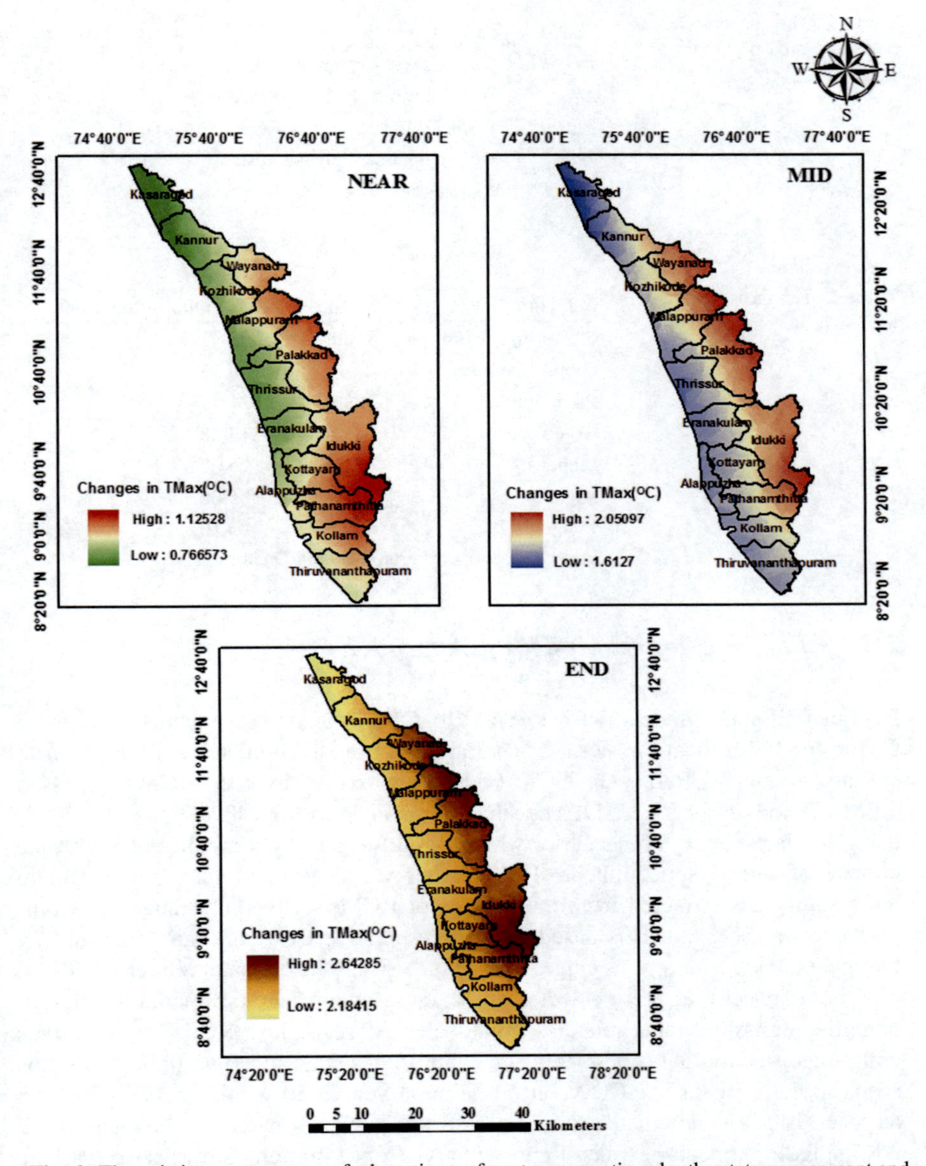

Fig. 6 The existing programmes of adaptation or forest conservations by the state government and department of forests in Kerala consists of 1. *Ente Maram** (My Tree), 2. *Njangalude maram**(Our Tree) 3. *Vazhiyora thanal** (Shading Trees) 4.*Haritha Theeram** (Trees on the coasts), 5. *Haritha Keralam** (Green Kerala) etc. (Fig. 8)

Table 1 The forest cover types and area of Kerala

Sl. No	Forest type	Area (lakh ha.)
1	Tropical wet evergreen forest	3.48
2	Tropical moist deciduous forests	4.1
3	Tropical dry deciduous forests	0.094
4	Mountain sub tropical forests	0.188
5	Plantations	1.538
	Total	9.4

Table 2 Kerala is bestowed with rich biodiversity

Category	Kerala	% to the Indian flora
Flowering plants	4500	25.71
Gymnosperms	4	6.25
Pteridophytes	236	21.45
Bryophytes	350	12.28
Lichens	520	26.00
Algae	325	5.00
Fungi	4800	33.10

Source http://www.kerenvis.nic.in/Database

2.3 *Climate Change Modeling Approach*

Regional Climate Models (RCMs) viz., RegCM4.0 of Abdus Salam International Centre for Theoretical Physics (ICTP), Italy was used to simulate the future climate scenarios under RCP 4.5. The GCM boundaries used to drive the RCMs were Had GEM- ES model for RegCM4. The simulation was runs for 128 years (1971–2098) using the latest RCP 4.5 scenario, which is said to be in good coherence with the climate of Indian subcontinent. The weather variables maximum and minimum temperature are extracted from the outputs of RCMs. A medium range emissions pathway was selected. It is based on a representative CO_2 concentration pathway that generates a radiative forcing of 4.5 Wm^{-2} (RCP 4.5) (Chaturvedi et al. 2012a, b). These trajectories were selected to be representative of the medium range of potential emissions and related climate change scenarios. RCP 4.5 represents emissions if action is taken to limit greenhouse gas emissions by 2050. Under this scenario the atmospheric CO2 concentration at year 2050 would be 478 ppm, and at year 2100 it is anticipated to rise to 538 ppm. Studies based on the latest IPCC AR5 emission trajectories [i.e., Representative Concentration Pathways (RCPs)] are only a handful on impact assessment, especially for the Indian scenarios (IPCC AR5 2013a, b).

From the daily outputs of RegCM4, mean annual estimations were made using Perl programme for climate analysis. The annual estimations were done for each grids and then averaged for the state to find the increase or decrease in trend of climate parameters till the end of 21st century.

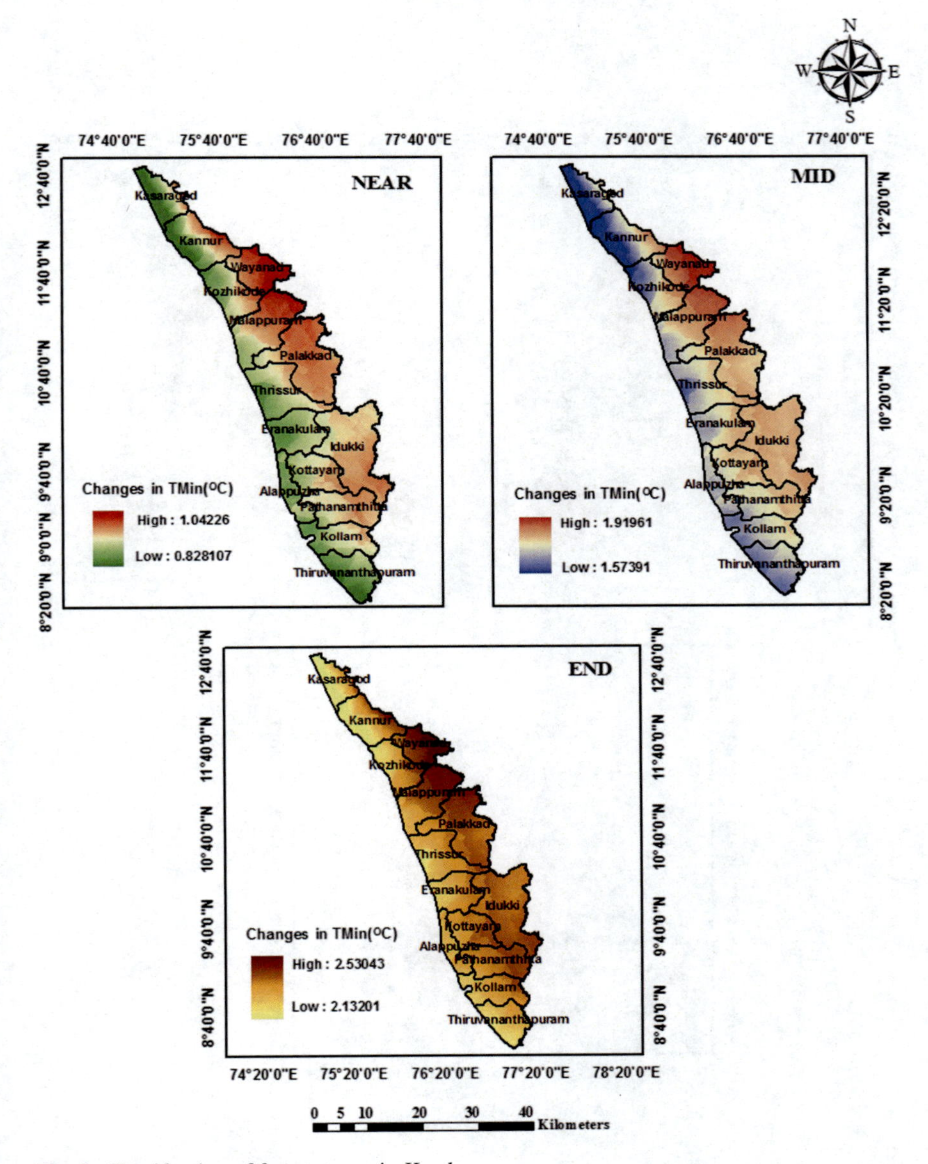

Fig. 7 Classification of forests types in Kerala

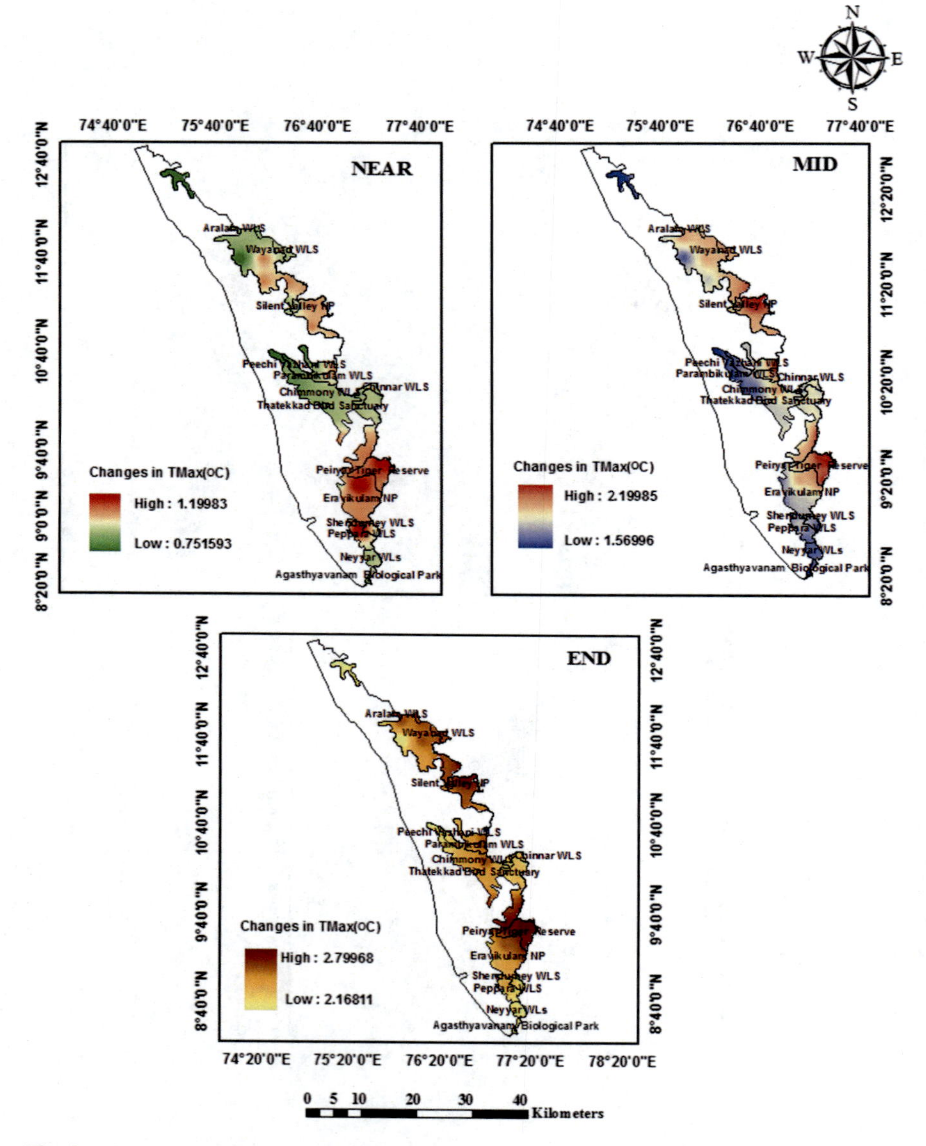

Fig. 8 The spatial extent of percentage of area under each forest circle in Kerala

3 Results and Discussion

3.1 *General Projections Warming Over the Districts of Kerala*

With reference to the baseline (1971–2000) simulations, monthly climatology were compared for understanding the future changes. Maximum temperature is likely to increase in this study (Fig. 7) as well as reported by Wiltshire et al. (2013). Similar kind of increased temperature projection for India was also observed by Chaturvedi et al. (2012a). In this study it was was noted that the rate of increase in minimum temperature is higher than that of maximum temperature and it was in agreement with Ramaraj et al. (2009) for Tamilnadu.

The projected warming during the day time was in the range of 0.7 °C in the western coastal districts and 1.1 °C in the hilly areas during the near century (2010–2040) period. Comparatively higher levels of warming was observed in the Pathanathitta, Kollam, Idukki, Palakkad, Malappuram and Wayanad districts, due to its rugged topography and land form features. The rise in warming was seen to be 1.6–2 °C during midcentury and 2.1 and 2.6 °C in the end of 21st century (Fig. 9).

However, there is clear spatial variation in the changes in the night time temperature over Kerala.

The projected night time warming was in the range of 0.8 °C in the western coastal districts and 1.04 °C in the hilly areas during the near century (2010–2040) period. Comparatively higher levels of warming was observed in the Wayanad Palakkad and Malappuram districts. The likely rise in warming was seen to be 1.5–1.9 °C during mid-century and 2.1 and 2.5 °C in the end of 21st century (Fig. 10).

3.2 *Exclusive Projections Warming for Forested Areas of Kerala*

Periyar Tiger Reserves, Silent Valley, Wayanad forested areas are projected to experience severe warming in future. These forested areas are going to experience severe warming of 2.7 °C during the day. This research also tries to understand the existing forest management practices undertaken jointly by the state government of Kerala and Department of Forests.

As far as the night time temperatures are concerned Wayanad, Silent Valley, areas are likely to experience the higher warming in future. These forested areas are going to experience severe warming of 2.5 °C during the night (Fig. 11).

As the model outcomes suggests potential for warming for kerala under mid emission trajectory RCP 4.5. In this context, the terrestrial ecosystems in the forested areas are likely to be changing in response to the climatic signals. Drier years and drier summers have often led to more large fires, many of which are more

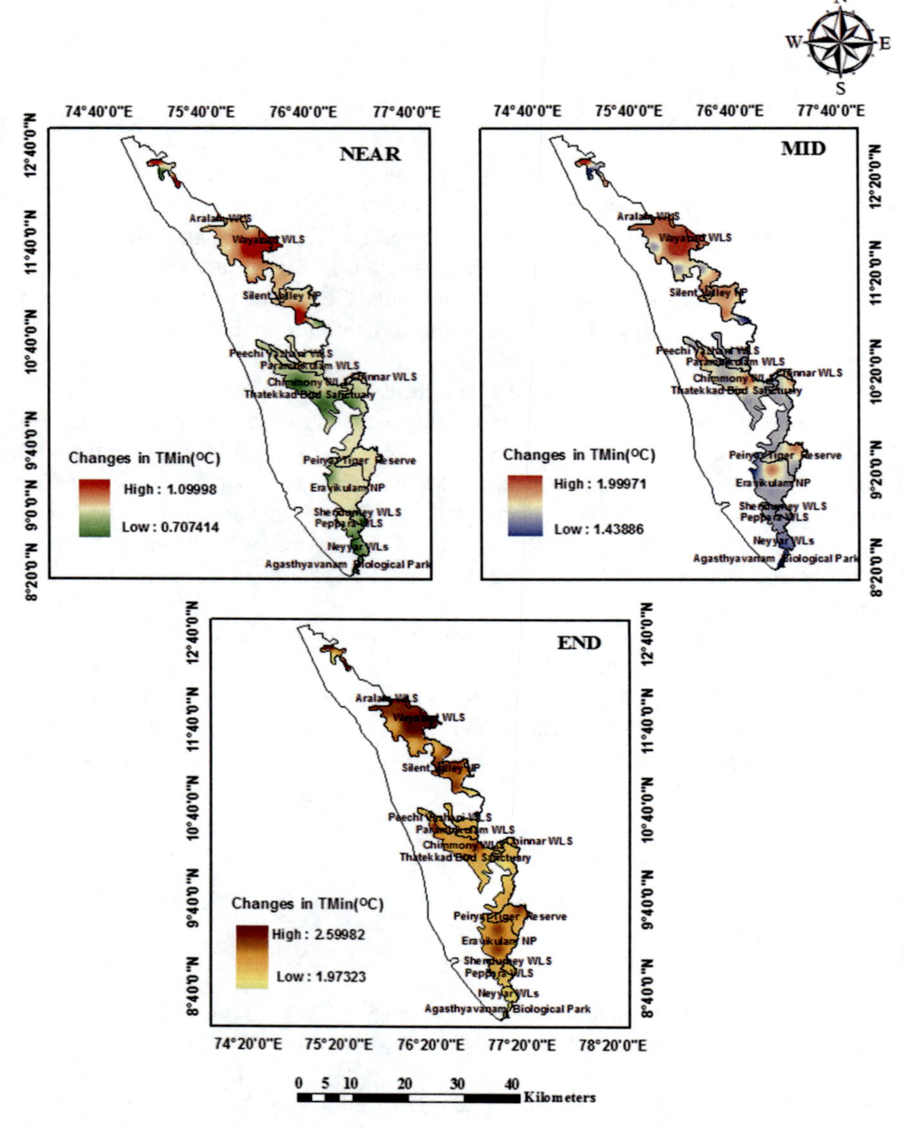

Fig. 9 Projected daytime temperature changes spanning three continuous time slices (near-2010–2040; mid-2041–2070, End Century-2071–2100)

severe. potential conversion of forests to shrubs in some places (FAO 2005a, b; Ramachandran et al. 2015; Meinshausen et al. 2011; FSI Report 1987–2009; Roa et al. 2009; UNCCD 2012). Envisioning future solutions will be facilitated through exploration of forest ecosystem dynamics across landscapes containing multiple

Fig. 10 Projected night time temperature changes spanning three continuous time slices (near-2010–2040 mid-2041–2070, end-2071–2100)

populations or patches and over long time periods containing many events and ecosystem response trajectories (FAO 2005a, b; GOI, NAPCC 2008). However, there is also recognition that increased dryness or length of dry spells could also have significant ecological consequences (Ramraj et al. 2009; Sabu et al. 2011). Afforestation and reforestation programmes require wider attention and focused actions (FAO 2005a) In climate change terms, this is called mitigation because better management results in lower emissions or in increased removal of carbon from the atmosphere, thus lowering the CO_2 in the atmosphere. At the same time, these same landscapes can provide benefits that increase resilience to climate change. Such adaptation co benefits include improved storage and release of water, maintained local and regional rainfall patterns and diversified economic opportunities through agricultural products, timber, non-timber products and tourism.

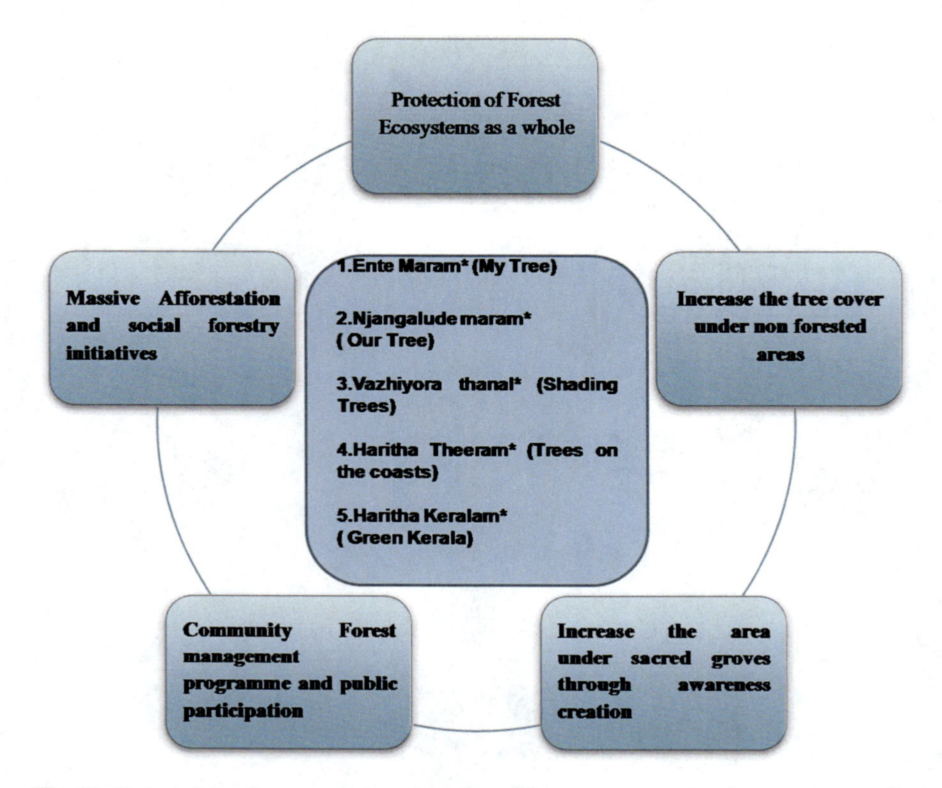

Fig. 11 Projected day time temperature changes over forest areas spanning three continuous time slices (near-2010–2040, mid-2041–2070, and End century-2071–2100)

These same landscapes can also provide services such as biodiversity maintenance and places where local people uphold cultural or religious values (Dillion et al. 2011; Pierce et al. 2004).

The geographic range of species will be a result of how populations of individual's respond. Not all species will be able to respond to changes in climate. Species that are unable to respond quickly enough will go extinct. The magnitude of extinctions could be immense due to climate change. For some areas and groups of plants and animals, 15–37% of species could be at risk of extinction by 2050 (Pounds et al. 2006). Changes in climate have overwhelmed species' abilities to respond at the same pace (Breashears et al. 2005; Westerling et al. 2006). Population decline and species loss are linked to the changing climate. Frogs stressed by warmer and drier climate are less able to resist fungal infection. As a result, populations decline and populations and species go extinct (Westerling et al. 2011; Morgan et al. 2008; Morrison et al. 2002; Easterling et al. 2000).

Scientific reports have already indicated that global warming is likely to pose a defining challenge on most of the economic sectors in India that are driven by climate such as water resources, agriculture and allied services, biodiversity, and forests (GOI 2008). Adapting to climate change reduces vulnerability by reducing risks and capitalising on benefits through maintaining social and ecological resilience (Nelson et al. 2007; Millar et al. 2007a, b). Kerala possess 95% of the flowering plants in the Western Ghats and contains 90% of the vertebrate fauna. Western Ghats in Kerala is one of the 34 world's hot spots of biodiversity. The latest estimates of future climate change predict that there will be an increment of 5–10 days in the consecutive dry days as per RCP 8.5 emission pathways for the period 2081–2100 for India. The latest simulation study reported by Chaturvedi et al. (2012a, b) indicated a rise in temperature in the range of 2.9–3.3 °C under RCP 4.5 and RCP 6 pathways. Maximum temperature is likely to increase in this study (Fig. 10) as reported by Wiltshire et al. (2013). Similar kind of increased projection for India was also observed by Chaturvedi et al. (2012a, b). Increasing extreme climate events, especially dryness pose serious soil degradations in the forest in terms of depletion in Organic Carbon. It is necessary to undertake necessary initiatives to preserve and enhance soil health of Western Coastal plains and Western Ghat region of India which has 8 and 6% of soil organic carbon and total carbon stock of the country respectively (Srinivas Roa et al. 2011; Spittlehouse and Stewart 2003).

4 Conclusions

There are lots of uncertainties attached to climate change projections and their plausible impacts. Impacts with respect to species distribution, frequency of extreme climate events, insect pest and diseases incidence, forest productivity, community livelihood etc. need to be understood in a better way. The anticipated future warming in the range of 2.7 °C at the eco sensitive areas of Periyar Tiger Reserves, Silent Valley, Wayanad showcases alarming situations if not taken atmost proactive care. A comparatively higher level of warming was observed in the Wayanad Palakkad and Malappuram districts. Further research involving Impact and vulnerability assessments using higher resolution projection data would provide a clearer picture at species level. Government of Kerala must focus on formulating enhanced research agendas and policies which are the need of the hour to understand how different predominant endemic species behave under the drastic or slow alterations in the climate and growing conditions in future. Micro level studies are the need of the hour to generate data as there exists spatial variability in climatic conditions and to design local forest divisional level planning. As the sustainable conservation initiatives are limited or happening at a slow pace in the state and hence it is important to increase the pace to exceed and overrule the extent of damage and fix priority areas of conservation.

References

Breshears DD, Cobb NS, Rich PM, Price KP, Allen CD, Balice RG, Romme WH, Kastens JH, Floyd ML, Belnap J, Anderson JJ, Myers OB, Meyer CW (2005) Regional vegetation die-off in response to global-change-type drought. Proc Nat Acad Sci USA 102:15144–15148

Chaturvedi RK, Joshi J, Jayaraman M, Bala G, Ravindranath NH (2012a) Multi model climate change projections for India under representative concentration pathways (RCPs): a preliminary analysis. Curr Sci 103:791–802

Chaturvedi RK, Joshi J, Jayaraman M, Bala G, Ravindranath NH (2012b) Multi model climate change projections for India under representative concentration pathways. Curr Sci 103(7): 1–12

Dillon GK, Holden ZA, Morgan P, Crimmins MA, Heyerdahl EK, Luce C (2011) Both topography and climate affected forest and woodland burn severity in two regions of the western US, 1984–2006, Ecosphere, in press

Easterling DR, Meehl GA, Parmesan C, Changnon SA, Karl TR, Mearns LO (2000) Climate extremes: observations, modeling, and impacts. Science 289:2068–2074

FAO (2005a) State of the world's forests. FAO, Rome

FAO (2005b) State of the world's forests 2005. Rome (also available at www.fao.org/docrep/007/y5574e/y5574e00.htm)

Food and Agriculture Organization (2005) Global forest resources assessment 2005 (Food and Agriculture Organization, Rome), Food and Agriculture Organization Forestry Paper 147

Forest survey of India (FSI) (1989–2009) State of forest report (1987–2007). Forest survey of India, Ministry of Environment and Forests, Dehra Dune

Gopalakrishnan R, Jayaraman M, Swarnim S, Chaturvedi RK, Bala G, Ravindranath NH (2010) Impact of climate change at species level: a case study of teak in India. Mitig Adapt Strategy Glob Change 16:199–209. gov/AR5/images/uploads/WGIIAR5-Chap19_FGDall.pdf., gov/AR5/images/uploads/WGIIAR5-Chap4_FGDall.pdf

Government of India (2008) National action plan on climate change (NAPCC). Govt. of India. Available via http://pmindia.nic.in/Pg01-52.pdf. Cited 26th Jan 2010

http://shodhganga.inflibnet.ac.in/bitstream/10603/5210/14/14_chapter%2010.pdf http://dyuthi.cusat.ac.in/jspui/bitstream/purl/2940/1/Dyuthi-T0931.pdf

IPCC (2007) Climate change 2007: impacts, adaptation and vulnerability.In: Parry ML, Canziani OF, Palutikof JP, van der Linden PJ and Hanson CE (eds) Contribution of Working Group II to the Fourth Assessment Report of the Intergovernmental Panel on Climate Change, p. 976. Cambridge, UK: Cambridge University Press

IPCC (2007) Climate Change 2007. The physical science basis, contribution of working group i to the fourth assessment report of the intergovernmental panel on climate change. Cambridge, University Press, Cambridge, UK, p 167

IPCC (2007) Climate change 2007: synthesis report; contribution of working groups I, II and III to the fourth assessment report of the intergovernmental panel on climate change. IPCC: Geneva, Switzerland, pp. 36–41

IPCC (2012) Emergent Risks and Key Vulnerabilities. Final Draft, Chapter 19 (online) http://ipccwg2

IPCC (2013) Terrestrial and Inland Water Systems. Final Draft, Chapter 4 (online) http://ipccwg2

IPCC (2013) Climate change 2013: the physical science basis. contribution of working group I to the fifth assessment report of the intergovernmental panel on climate change. Cambridge University Press, United Kingdom and New York, USA

Kumar KK, Kamala K, Rajagopalan B, Hoerling MP, Eischeid JK, Patwardhan SK, Srinivasan G, Goswami BN, Nemani R (2011) The once and future pulse of Indian monsoonal climate. Clim Dyn 36:2159–2170

Meinshausen M, Smith SJ, Calvin KV, Daniel JS, Kainuma JF, Lamarque M et al (2011) The RCP greenhouse gas concentrations and their extension from 1765 to 2300. Clim Change 109:213–241. https://doi.org/10.1007/s10584-011-0156-z

Millar CI, Stephenson NL, Stephens SL (2007) Climate change and forests of the future: managing in the face of uncertainty. Ecol Appl 17:2145–2151. pmid:18213958 https://doi.org/10.1890/06-1715.1

Millar CI, Stephenson NL, Stephens SL (2007b) Climate change and forests of the future: Managing in the face of uncertainty. Ecol Appl 17:2145–2151

Morgan P, Heyerdahl EK, Gibson CE (2008) Multi-season climate synchronized widespread forest fires throughout the 20th-Century. Northern Rocky Mountains. USA, Ecology 89:717–728

Morrison J, Quick MC, Foreman MGG (2002) Climate change in the Fraser River watershed: flow and temperature projections. J Hydrol 263:230–244

Nelson DR, Adger WN, Brown K (2007) Adaptation to environmental change: Contributions of a resilience framework. Ann Rev Environ Resour 32:395–419

Pierce JL, Meyer GA, Jull AJT (2004) Fire-induced erosion and millennial scale climate change in northern ponderosa pine forests. Nature 432:87–90

Pounds JA, Bustamante MR, Coloma LA et al (2006) Widespread amphibian extinctions from epidemic disease driven by global warming. Nature 439:161–167

Ramachandran A, Praveen D, Jaganathan R, Palanivelu K (2015) Projected and observed aridity and climate change in the east coast of south india under RCP 4.5. Sci World J 2015, Article ID 169761, 11: 2015. https://doi.org/10.1155/2015/169761

Ramaraj AP, Jagannathan R, Dheebakaran GA (2009) Impact of climate change on rice and groundnut yield using PRECIS regional climate model and DSSAT crop simulation model. ISPRS Archives XXXVIII-8/W3 Workshop Proceedings: Impact of Climate Change on Agriculture pp 143–146

Rao GSLHVP, Rao AVRK, Krishnakumar KN, Gopakumar CS (2009) Climate change projections and impacts on plantations in Kerala. In: Proceeding of climate change adaptation strategies in agriculture and allied sectors-invited papers. pp. 158

Rao SA, Chaudhari HS, Pokhrel S, Goswami BN (2010) Unusual central Indian drought of summer monsoon 2008:role of southern tropical Indian Ocean warming. Journal ofClimate 23(19):5163–5174

Ravindranath NH, Joshi NV, Sukumar R, Saxena A (2006) Impact of climate change on forest in India. Curr Sci 90(3):354–361

Ravindranath NH, Chaturvedi RK, Murthy IK (2008) Forest conservation, afforestation and reforestation in India: implications for forest carbon stocks. Curr Sci 95(2):216–222

Roa S, Venkeswaralu Ch, Srinivas B et al (2011) Soil carbon sequestration for climate change Mitigation and Food security, 24th November to 3ed December, 2011. Central Research Institute for Dryland Agriculture, Hyderabad, India, 322p

Sabu TK, Vinod KV, Latha M, Nithya S, Boby J (2011) Cloud forest dung beetles (Coleoptera: Scarabaeinae) in the Western Ghats, a global biodiversity hotspot in southwestern India. Tropical Conserv Sci 4(1):12–24. Available online: www.tropicalconservationscience.org

Spittlehouse DL, Stewart RB (2003) Adaptation to climate change in forest management. BC J Ecosys Manage 4:1–7

United Nations Convention to Combat Desertification (UNCCD) (2012) Report, 2012, http://www.unccd.int/ActionProgrammes/india-eng2001.pdf

Vandermeer J, Perfecto I (1997) The agroecosystem: a need for the conservation biologist's lens. Conserv Biol 11:1–3

Westerling AL, Hidalgo HG, Cayan DR, Swetnam TW (2006) Warming and earlier spring increases western US. For. Wildfire Act Sci 313:940–943. https://doi.org/10.1126/science.1128834

Westerling AL, Turner MG, Smithwick EAH, Romme WH, Ryan MG (2011) Continued warming could transform greater yellowstone fire regimes by mid-21st century. Proc Natl Acad Sci USA 108:13165–13170. https://doi.org/10.1073/pnas.1110199108

McCarty JP, Wolfenbarger LL, Wilson, JA (2009) Biological impacts of climate change. In: Encyclopedia of life sciences (ELS). Wiley, Chichester. https://doi.org/10.1002/9780470015902.a0020480

Wiltshire A, Kay G, Gornall J, Betts R (2013) The impact of climate, CO2 and population on regional food and water resources in the 2050s. Sustainability 5(5):2129–2151. https://doi.org/10.3390/su5052129

BioGeographic Areas of Kerala. http://forest.kerala.gov.in/images/abc/bgareaskerala.pdf

http://www.prokerala.com/kerala/forest.htm

Mediterranean Marginal Lands in Face of Climate Change: Biodiversity and Ecosystem Services

Helena Castro and Paula Castro

Abstract Mediterranean landscapes are the result of the interaction of a long history of anthropogenic disturbances (cultivation, grazing, timber and fuel wood) with natural disturbances (such as fire, floods and extreme droughts) and a variable climate. Often ecosystems in these landscapes are characterized by having soils that are marginal for production. On the other hand, these ecosystems often have multiple land uses, including livestock breeding, forestry and cultivation, creating ecosystems that are of great importance for their environmental and socio-economic value. Climate change is expected to modify patterns of precipitation with predictions for the Mediterranean pointing to enhanced drought and increased frequency of extreme events. The impact of rainfall variability on grassland productivity represents a topic of concern due to its relevance for agricultural activities such as livestock production a key economic activity in Mediterranean marginal lands. In this chapter we present a review of the effects of climate change on biodiversity and ecosystem services associated with agro-silvo-pastoral systems while also approaching the associated socio-economic and desertification issues. Finally we review studies on conventional and alternative management strategies in in search for strategies to cope with climate change.

1 Introduction

Current and historical human activity is an unavoidable subject when approaching the structure and function of ecosystems because it strongly influences modern vegetation patterns (Foster et al. 2003), particularly in the Mediterranean Basin where ecosystems and landscapes were transformed and redesigned by humans over several millennia. This, in combination with a variable climate and a high

H. Castro (✉) · P. Castro
CFE-Centre for Functional Ecology - Science for People & the Planet,
Department of Life Sciences, University of Coimbra, 3000-456 Coimbra, Portugal
e-mail: hecastro@ci.uc.pt

P. Castro et al. (eds.), *Climate Change-Resilient Agriculture and Agroforestry*,
Climate Change Management, https://doi.org/10.1007/978-3-319-75004-0_10

degree of spatial heterogeneity, has contributed to the high biodiversity of the Mediterranean Basin.

Mediterranean grasslands including, rangelands, pastures, meadows, and fodder crops (Varela and Robles-Cruz 2016) are important resources covering up to 48% of the Region. Extensive grazing systems represent one of the best adapted land uses in Mediterranean marginal lands and high biological diversity is often associated with these ecosystems. This review deals with extensive grazing lands associated to two main ecosystems: Montado and cereal steppes. The importance of these areas rests on both environmental as well as socio-economic values. They support outstanding biodiversity, form unique landscapes, are the source of high-quality food derived from animal production, sustain rural population, and constitute important areas for rural leisure and tourism.

In Mediterranean grasslands, the structure and composition of the vegetation, as well as the life history traits of many species show adaptations to human perturbations, such as fire, ploughing and grazing, that reflect the profound impact of anthropogenic activity in this region. The exceptional richness of annual plant species in the Mediterranean flora is to a large extent the result of long-standing human activities (Blondel and Aronson 1999).

The Mediterranean climate is characterized by the seasonality and unpredictability in precipitation. Climate change is expected to modify patterns of precipitation with predictions for the Mediterranean pointing to enhanced drought and increased frequency of extreme events (Sardans and Peñuelas 2013; Chelli et al. 2016). Several studies predict an increase in winter precipitation primarily in the form of high-intensity events, which together with decreased spring and autumn rainfall can have important consequences for ecosystem function (Miranda et al. 2011). It may speed up desertification process and reduce ecosystem productivity.

The impact of rainfall variability on grassland productivity represents a topic of concern due to its relevance for agricultural activities such as livestock production a key economic activity in Mediterranean marginal lands. In this chapter we present a review of the effects of climate change on biodiversity and ecosystem services associated with agro-silvo-pastoral systems while also approaching the associated socio-economic and desertification issues. Finally, we review studies on conventional and alternative management strategies in in search for strategies to cope with climate change.

2 Climate Change in the Mediterranean Region

Climate change has become an important issue of the 21st century deeply impacting biodiversity, habitats and ecosystems which in turn intensify the vulnerability of regions, economic sectors and communities (IPCC 2014; EEA 2013; EEA 2017). These impacts and vulnerabilities are of particularly prominence in the Mediterranean Region and arte expected to be worsened by climatic variation (EEA 2017; EEA, 2012; Ulbrich et al. 2012). In 2017 the European Environment Agency

(EEA 2017) reported the key observed and projected climate change and its impacts in southern Europe, which include: 1. Large increase in heat extremes, 2. Decrease in precipitation and river flow, 3. Increasing risk of droughts, 4. Increasing risk of biodiversity loss, 5. Increasing risk of forest fires, 6. Increased competition between different water users, 7. Increasing water demand for agriculture, 8. Decrease in crop yields, 9. Increasing risks for livestock production, 10. Increase in mortality from heat waves, 11. Expansion of habitats for southern disease vectors, 12. Decreasing potential for energy production, 13. Increase in energy demand for cooling, 14. Decrease in summer tourism and potential increase in other seasons, 15. Increase in multiple climatic hazards, 16. Most economic sectors negatively affected and 17. High vulnerability to spill over effects of climate change from outside Europe. Is thus mandatory that climate change is incorporated in ecosystem service assessments and in the planning procedures as well as in decision-making (Groves et al. 2012; Runting et al. 2017).

3 History and Main Trends of Mediterranean Grasslands in Southern Portugal

The agro-silvo-pastoral systems of the South of Portugal, similarly to others in the Mediterranean basin have been shaped by a long history of human and ecological disturbances (Serrão 1963; Blondel and Aronson 1999), which led to great transformations in the vegetation. In southern Portugal, these are thought to date from Palaeolithic times and to have intensified through time with increases in cultivated areas and with the demand of wood for naval construction in the XV and XVI centuries (Capelo et al. 2007). In the first half of the 20th century, the mechanization and intensification of cereal production in the undercover led to a wide scale destruction of the Montado, resulting in a decrease in tree cover and in extreme cases in the origin and expansion of the cereal steppe (Roxo et al. 1998; Marta-Pedroso et al. 2007). From 1980 to the present the Portuguese agricultural policy has been increasingly aligned with the European Common Agricultural Policy (CAP) (Guerra et al. 2016).

The cereal steppe is an open mosaic landscape, consisting mainly of cereal fields, stubble, ploughed and fallow land (usually grazed), and is based on extensive cultivation of cereals under a rotational scheme that lasts 5–7 years (Marta-Pedroso et al. 2007). The cereal steppes have an important conservation value as they provide habitat for many steppic birds with unfavorable conservation status such as the great bustard (*Otis tarda* L.), the little bustard (*Tetrax tetrax* L.), and the lesser kestrel (*Falco naumanni* Fleischer) constituting one of the last refuges for these species (Moreira et al. 2005). The Montado is characterized by sparse tree cover and a diversity of understory vegetation—ranging from shrub formations to grasslands. Currently, the existing Montado areas are cork oak, holm oak or a mixture of both species, which represent a cline from 5 to 65% oak tree cover, and coexist with pasture and crops in the undercover (Sá-Sousa 2014). In the undercover encroaching shrubs are cleared in at intervals 4–7 years to reduce the risk of wildfires and to promote the establishment of pasturelands or annual crops (Pinto-Correia 2000). In both cases the rotation creates a mosaic of land uses and habitat types with high conservation value. In both systems grazing plays a key role in the sustainability and in maintaining biodiversity. Traditionally, grazing was done by sheep, or mixed grazing with goat, but recently grazing by cattle has increased substantially, mainly due to funding schemes of the Common Agricultural Policy of the European Union which favour payments per cattle head (Ferraz de Oliveira et al. 2013). This is a strong pressure on these marginal grasslands. Grazing by wild species is also common in some Montado areas and is an important source additional income through hunting.

The trend in land cover change in these marginal lands reflects a polarization between extensification or abandonment of more peripheral and fragile areas and over-exploitation of the tree cover and/or intensification of activities in the undercover, such as overgrazing and mechanized ploughing (Pinto-Correia and Godinho 2013). In these systems, abandonment has a strong effect on vegetation composition with the replacement of herbaceous dominated communities with shrub dominated communities, resulting in a decrease in plant species richness (Castro et al. 2010). At the landscape level, the increasing abandonment may promote the appearance of large continuous areas of uniform flammable vegetation with the consequence of an increase in the number of fires, total surface burned and the distribution of areas affected by fire (Pausas 1999). Additionally, the complete abandonment of agro-pastoral uses implies the loss of a traditional landscape and many ecosystem functions as well as loss of habitat and species diversity (Moreira et al. 2005; Plieninger et al. 2014). The effects of abandonment on biodiversity in the Mediterranean basin are reviewed by Plieninger et al. (2013). On the other hand intensification may lead to the reduction in tree cover, leading to the disappearance of the forestry component, and the disappearance of the associated ecosystem services. Intensification may also cause soil erosion leading to the loss of valuable soil provided ecosystem services and ultimately to desertification.

4 Grassland Biodiversity and Services in a Changing Climate

Grassland composition, function and services can be strongly influenced by anthropic disturbance and expected climate changes may intensify the impacts of these disturbances. Mediterranean grasslands provide a range of ecosystem services including provisioning services, which is the most prominent service and has motivated their existence; supporting services, such as biodiversity conservation and habitat for wildlife; regulating services such as carbon fixation and prevention of erosion; and cultural services such as aesthetic and recreational services.

Provisioning services (i.e. grazing) produce a series of benefits, such as meat or milk, often with special sensorial and nutritive qualities linked to labels such as products with Denomination of Origin. In addition to the added value, these products may raise values linked to cultural heritage for consumers which may be important in connecting rural and urban populations (Varela and Robles-Cruz 2016).

Soil erosion is a severe problem in Southern Portugal. On the one hand, the plant cover provided by grasslands can reduce the likelihood of soil losses (Schnabel et al. 2009), on the other, overgrazing by livestock and soil mobilization can cause of soil erosion in the Mediterranean (Pinto-Correia and Godinho 2013).

Biodiversity has key role in the maintenance of ecosystem stability (Caldeira et al. 2005) and a key role in the support of ecosystem services. Plant diversity in Mediterranean grasslands results from the action of a multiplicity of factors including spatial variability, soil seed bank (Levassor et al. 1990), grazing intensity, inter-annual fluctuations in climatic conditions, and disturbance. Their vegetation is dominated by annual species, where most species germinate after the first significant autumn rains, flower in spring and die in summer. Mediterranean grasslands are characterized by high plant species diversity as indicated by various studies (e.g. Díaz-Villa et al. 2003; Castro et al. 2010; Castro et al. 2016), but the variation in yearly composition is also high with only part of the plant species being present consistently from year to year.

Inter-annual fluctuations in rainfall have a strong impact on plant community composition as shown by multiple studies performed in Mediterranean grasslands (e.g. Castro et al. 2016; López-Carrasco et al. 2015; Henkin et al. 2010; López-Sánchez et al. 2016). The effect of meteorological fluctuations depends on the total rainfall during the growing season (the more water available, the greater the species richness), and of the autumn rainfall distribution (the less drought periods between rain episodes, the greater the richness). Data comparing a wet and a dry year in two Mediterranean grasslands shows a decrease on overall species richness (Fig. 1) and plant cover, as well as that functional groups are affected differently (Fig. 2).

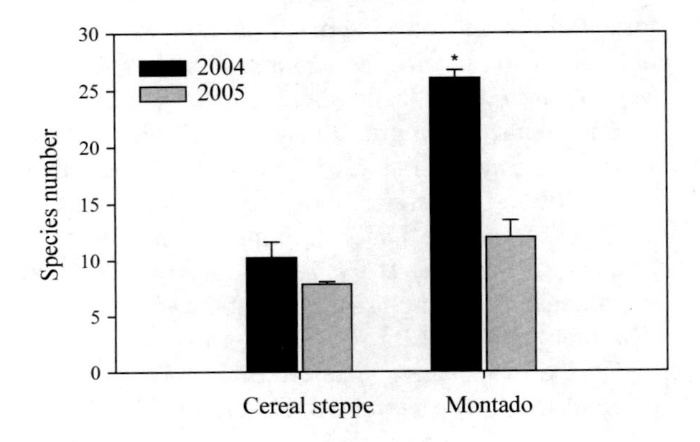

Fig. 1 Plant species richness per 0.25 m^2 in two Mediterranean grasslands in a wet year (2004) and a dry year (2005). Significant differences between years, as determined by a t-test, are indicated by *

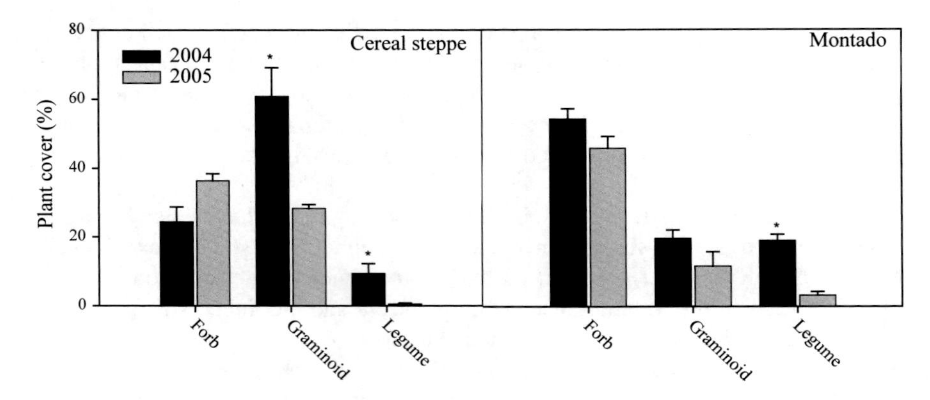

Fig. 2 Cover of plant functional groups in two Mediterranean grasslands in a wet year (2004) and a dry year (2005). Significant differences between years, as determined by a t-test, are indicated by *

Legumes were affected negatively by the dry year, with both a decrease in species number and cover. This was also observed in other studies in Mediterranean grasslands (e.g. Castro et al. 2016; Henkin et al. 2010). Overall, grasses and forbs seem to be less affected by fluctuations in rainfall. In addition to rainfall, increased yearly temperatures negatively affect plants by accelerating their development and reducing the time available for seed dispersal (Hernández and Pastor 2006).

In these ecosystems, mostly composed of annual species, the status of the seed bank is directly linked to the annual fertility of adult plants, ultimately threatening the stability of the population if many episodes interrupt seed renewal (e.g. prolonged droughts, droughts early in spring, overgrazing) (Jongen et al. 2013).

Grazing is another major factor affecting plant structure and diversity in Mediterranean grasslands (Noy-Meir et al. 1989). Large herbivores affect species richness and diversity by selective defoliation, trampling, creating nutrient patches and influencing patterns of litter inputs. In Mediterranean grasslands these small-scale effects on species richness are generally positive (Sternberg et al. 2000). Therefore, grazing may not be a disturbance in these grasslands. On the contrary, herbivore elimination may constitute a disturbance because it can lead to a transformation of the ecosystem (Montalvo et al. 1993). Several studies show the importance of grazing as a management tool for maintaining plant diversity in Mediterranean grasslands (e.g. Golodets et al. 2011).

The interaction of grazing and rainfall plays a key role in the sustainability these systems. Carmona et al. (2012) examined the combined effect of grazing and water availability on the diversity of Mediterranean grasslands and concluded that water availability was the major determinant of both taxonomical and functional compositions. However, the effect of grazing was also significant, supporting the importance of grazing in influencing the diversity and composition of these systems (Carmona et al. 2012).

On the other hand, Mediterranean grasslands display a remarkable capacity to respond to climate perturbations due to the different strategies of the component species, which allow them to survive in a variable and highly unpredictable environment (Merou et al. 2013). In line with this Carmona et al. (2012) found that, despite great interannual variations in species diversity, functional diversity was rather stable over time and space, showing the temporal functional stability of the system. However, the same authors also found that under more limited water availability, grazing intensification reduced the functional diversity of these grasslands (Carmona et al. 2012). The predicted reduction in rainfall in the Mediterranean area increases the risk of high grazing levels causing dramatic declines in the functional diversity of Mediterranean grasslands, probably compromising their stability and resilience over time, confirming that the interaction of grazing and rainfall plays a key role in the sustainability these systems.

5 Primary Productivity, Food Quality and Implications for Livestock

Biomass production in grasslands is a key component of food provision for domestic herbivores and is known to depend on climate, resource availability, and on the functional characteristics of communities (Chollet et al. 2014). Biomass production in grasslands associated to cereal steppe and montado areas, in particular, is constrained by rainfall, duration of the growth season and low soil fertility. Consequently, they are characterized by large spatial and temporal variation in the availability and quality of plant forage, which affects grassland productivity and feed value causing variations in animal body conditions (Scocco et al. 2016).

Biomass production is affected by rainfall as shown in Fig. 3, which compares biomass data from wet (2004) and a dry (2005) year in two Mediterranean grasslands. Similar results were also observed by other studies in Mediterranean grasslands in Southern Portugal (e.g. Henkin et al. 2010; López-Sánchez et al. 2016; Vázquez-De-Aldana et al. 2008; Caldeira et al. 2005). Variability of herb biomass is often related to changes in the cover of dominant species, predominantly legumes, as shown by the very low legume biomass yield under dry conditions in the study by López-Carrasco et al. (2015).

Plant food resource quality is related to species composition, namely abundance of legumes or grasses and the chemical and physical characteristics of each species (Vázquez-de-Aldana et al. 2000). Forage quality depends on nutrient concentration, which determines digestibility, partitioning of metabolized products in the digestive tract and forage intake (Dumont et al. 2015). Chemical and physical characteristics vary along the growing season with protein and mineral contents and dry matter digestibility decreasing (Corona et al. 1998). The effect of climate change on physical and chemical characteristics of forages was reviewed by Dumont et al. (2015). Rainfall and temperature affect both the quantity and quality of forage. Water stress leads to an average increase forage N concentration and to a decrease in the plant cell-wall, as well as, to an average increase in digestibility, even though

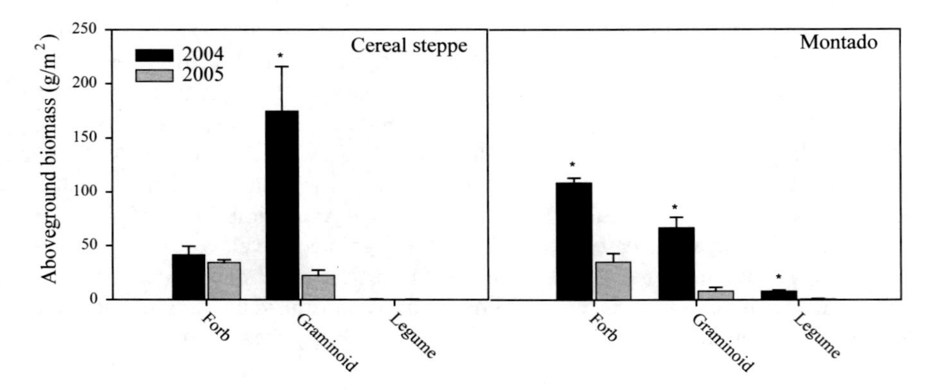

Fig. 3 Biomass of plant functional groups in two Mediterranean grasslands in a wet year (2004) and a dry year (2005). Significant differences between years, as determined by a t-test, are indicated by *

there was high variation among experiments (Dumont et al. 2015). Protein concentration was significantly and negatively correlated with the annual precipitation in the study by Vázquez-De-Aldana et al. (2008). Variations in forage N content can also be expected if legume abundance varies, as legumes are generally richer in N than other functional groups (Dumont et al. 2015). Legume component in pasture is an important factor in the protein concentration of herbage, since this botanical component has much greater protein concentration than grasses and forbs (Vázquez-de-Aldana and Pérez-Corona 2000). As referred above legume cover and biomass is related to rainfall.

In addition to influencing the total biomass, drought can accelerate plant maturation or even lead to tissue senescence which strongly decreases forage quality (Scocco et al. 2016).

Local topographic differences such as slope location (upper vs. lower location) can influence time plant maturation, biomass and nutritional quality. Lower locations tend to be show later maturation and have changes in biomass and quality less affected by fluctuations in rainfall than upper areas (Vázquez-de-Aldana and Pérez-Corona 2000; Vázquez-De-Aldana et al. 2008).

6 Strategies to Cope with Climate Change

Vulnerability to drought in pastures in semi-arid areas can lead to considerable socio-economic and environmental losses in the absence of mitigation and adaptation strategies.

Climate change is expected to modify patterns of precipitation with predictions for the Mediterranean pointing to enhanced drought and increased frequency of extreme events (Sardans and Peñuelas 2013; Chelli et al. 2016), which may be a major threat to grazing activities. These activities play a key role in the sustainability and in maintaining biodiversity of highly valued silvo-pastoral ecosystems, such as Montados and cereal steppes, which provide many environmental, cultural and economic services.

A study by Iglesias et al. (2016) addressing drought risk in the Spanish Dehesa showed that that main feeding costs are incurred during the summer months of July, August and September while the higher variability happens in February–March, May–June and October–November. The same study also indicates a decupling between climate risk and economic consequences, with climate risks concentrating at the beginning of spring and autumn, and economic risk usually occurring later and lasting longer (Iglesias et al. 2016).

Climate change is expected to modify patterns of precipitation with predictions for the Mediterranean pointing to enhanced drought and increased frequency of extreme events (Sardans and Peñuelas 2013; Chelli et al. 2016). This can negatively affect the sustainability of these ecosystems by affecting rangeland vegetation quality and productivity, the two primary factors affecting the carrying capacity of the rangeland for livestock, as well as affecting the biodiversity and underlying services. In this systems understanding and valuing the social component is essential for sustainable management (Louhaichi et al. 2016). Similarly it is also important to understand that although drought may be perceived as a climate threat, its effects may be worsened or mitigated by the interaction of various environmental and socioeconomic factors (Iglesias et al. 2016). Grazing intensity and soil mobilization in areas under cultivation or shrub clearing, which increase the risk of soil erosion and ecosystem desertification are two main concerns, that can in worsened by climate change. As pointed before, the interaction of human activity and rainfall plays a key role in the sustainability these systems. Below we present some measures that have been employed or undergone studies with the objective of achieving sustainability and coping with climate change.

The system of 'sown biodiverse permanent pastures rich in legumes', a semi-intensive system for meat production, that started being developed in Portugal in the second half of the 1960s has been presented as a being an economic and ecological win-win solution that addresses many of the causes of land degradation in the Mediterranean while also recovering soil condition, ecosystem functions and services (Teixeira et al. 2014). According to (Teixeira et al. 2014) these systems have higher productivity than natural pastures, allowing a sustainable increase in animal carrying capacity, contribute to the increase in soil organic matter, which together with avoided soil mobilization due to seed bank persistence, decreases sediment loss and soil erosion.

Stocking rates can have important implications in grassland sustainability as shown by several studies. In this context there is the interesting study by Iglesias et al. (2016) which shows that, under drought conditions, lowering the stocking rate may significantly reduce vulnerability and decrease the chance of worst outcomes, but also entails opportunity costs in favourable weather circumstances. This may lead to farmers' reluctance to adopting this type of strategy (Iglesias et al. 2016).

A study by Escribano et al. (2012) comparatively evaluating conventional and organic beef production in the Spanish dehesa concluded that organic farms are more sustainable, obtaining higher scores than conventional farms in stability and self-reliance, and similar scores in productivity, adaptability, and equity. Organic dehesa beef farms seem to have low dependence on external products and services and to be more adapted to their environment (lower stocking rates for an optimal use of the system's feedstuff production) (Escribano et al. 2012).

Extensive animal production with pastures and forages integrated in multiple use of forest are part of sustainable rural development (Sequeira 1997). These systems increase security against price change, increase farm income, employment stability and qualification, combat depopulation, and especially combat desertification. The perennial herbaceous component of these communities can buffer against drastic changes in community structure potentially decreasing vulnerability to climate change (Henkin et al. 2010). The presence of trees can influence herbaceous biomass and diversity. Overall, herbaceous biomass is higher outside the canopy, except in wet years, and forb and legume biomass are higher outside while grass biomass is higher under the canopy in wet years (López-Carrasco et al. 2015). Over the three years, there was a consistent trend in species richness, with low values beneath the crowns and higher values away from the oaks (López-Carrasco et al. 2015). New methods of reclaim eroded soils by sewage sludge mud subsoil injection, retaining water runoff are developed to avoid erosion, fertility and water loss (Sequeira 2002a, b; Sequeira 2004).

7 Conclusion

The Mediterranean Region is one of the most vulnerable regions in the world to the impacts of global warming. Critical factors as rainfall and temperature pressure the dynamic and functioning of grassland ecosystems and the expected increases in drought events and severity will constrain the conservation of natural resources and place the sustainability of forage production and other ecosystems services at risk. These systems play important ecological, economic and social roles and expected climate change scenarios deserve special attention from stakeholders, scientific community and decision-makers.

References

Blondel J, Aronson J (1999) Biology and wildlife of the Mediterranean region. Oxford University Press, Oxford

Caldeira MC, Hector A, Loreau M, Pereira JS (2005) Species richness, temporal variability and resistance of biomass production in a Mediterranean grassland. Oikos 110:115–123

Capelo J, Catry F, Silva JS (2007) Biologia, Ecologia e distribuição da azinheira. Lisboa, pp 119–129

Carmona CP, Azcárate FM, de Bello F et al (2012) Taxonomical and functional diversity turnover in Mediterranean grasslands: interactions between grazing, habitat type and rainfall. J Appl Ecol 49:1084–1093. https://doi.org/10.1111/j.1365-2664.2012.02193.x

Castro H, Lehsten V, Lavorel S, Freitas H (2010) Functional response traits in relation to land use change in the Montado. Agric Ecosyst Environ 137:183–191. https://doi.org/10.1016/j.agee.2010.02.002

Castro H, Barrico L, Rodríguez-Echeverría S, Freitas H (2016) Trends in plant and soil microbial diversity associated with Mediterranean extensive cereal–fallow rotation agro-ecosystems. Agric Ecosyst Environ 217:33–40. https://doi.org/10.1016/j.agee.2015.10.027

Chelli S, Canullo R, Campetella G et al (2016) The response of sub-Mediterranean grasslands to rainfall variation is influenced by early season precipitation. Appl Veg Sci 19:611–619. https://doi.org/10.1111/avsc.12247

Chollet S, Rambal S, Fayolle A et al (2014) Combined effects of climate, resource availability, and plant traits on biomass produced in a Mediterranean rangeland. Ecology 95:737–748. https://doi.org/10.1890/13-0751.1

Corona MEP, De Aldana BRV, Criado BG, Ciudad AG (1998) Variations in nutritional quality and biomass production of semiarid grasslands. J Range Manag 51:570–576. https://doi.org/10.2307/4003378

Díaz-Villa MD, Marañón T, Arroyo J, Garrido B (2003) Soil seed bank and floristic diversity in a forest-grassland mosaic in southern Spain. J Veg Sci 14:701–709. https://doi.org/10.1111/j.1654-1103.2003.tb02202.x

Dumont B, Andueza D, Niderkorn V et al (2015) A meta-analysis of climate change effects on forage quality in grasslands: specificities of mountain and mediterranean areas. Grass Forage Sci 70:239–254. https://doi.org/10.1111/gfs.12169

EEA (European Environment Agency) (2012) Climate change, impacts and vulnerability in Europe 2012: an indicator-based report. Office for official publications of the European union, Luxembourg. EEA report no 12/2012

EEA (European Environment Agency) (2013) Adaptation in Europe. Addressing risks and opportunities from climate change in the context of socio-economic developments. Publications office of the European union, Luxembourg. EEA report no 3/2013

EEA (European Environment Agency) (2017). Climate change, impacts and vulnerability in Europe 2016. An indicator-based report. Publications office of the European union, Luxembourg. EEA report no 1/2017

Escribano AJ, Mesias FJ, Gaspar P et al (2012) Sustainability of organic and conventional beef cattle farms in SW Spanish rangelands ('Dehesas'): a comparative study. 10TH Eur IFSA symp prod reprod farming syst new modes Organ sustain food syst tomorrow. https://doi.org/10.13140/rg.2.1.4293.4889

Ferraz de Oliveira MI, Lamy E, Bugalho MN et al (2013) Assessing foraging strategies of herbivores in Mediterranean oak woodlands: a review of key issues and selected methodologies. Agrofor Syst 87:1421–1437. https://doi.org/10.1007/s10457-013-9648-3

Field CB, Barros VR, Dokken DJ et al (eds) (2014) Climate Change 2014. Impacts, adaptation, and vulnerability. part a: global and sectoral aspects. IPCC (Intergovernmental Panel on Climate Change). Cambridge University Press, Cambridge

Foster D, Swanson F, Aber J et al (2003) The importance of land-use legacies to ecology and conservation. Bioscience 53:77–88

Golodets C, Kigel J, Sternberg M (2011) Plant diversity partitioning in grazed Mediterranean grassland at multiple spatial and temporal scales. J Appl Ecol 48:1260–1268. https://doi.org/10.1111/j.1365-2664.2011.02031.x

Groves CR, Game ET, Anderson MG et al (2012) Incorporating climate change into systematic conservation planning. Biodivers Conserv 21:1651–1671. https://doi.org/10.1007/s10531-012-0269-3

Guerra CA, Metzger MJ, Maes J, Pinto-Correia T (2016) Policy impacts on regulating ecosystem services: looking at the implications of 60 years of landscape change on soil erosion prevention in a Mediterranean silvo-pastoral system. Landsc Ecol 31:271–290. https://doi.org/10.1007/s10980-015-0241-1

Henkin Z, Perevolotsky A, Sternberg M (2010) Vulnerability of Mediterranean grasslands to climate change: what can we learn from a long-term experiment? In: Porqueddu C, Ríos S (eds) The contributions of grasslands to the conservation of Mediterranean biodiversity. CIHEAM, Zaragoza, pp 167–174

Hernández AJ, Pastor J (2006) Mediterranean grasslands : ecological observations related to the climate of the past 55 years. In: Lloveras J, González-Rodrígues A, Vázquez-Yañez O et al (eds) Sustainable grassland productivity. 21st General Meeting of the European Grassland Federation, pp 790–792

Iglesias E, Báez K, Diaz-Ambrona CH (2016) Assessing drought risk in mediterranean dehesa grazing lands. Agric Syst 149:65–74. https://doi.org/10.1016/j.agsy.2016.07.017

Jongen M, Unger S, Fangueiro D et al (2013) Resilience of montado understorey to experimental precipitation variability fails under severe natural drought. Agric Ecosyst Environ 178:18–30. https://doi.org/10.1016/j.agee.2013.06.014

Levassor C, Ortega M, Peco B (1990) Seed bank dynamics of mediterranean pastures subjected to mechanical disturbance. J Veg Sci 1:339–344

López-Carrasco C, López-Sánchez A, San Miguel A, Roig S (2015) The effect of tree cover on the biomass and diversity of the herbaceous layer in a Mediterranean dehesa. Grass Forage Sci 70:639–650. https://doi.org/10.1111/gfs.12161

López-Sánchez A, San Miguel A, Dirzo R, Roig S (2016) Scattered trees and livestock grazing as keystones organisms for sustainable use and conservation of Mediterranean dehesas. J Nat Conserv 33:58–67. https://doi.org/10.1016/j.jnc.2016.07.003

Louhaichi M, Clifton K, Kassam SN, Werner J (2016) Overlooked benefits and services of grasslands to support policy reform. In: Kyriazopoulos A, López-Francos C, Porqueddu C, Sklavou P (eds) Ecosystem services and socio-economic benefits of Mediterranean grasslands, pp 301–312

Marta-Pedroso C, Domingos T, Freitas H, de Groot RS (2007) Cost-benefit analysis of the zonal program of castro verde (Portugal): highlighting the trade-off between biodiversity and soil conservation. Soil Tillage Res 97:79–90

Merou TP, Tsiftsis S, Papanastasis VP (2013) Disturbance and recovery in semi-arid Mediterranean grasslands. Appl Veg Sci 16:417–425. https://doi.org/10.1111/avsc.12013

Miranda JD, Armas C, Padilla FM, Pugnaire FI (2011) Climatic change and rainfall patterns: effects on semi-arid plant communities of the Iberian Southeast. J Arid Environ 75:1302–1309. https://doi.org/10.1016/j.jaridenv.2011.04.022

Montalvo J, Casado MA, Levassor C, Pineda FD (1993) Species diversity patterns in Mediterranean grasslands. J Veg Sci 4:213–222. https://doi.org/10.2307/3236107

Moreira F, Beja P, Morgado R et al (2005) Effects of field management and landscape context on grassland wintering birds in Southern Portugal. Agric Ecosyst Environ 109:59–74

Noy-Meir I, Gutman M, Kaplan Y (1989) Responses of mediterranean grassland plants to grazing and protection. J Ecol 77:290–310

Pausas JG (1999) Mediterranean vegetation dynamics: modelling problems and functional types. Plant Ecol 140:27–39

Pinto-Correia T (2000) Future development in Portuguese rural areas: how to manage agricultural support for landscape conservation? Landsc Urban Plan 50:95–106

Pinto-Correia T, Godinho S (2013) Changing agriculture—changing landscapes: what is going on in the high valued montado. In: Agriculture in mediterranean Europe: between old and new paradigms. Emerald group, pp 75–90

Plieninger T, Gaertner M, Hui C, Huntsinger L (2013) Does land abandonment decrease species richness and abundance of plants and animals in Mediterranean pastures, arable lands and permanent croplands? Environ Evid 2:3. https://doi.org/10.1186/2047-2382-2-3

Plieninger T, Hui C, Gaertner M, Huntsinger L (2014) The impact of land abandonment on species richness and abundance in the Mediterranean Basin: a meta-analysis. PLoS One. https://doi.org/10.1371/journal.pone.0098355

Roxo MJ, Mourão JM, Casimiro PC (1998) Políticas agrícolas, mudanças de uso do solo e degaradção dos recusros naturais

Runting RK, Bryan BA, Dee LE et al (2017) Incorporating climate change into ecosystem service assessments and decisions: a review. Glob Chang Biol 23:28–41. https://doi.org/10.1111/gcb.13457

Sardans J, Peñuelas J (2013) Plant-soil interactions in Mediterranean forest and shrublands: impacts of climatic change. Plant Soil 365:1–33

Sá-Sousa P (2014) The Portuguese montado: conciliating ecological values with human demands within a dynamic agroforestry system. Ann For Sci 71:1–3. https://doi.org/10.1007/s13595-013-0338-0

Schnabel S, Gómez Gutiérrez A, Lavado Contador JF (2009) Grazing and soil erosion in dehesas of SW Spain. Adv Stud Desertif 725–728 (732)

Scocco P, Piermarteri K, Malfatti A et al (2016) Effects of summer rainfall variations on sheep body state and farming sustainability in sub-Mediterranean pastoral systems. Spanish J Agric Res 14:4–7. https://doi.org/10.5424/sjar/2016143-9230

Sequeira EM (1997) As pastagens, a nova PAC e o Ambiente no Alentejo. Pastagens e Forragens 18:49–74

Sequeira EM (2002a) O uso das lamas de ETAR no combate à Desertificação—O caso do "Projecto Piloto de Combate à Desertificação" da LPN. Apresentado no Painel 3- Valorização Agrícola de Lamas do Simposium Gestão e Valorização de Lamas de ETA's e ETAR's em Portugal, organizado pela APDA e APEA no Forum Lisboa, em 27 e 28 de Outubro

Sequeira EM (2002b) O "Projecto Piloto de Combate à Desertificação" da LPN. O uso de lamas de ETAR no combate à Desertificação e na recuperação de solos degradados. Apresentado no Encontro Anual da Sociedade Portuguesa da Ciência do Solo—Sistemas de Uso da Terra, Ordenamento do Território e Ambiente, na Escola Superior Agrária de Ponte de Lima, de 5 a 7 de Setembro

Sequeira EM (2004) Um exemplo português de recuperação da "terra"—Campo Branco". In: Louro V (ed) Desertificação. Sinais, Dinâmicas e Sociedade Instituto Piaget, pp 153–164

Serrão J (1963) Dicionário de historia de Portugal. Iniciativas Editoriais, Lisboa

Sternberg M, Gutman M, Perevolotsky A et al (2000) Vegetation response to grazing management in a Mediterranean herbaceous community: a functional group approach. J Appl Ecol 37:224–237

Teixeira RFM, Proença V, Valada T et al (2014) Sown biodiverse pastures as a win-win approach to reverse the degradation of Mediterranean ecosystems. EGF 50 Futur Eur grasslands Proc 25th Gen Meet Eur Grassl Fed Aberystwyth, Wales, 7–11 Sept 2014 258–260

Ulbrich U, Lionello P, Belušic D, et al (2012). Climate of the Mediterranean: synoptic patterns, temperature, precipitation, winds, and their extremes. In: Lionello P (ed) The climate of the mediterranean region: from the past to the future. Elsevier, pp 301–346

Varela E, Robles-Cruz AB (2016) Ecosystem services and socio-economic benefits of Mediterranean grasslands. In: Kyriazopoulos A, López-Francos A, Porqueddu C, Sklavou P (eds) Ecosystem services and socio-economic benefits of Mediterranean grasslands. CIHEAM, pp 13–28

Vázquez-de-Aldana García-Ciudad, Pérez-Corona García-Criado (2000) Nutritional quality of semi-arid grassland in western Spain over a 10-year period: changes in chemical composition of grasses, legumes and forbs. Grass Forage Sci 55:209–220. https://doi.org/10.1046/j.1365-2494.2000.00217.x

Vázquez-De-Aldana BR, García-Ciudad A, García-Criado B (2008) Interannual variations of above-ground biomass and nutritional quality of Mediterranean grasslands in Western Spain over a 20-year period. Aust J Agric Res 59:769–779. https://doi.org/10.1071/AR07359

Sustainable Food Systems in Culturally Coherent Social Contexts: Discussions Around Culture, Sustainability, Climate Change and the Mediterranean Diet

F. Xavier Medina

Abstract In the developed countries, the large number of industrial processes and transformations of all kinds that food goes through before reaching the consumer generates in the latter a blind mistrust towards it. Actually, we can find an increasingly important movement from civil society, asking for more attention to be paid to local food and the sustainability of ecosystems and landscapes, in a context where climate change troubles are taking a leading role. In the Mediterranean basin, and from a local point of view and as a model of proximity consumption, Mediterranean food and diet can be a sustainable resource for the Mediterranean area. But this challenge requires a big effort and a very active and committed role of the public sector, combined with the private action.

1 Introduction

Food systems are changing rapidly all around the world. In the developed countries, well-provided with food, we can find a larger production and distribution of products at an industrial level and a more fluid access to a large amount of food by the public, at much more accessible prices. In this sense, the large number of industrial processes and transformations of all kinds which food goes through before reaching the consumer generates in the latter a blind mistrust towards it (Medina 2015). More and more often, people pursue going to concepts such as "sustainable", "traditional", "local", "organic" "bio"..., at the same time that green and ethical information schemes could become much more in line with the reflective nature of *green*, political consumers (Boström and Klintman 2009).

On the other hand, we know that there is growing evidence of the diets cost for the environment (including climate change), society and public health nutrition (Bottalico et al. 2016). The sustainability of dietary patterns has emerged in the

F. Xavier Medina (✉)
Universitat Oberta de Catalunya (UOC), Barcelona, Spain
e-mail: fxmedina@uoc.edu

P. Castro et al. (eds.), *Climate Change-Resilient Agriculture and Agroforestry*,
Climate Change Management, https://doi.org/10.1007/978-3-319-75004-0_11

last years as a public health nutrition challenge (Buttriss and Riley 2013; O'Kane 2012) as well as within the international debate on sustainability (Burlingame and Dernini 2011), food security and nutrition (Berry et al. 2015; FAO 2013a, b; Garnett 2013) and climate change (FAO 2016).

Finally, we also know that food and eating behaviours in general fall within the framework of the societies that produce and recreate them, and therefore within specific sociocultural systems (Medina 1996). Different (and more accelerated than ever) socio-cultural and economic factors such as urbanization, globalization, production, changes in gender or intergenerational relationships, organization of working time… have had important effects on lifestyles, leading also important changes in food patterns and behaviour.

This international debate has also emerged in the Mediterranean area, where one of the most important challenges is still food and nutrition security (FAO 2012a, b), but also sustainability (Burlingame and Dernini 2011) or cultural coherency and food heritage (Medina 2009). Enhancing the transition towards more sustainable food systems in the Mediterranean area requires a development of holistic approaches within different spheres and arenas of agriculture, economy, culture and lifestyle, environment, climate change, nutrition and health (Bottalico et al. 2016). This is crucial for designing cross-sectorial policy instruments allowing the improvement of the sustainability of the diets and food systems (Adinolfi et al. 2015).

1.1 *What Is a* Sustainable Diet*?*

The incorporation of sustainability issues into the International agri-food and nutritional agenda has been increasingly discussed over the last decades. The concept of sustainable diets acknowledges the interdependencies of food production and consumption with food requirements, and nutrient recommendations (Dernini et al. 2016), and at the same time, expresses the notion that food (including production, distribution, and consumption, social and cultural aspects, health, or economy, among others) cannot work separately from that of ecosystem.

As FAO has highlighted in today's most accepted definition of sustainable diets, established after the International Scientific Symposium on Biodiversity and Sustainable Diets united against Hunger: "Sustainable diets are protective and respectful of biodiversity and ecosystems, culturally acceptable, accessible, economically fair and affordable; nutritionally adequate, safe and healthy; while optimizing natural and human resources" (Burlingame and Dernini 2011). After this main definition, some authors, like Jonston et al. (2014) adds that sustainable diets must be also culturally sensitive and acceptable.

As in different other fields and items, cultural aspects have traditionally been neglected, observed only as subservient or complementary to other, more important items. Even after the declaration by UNESCO of the Mediterranean Diet as intangible cultural heritage of the Humanity, definitions of diet, sustainable diets, or the Mediterranean Diet—even the more open ones; even those drafted by supranational institutions—continue to relegate to the background those aspects more closely linked to culture. In this chapter, we will focus on the social and cultural perspective of food and its relationship with diets, territories, sustainability and climate change.

1.2 Climate Change and Sustainable Food Production

Experts warn that climate change is real and it is already happening. Since the 1950s, the global temperature of the atmosphere and oceans has increased, the volume of snow and ice have diminished, sea level has risen and heavy rains have increased (IPCC 2014). The agricultural production is being negatively affected by higher temperatures, greater frequency of heat waves, changes in the frequency and quantity of rainfall, frequency of droughts, rising sea levels and salinization of agricultural land and aquifers. Therefore, it will become more difficult and more expensive to grow, raise animals, managing forests and fish in the same places as before (FAO 2016).

The models predict that climate change, combined with the expected increase of food request, will affect food security, reducing biodiversity and the availability of water, and loosing nutritional value of some crops due to the impoverishment of the soil (IPCC 2014).

Together with a greater awareness of this effect, the interest in sustainable diets has raised. Already in 2008, the Report of the FAO Regional Conference for Europe made important statements about sustainable diets: "that the goal of increased global food production, including biofuels, should be balanced against the need to protect biodiversity, ecosystems, traditional foods and traditional agricultural practices". On the other hand and also under the umbrella of the FAO, in 2010, a common scientific position was reached on the definition of "sustainable diets": "*Sustainable diets are those diets with low environmental impacts which contribute to food and nutrition security and to healthy life for present and future generations. Sustainable diets are protective and respectful of biodiversity and ecosystems, culturally acceptable, accessible, economically fair and affordable; nutritionally adequate, safe and healthy; while optimizing natural and human resources*" (FAO and Bioversity 2012).

Supply chains are complex and varied, and food supply chains are especially challenging because of seasonality, freshness, spoilage, and sanitary considerations. As Wakeland et al. (2012: 233) pointed out, the transportation-related carbon footprint varies from a few percent to more than half of the total carbon footprint associated with food production, distribution, and storage, measuring transportation-related carbon footprint involves careful choice of the scope of the analysis, and there is much uncertainty in the results.

Nevertheless, reducing food losses and waste, and promoting a transition to more sustainable diets, can also deliver emissions reductions and contribute to global food security (FAO 2016). The footprint of local production and consumption seems to be low, less aggressive with the environment and acting in favour the social, cultural and economic development of the local societies.

1.3 The Case of the Mediterranean Diet

Within the international debate on a shift towards more sustainable food systems and diets, interest in the Mediterranean diet as a model of a sustainable dietary pattern seems to be increased. The notion of the Mediterranean diet has undergone a progressive evolution over the past 60 years—from a coronary healthy dietary pattern to the present model of a sustainable diet (Dernini et al. 2016).

Even if many studies have shown that the Mediterranean Diet seems to have a lower environmental impact than other dietary patterns (Tilman and Clark 2014; Heller et al. 2013) in the Mediterranean area, and some experts agrees also that the Mediterranean Diet as a model should provide more environmental benefits in the Mediterranean region, characterized by an increasing water scarcity (Capone et al. 2013), it is also true that Mediterranean diets are not homogeneous, and of course there are not all the same in what regards environmental sustainability, and the differences between meat/fish-based and plant-based diets.

The urban-rural dichotomy or the socioeconòmic differences among the countries of the region plays also an important role to have in mind. The water footprint of consumption, for example, varies greatly among the different countries of the region. About 91% of the regional water footprint of consumption is due to agricultural products consumption (Hachem et al. 2016). The increase in food demand will have effects on the volumes of water used for irrigation. Meat, dairy products and wheat represent more than a half of the water footprint of food supply in Mediterranean countries (Lacirignola et al. 2014). Nevertheless, Roberto Capone et al. (2013) analysed the case study of the environmental cost for Italy, in terms of water consumption, of non-adherence to the Mediterranean dietary pattern by comparing the estimated water footprint of the traditional diet and that of the current dietary pattern: the result was that the latter is about 70% higher than that of the "ideal" diet.

But even if ideally acceptable, this intended sustainability has not only an environmental dimension, but also social and cultural. A seasonal consumption of fresh and local products, diversity of foods, traditional dishes, conviviality, represent the cornerstone of conserving the Mediterranean diet heritage. Until the present, the Mediterranean diet has been observed as a healthy model of medical behaviour. Nevertheless, and after its declaration as a Cultural Heritage of Humanity at UNESCO, the Mediterranean diet is actually being (and must be) observed as a part of Mediterranean culture, and opening the concept as an equivalent of the Mediterranean cultural food system or Mediterranean culinary system (Medina 2009, 2015; Dernini et al. 2016).

The Mediterranean Diet was inscribed in 2010 on the UNESCO's Representative List of Intangible Cultural Heritage of Humanity as "*a set of skills, knowledge, practices and traditions ranging from the landscape to the table, including crops, harvesting, fishing, conservation, processing, preparation and, particularly, consumption of food […]. This unique lifestyle determined by the Mediterranean climate and space, is also shown through the associated festivities and celebrations*" (Mediterranean Diet 2010). Besides, through its social and cultural functions as well as its significance, it embodies landscapes, natural resources and associated occupations as well as the fields of health, welfare, creativity, intercultural dialogue and at the same time values such as hospitality or conviviality, sustainability or biodiversity (Serra-Majem and Medina 2016).

After this definition, the Mediterranean diet is a concept that tries to embrace biodiversity, sustainability, quality, palatability, health and cultural heritage. Safeguarding the Mediterranean diet should be the driving force behind responsible sustainable consumption (González Turmo and Medina 2012). From a local, Mediterranean point of view, and as a model of proximity consumption, Mediterranean foods and diets can be a sustainable development resource for the Mediterranean area (Medina 2011).

Consumption, as part of the Mediterranean diet, cannot be separated from production, distribution or other social and cultural factors that have built historically around food in the Mediterranean region. In this sense, the Mediterranean diet is not simply a set of healthy nutrients, but a complex web of cultural aspects that depend on each other and lead from nutrition to the economy, through law, history, politics or religion. This point of view has to be highlighted in future discussions about the Mediterranean diet, its challenges and its future perspectives.

1.3.1 Culturally "Coherent", More Than Culturally "Acceptable"

As in different other fields and items, cultural aspects have traditionally been neglected, observed only as subservient or complementary to other, more important items. Even after the declaration by UNESCO of the Mediterranean Diet as intangible cultural heritage of the Humanity, definitions of *diet, sustainable diets*, or the *Mediterranean Diet*—even the more open ones; even those drafted by supranational institutions—continue to relegate to the background those aspects more closely linked to culture. In this sense: Sustainable diets are protective and respectful of biodiversity and ecosystems, *culturally acceptable*,[1] accessible, economically fair and affordable; nutritionally adequate, safe and healthy; while optimizing natural and human resources (FAO 2011; Burlingame and Dernini 2011). Jonston et al. (2014) adds that food must be *culturally sensitive and acceptable*.

But, more than "acceptable", food (but not only) must be culturally "coherent" (Medina 2015). The Cambridge Dictionary (online edition) defines "acceptable" as:

[1]Our italics.

"satisfactory and able to be agreed to or approved of", or simply "just good enough, but not very good". In this sense, if something must be agreed or approved, usually it is because it is not taking part of the system itself, and must be *accepted* from the outside. Or, as the second of the meanings cited above explains very well, it is "just good enough, but not very good".

On the other hand, something "coherent": "it is clear and carefully considered, and each part of it connects or follows in a natural or reasonable way". In this sense, many things may be *acceptable*, but very few are *coherent*. From a local and sustainable point of view, in addition, betting on the cultural "acceptability" of a food can open up too much the spectrum of what is acceptable as edible. But its cultural coherence within a system appeals to other aspects that have nothing to do with what is simply acceptable.

2 Conclusion

The Mediterranean Diet is equivalent of Mediterranean culinary system. A broader knowledge and promotion of this reality as a system would make a significant contribution to greater sustainability and footprint of Mediterranean food production and consumption in the Mediterranean area, in addition to other possible and well-known benefits.

Local production and consumption seems to have a lower environmental footprint and acts in favour the social, cultural and economic development of the local societies. In this sense, local production and consumption act in favour of a greater sustainability and a lower environmental footprint, but also in favor of a greater conservation of heritage and cultural coherency (from the field to the plate).

In this sense, we reviewed the definition of the Mediterranean Diet as a whole, inclusive and interdependent subject. From this point of view, none of the particular elements that compose this heritage must be considered separately or individually: from the production to the consumption; from the origin to the eventual recycling.

From a local Mediterranean point of view and as a model of proximity consumption, Mediterranean food and diet can be a sustainable resource for the Mediterranean area.

References

Adinolfi F, Capone R, El Bilali H (2015) Assessing diets, food supply chains and food systems sustainability: towards a common understanding of economic sustainability. In: Meybeck A, Redfern S, Paoletti F, Strassner C (eds) Proceedings of international workshop "Assessing sustainable diets within the sustainability of food systems—Mediterranean Diet, organic food: new challenges. 15–16 Sept 2014, Rome. FAO, Rome, pp 167–175. Available at http://www.fao.org/3/a-i4806e.pdf#page=221

Berry E, Dernini S, Burlingame B, Meybeck A, Conforti P (2015) Food security and sustainability: can one exist without the other? Public Health Nutr 18(13):2293–2302

Boström M, Klintman M (2009) The green political food consumer. Anthropology of food, S5, September 2009. Available at http://aof.revues.org/index6394.html. Accessed 15 May 2017

Bottalico F, Medina FX, Capone R, El Bilali H, Debbs Ph (2016) Erosion of the Mediterranean Diet in Apulia Region, south-eastern Italy: socio-cultural and economic dynamics. J Food Nutr Res 4(4):258–266 (10.12691)

Burlingame B, Dernini S (2011) Sustainable diets: the Mediterranean Diet as an exemple. Public Health Nutr 14(12A):2285–2287

Buttriss J, Riley H (2013) Sustainable diets: Harnessing the nutrition agenda. Food Chem 140:402–407

Capone R, El Bilali H, Debs P, Cardone G, Berjan S (2013) Nitrogen fertilizers in the Mediterranean Region: use trends and environmental implications. In: Fourth international scientific symposium Agrosym 2013, Jahorina, Bosnia and Herzegovina, Faculty of Agriculture, University of East Sarajevo, pp 1143–1148

Dernini S, Berry E, Serra Majem L, La Vecchia C, Capone R, Medina FX, Aranceta J, Belahsen R, Burlingame B, Calabrese G, Corella D, Donini LM, Meybeck A, Pekcan AG, Piscopo S, Yngve A, Trichopoulou A (2016) Med Diet 4.0. The Mediterranean diet with four sustainable benefits. Public Health Nutr 20(7):1322–1330

Food and Agriculture Organization of the United Nations (2011) International Scientific Symposium. Biodiversity and Sustainable Diets United Against Hunger. FAO. Rome. Source http://www.fao.org/ag/humannutrition/28506-efe4aed57af34e2dbb8dc578d465df8b.pdf. Accessed 10 Apr 2017

Food and Agriculture Organization of the United Nations (2012a) Greening the economy with agriculture. Working paper 4: utilization. Improving food systems for sustainable diets in a green economy. FAO, Rome. www.fao.org/docrep/015/i2745e/i2745e00.pdf

Food and Agriculture Organization of the United Nations (2012b) Towards the future we want. End hunger and make the transition to sustainable agricultural and food systems. FAO, Rome. http://www.fao.org/docrep/015/an894e/an894e00.pdf

FAO and Bioversity (2012) Sustainable diets and biodiversity. Directions and Solutions for Policy, Research and Action. FAO, Rome. http://www.fao.org/docrep/016/i3004e/i3004e00.htm

Food and Agriculture Organization of the United Nations (2013a) The State of Food and Agriculture. Food systems for better nutrition. FAO, Rome

Food and Agriculture Organization of the United Nations (2013b) Food security and nutrition in the Southern and Eastern Rim of the Mediterranean Basin. FAO, Cairo

Food and Agriculture Organization of the United Nations (2013c) Tackling climate change through livestock. A global assessment of emissions and mitigation opportunities. Rome: FAO. Source http://www.fao.org/3/8d293990-ea82-5cc7-83c6-8c6f461627de/i3437e.pdf. Accessed 25 Apr 2017

Food and Agriculture Organization of the United Nations (2016) The state of food and agriculture 2016. Climate Change, Agriculture and Food Security. Avalilabre at http://www.fao.org/3/a-i6030e.pdf. Accessed 18 May 2017

Garnett T (2013) Food sustainability: problems, perspectives and solutions. Proc Nutr Soc 72:29–39

González Turmo I, Medina, FX (2012) Défis et responsabilités suite à la déclaration de la diète méditerranéenne comme patrimoine culturel immatériel de l'humanité (Unesco), in Revue d'Ethnoécologie, 2. https://ethnoecologie.revues.org/957. Accessed 3 Apr 2017

Hachem F, Capone R, Dernini S, Yannakoulia M, Hwalla H, Kalaitzidis Ch (2016) The Mediterranean Diet: a sustainable consumption pattern. MedTerra 2016. Bari, CIHEAM

Heller MC, Keoleian GA, Willett WC (2013) Toward a life cycle-based, diet-level framework for food environmental impact and nutritional quality assessment: a critical review. Env Sci Tech 47:12632–12647

IPCC (2014) Climate change 2014: synthesis report. Contribution of Working Groups I, II and III to the Fifth Assessment Report of the Intergovernmental Panel on Climate Change

Jonston JL, Fanzo JC, Cogill B (2014) Understanding sustainable diets: a descriptive analysis of the determinants and processes that influence diets and their impact on health, food security, and environmental sustainability. Adv Nutr July Adv Nutr 5:418–429

Lacirignola C, Dernini S, Capone R, Meybeck A, Burlingame B, Gitz V, El Bilali H, Debs P, Belsanti V (eds) (2012) Vers l'élaboration de recommandations pour améliorer la durabilité des régimes et modes de consommation alimentaires: la diète méditerranéenne comme étude pilote. Bari, CIHEAM and FAO (www.iamm.ciheam.org/ress_doc/opac_css/doc_num.php?explnum_id=9369)

Medina FX (1996) Alimentación, dieta y comportamientos alimentarios en el context mediterráneo. In: Medina FX (ed) La alimentación mediterránea. Historia, cultura, nutrición. Icaria, Barcelona, pp 21–44

Medina FX (2009) Mediterranean diet, culture and heritage: challenges for a new conception. Pub Health Nutr 12(9A):1618–1620

Medina FX (2011) Food consumption and civil society: Mediterranean diet as a sustainable resource for the Mediterranean area. Pub Health Nutr 14(12A):2346–2349

Medina FX (2015) Assessing sustainable diets in the context of sustainable food systems: socio-cultural dimensions. In: Meybeck A, Redfern S, Paoletti F, Strassner P (eds) Assessing sustainable diets within the sustainable food systems. Mediterranean Diet, organic food: new challenges. Roma, FAO, CREA and International Research Network for Food Quality and Health

Mediterranean Diet (2010) Transnational nomination. Greece, Italy, Morocco, Spain

O'Kane G (2012) What is the real cost of our food? Implications for the environment, society and public health nutrition. Public Health Nutr 15(02):268–276

Serra-Majem L, Medina FX (2016) The Mediterranean diet as an intangible and sustainable food culture. In: Preedy VR, Watson DR (eds) The Mediterranean diet: an evidence-based approach. Academic Press-Elsevier, London, pp 37–46

Tilman T, Clarck M (2014) Global diets link environmental sustainability and human health. Nature 515:518–522

Wakeland W, Cholette S, Venkat K (2012) Food transportation issues and reducing carbon footprint. In: Boye J, Arcand Y (eds) Green technologies in food production and processing. Food Engineering Series. Springer, Boston

Multifunctional Urban Agriculture and Agroforestry for Sustainable Land Use Planning in the Context of Climate Change in Serbia

Jelena Živanović Miljković and Tijana Crnčević

Abstract Faced with the various developments and modern lifestyle trends of urban dwellers toward healthy food and closer contact with nature, people in urban areas have evolved extensive forms of urban agriculture and agroforestry to meet their needs. Urban agriculture and agroforestry systems, as a part of multifunctional landscapes, have various positive effects (they support urban climate amelioration and short food chains, provide fresh food, contribute to urban economic growth, conserve biodiversity, and have therapeutic qualities). However, competing land use needs and the high value of land in urban areas are the major challenges for the development of urban agriculture and agroforestry. In this chapter, the authors analyse urban agriculture and agroforestry as multifunctional and sustainable land use options for urban areas in Serbia. Through selected case studies, the authors give an overview of the physical, planning and institutional capacities for urban agriculture and agroforestry development in Serbia. The authors also identify trends in urban agriculture and agroforestry within the context of the impact of climate change on multifunctional landscapes.

Keywords Urban agriculture and agroforestry · Multifunctional land use
Climate change · Sustainable land use planning · Serbia

1 Introduction

Multifunctional agriculture (MFA) has become an important topic in international political and scientific discourse, extensively propounded in Europe. As a land use, agriculture is multifunctional, offering both commodity (food and fibre) and non-commodity outputs, such as environmental and rural amenities, food security and contribution to rural viability (Maier and Shobayashi 2001). Multifunctional

J. Živanović Miljković (✉) · T. Crnčević
Institute of Architecture and Urban & Spatial Planning of Serbia,
Bulevar Kralja Aleksandra 73/II, 11000 Belgrade, Serbia
e-mail: jelena@iaus.ac.rs

P. Castro et al. (eds.), *Climate Change-Resilient Agriculture and Agroforestry*,
Climate Change Management, https://doi.org/10.1007/978-3-319-75004-0_12

land use is related to multiple concepts, with strong links to the concept of sustainability (Maier and Shobayashi 2001; OECD 2001; Piorr et al. 2006; Turpin et al. 2010, etc.). While sustainability is a resource-oriented, long-term and global concept, multifunctionality is an economic activity-oriented concept that refers to specific properties of the production process [emphasising joint production (e.g. Turpin et al. 2010; OECD 2001)] and its multiple outputs (Maier and Shobayashi 2001). Hence, the multifunctional role of agriculture is understood as the range of associated environmental, economic and social functions of agriculture, as following the concept of sustainability (Silber and Wytrzens 2006: 31; Piorr et al 2006: 48; Zasada 2011, etc.).

Urban agriculture (UA) is a broad term which is defined as the growing, processing and distribution of food and other products through intensive plant cultivation (nonfood plant and tree crops) and animal husbandry in and around cities (Urban Agriculture Committee of the Community Food Security Coalition 2003: 3; Mougeot 2006: 4), e.g. on urban and peri-urban land. That is why the main challenge and opportunity in recent planning practice has been to design urban agriculture spaces to be multifunctional, matching the specific needs and preferences of local residents, while also protecting the environment (Lovell 2010). In that sense, UA should be viewed as a tool[1] contributing to sustainable urban development (Mougeot 2006: 10; Zivanovic Miljkovic et al. 2012). However, despite the growing interest in UA because of its enormous benefits for individuals and communities, it has been documented that urban agriculture is largely ignored in urban and regional planning (Lovell 2010) and agricultural policy (FAO 2008).

On the other hand, *agroforestry* (AF) combines the best practices of tree growing and agricultural systems, resulting in more sustainable use of land. It is defined as the collective name for land use systems and technologies where woody perennials (trees, shrubs, palms, bamboos, etc.) are deliberately used on the same land management units as agricultural crops and/or animals, in some form of spatial arrangement or temporal sequence (Nair 1994: 14; FAO 2013). Three main types of AF are recognized: (1) agrisilvicultural systems which represent a combination of crops and trees; (2) silvopastoral systems where forestry and the grazing of domesticated animals on pastures, rangelands or on-farm is combined; and (3) agrosylvopastoral systems where trees, animals and crops are integrated.

The integration of agriculture and forestry can provide numerous benefits, because the land can simultaneously be used for many purposes and also make easier the transition from one type of crop to another as the market demands (Drazic et al. 2013: 867). AF provides various benefits: (a) helps protect and sustain agricultural productive capacity; (b) ensures food diversity and seasonal nutritional security; (c) diversifies rural incomes and enhances productivity; (d) strengthens resilience to climatic fluctuations; (e) helps perpetuate local knowledge and social

[1]In Africa, UA is largely seen as a socio-economic problem, resulting from a failure to adequately address rural development needs (Mougeot 2006), while in most other examples it is viewed as a sustainable land use solution.

and cultural values; (f) often contributes to climate change mitigation; (g) supports biodiversity within the ecosystem; (h) contributes to the landscape characteristic, and this increase provides opportunities for recreational uses (Nair 1994: 153; Kandji et al. 2006: 23; Verchot et al. 2007: 901; FAO 2013; Drazic et al. 2013).

Multifunctional landscape approaches are increasingly applied to agroecosystems, but also in sustainable planning of urban ecosystems (Lovell and Taylor 2013). Urban agriculture and agroforestry systems, as a part of multifunctional landscapes (Silber and Wytrzens 2006; Lovell and Taylor 2013, etc.), provide various positive effects such as food security, economic benefits, health and wellness, environmental benefits, neighborhood revitalization, fresh food, and short food chains.

Urban and especially peri-urban agriculture is confronted with disadvantages (e.g. increased competition for land) as well as advantages (e.g. larger market) caused by urbanization (Van Huylenbroeck et al. 2005). There is a high degree of land use transition and conversion for urban purposes (housing, commercial development), so land use planning represents the mechanism that has an impact on the reduction of pressure on land resources[2] (Zivanovic Miljkovic et al. 2012). Hence, Lovell (2010: 2511) argues for integrating UA directly into the planning of green infrastructure (GI) in cities,[3] even though UA features (e.g. allotment gardening, community gardening) are not always given the same level of importance as other open green space (Lovell 2010).[4]

Further, urban agriculture and agroforestry have to compete in the land market with other non-agricultural land uses[5] that provide greater profit for the landowner. Therefore, land speculation is very common in urban and peri-urban areas, where there is a strong financial incentive for farmers to sell land for purposes of urban development (Zasada 2011: 640).

With these issues in mind, in this chapter the authors perceive and document the current state of physical, planning and institutional capacities for urban agriculture and agroforestry development in Serbia through the prism of two key research concepts—multifunctionality and sustainability. At the same time, the authors perceive trends in urban agriculture and agroforestry in the context of climate change-impacts on their role in multifunctional landscapes. In three case studies the authors show UA and AF application in Serbia and stress the positive role of urban

[2]Advanced and integrated land use, planning and natural resource management have a critical role in reducing non-adequate soil use (Zivanovic Miljkovic 2008).

[3]Lovell and Taylor (2013: 1488) document that the loss of agriculture will place greater pressure on urban green spaces, so urban GI will have a critical role to play in various functions, e.g. 'ecosystem services'.

[4]As a result, although there are thousands of community gardens across North America, only a few cities include them in their urban development plans and fewer still protect these features through zoning (Hou et al. 2009 according to Lovell 2010).

[5]In Denmark, for example, protection against land use change depends on the status of allotment gardens because legal regulation ensures that the majority of Denmark's allotment gardens, mainly state-owned, are permanent (Drilling et al. 2016).

planning in such applications. Hence, this chapter presents a complementing research of UA and AF from a spatial/urban planning viewpoint, which can be considered as its main contribution.

1.1 Methodological Approach

Based on background analysis and review of the relevant literature, the authors give an overview of viewpoints and definitions of the multifunctionality and sustainability of agriculture and agroforestry. Next, the general context of agriculture and agroforestry in terms of climate change in Serbia is reviewed. The authors apply a comprehensive and in-depth analysis of the national institutional framework of agriculture and agroforestry policies in Serbia, with special reference to aspects of climate change. This approach is based on comparative and critical law analysis, as well as on the current strategic documents of national importance that direct the planning practice. This is followed by a review of the multifunctional and sustainable role of agriculture and agroforestry within existing planning practice in Serbia. Although sustainable land use has a good institutional background, this overview paper indicates legal constraints that do not fully recognize UA and AF as sustainable and multifunctional land use options in Serbia. At the same time, the authors stress the positive role of spatial/urban planning in promoting UA and AF as sustainable land use.

2 General Context of Agriculture and Agroforestry in Terms of Climate Change in Serbia

Although AF has a long tradition—over 6000 years, and is seen today as a measure that falls within environmental protection, the implementation of sustainable development and climate change, it is insufficiently applied in Serbia and its region (Drazic et al. 2013). In Serbia, AF is not applied in a systematic, organized way or across large areas, but sporadically, intuitively and by smaller households (Ibid., 868), mainly due to an inadequate and, primarily, legal and planning basis. Thus, in the province of Vojvodina AF wind protection belts are present whose role is, primarily, to prevent the dispersal of land and prevent wind erosion while, at the same time, to improve the quality and yield of crops.

Regarding the development of renewable energy sources (RES) the main preconditions are given within the *National Action Plan for renewable energy sources of the Republic of Serbia* (NAPRES) and further in the *National Strategy for the inclusion of the Republic of Serbia in the Clean Development Mechanism* (CDM)—*waste management, agriculture and forestry*, which introduce the framework for projects for, among others, biomass utilization. Biomass as RES defines it is the "biodegradable fraction of products, waste and residues of biological origin from agriculture (including vegetal and animal substances), forestry and related industries, as well as the biodegradable fraction of industrial and municipal waste" (article 2 of the *Law of Energy*).

It is important to stress that "this type of biomass is not specifically considered and singled out as the potential of biomass from forestry and agriculture" within official data regarding the potentials of Republic of Serbia in RES (Crncevic et al. 2016a: 231). Further, the legal framework (*Law on Energy*) does not provide adequate conditions for their cultivation (Ibid.). However, the potential for biomass production—agroforestry is seen within unused arable land—250,000 ha or 4.9% of the total arable land for which there is no special plan to be used for energy purposes (RS 2013). Also, raising forest plantations of fast growing species is found to be the most frequently applied measure within for re-cultivation of degraded land of mining or other activities, such as are ongoing projects—growing of poplar and willow within the process of re-cultivation of "Kolubara"mining basin or reed Miscanthus (Miscanthus giganteus) on landfill of ash and slag in Veliki Crljeni within thermal power plant "Nikola Tesla B", etc. (RB Kolubara 2012; Jurekova and Drazic 2011, etc.).

The significance of AF in the context of climate change[6] is of great importance. Thus, the establishment of protective field belts could reduce treated areas and thus influence the reduction of fuel consumption, pesticides and fertilizers, so that from an economic standpoint it is cost-effective, providing higher yields and lower costs. It should also be noted that AF can be classified within CDM projects—in Serbia approved projects include new afforestation of bare land, production of biomass, etc. (E-kapija 2010).

Climate changes that for Serbia include an increase in temperature as well changes within the precipitation regime will cause a more frequent occurrence of drought so that the sites with the most favorable bioclimatic conditions in the period 2071–2100 will correspond to the sites that had the least favorable conditions in the 20th century (RS 2017). With the aim of adapting forest ecosystems, as adaptation measure, among others, are changes promoted within management practices and raising public awareness of the multiple ecosystem services provided by forests and, as well, on their multifunctionality (Ibid.).

The main preconditions for the inclusion of the climate change issue within spatial and urban planning are given by the *Spatial Plan of the Republic of Serbia* followed by the set of laws covering planning (*Law on planning and construction*), environmental and nature protection (*Law on Environmental Protection*, *Law on*

---

[6]In Serbia, special efforts are directed towards the harmonization of policy development with respect to the issue of climate change. As a signatory of the established global framework—the Kyoto Protocol, The United Nations Framework Convention on Climate Change (UNFCCC), as well the most recent Paris Agreement as a result of the 21st Conference of the Parties of the United Nations Framework Convention on Climate Change (UNFCCC) (Serbia is among 194 countries that have signed, although it has still not ratified) (UNFCCC 2017) necessary reforms are being implemented continuously.

Nature Protection, *Law on Strategic Environmental Assessment*, *Law on Environmental Impact Assessment*).

It should be stressed that Serbia, following the Kyoto Protocol, as a member of the group of non-developing countries (non-Annex I countries) can use one of the three marker-based mechanisms—CDM without obligation for greenhouse gas reduction. Within the CDM project, so far, the most represented is the project for biomass production (E-kapija 2010).

Regarding climate change, the strengths that are stressed within current planning practice are the presence of measures such as developing ecological networks, increasing protected areas, ecosystem protection, monitoring invasive species, and planning measures for their suppression, while the main limitations are found within missing data regarding the impacts of climate change (Crncevic 2013; Crncevic et al. 2016b).

3 Institutional Framework for Sustainable Land Use in Serbia

The Republic of Serbia, in accordance with its European integration process, continuously carries out activities aimed at the innovation of the strategic, legislative and planning framework. The national development policy defined by strategies, aims to establish the basic conditions for the implementation of the global adopted strategy of sustainable development—*National Strategy for Sustainable Development* and *Action Plan for the period 2009–2017*, *National Strategy for Sustainable Use of Natural Resources and Properties, Biodiversity Strategy for the period 2011 to 2018*, as well as environmental protection—*National program of Environmental Protection for the period 2010–2019* and climate change adaptation—*National Strategy for inclusion of Serbia within the clean development mechanism* (CDM), and others. Coordinated operation of agricultural policy and environmental policy in the field of sustainable land use,[7] as well as the inclusion of these actions in spatial development policy is characterized as necessary.

3.1 Agriculture

As defined by the *Law on Agricultural Land*, agricultural land is a *resource of public interest* to the Republic of Serbia, which is used for agricultural production.

[7]As a multiple resource, land is managed through *land policy*, concerning key issues regarding sustainable *use*, *regulation* (land use planning), security and equitable distribution of *land rights* and *access to land*, including the forms of *tenure* under which it is held (Zivanovic Miljkovic and Popovic 2014).

The *Spatial Plan of the Republic of Serbia*, as the primary planning document that sets out a long-term basis for the organization, development, use and protection of space, recognizes the sustainable use and protection of natural resources as the main objectives of planned development of the country. Among these objectives, the priority is the protection of agricultural land and the conservation of its biodiversity for food production, which is threatened by the expansion of settlements, mining activities, industry, and greenfield investments. The key measure is to prevent uncontrolled and irrational coverage of quality agricultural land buildings, primarily control of the scope and structure of a greenfield investment.

The protection of agricultural land as a basic natural resource for food production is a priority given by all strategic documents. The *Law on Agricultural Land* prohibits the use of arable land of the highest quality for non-agricultural purposes, except in cases stipulated by law and in cases when public interest is determined (e.g. when an urban plan has such stipulations), with the payment of compensation for land use changes.

The strategic objectives of sustainable land use are defined by the *National Strategy for Sustainable Development of the Republic of Serbia* and relate to the prevention of further loss of land and the preservation and improvement of its quality, as well as the prevention of environmental degradation and changes in land use and agricultural land. Efficient land management and the increase of the availability of land resources are the key elements of the agricultural policy of Serbia and the priority areas of policy changes in the future. The *Strategy for Agriculture and Rural Development of the Republic of Serbia 2014–2024* advocates for a higher degree of agricultural land use, more efficient land use of poor quality or non-arable agricultural land, and the controlled re-appropriation of agricultural land.

In terms of management of agricultural land in *private ownership*, it is provided that the owner or user of agricultural land is required to cultivate arable land regularly and to implement appropriate measures. Arable land that is not cultivated in the preceding vegetation period may be leased to the competent Ministry for a period of up to three years, with payment of rent to the land owner, and after deduction of the costs of the proceedings. The main objectives of agricultural land in *public ownership* as the efficient use of the land in accordance with the principles of sustainability, increasing possession of farms and the enlargement of the parcels. Agricultural land in public ownership is managed by the competent Ministry, it cannot be sold, but it can be used in procedures such as: (a) the lease for a period of one to 40 years; (b) making it available for use at no charge; (c) transfer of the ownership right between public agencies, organizations and public enterprises.

According to the Republic of Serbia Treasury data, in 2012 in the register of agricultural holdings 2480 million hectares were registered, of which the leased was 773,603 ha. Of this, about 40% of agricultural land is state owned. On the basis of these data, *The Strategy of Agriculture and Rural Development of the Republic of Serbia* (2014: 18) assesses that the market of agricultural land, in terms of the lease, is active.

The *Law on Incentives in Agriculture and Rural Development* predicts agro-environmental measures will support and obliges users of incentives to respect regulations governing environmental quality standards and the protection of public health, animal and plant health, animal welfare and agricultural land.

3.2 Agroforestry and Green Infrastructure

The *Forestry Development Strategy* promotes the role of forests in the context of climate change mitigation and supports their continuous improvement. Recognizing the multifunctionality of forests, the Strategy encourages the promotion of cooperation with other sectors (such as agriculture and tourism), as well with the wider public with an aim of using other resources of forest areas and, especially, highlights the importance of intensive forest plantations as renewable energy sources.

Regarding agroforestry, it should be stated that within the legal framework the main preconditions are given within the *Law on Forests*. However, within the Law the term agroforestry is not applied. AF support is contained in Article 5 where "under trees forest nurseries within the forest complex and seed orchards are consided, as well as protective belts of trees with an area greater than 5 acres". Further, the Law indirectly provides the main preconditions for the production of fast growing forests by establishing the functions and purposes of forests—economic and forests with special purposes. Article 33 allows "for the planting of forest tree species for short production within runs of up to ten years and intended for production of certain assortments and special projects for the planting of short production cycles". Also, Article 52 states that razing, browsing or acorn collecting in the woods can be done "only with the permission of the owner, i.e. the forest user, who can give license only if grazing, browsing or acorn collecting are included within forest management plans and if the forest is not in the recovery phase."

Further, indirectly agroforestry, the actual growing of energy crops is promoted within:

(1) the *Spatial Plan of the Republic of Serbia* by promoting RES and defining the obligation for developing a database for regional plans of local potential within RES; more than 200,000 ha of land near rivers and canals are found to be suitable for energy crops; (2) the *Law on Planning and Construction* by promoting optimal use of renewable energy sources; and (3) the *Law on Agricultural Land* as a measure for protection of land from erosion allows the raising of protective forest belts on agricultural land, even on land of high quality.

3.3 Resume

Previous analyses show that in Serbia a regulatory framework for the planning, use and protection of agricultural land is established. Still, there is not an appropriate legal base for the planning of GI in Serbia. The basic law—*the Draft law on the Protection and Improvement of Green Areas* has not been entered into procedure yet. The existing framework for the planning of public green areas within urban and spatial planning includes primarily the *Law on the Spatial Plan of the Republic of Serbia* and the *Law on Environmental Protection* while the *Law on Planning and Construction* establishes the basis, primarily, in the context of planning urban infrastructure. Despite lack of a legal framework, numerous examples of good practices that affirm the role and importance of GI are present (Crncevic et al. 2015; Maric et al. 2015, Crncevic and Sekulic 2012; Manic et al. 2011; Crncevic and Bakic 2008; etc.).

4 Sustainable Land Use Application at Various Levels—Examples of Multifunctional Urban Agriculture and Agroforestry in Serbia

4.1 Public and Institutional Green Spaces

An example of good practice in terms of innovation of GI planning is the *Plan of Detailed Regulation of Block 23 (settlement Rasadnik 2)* at Bela Crkva, which promotes the development of agriculture within green areas of mixed and multi-family housing. The organizational concept provides an open block, with existing multi-family buildings and a kindergarten that are freely placed in space and separated by free and green areas. The greening of the area is based on respecting the setting of the plan at a higher level, in this case *the Master Plan of Bela Crkva*, which outlines the interconnection of green areas and their equally distribution within the village. The scope of the plan takes into account the established system of the green areas at the level of the Master plan, the green areas of limited use—a green area within the complex of the kindergarten and green areas among housing structures, as well as public green areas (street greenery). Following the basic concept promoted by the Master Plan, greening is planned through the participation of autochthonous woody species at 50–80%. Free surfaces—green areas within housing areas, owned by the Municipality of Bela Crkva, allows the local population to use it for agricultural production so that, within the unplanned and undeveloped land, small family vegetable gardens are partly represented (Fig. 1).

The plan provides the further use of this land for the purpose of agricultural production as experimental gardens. In this regard, taking into account that this is a state-owned land, there is planned revitalization and reconstruction of the area where the Municipality will lease land to interested residents of the block. This plan, as stated, “shows the specific application of greenery, its commercial function as farmland—kitchen gardens, orchards and nurseries in the block” (Manic et al. 2011: 74).

Fig. 1 Agriculture on public plots in Bela Crkva (photo by B. Manic)

Such a planning approach represents a significant step forward in terms of innovation of the usual planning practice system of green areas/GI. Taking into account the fact that the current legal framework does not establish a GI planning framework and that urban agriculture is "not recognized" within the system, through this approach and with such planning solutions, planning, and thus the legal base and implementation framework are provided, since the adoption of the plan results in normative obligations.

4.2 *Urban Agriculture and Agroforestry as Part of the Green Infrastructure*

The example of the town of Bor is unique as the area is characterized by its Mining Metallurgical Basin (MMB), which covers the north-eastern part of the city and is also the area with the largest open pit mines in Serbia. In an earlier period, within the area of the city of Bor, the system of green areas was not established so their development was based on the "intensive greening of open spaces and special traffic routes and the formation of protective green belts at the contact between housing and environmentally aggressive production." Within current spatial and urban planning documents, the system of green spaces-green infrastructure in Bor has been established, including urban green areas within city building land and other green areas outside the boundary of city building land.

The *Master Plan of the City of Bor* covers 1337.90 ha of the outter building land is about 280 ha of agricultural land and 412 ha of forest land. Within other green spaces there are woody, shrubby and herbaceous vegetation. Taking into account the planned expansion of commercial, residential and business zones, there will be a reduction of the existing surface area of forest and agricultural land. The basic limitation for the sustainable use of forest and forest land is, *inter alia*, the high extent of damage to the forest fund, the land, wildlife and ecosystems by the former development of MMB, as well as the continuing tendency of drying forests due to air pollution, the development of plant diseases, the harmful insect populations and the strengthened impact of other threatening factors, including climate change. In addition, there is uncertainty as to the real sources of funding due to the lack of an efficient system of institutional and financial support for multi-purpose use of the total forest resources at the local, regional and national level.

The plan of the use of forest land is based on the concept of integrating forestry within the environmental policy (Fig. 2). Implementing this concept requires taking measures that result in improving the condition of existing forests and the planting of new forests on lands that are overgrown with shrub and other vegetation with poor protection and economic power, as well by the establishment of plantation for biofuel production, and in part on agricultural land with major limitations for regular cultivation.

The Plans promote biomass plantations within the area outside the border of the building area, covering approximately 30 ha (9% of the total surface of the land outside the borders of building land). It should be stressed that these areas are planned (reserved) locations for development and that further developments require additional project development, such as conducting basic soil studies and remediation of contaminated land.

This case study shows that by providing a planning framework for biomass production agroforestry production has the potential to offer certain incentives to local governments regarding further development and promoting agroforestry production.

4.3 Private Parcels in Peri-urban Area

Urban and peri-urban gardening is not a new trend in Serbia. Various forms of urban gardens existed during the 20th century, usually private and family run, mostly for growing flowers and medicinal and spice herbs, but also fruits trees and vegetables. Mass apartment construction in the period of socialism did not favor establishing gardens within the scope of the block settlements, so wealthier citizens chose to grow their vegetables and flowers in the backyards of their weekend houses, which were built in ecological oases in the peri-urban area (Popovic and

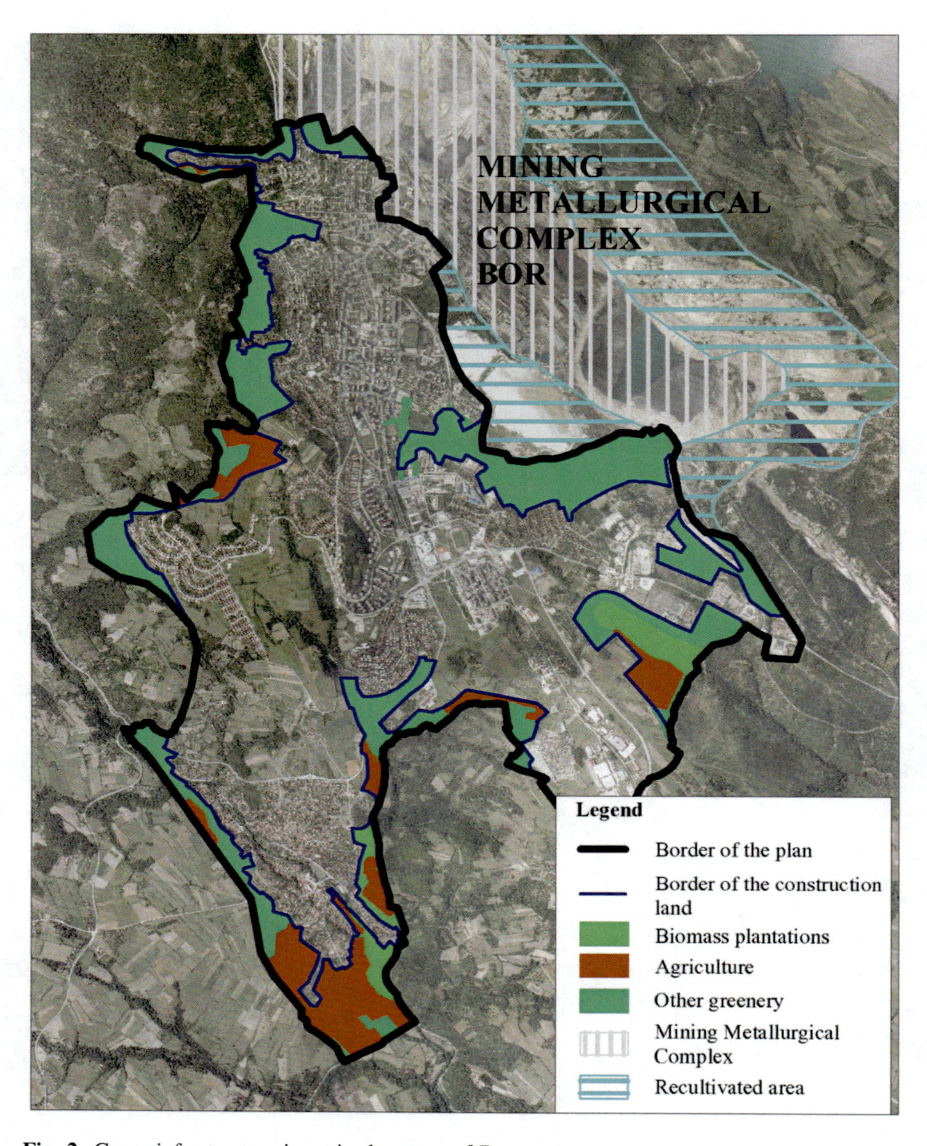

Fig. 2 Green infrastructure in peri-urban area of Bor

Zivanovic Miljkovic 2013). In the period of post-socialist transition, since the 1990s, due to lack of money, residents illegally occupied areas of urban construction land that had not been put to use, and transformed them into small allotments, where they grew basic vegetables without soil quality control or basic infrastructure requirements (Ibid.). Since there are no legal regulations on urban gardens (community gardens, allotment gardens), at either the national and local level, their status is undefined as to whether they are illegal or the initiative of local residents.

In the city of Belgrade, where strong land competition is present, private initiative has taken advantage of various benefits offering community gardening and provided private gardening space for interested residents. In 2012 the project "Bastaliste" in the Slanci settlement was started (10 km from the city centre), which is the first community garden in Serbia.[8] Within an area of 8.8 ha there are 22 plots where tenants cultivate and grow vegetables under the principles of organic production and permaculture (Bastaliste 2017).

To allow for the rational and appropriate use of agricultural land, the *Strategy of Agricultural Development of the City of Belgrade until 2015* established the leasing free of charge of the small state owned plots as low-income family farms. The strategic planning document, the *Master Plan of Belgrade* 2021 envisages development of a program for the establishment of allotment gardens, divided into small plots for several years' use by individuals, under the scope of the agricultural and rural enclaves around the city center and the other predominantly residential areas. The results on the operationalization and implementation of these commitments have not been registered so far (Popovic and Zivanovic Miljkovic 2013).

5 Conclusions

Starting from the fact that the literature overview has documented the various positive functions provided by UA and AF, in this chapter the authors primarily consider urban agriculture and agroforestry as multifunctional and sustainable land use options and present some possibilities for planning of urban and peri-urban areas in Serbia.

Although sustainable land use has a good institutional background in Serbia, in this review chapter the authors indicate the legal constraints that do not fully recognize urban agriculture and urban agroforestry as sustainable and multifunctional land use options.

Still, urban planning, as a public mechanism that has impacts on sustainable land use, could have a positive role in considered issues. Selected case studies have shown that UA and AF can be integrated in many different forms. The lack of legal regulations related to urban gardens, leads to the fact that only private initiative will implement their multifunctionality. The examples also show that urban planning has a decisive role in avoiding spatial conflicts in urban and peri-urban areas by integrating urban agriculture and agroforestry issues into plans. Hence, processes of spatial and urban planning should continue to consider the value of urban gardening

[8]Interest in the community garden concept is spreading to other cities in Serbia. The city of Sabac is the only city in the Republic of Serbia with community gardens on a public plot. Leasing agreements for plots are available for three years, with the possibility of extension (City of Sabac 2016).

and AF, even if they are economically less valuable, primarily in terms of food production, as well as climate amelioration.

Acknowledgements The paper was prepared as a result of work on scientific projects: "The role and implementation of the national spatial plan and regional development documents in renewal of strategic research, thinking and governance in Serbia" (No. III 47014) and "Sustainable spatial development of the Danube region in Serbia" (No. TR 36036), financed by the Ministry of Education, Science and Technological Development of the Republic of Serbia in the period 2011–2019.

References

Bastaliste (2017) First community garden in Serbia. http://bastaliste.org/about. Accessed 20 May 2017

City of Sabac (2016) First year of the urban garden. http://sabac.rs/aktuelnosti/vesti/prva-godina-gradske-baste.htm. Accessed 20 May 2017

Crncevic T (2013) Planning and protection of nature, natural values and landscape in the context of climate change in Republic of Serbia—contribution to the development of methodological framework. Speacial editions, No 72. IAUS, Belgrade (in Serbian)

Crncevic T, Bakic O (2008) The system of green surfaces in spas with special reference to the case studies: Vrnjacka, Kanjiza and Pribojska spa. Spatium 17(18):92–97

Crncevic T, Sekulic M (2012) Green Roofs in the context of climate change—overview of new experiences. Arhitektura i urbanizam 36:57–67 (in Serbian)

Crncevic T, Manic B, Maric I (2015) Green walls of urban spaces in the context of climate change—an overview of the actual planning frameworks and experiences. Arhitektura i urbanizam 41:40–48 (in Serbian)

Crncevic T, Jokic V, Bezbradica L (2016a) Planning energy crops in Serbia with special reference to the fast growing forest. In: Stevic Z (ed) Proceedings 4th International Conference on Renewable Electrical Power Sources, Belgrade, October 2016, Union of Mechanical and Electrotechnical Engineers and Technicians of Serbia (SMEITS) Society for Renewable Electrical Power Sources, pp 229–237

Crncevic T, Dzelebdzic O, Milijic S (2016b) Planning and climate change: a case study on the spatial plan of the danube corridor through Serbia. In: Leal Filho W et al (eds) Implementing climate change adaptation in cities and communities, climate change management, Springer International Publishing Switzerland, pp 161–177

Drazic D, Cule N, Veselinovic M, Rakonjac Lj, Bojovic S, Mitrovic S, Todorovic N (2013) Agroforestry—possibilities of multifunctional land use. In: Kovacevic D (ed) Book of proceedings IV international symposium "Agrosym 2013", Jahorina, pp 867–872

Drilling M, Giedych R, Ponizy L (2016) The idea of allotment gardens and the role of spatial and urban planning. In: Bell S, Fox-Kämper R, Keshavarz N, Benson M, Caputo S, Noori S, Voigt A (eds) Urban allotment gardens in europe. Routledge, New York, pp 35–61

E-kapija (2010) CDM projects in Serbia. http://www.ekapija.com/website/sr/company/photoArticle.php?id=349989&path=ekologija_150414.j. Accessed 15 Feb 2017

FAO (2008) Urban agriculture for sustainable poverty alleviation and food security. WB paper, http://www.fao.org/fileadmin/templates/FCIT/PDF/UPA_-WBpaper-Final_October_2008.pdf. Accessed 20 May 2017

FAO (2013) Advancing agroforestry on the policy agenda a guide for decision-makers. By G. Buttoud, in collaboration with O. Ajayi, G. Detlefsen, F. Place & E. Torquebiau. Agroforestry Working Paper no 1. FAO, Rome

Jurekova Z, Drazic G (ed) (2011) External and internal factors influencing the growth and biomass production of short rotation woods genus Salix and perennial grass Miscanthus. Faculty of applied ecology FUTURA, Singidunum University Belgrade. Available via https://is.uniag.sk/dok_server/slozka.pl?id=8311;download=6339;z=1. Accessed 20 Dec 2016

Kandji ST, Verchot LV, Mackensen J (2006) Climate change and variability in southern africa: impacts and adaptation in the agricultural sector. UNEP, Nairobi

Lovell ST (2010) Multifunctional urban agriculture for sustainable land use planning. Sustainability 2:2499–2522

Lovell ST, Taylor JR (2013) Supplying urban ecosystem services through multifunctional green infrastructure in the United States. Landscape Ecol 28:1447–1463

Maier L, Shobayashi M (2001) Multifunctionality. Towards an analitycal framework. OECD, Paris

Manic B, Crncevic T, Nikovic A (2011) The role of green spaces within the spatial-functional concept of block 23 in Bela Crkva. Arhitektura i urbanizam 33:67–74 (in Serbian)

Maric I, Crncevic T, Cvejic J (2015) Green infrastructure planning for cooling urban communities: overview of the contemporary approaches with special reference to Serbian experiences. Spatium 33:55–61

Mougeot LJA (2006) Growing better cities: Urban agriculture for sustainable development. International Development Research Centre, Ottawa

Nair RPK (1994) An introduction to agroforestry. Kluwer Academic Publishers, Dordrecht

OECD (2001) Multifunctionality0 Towards an analitycal framework. OECD, Paris

Piorr A, Uthes S, Waarts Y, Sattler C, Happe K, Müller K (2006) Making the multifunctionality concepts operational for impact assessment. In: Meyer BC (ed) Sustainable land use in intensively used agricultural regions. Wageningen, Alterra, pp 47–54

Popovic V, Zivanovic Miljkovic J (2013) Community gardening and urban permaculture design. In: Cvijanović D, Subić J, Vasile AJ (eds) International scientific conference sustainable agriculture and rural development in terms of the republic of Serbia strategic goals realization within the Danube region—achieving regional competitiveness, thematic proceedings. Institute of Agricultural Economics, Belgrade, pp 1265–1282

RB Kolubara (2012) First planted fast-growing forests. www.rbkolubara.rs/list-kolubara/?p=7186. Accessed 22 Dec 2016

Silber R, Wytrzens HK (2006) Supporting multifunctionality of agriculture in intensively used urban regions—a case study in Linz/Urfahr (Upper Austria). In: Meyer BC (ed) Sustainable land use in intensively used agricultural regions. Alterra Reports No. 1338, Landscape Europe, Wageningen

Turpin N, Stapleton L, Perret E, van der Heide CM, Garrod G, Brouwer F, Voltr V, Cairol D (2010) Assessment of multifunctionality and jointness of production. In: Brouwer FM, van Ittersum M (eds) Environmental and agricultural modelling: integrated approaches for policy impact assessment. Springer, Dordrecht, pp 11–36

UNFCCC (The United Nations Framework Convention on Climate Change) (2017) Paris agreement—status of ratification. Alailable via: http://unfccc.int/paris_agreement/items/9485.php. Accessed 15 Aug 2016

Urban Agriculture Committee of the Community Food Security Coalition (CFSC) (2003) Urban agriculture and community food security in the United States: farming from the city center to the urban fringe. Available at: http://community-wealth.org/sites/clone.community-we alth.org/files/downloads/report-brown-carter.pdf. Accessed 20 Dec 2016

Van Huylenbroeck G, Van Hecke E, Meert H, Vandermeulen V, Verspecht A, Vernimmen T, Boulanger A, Luyten S (2005) Development strategies for multifunctional agriculture in peri-urban areas. Available via: http://www.belspo.be/belspo/organisation/publ/pub_ostc/cpagr/rappcp18r_en.pdf. Accessed 15 Sep 2016

Verchot LV, Van Noordwijk M, Kandji S, Tomich T, Ong C, Albrecht A, Mackensen J, Bantilan C, Anupama KV, Palm C (2007) Climate change: linking adaptation and mitigation through agroforestry. Mitig Adapt Strat Glob Change 12:901–918
Zasada I (2011) Multifunctional peri-urban agriculture—a review of societal demands and the provision of goods and services by farming. Land Use Policy 28(4):639–648
Zivanovic Miljkovic J (2008) Some measures for soil regulation in Belgrade peri-urban zone. Spatium 17(18):68–71
Zivanovic Miljkovic J, Crncevic T, Maric I (2012) Land use planning for sustainable development of peri-urban zones. Spatium 28:15–22
Zivanovic Miljkovic J, Popovic V (2014) Land use regulation and property rights regime over land in Serbia. Spatium 32:22–27

National regulative, strategies and plans

Biodiversity Strategy for the period 2011 to 2018, Official Gazette of RS, No 13/11
Forestry Development Strategy, Official Gazette of RS, No 59/2006
Law on Agricultural Land, Official Gazette of the Republic of Serbia, No. 62/2006, 65/2008—other law, 41/2009, 112/2015
Law on Energy, Official Gazette of the RS, No 145/2014
Law on Environmental Impact Assessment, Official Gazette of RS no. 135/2004.36/2009
Law on Environmental Protection, Official Gazette of RS no. 135/2004… 14/2016
Law on forests, Official Gazette of the RS no. 30/2010, 93/2012, 89/2015
Law on Incentives in Agriculture and Rural Development, Official Gazette of the Republic of Serbia, No. 10/2013, 142/2014, 103/2015
Law on Nature Protection, Official gazette of RS no. 36/2009, 88/2010, 91/2010—correction and 14/2016
Law on planning and contruction, Official Gazette of the RS no.72/09, 81/09—correction, 64/10, 24/11,121/12, 42/13, 50/13, 98/13, 132/14 and 145/14
Law on Spatial plan of the Republic of Serbia, Official Gazette of the Republic of Serbia, No. 88/10
Law on Strategic Environmental Assessment, Official Gazette of RS no. 88/2010
Master plan of Belgrade 2021, Official Gazette of the City of Belgrade, No. 27/03, 25/05, 34/07, 63/09 and 70/14
Master Plan of the City of Bor, Official Gazette of the Bor municipality no. 20/2015 and 21/2015
National Strategy for Suistanable Development of the Republic of Serbia, Action plan for the period 2009–2017, Official Gazette of RS, No 57/2008
National Strategy for Sustainable Use of Natural resourses and properties, Official Gazette of RS, No 33/2012
National Strategy for the inclusion of the Republic of Serbia in the Clean Development Mechanism (CDM)—waste management, agriculture and forestry, Official Gazette of the RS, No 8/2010
National Sustainable Development Strategy of the Republic of Serbia, Official Gazette of the Republic of Serbia, No. 57/08
Plan of Detailed Regulation of Block 23 (setllement Rasadnik 2) in Bela Crkva, Official Gazette of the municipalty Bela Crkva, No 7/2011
RS (Republic of Serbia) (2013) National action plan for renewable energy sources of the republic of Serbia (NAPRES). Available via: http://energetskiportal.rs/dokumenta/Strategije/Nacionali%20akcioni%20plan%20za%20obnovljive%20izvore%20energije.pdf. Accessed 15 Sep 2016
RS (Republic of Serbia) (2017) Second report of republic of serbia after united nations framework convention on climate change. Available via: http://www.klimatskepromene.rs/wp-content/uploads/2017/04/SNC_na-misljenje.pdf. Accessed 15 Apr 2017

Strategy of Agricultural Development of the City of Belgrade unitil 2015. Available via: http://www.beograd.rs/lat/gradska-vlast/1613406-strategija-razvoja-poljoprivrede-grada-beograda-do-2015/. Accessed 20 Nov 2013
Strategy of agriculture and Rural Development 2014-2020, Official Gazette of the Republic of Serbia, No. 85/2014

Jelena Živanović Miljković holds B.Sc., M.Sc. and Ph.D. degrees in Spatial Planning from the University of Belgrade, Faculty of Geography. She works as a Research Fellow at the Institute of Architecture and Urban & Spatial Planning of Serbia. Her research interests are focused on spatial, urban and peri-urban planning, land use planning, land policy, land development, and land property rights. She has published numerous scientific papers on these topics, both as author and co-author. She is a member of the International Academic Association on Planning, Law and Property Rights (PLPR).

Tijana Crnčević is currently a Senior Research Fellow at the Institute of Architecture and Urban & Spatial Planning of Serbia. Her main research interests include landscape and green infrastructure planning, environmental management and sustainable development. She holds a Ph.D. in Environmental Management from the University of Belgrade, Serbia, a Master of Science in Environmental Assessment and Management and a Certificate in Strategic Environmental Assessment from the School of Planning, Oxford Brookes University, Oxford, United Kingdom and the diploma of Graduate Engineer of Forestry, Specialist in Landscape Architecture from the University of Belgrade, Forestry Faculty, Department of Landscape Architecture. She has over 15 years of experience in planning and research in the field of spatial and urban planning.

Alien Plant Species: Environmental Risks in Agricultural and Agro-Forest Landscapes Under Climate Change

Joana R. Vicente, Ana Sofia Vaz, Ana Isabel Queiroz, Ana R. Buchadas, Antoine Guisan, Christoph Kueffer, Elizabete Marchante, Hélia Marchante, João A. Cabral, Maike Nesper, Olivier Broennimann, Oscar Godoy, Paulo Alves, Pilar Castro-Díez, Renato Henriques and João P. Honrado

Abstract Alien plant species have been essential for farming and agro-forestry systems and for their supply of food, fiber, tannins, resins or wood from antiquity to the present. They also contributed to supporting functions and regulating services (water, soil, biodiversity) and to the design of landscapes with high cultural and scenic value. Some of those species were intentionally introduced, others arrived accidentally, and a small proportion escaped, naturalized and became invasive in natural ecosystems—these are known as invasive alien species (IAS). Here, invasive means that these species have some significant negative impact, either by spreading from human-controlled environments (e.g. fields, gardens) to natural ecosystems, where they can cause problems to native species, or to other production systems or urban areas, impacting on agricultural, forestry activities or human

J. R. Vicente (✉) · A. S. Vaz · A. R. Buchadas · P. Alves · J. P. Honrado
Research Network in Biodiversity and Evolutionary Biology, Research Centre in Biodiversity and Genetic Resources (InBIO-CIBIO), Campus Agrário de Vairão, Rua Padre Armando Quintas, Vairão PT4485-661, Portugal
e-mail: jsvicente@fc.up.pt

J. R. Vicente · A. S. Vaz · J. P. Honrado
Faculty of Sciences, University of Porto, Rua do Campo Alegre, s/n, Porto 4169-007, Portugal

J. R. Vicente · J. A. Cabral
Laboratory of Applied Ecology, CITAB—Centre for the Research and Technology of Agro-Environment and Biological Sciences, University of Trás-os-Montes e Alto Douro, Vila Real, Portugal

A. I. Queiroz
IHC-FCSH, NOVA de Lisboa, Avenida de Berna 26C, Lisbon 1069-061, Portugal

A. Guisan · O. Broennimann
Department of Ecology and Evolution, Biophore, University of Lausanne, 1015 Lausanne, Switzerland

P. Castro et al. (eds.), *Climate Change-Resilient Agriculture and Agroforestry*, Climate Change Management, https://doi.org/10.1007/978-3-319-75004-0_13

health. Socio-environmental impacts associated with plant invasions have been increasingly recognized worldwide and are expected to increase considerably under changing climate or land use. Early detection tools are key to anticipate IAS and to prevent and control their impacts. In this chapter, we focus on crop and non-crop alien plant species for which there is evidence or prediction of invasive behaviour and impacts. We provide insights on their history, patterns, risks, early detection, forecasting and management under climate change. Specifically, we start by providing a general overview on the history of alien plant species in agricultural and agroforestry systems worldwide (Sect. 1). Then, we assess patterns, risks and impacts resulting from alien plants originally cultivated and that became invasive outside cultivation areas (Sect. 2). Afterwards, we provide several considerations for managing the spread of invasive plant species in the landscape (Sect. 3). Finally, we discuss challenges of alien plant invasions for agricultural and agroforest systems, in the light of climate change (Sect. 4).

Keywords Ecosystem service · Impact assessment · Introduction history
Plant invasions · Predictive modelling · Remote sensing

A. Guisan · O. Broennimann
Institute of Earth Surface Dynamics, Geopolis, University of Lausanne, 1015 Lausanne, Switzerland

C. Kueffer
Department of Environmental Systems Science, Institute of Integrative Biology, ETH Zurich, 8092 Zurich, Switzerland

E. Marchante · H. Marchante
Department of Life Sciences, Centre for Functional Ecology, University of Coimbra, Calçada Martim de Freitas, Coimbra 3000-456, Portugal

H. Marchante
Escola Superior Agrária, Instituto Politécnico de Coimbra, Bencanta, 3045-601 Coimbra, Portugal

M. Nesper
Ecosystem Management, Institute of Terrestrial Ecosystems, ETH Zurich, Universitaetstrasse 16, 8092 Zurich, Switzerland

O. Godoy
Instituto de Recursos Naturales y Agrobiología de Sevilla (IRNAS-CSIC), Av. Reina Mercedes 10, 41080 Seville, Spain

P. Castro-Díez
Department of Life Sciences, Faculty of Sciences, University of Alcalá, 28805 Alcalá de Henares, Madrid, Spain

R. Henriques
Departamento de Ciências da Terra, Instituto de Ciências da Terra, Universidade do Minho, ICT/CCT/UM), Braga, Portugal

1 Introduction

For millennia the introduction of alien species has been part of human culture and subsistence. The intentional cultivation of alien plant species has enhanced the local production of food, fiber and fuel throughout the history of agriculture and agroforestry. Over time, this practice has inevitably lead to the selection of crop and tree cultivars that could adapt to and thrive in new areas with compatible environmental conditions, even if these were often distinct from those observed in the native range (Davis and Landis 2011). Some currently traded crops were once widespread weeds in disturbed ecosystems, characterised by particular features that make them well-adapted and fast-growing plants outside their native range (Doebley 2006). Similarly, a broad range of tree species that currently form the basis of many forestry enterprises have been planted for centuries to obtain timber and other products and services (e.g. erosion prevention; (Richardson 2011).

These long-term processes of human introductions and alien species adaptations to novel environments have thus shaped many contemporary agricultural and agroforest landscapes. However, current fluctuations in market demands have been pushing the trade of alien crops beyond the limits of local-scale economies (Davis and Landis 2011). Also, the increasing market competition for faster and higher production of commercial wood products led to large scale afforestation with alien tree species (Brundu and Richardson 2016). As a result, modern agriculture and agroforestry are grounded on the regional maintenance and international trade of many alien crops and trees (Davis and Landis 2011; Richardson 2011).

Depending on several ecological, economic, social and political factors, modern agriculture, agroforestry and the related trade can promote plant invasions, i.e. the spread of alien crops and trees outside their cultivation areas, leading to major impacts on the environment and society (Richardson and Rejmánek 2011). Besides threatening native biodiversity and ecosystems (Pysek et al. 2008; Simberloff et al. 2013), these plant invasions have been associated to drastic changes in the state of ecosystems and of their associated social systems, with various consequences for human culture, health and economy (Pejchar and Mooney 2009; Vaz et al. 2017).

Since invasive plants are particularly relevant in agroforestry, many cultivated trees rank among the most problematic invasive species world-wide (Richardson and Rejmánek 2011), leading to decreases in the production of forest resources and respective revenues (Garcia et al. 2010; Garcia-Llorente et al. 2011). There is also an increasing attention devoted to problematic trends of plant invasions in agricultural systems (Chytrý et al. 2009). Alien agricultural crops are introduced more frequently and at a higher abundance to new ranges, leading to higher propagule pressure, which increases invasion risk in land surrounding agricultural landscapes. Associated with these introductions there is also an increased unintentional transfer of pathogens, weeds and insect pests (Davis and Landis 2011; Garcia et al. 2010).

Although many currently traded crops and trees do not show an invasion potential or risk, some of these species may change their behaviour under climate change, for instance through adaptive physiological responses (Davis and Landis

2011; Krumm and Vítková 2016). Other species will profit from new climates, resulting in new invasions (Hannah et al. 2002) with consequences to human societies, economies and the environment. How problematic those new invasions driven by climate change will be depends on the characteristics of the species and on the effectiveness of invasive plant management in agricultural and agroforestry systems and in the surrounding landscapes. It is now widely recognised that the most time- and cost-efficient management option for potentially invasive species is to prevent the introduction and spread of those species as early as possible (Vicente et al. 2016; Hulme 2006; Davies and Sheley 2007).

Therefore, predicting and identifying future potentially invasive species, and anticipating where new invasions will most likely occur (Theoharides and Dukes 2007) under scenarios of climate and landscape change, is a major challenge in invasion science. Novel and more robust modelling tools (e.g., by means of species distribution models; (Kolar and Lodge 2001; Shmueli 2010) are being developed in the context of invasion risk assessment. These tools open promising avenues to minimize undesirable consequences of potential future invasions, thereby supporting efficient management that secures native biodiversity and promotes ecosystem functioning and the resulting services (MA 2005) Theoharides and Dukes 2007).

This chapter is organised as follows: it starts by providing a general overview on the history of alien plant species in agricultural and agroforestry systems worldwide (Sect. 1). Then, it focuses on the patterns, risks and impacts derived from alien plants that were originally cultivated but became invasive outside cultivation areas (Sect. 2). Subsequently, the chapter provides considerations for managing the spread of invasive plant species in the broader landscape (Sect. 3). Finally, this chapter discusses current and future challenges of alien plant invasions for agricultural and agroforest systems, in the light of climate change (Sect. 4), encouraging the wide application of predictive models from which management strategies can be designed to restore ecosystems structure, functions and services.

2 A Brief Historical Perspective on the Introduction of Alien Plant Species

Agriculture arose approximately 10,000 years ago. From the main centers of origin of Neolithic Agriculture practices, it has spread to most other regions of the world, transporting plants with edible seeds, fruits, leaves, stems and roots to new ranges (Mazoyer and Roudart 2006). In these new areas, introduced species together with domesticated local species and varieties ensured the survival of settlers. Agroforestry, i.e., the practice of cultivating tree species and agricultural crops in intimate combination, was an integrated land use that showed several advantages with the ultimate objective of food production in several territories since ancient times (Smith 2010). People cultivated food crops and created pastures on the cleared areas inside woodlands, where the soils were deep and rich in humus. Native and alien trees and shrubs (e.g. brooms, reeds, Indian-figs and Iron trees) were also used as fences to protect crops and cattle from strong winds and sunstrokes.

Until Modern times, agriculture or agroforestry diversification developed slowly but steadily. In the 18th and mainly in the 19th century, cumulative factors contributed to the globalization of these activities. Portugal, Spain, France, Netherlands and the United Kingdom were then European empires and developed colonial science and agriculture transferring live beings to and from their colonies in the four other continents and islands (Crosby 2004; Grove 1995; Murphy 2007; Rotherham 2011). The emergence of the concept of acclimatization led to "an intended and 'scientific' mediated transplantation of organisms" (Osborne 2000). Acclimatization societies, frequently supported by the states, promoted the exchange of plants for horticulture, satisfying aesthetical pleasing through collecting and gardening, and adapting new species to create more diverse, resilient and productive farming systems. Increasing transportation of people and livelihoods, and the booming transatlantic trade, also facilitated intentional and accidental introductions. Additionally, "philosophical travels" were made by naturalists in charge of finding, collecting and transporting species that could provide cultural and economic benefits in the arrival areas. Living plants and seeds were then cultivated in botanical gardens and later, if successful, distributed by farmers.

Towards the end of the 19th century, while agronomy became a science, and scarcity and hunger were a reality, farmlands and cultivated forests expanded with a whole new species composition (field crops and trees). Among intentionally introduced alien species, and others that arrived in association with them, a few escaped from gardens, fields and greenhouses and became invasive.

3 Invasive Plant Species Originally Cultivated in Production Systems and Their Impacts Outside Crop Areas

3.1 The Most Wide-Spread Plant Invasions

Biological invasions by weeds are intimately related with human history, as humans have not only enabled transportation of organisms from one site to another, they have also imprinted changes in the landscape that have facilitated the establishment of those organisms. When human settlements began to thrive, there was a huge increase of area occupied by crops and pastures at the expense of the natural forests, which originated the landscapes that we know today (Meeus et al. 1990). Weeds were the first plants introduced outside their native area, a process that started in ancient historical periods, making it difficult to determine the origin of some of those species. The first flow of weeds presumably occurred between Asia and

Europe, although there are indications of plant movements in the American Continent because of the intense agricultural activity practiced by the pre-Columbian cultures (Elton 2000). At the end of the 15th century, the exploration and colonization of distant areas of the globe began to substantially increase the number of naturalized species able to colonize areas outside their native range (Cousens and Mortimer 1995).

Some plant species that colonize agricultural fields are among the most common invasive species worldwide. Weed species from tropical areas such as C4 grasses are successful in many agricultural areas around the globe because of their drought tolerance and efficient photosynthetic pathway (Maillet and Lopez-Garcia, 2000). Some of these weed species with shorter lifespans have evolved to support environments other than their native areas, such as some C4 grasses (e.g. *Echinochloa crus-galli*) that are colonizing areas further north, as a result of the increased catalytic efficiency of some enzymes (Lee 2002).

Other species introduced by their interest for humans are nowadays a nuisance for agriculture and silviculture. Among those are the species of genus *Opuntia*, from America, and of genus *Acacia*, from Australia (Thuiller et al. 2006). In the majority of areas of the world, most weeds in a region are not native but were introduced by humans, often deliberately (Mack and Lonsdale 2001). And although there are common invasive weeds across the globe (e.g. *Acacia mearnsii*, *Arundo donax*, *Eichornia crassipes*, *Opuntia stricta*), others are only considered a problem in certain parts of the world (e.g. *Ulex europaeus*, *Euphorbia esula*), highlighting human facilitation as a major driver (Lowe et al. 2000).

3.2 Impacts on Ecosystem Services

Ecosystem services (ES) are the material and non-material contributions that ecosystems bring to human development and wellbeing (MA 2005). There are different frameworks to categorise ES, such as the *Millennium Ecosystem Assessment* (MA 2005), *The Economics of Ecosystems and Biodiversity* (TEEB 2010) and the *Common International Classification of Ecosystem Services* (Haines-Young and Potschin 2013).

These frameworks usually recognise three ES categories—provisioning, regulating and maintenance, and cultural services (MA 2005; TEEB 2010). Provisioning services include material and energy outputs from ecosystems (e.g., timber, food, water for consumption), whereas regulating and maintenance services comprise the benefits that derive from regulatory ecosystem processes (e.g., soil stabilization, climate regulation, pollination). Cultural services comprise the intangible benefits from ecosystems such as spiritual and cognitive enrichment, aesthetics, and recreation (Haines-Young and Potschin 2013; MA 2005).

In agricultural and agroforestry systems worldwide, the intentional cultivation of alien plant species has for long been a practice to obtain specific provisioning services, e.g. timber, pulp, forage or tannins (Davis and Landis 2011). Other alien plant species have further been cultivated to maximize regulating services (e.g. soil nitrogen fixation, land stabilisation, erosion protection) besides provisioning services (Richardson 2011). At a minor scale, in particular when planted in agroforestry systems, alien plant species can further contribute to cultural services, supporting landscape amenity and recreation (Foxcroft et al. 2008; Vaz et al. 2018).

However, some cultivated alien plant species can establish and spread beyond cultivation sites, i.e. they become invasive (Brundu and Richardson 2016). These invasive alien plant species (IAS) can alter ecosystem processes, in particular by reducing native biodiversity and competing with service-providing species (Levine et al. 2003; Simberloff et al. 2013; Vilà and Hulme 2017). For instance, IAS can indirectly affect provisioning services, by minimising the availability or reducing the quality of resources that humans use outside cultivation areas. Notable examples include the depletion of water for human consumption through several *Prosopis* and *Acacia* species, or livestock poisoning by *Prosopis juglans* and *Sorghum halapense* (Carruthers et al. 2011; Davis and Landis 2011; Richardson 2011). IAS can also change regulating services, by altering and disrupting ecosystem functions and processes outside production sites. Examples include the amplification of fire regimes by *Pinus radiata*, the reduction of stream flow by *Prosopis juliflora and Arundo donax*, and altered nutrient cycles by *Sorghum halepense* (Davis and Landis 2011; Dickie et al. 2014; Richardson et al. 2014).

Although IAS can have straightforward effects on provisioning and regulating services, IAS effects on cultural services may be perceived differently by different stakeholders (Vaz et al. 2018). For example, *Acacia dealbata* and a few *Pinus* species are by some people recognised for their ornamental (e.g. due to attractive blooms) and recreation (e.g. because of shade provision) values, respectively (Dickie et al. 2014; Vaz et al. 2017). At the same time, several *Eucalyptus* and *Acacia* species are perceived by other people as undesirable for aesthetic (e.g. invaded landscapes considered as monotonous) or spiritual (invaded areas associated to the lack of a wilderness character) values, respectively (Carruthers et al. 2011; Kueffer and Kull 2017; Vaz et al. 2018).

4 Managing the Spread of Invasive Plant Species in the Broader Landscape

4.1 Strategies to Tackle Invasive Plant Species

Given the economic, ecological and social impacts of widely dispersed invasive alien plant species (IAS), it is crucial to focus the existing (often limited) resources in management strategies that allow tackling them more successfully. In this context, it is vital to prioritize the prevention of new (potentially) IAS and the early-detection of new outbreaks in order to promote effective and timely management that limits new invasions.

Preventing the introduction of IAS into new areas is surely one of the most cost-effective management strategies (Pysek and Richardson 2010). In the context of this chapter, this may include preventing the spread of alien species grown in agricultural and agro-forest landscapes into surrounding areas; these invasion events may increase in the future as some species are expected to change their behaviour under climate change (Hellmann et al. 2008). Different actions can be considered, such as the implementation of crop monitoring protocols to detect alien plants starting to establish and escape, surveillance of dispersal routes and vectors to intercept new spreading events, raising awareness of different stakeholders and communities, etc.

Early Detection and Rapid Response (EDRR) are also crucial to minimize the impacts of IAS, being highly cost-effective when compared to management costs in later phases, and thus justifying a strong investment to improve protocols and techniques available. These include space- and air-borne remote sensing platforms for species monitoring and surveillance, species distribution modelling to map areas with higher potential for invasion, and the development of species identification tools and protocols. Although the extension of surveillance areas and the large diversity of potentially IAS make EDRR difficult (Pysek and Richardson 2010), in the context of cultivated areas, potential IAS are frequently identified and easier to target, and the surrounding area may be more restricted, turning early detection more feasible. Early detection may prioritize species or areas (e.g., habitats with higher conservation value surrounding crops, areas prone to invasion, species with higher invasive risk) in order to increase its effectiveness. Rapid response should include a rapid and detailed inventory of the populations occurring in all the surrounding area in a way to prevent small scattered foci from passing unnoticed, thus precluding eradication. Citizens or specific stakeholders previously trained can also be important allies in EDRR and associated surveillance and monitoring schemes (Holcombe and Stohlgren 2009).

IAS initially used as crops, but that nowadays promote negative impacts demanding their control, can be tackled through eradication, containment, control or mitigation, depending on the invaded extension and availability of resources. Associated with any of these options, it is essential to consider the monitoring of results (to learn from experience and adapt follow-up interventions), the persistence of management interventions, and eventually actions to actively restore the invaded areas. Although all these strategies are frequently very costly, complex and time-consuming, well-planned interventions based on strong technical and scientific knowledge, planned for medium-long term and involving the different stakeholders and their interests, allow reaching good levels of success. Nevertheless, when resources are insufficient, managers should pragmatically analyse if control is cost-efficient or if inaction may be preferable.

4.2 Modelling and Detection Tools to Support Management

4.2.1 Remote Sensing Based on Airborne Sensors

In recent years, the development of unmanned aerial vehicles (UAVs) has been an important contribution to the study of the ecological dynamics of invasive species. These airborne systems have become more reliable and functional, being able to operate with low planning needs and in increasingly demanding meteorological and terrain conditions. Although they do not have the capacity to cover large areas, as satellite systems or classical aerial photography, their implementation is inexpensive and, with the right combination of aircraft and sensors, the final results can be very useful.

UAVs have all the advantages of an aerial point of view to the landscape, and since they fly significantly lower than conventional flight systems, they can be used to obtain aerial photography with a very high resolution (Gonçalves and Henriques 2015). Typically, a small fixed wing or multicopter UAV system can cover areas between 80 and 200 ha with a single battery and with about 3–6 cm/pixel of ground sampling distance (Colomina and Molina 2014). The airborne sensors have also evolved. Nowadays it is possible to use several small RGB, multispectral and hyperspectral cameras available on the market, allowing the use of aerial imagery obtained with these systems with any remote sensing computational tool for vegetation study purposes (Ahmed et al. 2017).

A major advantage of these systems is that the collection of aerial imaging can be done simultaneously with georeferenced ground fieldwork for species identification. This ensures that the remote detection final results are more robustly validated. Besides the high spatial resolution, these systems also ensure a high temporal resolution of the monitored area, by repeating flights with the same flight plan. Accurate multispectral orthophotos and three-dimensional digital surface models are usually the most common products that are obtained with these systems. These products can be processed by the most common remote sensing software.

4.2.2 Habitat Distribution Models

Habitat suitability models (HSM; Guisan et al., in press) represent important tools for anticipating plant invasions and spread (Gallien et al. 2012; Peterson 2003), and thus for the management of invasive alien plants (Guisan et al. 2013; Meier et al. 2014; Richter et al. 2013; Venette et al. 2010), especially if based on ensemble modelling (Bellard et al. 2013; Stohlgren et al. 2010). In particular, HSMs could be embedded in adaptive management frameworks connected to national and international biodiversity databases (e.g. GBIF), to direct field monitoring and eradication efforts (Honrado et al. 2016; Venette et al. 2010). Their data results could in turn update the models in an iterative and adaptive way, as already proposed for the management of rare species (Guisan et al. 2006).

Early efforts applying HSMs to anticipate invasions proposed to fit models in the native range and project them into the invaded range (Peterson 2003), but issues were raised later about the possibility that species' ecological niches, on which these models are based, may change for some species after invasion, hampering the proper application of these models (Early and Sax 2014; Guisan et al. 2014) and references therein). This supports the use of data from both the native and invaded (exotic) range for building HSMs, which can then be applied more accurately in both ranges (Beaumont et al. 2009; Broennimann and Guisan 2008). Another important argument in favor of such combined ranges approach is that it makes more likely that the full ecological niche is captured in the HSM, whereas using local data (e.g., only in the invaded range) bears the risk of fitting truncated response curves and making wrong predictions (Petitpierre et al. 2016; Thuiller et al. 2004).

In general, the study of biological invasions will require hierarchical modelling approaches combining multiple scales (Pysek and Hulme 2005). Future modelling directions include (i) adding more dynamic components, such as dispersal and population dynamics, in HSMs though the further development of hybrid models (Ferrari et al. 2014; Gallien et al. 2010; Pysek and Hulme 2005); (ii) better understanding the dynamics of niche filling and expansion during invasions (Broennimann et al. 2014); (iii) improving the environmental predictors used in HSMs, such as new remotely sensed images (Andrew and Ustin 2009; Bradley 2014), as these are also typically used in more general monitoring programs (e.g. GEOBON); and (iv) fitting invasive species models in a hierarchical manner, combining the strength of local data at fine resolution with global data (e.g., to fit the whole species climatic niche) (Petitpierre et al. 2016).

4.2.3 Dynamic Ecological Models

Dynamic models are focused on ecological processes, which differ from static models by explicitly incorporating time-dependent changes in the state of a system (Jørgensen 2008). These models include, among others, biogeochemical dynamics (e.g. Soetaert et al. 2000), population dynamics (e.g. Kriticos et al. 2009), individual-based models (e.g. Nehrbass and Winkler 2007), and cellular automata systems (e.g., Crespo-Pérez et al. 2011).

The ability to anticipate invasive plant species trends and distribution patterns as ecological indicators under different socio-ecological scenarios depends on the accuracy to predict their invasiveness and the invasibility of threatened habitats (Peterson 2003). Therefore, biological invasion studies have been improved by creating dynamic models that simultaneously attempt to capture the structure and functioning of vulnerable systems, in which the potential invasive trajectory can be predicted a priori (Santos et al. 2011).

Although static Habitat Distribution Models have been widely applied in predicting invasive plant species distributions under global change scenarios (e.g., Vicente et al. 2016), their deterministic assumptions and the lack of integration with dynamic and/or stochastic processes limit their accuracy in capturing ecological responses to more local changes (Kandziora et al. 2013). In contrast, dynamic modelling can be adjusted to the local management requirements (objectives and parameters) and capable of responding with flexibility to specific contexts, guiding current decision-making (Cuddington et al. 2013). Therefore, dynamic models can complement static models to design and test specific cost-effective local conservation measures in the scope of invasive plant species broader management (Bastos et al. 2012). In fact, the combination of empiric, mechanistic and correlative modelling approaches is a promising research field, known as hybrid multi-modelling (Box 1), that enables to improve the understanding of the main drivers implicated on the spatial-dynamic patterns of invasive plant species (Buchadas et al. 2017).

Box 1—Dynamic models for supporting invasion management: a review

Dynamic models are preferential tools to guide management decisions (Cuddington et al. 2013). Based on a literature review, in which keywords related with "Invasion", "Management" and "Dynainic Modelling" were searched using ISI WOK (http://webofknowledge.com) (Buchadas et al. 2017), 59 publications on plant invasions in agricultural and agro-forest landscapes were considered. To investigate the extent and how dynamic modelling has been applied on management, we reviewed each one of the 59 publications to identify their modelling approach (i.e., purely dynamic vs. hybrid models; Fig. 1), management type (i.e., active management targeting later stages of the invasion process, vs. passive management targeting early invasion stages), and if climate change was considered.

Dynamic models were more often applied in active management actions, such as invasion control. Contrastingly, hybrid models were more often applied in passive management, e.g. prevention and early-detection. Only 5% of publications considered climate change, which applied hybrid approaches

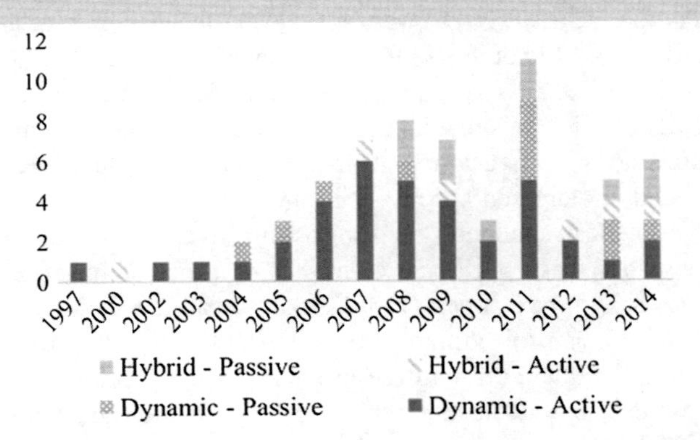

Fig. 1 Temporal evolution of the application of dynamic modelling approaches, dynamic or hybrid (static-dynamic) to support active or passive management actions on plant invasions in agricultural and agro-forest landscapes

for passive management. The ability of dynamic models to mimic demographic processes, such as dispersal and growth, is of key importance for active management measures. Still, the more recently applied, hybrid models have the capacity to deliver spatial outputs which are effective to support management at early invasion stages (from prevention to early detection), and to improve the monitoring and impact assessment of invasion processes. Furthermore, hybrid models may be relevant when considering the role of climate change, due to their capacity to effectively extrapolate invasion process beyond uncertain environmental conditions. Therefore, further use of hybrid approaches to study invasions management while considering climate change may constitute a research pathway to support invasion management in agricultural and agroforestry systems.

5 Challenges of Alien Plant Invasions in the Light of Climate Change

5.1 Forecasting

Because biological invasions are ongoing processes, anticipating their spread requires advanced predictive tools embedded in adaptive and iterative modelling frameworks (Baxter and Possingham 2011; Uden et al. 2015). These typically combine data from the native and invaded ranges (Broennimann and Guisan 2008; Beaumont et al., 2009) to limit effects of niche changes between ranges (Guisan et al. 2014). It is however not enough to anticipate spread under current climate, as predictions may differ in future climates (Bellard et al. 2013), as shown for various mountains (Petitpierre et al. 2016), for North America (Bradley et al. 2010) or for South-Africa (Parker-Allie et al. 2009). Additionally, changes in other factors, such as land use (Vicente et al. 2011) or global trade (Seebens et al. 2015), may also influence predictions. Models that incorporate climate or land use changes scenarios may also anticipate how invasions will threaten protected areas (Kleinbauer et al. 2010) or rare and endangered species (Vicente et al. 2011).

Perhaps the greatest challenge for forecasting invasions in future conditions is when novel situations—new climate or new land uses, for instance—will characterize some areas in future times (van Klinken et al. 2009; Vicente et al. 2011; Webber et al. 2011), as projections there will be entailed with greatest uncertainty. Another important challenge for future projections is to find robust ways to build hierarchical models, based on hierarchical studies of biological invasions (Pysek and Hulme 2005), that combine models of entire species' climatic niches at the global scale with local models incorporating the species requirements in the area of interest (Petitpierre et al. 2016).

5.2 Predicting Impacts on Ecosystem Services

Detailed knowledge of how climate change will mediate IAS escaped from agriculture and agroforestry systems, and how these species will affect ecosystem services, is lacking. Still, there is a clear recognition that climate change can alter habitat conditions that facilitate or hinder IAS performance, and thereby the effects of IAS on ecosystem services (Brundu and Richardson 2016; Krumm and Vítková 2016; Nie et al. 2017).

It has been shown that agricultural and agroforestry IAS, such as *Acacia dealbata*, *Robinia pseudoacacia*, *Salix alaxensis* and some *Lupinus* species, will potentially spread into areas that are currently unsuitable, due to more favourable future climate conditions (Gassó et al. 2012; Kleinbauer et al. 2010; Pickart et al. 1998; Vicente et al. 2013; Wasowicz et al. 2013). Accordingly, the impacts that these IAS currently have on ecosystem services (e.g. increased fire regimes, water depletion) will likely be altered (Hellmann et al. 2008; Nie et al. 2017).

Future research should prioritise studies on the interactions between IAS, ecosystem services, and management practices in agriculture and agroforestry systems under different climate change scenarios (Richardson et al. 2010). Such interactions should be investigated considering the complex and context-specific interlinks between ecological impacts, social valuations and perceptions, and management decisions (Dickie et al. 2014; Vaz et al. 2017).

5.3 Adaptive Management

With changing climate and with the foreseeable introduction of new alien species, the occurrence of new IAS with novel traits that behave differently under new environmental conditions (including e.g., new interactions pests and diseases; (Pauchard et al. 2016) will likely increase. Thus, management will have to be adaptive to successfully tackle species for which reliable observational data from the past do not exist (ghost of invasion past problem, (Kueffer et al. 2013). Dealing with this challenging situation will only be possible through a flexible approach that allows management priorities and strategies to be continuously adapted, considering the response of ecosystems to management actions. Adaptation needs to be based not only on new research, but also on the ways ecosystem services are perceived and valued by local stakeholders, considering the social, technological, economic, environmental and political contexts (Foxcroft and McGeoch 2011). To foster such an approach, continuous knowledge exchange between practitioners, scientists and other stakeholders is needed, data flows supported by adaptive monitoring must be enabled (Honrado et al. 2016), and legal and institutional frameworks must be adapted so that they become more flexible and focused on anticipation and early response. There is also a need to develop a more balanced approach towards evaluating invasive species and their effects on ecosystems based not only on their negative impacts but also on their positive roles in socio-ecological systems (Vaz et al. 2017).

Box 2—Alien tree species in coffee agroforestry systems—implications for ecosystem functions anil services

Coffee (*Coffea canephora* and *C. arabica*) production depends on shading. Traditionally coffee is produced under a native multispecies tree canopy, however, increasingly monospecific shade tree layers composed of ail alien species (e.g. *Grevillea robusta*) are used. Ecosystem functioning and services differ substantially between native multispecics and alien monospecific agroforestry systems, as has for instance been shown in a long-term study in Kodagu (Western Ghats, India) (Garcia et al. 2010, Krishnan et al. 2011; Boreux et al. 2013, Nesper et al. 2017). The presence of insect pollinators and their visitation rates of coffee are affected by canopy composition. While habitat quality, tor instance tor bees such as *Apis dorsata*, is improved in a diverse agroforest (Krishnan et al. 2011), pollination can sometimes be reduced under a multispecies shade cover; possibly due to floral competition with coffee plants (Boreux et al. 2013). Lower predator abundance and/or diversity might be the cause for increased in testation by the coffee pest *Hypothenemus hampei* under an alien *G. robusta* canopy (Nesper et al. 2017). Further, there is a tendency towards reduced nutrient availabilities under *G. robusta* due to the low litter quality of this alien tree (Nesper et al. in review). There are also indications that microclimate and leaf phenology peaks become more extreme under an alien canopy (Nesper et al. in review), which might have important implications for resilience against climate change in a seasonal dry climate. Effects on ecosystem services can result in reduced coffee production and quality under alien shade trees (Nesper et al. 2017). These effects are partly due to the presence of a particular alien tree species, but result partly also from the replacement of old-grown multispecies canopies.

6 Conclusions

This chapter provided a general overview on the history, patterns, risks and impacts of alien plant species in agricultural and agroforestry systems. The major highlights are that, since pre-Columbus times, plant species have been moved across the world. Still, the increasing transportation of people and livelihoods, and the booming transatlantic trade during the 18th and 19th centuries facilitated intentional and accidental introductions of alien plant species. A proportion of these species has become invasive, namely *Acacia mearnsii, Arundo donax, Eichornia crassipes*, and

Opuntia stricta, that were able to expand beyond cultivation sites. These alien invasive species have many documented impacts in biodiversity and on provisioning, regulating and cultural ecosystem services. Management strategies, such as early detection and/or rapid response actions, are thus relevant for preventing undesirable impacts from these species. Modelling and detection tools are being increasingly applied to support effective management. These tools comprise, for example, the remote sensing information captured by airborne systems, the habitat distribution models, and the dynamic models. These approaches are even more relevant in the context of climate change. Specifically, they can be implemented to forecast and anticipate invasion processes, as well as to predict and prevent impacts on ecosystem services. These emergent are thus key tools for achieving adaptive management, allowing institutional frameworks to become well-adapted, -flexible and -focused on the anticipation and early response to invasions, their costs and benefits in socio-ecological systems.

Acknowledgements Joana R. Vicente was supported by POPH/FSE and FCT (Post-Doc grant SFRH/BPD/84044/2012). Ana Sofia Vaz was supported by FSE/MEC and FCT (Ph.D. grant PD/BD/52600/2014). Ana Isabel Queiroz supported by FCT—the Portuguese Foundation for Science and Technology [UID/HIS/04209/2013 and IF/00222/2013/CP1166/CT0001]. This work received financial support from the European Union (FEDER funds POCI-01-0145-FEDER-006821) and National Funds (FCT/MEC, Fundação para a Ciência e Tecnologia and Ministério da Educação e Ciência) under the Partnership Agreement PT2020 UID/BIA/50027/2013.

References

Ahmed OS, Shemrock A, Chabot D, Dillon C, Williams G, Wasson R, Franklin SE (2017) Hierarchical land cover and vegetation classification using multispectral data acquired from an unmanned aerial vehicle. Int J Remote Sens 38:2037–2052. https://doi.org/10.1080/01431161.2017.1294781

Andrew ME, Ustin SL (2009) Habitat suitability modelling of an invasive plant with advanced remote sensing data. Diversity Distrib 15:627–640. https://doi.org/10.1111/j.1472-4642.2009.00568.x

Bastos R et al (2012) Testing a novel spatially-explicit dynamic modelling approach in the scope of the laurel forest management for the endangered Azores bullfinch (Pyrrhula murina) conservation. Biol Conserv 147:243–254. https://doi.org/10.1016/j.biocon.2012.01.009

Baxter PWJ, Possingham HP (2011) Optimizing search strategies for invasive pests: learn before you leap. J Appl Ecol 48:86–95. https://doi.org/10.1111/j.1365-2664.2010.01893.x

Beaumont LJ, Gallagher RV, Thuiller W, Downey PO, Leishman MR, Hughes L (2009) Different climatic envelopes among invasive populations may lead to underestimations of current and future biological invasions. Diversity Distrib 15:409–420. https://doi.org/10.1111/j.1472-4642.2008.00547.x

Bellard C, Thuiller W, Leroy B, Genovesi P, Bakkenes M, Courchamp F (2013) Will climate change promote future invasions? Glob Change Biol 19:3740–3748. https://doi.org/10.1111/Gcb.12344

Boreux V, Krishnan S, Cheppudira KG, Ghazoul J (2013) Impact of forest fragments on bee visits and fruit set in rain-fed and irrigated coffee agro-forests. Agric Ecosyst Environ 172:42–48. https://doi.org/10.1016/j.agee.2012.05.003

Bradley BA (2014) Remote detection of invasive plants: a review of spectral, textural and phenological approaches. Biol Invasions 16:1411–1425. https://doi.org/10.1007/s10530-013-0578-9

Bradley BA, Wilcove DS, Oppenheimer M (2010) Climate change increases risk of plant invasion in the Eastern United States. Biol Invasions 12:1855–1872. https://doi.org/10.1007/s10530-009-9597-y

Broennimann O, Guisan A (2008) Predicting current and future biological invasions: both native and invaded ranges matter. Biol Lett 4:585–589. https://doi.org/10.1098/rsbl.2008.0254

Broennimann O, Mráz P, Petitpierre B, Guisan A, Müller-Schärer H (2014) Contrasting spatio-temporal climatic niche dynamics during the eastern and western invasions of spotted knapweed in North America. J Biogeogr 41:1126–1136. https://doi.org/10.1111/jbi.12274

Brundu G, Richardson DM (2016) Planted forests and invasive alien trees in Europe: a code for managing existing and future plantings to mitigate the risk of negative impacts from invasions. In: Daehler CC, van Kleunen M, Pyšek P, Richardson DM (eds) Proceedings of 13th international EMAPi conference, Waikoloa, Hawaii, 2016. NeoBiota, pp 5–47

Buchadas A et al (2017) Dynamic models in research and management of biological invasions. J Environ Manage 196:594–606. https://doi.org/10.1016/j.jenvman.2017.03.060

Carruthers J, Robin L, Hattingh JP, Kull CA, Rangan H, van Wilgen BW (2011) A native at home and abroad: the history, politics, ethics and aesthetics of acacias. Diversity Distrib 17:810–821. https://doi.org/10.1111/j.1472-4642.2011.00779.x

Chytrý M, Pyšek P, Wild J, Pino J, Maskell LC, Vilà M (2009) European map of alien plant invasions based on the quantitative assessment across habitats. Divers Distrib 15:98–107. https://doi.org/10.1111/j.1472-4642.2008.00515.x

Colomina I, Molina P (2014) Unmanned aerial systems for photogrammetry and remote sensing: a review. ISPRS J Photogramm Remote Sens 92:79–97. https://doi.org/10.1016/j.isprsjprs.2014.02.013

Cousens R, Mortimer M (1995) Dynamics of weed populations. Cambridge University Press, Cambridge, UK

Crespo-Pérez V, Rebaudo F, Silvain J-F, Dangles O (2011) Modeling invasive species spread in complex landscapes: the case of potato moth in Ecuador. Landscape Ecol 26:1447–1461. https://doi.org/10.1007/s10980-011-9649-4

Crosby AW (2004) Ecological imperialism, 2nd edn. Cambridge University Press, New York

Cuddington K, Fortin MJ, Gerber LR, Hastings A, Liebhold A, O'Connor M, Ray C (2013) Process-based models are required to manage ecological systems in a changing world. Ecosphere 4:1–12. https://doi.org/10.1890/ES12-00178.1

Davies KW, Sheley RL (2007) A conceptual framework for preventing the spatial dispersal of invasive plants. Weed Sci 55:178–184. https://doi.org/10.1614/WS-06-161

Davis AD, Landis DA (2011) Agriculture. In: Simberloff D, Rejmánek M (eds) Encyclopeia of biological invasions. University of California Press, Berkeley and Los Angeles, pp 7–10

Dickie I et al (2014) Conflicting values: ecosystem services and invasive tree management. Biol Invasions 16:705–719. https://doi.org/10.1007/s10530-013-0609-6

Doebley J (2006) Unfallen grains: how ancient farmers turned weeds into crops. Science 312:1318. https://doi.org/10.1126/science.1128836

Early R, Sax DF (2014) Climatic niche shifts between species' native and naturalized ranges raise concern for ecological forecasts during invasions and climate change. Glob Ecol Biogeogr 23:1356–1365. https://doi.org/10.1111/geb.12208

Elton CS (2000) The ecology of invasions by animals and plants. University of Chicago Press, Chicago

Ferrari JR, Preisser EL, Fitzpatrick MC (2014) Modeling the spread of invasive species using dynamic network models. Biol Invasions 16:949–960. https://doi.org/10.1007/s10530-013-0552-6

Foxcroft LC, McGeoch M (2011) Implementing invasive species management in an adaptive management framework. Koedoe 53. https://doi.org/10.4102/koedoe.53i2.1006

Foxcroft LC, Richardson DM, Wilson JRU (2008) Ornamental plants as invasive aliens: problems and solutions in Kruger national park South Africa. Environ Manage 41:32–51. https://doi.org/10.1007/s00267-007-9027-9

Gallien L, Douzet R, Pratte S, Zimmermann NE, Thuiller W (2012) Invasive species distribution models—how violating the equilibrium assumption can create new insights. Glob Ecol Biogeogr 21:1126–1136. https://doi.org/10.1111/j.1466-8238.2012.00768.x

Gallien L, Munkemuller T, Albert CH, Boulangeat I, Thuiller W (2010) Predicting potential distributions of invasive species: where to go from here? Diversity Distrib 16:331–342. https://doi.org/10.1111/j.1472-4642.2010.00652.x

Garcia CA et al (2010) Biodiversity conservation in agricultural landscapes: challenges and opportunities of coffee agroforests in the Western Ghats. India Conservation Biology 24:479–488. https://doi.org/10.1111/j.1523-1739.2009.01386.x

Garcia-Llorente M, Martin-Lopez B, Nunes PA, Gonzalez JA, Alcorlo P, Montes C (2011) Analyzing the social factors that influence willingness to pay for invasive alien species management under two different strategies: eradication and prevention. Environ Manage 48:418–435. https://doi.org/10.1007/s00267-011-9646-z

Gassó N, Thuiller W, Pino J, Vilà M (2012) Potential distribution range of invasive plant species. Spain NeoBiota 12:25–40. https://doi.org/10.3897/neobiota.12.2341

Gonçalves JA, Henriques R (2015) UAV photogrammetry for topographic monitoring of coastal areas. ISPRS J Photogram Remote Sens 104:101–111. https://doi.org/10.1016/j.isprsjprs.2015.02.009

Grove RH (1995) Green imperialism. Colonial expansion, tropical islands Edens and the origins of environmentalism, 1600–1860. Cambridge University Press, Cambridge

Guisan A et al (2013) Predicting species distributions for conservation decisions. Ecol Lett 16:1424–1435. https://doi.org/10.1111/Ele.12189

Guisan A, Broennimann O, Engler R, Vust M, Yoccoz NG, Lehmann A, Zimmermann NE (2006) Using niche-based models to improve the sampling of rare species. Conserv Biol 20:501–511. https://doi.org/10.1111/j.1523-1739.2006.00354.x

Guisan A, Petitpierre B, Broennimann O, Daehler C, Kueffer C (2014) Unifying niche shift studies: insights from biological invasions. Trends Ecol Evol 29:260–269. https://doi.org/10.1016/j.tree.2014.02.009

Haines-Young R, Potschin M (2013) In: Common international classification of ecosystem services (CICES): consultation on version 4, Aug–Dec 2012. EEA Framework contract No EEA/IEA/09/003

Hannah L, Midgley GF, Millar D (2002) Climate change-integrated conservation strategies. Glob Ecol Biogeogr 11:485–495. https://doi.org/10.1046/j.1466-822X.2002.00306.x

Hellmann JJ, Byers JE, Bierwagen BG, Dukes JS (2008) Five potential consequences of climate change for invasive species. Conserv Biol 22:534–543. https://doi.org/10.1111/j.1523-1739.2008.00951.x

Holcombe T, Stohlgren TJ (2009) Detection and early warning of invasive species. In: Clout MN, Williams PA (eds) Invasive species management: a handbook of principles and techniques. Oxford University Press, Oxford, pp 36–46

Honrado JP, Pereira HM, Guisan A (2016) Fostering integration between biodiversity monitoring and modelling. J Appl Ecol 53:1299–1304. https://doi.org/10.1111/1365-2664.12777

Hulme PE (2006) Beyond control: wider implications for the management of biological invasions. J Appl Ecol 43:835–847. https://doi.org/10.1111/j.1365-2664.2006.01227.x

Jørgensen SE (2008) Overview of the model types available for development of ecological models. Ecol Model 215:3–9. https://doi.org/10.1016/j.ecolmodel.2008.02.041

Kandziora M, Burkhard B, Müller F (2013) Interactions of ecosystem properties, ecosystem integrity and ecosystem service indicators—a theoretical matrix exercise. Ecol Ind 28:54–78. https://doi.org/10.1016/j.ecolind.2012.09.006

Kleinbauer I, Dullinger S, Peterseil J, Essl F (2010) Climate change might drive the invasive tree Robinia pseudacacia into nature reserves and endangered habitats. Biol Cons 143:382–390. https://doi.org/10.1016/j.biocon.2009.10.024

Kolar CS, Lodge DM (2001) Progress in invasion biology: predicting invaders. Trends Ecol Evol 16:199–204. https://doi.org/10.1016/S0169-5347(01)02101-2

Krishnan S, Chengappa SK, Ghazoul J (2011) Bee diversity and the extent of forests in the context of the wider landscape matrix. Zurich

Kriticos DJ, Watt MS, Withers TM, Leriche A, Watson MC (2009) A process-based population dynamics model to explore target and non-target impacts of a biological control agent. Ecol Model 220:2035–2050. https://doi.org/10.1016/j.ecolmodel.2009.04.039

Krumm F, Vítková L (2016) Introduced tree species in European forests: opportunities and challenges. European Forest Institute, Germany

Kueffer C, Kull C (2017) Non-native species and the aesthetics of nature. In: Hulme P, Vilà M, Ruiz G (eds) Impact of biological invasions on ecosystem services. Springer, Berlin

Kueffer C, Pysek P, Richardson DM (2013) Integrative invasion science: model organisms, multi-site studies, unbiased meta-analysis, and invasion syndromes (Tansley review). New Phytol 200:615–633. https://doi.org/10.1111/nph.12415

Lee CE (2002) Evolutionary genetics of invasive species. Trends Ecol Evol 17:386–391. https://doi.org/10.1016/S0169-5347(02)02554-5

Levine JM, Vilà M, D'Antonio CM, Dukes JS, Grigulis K, Lavorel S (2003) Mechanisms underlying the impacts of exotic plant invasions. Proc R Soc B: Biol Sci 270:775–781. https://doi.org/10.1098/rspb.2003.2327

Lowe S, Browne M, Boudjelas S, De Poorter M (2000) 100 of the world's worst invasive alien species: a selection from the global invasive species database, vol 12. Invasive Species Specialist Group, Auckland

MA (2005) Ecosystems and human well-being: synthesis (millennium ecosystem assessment). Island Press, Washington, DC

Mack RN, Lonsdale WM (2001) Humans as global plant dispersers: getting more than we bargained for: current introductions of species for aesthetic purposes present the largest single challenge for predicting which plant immigrants will become future pests. Bioscience 51:95–102. https://doi.org/10.1641/0006-3568(2001)051%5b0095:HAGPDG%5d2.0.CO;2

Maillet J, Lopez-Garcia C (2000) What criteria are relevant for predicting the invasive capacity of a new agricultural weed? The case of invasive American species in France. Weed Res-Oxford 40:11–26. https://doi.org/10.1046/j.1365-3180.2000.00171.x

Mazoyer M, Roudart L (2006) A history of world agriculture: from the neolithic age to current (trans: Membrez JH). Earthscan, London

Meeus JHA, Wijermans MP, Vroom MJ (1990) Agricultural landscapes in Europe and their transformation. Landscape Urban Plann 18:289–352. https://doi.org/10.1016/0169-2046(90)90016-U

Meier ES, Dullinger S, Zimmermann NE, Baumgartner D, Gattringer A, Hulber K (2014) Space matters when defining effective management for invasive plants. Diversity Distrib 20:1029–1043. https://doi.org/10.1111/Ddi.12201

Murphy DJ (2007) Imperial botany and the early scientific breeders. In: Murphy DJ (ed) People, plants and genes. The story of crops and humanity. Oxford University Press, Oxford, pp 247–260

Nehrbass N, Winkler E (2007) Is the Giant Hogweed still a threat? An individual-based modelling approach for local invasion dynamics of Heracleum mantegazzianum. Ecol Model 201:377–384. https://doi.org/10.1016/j.ecolmodel.2006.10.004

Nesper M, Kueffer C, Krishnan S, Kushalappa CG, Ghazoul J (2017) Shade tree diversity enhances coffee production and quality in agroforestry systems in the Western Ghats. Agric Ecosyst Environ 247:172–181. https://doi.org/10.1016/j.agee.2017.06.024

Nie M, Shang L, Liao C, Li B (2017) Changes in primary production and carbon sequestration after plant invasions. In: Vilà M, Hulme PE (eds) Impact of biological invasions on ecosystem services. Springer, Cham, Switzerland

Osborne MA (2000) Acclimatizing the world: a history of the paradigmatic colonial science. Osiris 15:135–151

Parker-Allie F, Musil CF, Thuiller W (2009) Effects of climate warming on the distributions of invasive Eurasian annual grasses: a South African perspective. Clim Change 94:87–103. https://doi.org/10.1007/s10584-009-9549-7

Pauchard A et al (2016) Non-native and native organisms moving into high elevation and high latitude ecosystems in an era of climate change: new challenges for ecology and conservation. Biol Invasions 18:345–353. https://doi.org/10.1007/s10530-015-1025-x

Pejchar L, Mooney HA (2009) Invasive species, ecosystem services and human well-being. Trends Ecol Evol 24:497–504. https://doi.org/10.1016/j.tree.2009.03.016

Peterson AT (2003) Predicting the geography of species' invasions via ecological niche modeling. Q Rev Biol 78:419–433. https://doi.org/10.1086/378926

Petitpierre B, McDougall K, Seipel T, Broennimann O, Guisan A, Kueffer C (2016) Will climate change increase the risk of plant invasions into mountains? Ecol Appl 26:530–544. https://doi.org/10.1890/14-1871

Pickart AJ, Miller LM, Duebendorfer TE (1998) Yellow Bush Lupine invasion in Northern California Coastal Dunes I. Ecological impacts and manual restoration techniques. Restor Ecol 6:59–68. https://doi.org/10.1046/j.1526-100x.1998.00618.x

Pysek P, Hulme PE (2005) Spatio-temporal dynamics of plant invasions: linking pattern to process. Ecoscience 12:302–315. https://doi.org/10.2980/i1195-6860-12-3-302.1

Pysek P, Richardson DM (2010) Invasive species, environmental change and management, and health. Ann Rev Environ Resour 35:25–55. https://doi.org/10.1146/annurev-environ-033009-095548

Pysek P, Richardson DM, Pergl J, Jarosik V, Sixtova Z, Weber E (2008) Geographical and taxonomic biases in invasion ecology. Trends Ecol Evol 23:237–244. https://doi.org/10.1016/j.tree.2008.02.002

Richardson DM (2011) For Agrofor. In: Simberloff D, Rejmánek M (eds) Encyclopeia of biological invasions. University of California Press, Berkeley and Los Angeles, pp 241–248

Richardson DM, Hui C, Nuñez MA, Pauchard A (2014) Tree invasions: patterns, processes, challenges and opportunities. Biol Invasions 16:473–481. https://doi.org/10.1007/s10530-013-0606-9

Richardson DM, Iponga DM, Roura-Pascual N, Krug RM, Milton SJ, Hughes GO, Thuiller W (2010) Accommodating scenarios of climate change and management in modelling the distribution of the invasive tree Schinus molle in South Africa. Ecography 33:1049–1061. https://doi.org/10.1111/j.1600-0587.2010.06350.x

Richardson DM, Rejmánek M (2011) Trees and shrubs as invasive alien species—a global review. Diversity and Distributions 17:788–809. https://doi.org/10.1111/j.1472-4642.2011.00782.x

Richter R, Berger UE, Dullinger S, Essl F, Leitner M, Smith M, Vogl G (2013) Spread of invasive ragweed: climate change, management and how to reduce allergy costs. J Appl Ecol 50:1422–1430. https://doi.org/10.1111/1365-2664.12156

Rotherham I (2011) The history and perception of animals and plant invasions—the case of acclimatization and wild gardens. In: Rotherham ID, Lambert RA (eds) Invasive and introduced plants and animals: human perceptions, attitudes and approaches to management. Earthscan, London

Santos M, Freitas R, Crespí AL, Hughes SJ, Cabral JA (2011) Predicting trends of invasive plants richness using local socio-economic data: an application in North Portugal. Environ Res 111:960–966. https://doi.org/10.1016/j.envres.2011.03.014

Seebens H et al (2015) Global trade will accelerate plant invasions in emerging economies under climate change. Glob Change Biol 21:4128–4140. https://doi.org/10.1111/gcb.13021

Shmueli G (2010) To explain or to predict? Stat Sci 25:289–310. https://doi.org/10.1214/10-STS330

Simberloff D et al (2013) Impacts of biological invasions: what's what and the way forward. Trends Ecol Evol 28:58–66. https://doi.org/10.1016/j.tree.2012.07.013

Smith J (2010) The history of temperate agroforestry. Progressive Farming Trust Limited, Berkshire

Soetaert K, Middelburg JJ, Herman PMJ, Buis K (2000) On the coupling of benthic and pelagic biogeochemical models. Earth Sci Rev 51:173–201. https://doi.org/10.1016/S0012-8252(00)00004-0

Stohlgren TJ, Ma P, Kumar S, Rocca M, Morisette JT, Jarnevich CS, Benson N (2010) Ensemble habitat mapping of invasive plant species. Risk Anal 30:224–235. https://doi.org/10.1111/j.1539-6924.2009.01343.x

TEEB (2010) The economics of ecosystems and biodiversity: mainstreaming the economics of nature: a synthesis of the approach, conclusions and recommendations of TEEB. Malta

Theoharides KA, Dukes JS (2007) Plant Invasion across space and time: factors affecting nonindigenous species success during four stages of invasion. New Phytol 176:256–273. https://doi.org/10.1111/j.1469-8137.2007.02207.x

Thuiller W, Brotons L, Araújo MB, Lavorel S (2004) Effects of restricting environmental range of data to project current and future species distributions. Ecography 27:165–172

Thuiller W, Richardson DM, Rouget M, Procheş Ş, Wilson JR (2006) Interactions between environment, species traits, and human uses describe patterns of plant invasions. Ecology 87:1755–1769. https://doi.org/10.1111/j.0906-7590.2004.03673.x

Uden DR, Allen CR, Angeler DG, Corral L, Fricke KA (2015) Adaptive invasive species distribution models: a framework for modeling incipient invasions. Biol Invasions 17:2831–2850. https://doi.org/10.1007/s10530-015-0914-3

van Klinken RD, Lawson BE, Zalucki MP (2009) Predicting invasions in Australia by a Neotropical shrub under climate change: the challenge of novel climates and parameter estimation. Glob Ecol Biogeogr 18:688–700. https://doi.org/10.1111/j.1466-8238.2009.00483.x

Vaz AS et al (2017) Integrating ecosystem services and disservices: insights from plant invasions. Ecosyst Serv 23:94–107. https://doi.org/10.1016/j.ecoser.2016.11.017

Vaz AS et al (2018) An indicator-based approach to analyse the effects of non-native tree species on multiple cultural ecosystem services. Ecol Ind 85:48–56. https://doi.org/10.1016/j.ecolind.2017.10.009

Venette RC et al (2010) Pest risk maps for invasive alien species: a roadmap for improvement. Bioscience 60:349–362. https://doi.org/10.1525/bio.2010.60.5.5

Vicente J, Randin CF, Goncalves J, Metzger MJ, Lomba A, Honrado J, Guisan A (2011) Where will conflicts between alien and rare species occur after climate and land-use change? A test with a novel combined modelling approach. Biol Invasions 13:1209–1227. https://doi.org/10.1007/s10530-011-9952-7

Vicente JR et al (2013) Using life strategies to explore the vulnerability of ecosystem services to invasion by alien plants. Ecosystems 16:678–693. https://doi.org/10.1007/s10021-013-9640-9

Vicente JR et al (2016) Cost-effective monitoring of biological invasions under global change: a model-based framework. J Appl Ecol 53:1317–1329. https://doi.org/10.1111/1365-2664.12631

Vilà M, Hulme PE (2017) Impact of biological invasions on ecosystem services. Invading nature—Springer Series in invasion ecology, vol 12. Springer, Switzerland

Wasowicz P, Przedpelska-Wasowicz EM, Kristinsson H (2013) Alien vascular plants in Iceland: diversity, spatial patterns, temporal trends, and the impact of climate change. Flora—Morphol Distrib Funct Ecol Plants 208:648–673. https://doi.org/10.1016/j.flora.2013.09.009

Webber BL et al (2011) Modelling horses for novel climate courses: insights from projecting potential distributions of native and alien Australian acacias with correlative and mechanistic models. Diversity Distrib 17:978–1000. https://doi.org/10.1111/j.1472-4642.2011.00811.x

Remote Sensing of Droughts Impacts on Maize Prices Using SPOT-VGT Derived Vegetation Index

John A. Ogbodo, Ejiet John Wasige, Sakirat M. Shuaibu, Timothy Dube and Samuel Emeka Anarah

Abstract Maize production in Kenya is usually affected by climate variability. Climate variability, further has implications for maize prices and national food security. The main objective of this study was to determine temporal fluctuations of maize prices in five (5) markets in Kenya; using NDVI values from SPOT-VEGETATION imagery of 1998–2010. The results show a weak relationship between maxNDVI and Kenyan maize wholesale price. Out of the five (5) markets analysed, only Kisumu ($r^2 = -0.11$) shows a negative regression value; whereas, Nairobi ($r^2 = 0.29$), Mombasa ($r^2 = 0.27$), Nakuru ($r^2 = 0.44$) and Eldoret ($r^2 = 0.05$) portray a positive relationship. Overall, the findings of this study indicate that maize prices were high during drought periods (i.e. negative anomalies) and low during wet seasons (i.e. positive anomalies). The findings of this work underscores the potential for maize price monitoring using satellite derived vegetation indices, such as the normalised difference vegetation index towards providing

J. A. Ogbodo (✉)
Department of Forestry and Wildlife Management,
Nnamdi Azikiwe University, P.M.B 5025, Awka, Nigeria
e-mail: jaogbodo@gmail.com
URL: https://steifoundation.org/

J. A. Ogbodo
STEi Foundation - Sustainable TransEnvironment International Foundation,
Makurdi, Nigeria

E. J. Wasige
Department of Environmental Management, Makarere University,
Kampala, Uganda

S. M. Shuaibu
University of Twente, No. 5/29 Santley Crescent, Kingswood,
NSW 2747, Australia

T. Dube
Department of Earth Sciences, University of Western Cape, Private Bag X17,
Bellville 7535, South Africa

S. E. Anarah
Department of Agricultural Economics and Extension, Nnamdi Azikiwe University,
P.M.B 5025, Awka, Nigeria

P. Castro et al. (eds.), *Climate Change-Resilient Agriculture and Agroforestry*,
Climate Change Management, https://doi.org/10.1007/978-3-319-75004-0_14

valuable inputs to the food security modelling community. We however, recommend that, in the future, there is need to integrate a cumulative vegetation index (CVI) method to reduce any differences that could exist during analysing maize growing stage using vegetation index remote sensing techniques.

Keywords Drought · Food insecurity · maxNDVI · Stepwise regression
SPOT vegetation · Wholesale maize price

1 Introduction

In Kenya, maize (*Zea mays*) is a staple and an important cereal food crop. Hence, *food insecurity* is synonymous to poor maize production and its scarcity in Kenyan markets. In other words, whenever maize production is impaired, majority of the Kenyans would become food insecure (FEWSNET 2010, 2011, 2013). The Food and Agriculture Organization (FAO) defined *food insecurity* as a situation that exists when people or populations lack secure access to sufficient amounts of safe and nutritious food, for normal growth and development; and for an active and healthy life (FAO-WFP 1996). According to Grace et al. (2014), episodes of household food insecurity occur when access to food decreases (i.e. cost increases while income or entitlements do not increase), when food availability decreases (i.e. crop production decreases). In Kenya, many poor people rely on rainfed, locally produced food for the majority of their caloric intake. Also, Kenyans maize production is usually affected by variations in rainfall, which in turn, have implications on national maize production rates and pricing system (Grace et al. 2014; FEWSNET 2013; FAO 2011). Usually, shifts in climate and weather patterns dramatically reduce agricultural productivity (Ogbodo et al. 2018; Livingston et al. 2011). This reduction in maize productivity reduces overall food availability and increases local food prices, due to rising demand and diminished supply. This ultimately impacts food accessibility, thus, putting millions of Kenyans at risk of malnutrition; as in the examples of years 2002 and 2009. In 2009, the expected yield per acre of maize was less than the normal yield, as a result of drought (Smale et al. 2011). As a consequence, approximately, one third (1/3) of every Kenya households became food insecure, when the production output of a staple crop as maize turns out to be poor.

Zea mays is a staple and an important cereal food crop in Kenya (FEWSNET 2013). Area of Kenyan agricultural land under maize cultivation was estimated at 1.6 MH (Deenapanray and Tan 2011; EPZ-Kenya 2005). Erenstein et al. (2011) expressed that, maize cultivation is highly dependent on annual rainfall. FEWSNET (2013) and Kangethe (2011), further affirmed that, maize is a rain-fed crop, which is cultivated across a range of Kenyan soil types, moisture regimes, latitudes, slopes and altitudes. Maize is primarily produced for home consumption and for local markets by small-scale farmers (Short et al. 2012; EPZ-Kenya 2005). Small-scale farmers contribute about 75% of the overall maize production in Kenya (Nyoro et al. 2004; Nyoro 2002).

Prolonged droughts in Kenya have become common in the recent years, particularly since 2000. According to FEWSNET (2009), the impact of drought had affected larger areas in that country. The drought of 1999/2001 was regarded as the "worst ever" in Kenya's history. Therefore, drought is, one of the impacts of climate change that threatens the livelihoods of most communities in Kenya (Mariara and Karanja 2007). In years past, agricultural crops including maize, failed in one out of every three (3) seasons due to the frequent reoccurrence of agricultural droughts (Rojas et al. 2011; Bassi 2010). Mishra and Singh (2010), defined *agricultural drought* as a period with declining soil moisture which consequently leads to crop failures in areas where it is not possible to adopt irrigational measures. Agricultural drought usually diminishes water supplies, reduces crop productivity and most often results to a widespread famine. A reduction in maize productivity can further lead to a reduction in the overall maize availability, and an increase in Kenya wholesale prices for maize meals paid by consumers due to rising demand and diminished supply (Jayne et al. 2008). This scenario has negative impacts on food affordability, and it usually puts millions of Kenyans at risk of malnutrition (Grace et al. 2014). Therefore, in an effort to improve scientific understanding of food insecurity and evaluate the usefulness of pricing data as an indicator of food insecurity (FAO 2014); there is a need for agricultural drought assessment studies using novel remote sensing techniques.

Remote-sensing data acquired by satellite sensors have a wide scope in agricultural applications owing to their synoptic and repetitive coverage. Generally, vegetation indices derived from geospatial (i.e. remote sensing and geographic information systems) methods are considered as potential tools for improving vegetation health simulations in real-time (Gitelson et al. 2002). Some past scientific studies have successfully applied vegetation indices remote sensing techniques to crop yields estimations (Shuaibu et al. 2016; Bolton and Friedl 2013; Vrieling et al. 2011). However, in addition to vegetation indices, rainfall distribution in space and time are also incorporated into remote sensing/GIS models on crop yields estimations (Andrea et al. 2011; Rojas et al. 2011; Dinku et al. 2007). NDVI (Normalized Difference Vegetation Index) remote sensing is another geospatial approach, which can combine satellite images and rainfall data in assessing drought —prone agricultural areas, crop failures and poor pasture conditions (Mutanga and Skidmore 2004; FEWSNET 2010; Dinku et al. 2007).

Thus, NDVI has been the most frequently used vegetation index in agro-meteorological analysis (Andrea et al. 2011; Rojas et al. 2011). It is defined as: NDVI = (NIR − RED)/(NIR + RED). Where, *NIR* stands for reflectance (%) in the *near infrared* band; and *RED* is *reflectance (%)* in the *red* band of an electromagnetic spectrum (Shuaibu et al. 2016). It is easy to understand the vegetation index, when the characteristics of absorption and reflection of the radiation by green leaves, are studied. The chlorophyll of plants absorbs the majority of radiation in the visible part of the electromagnetic spectrum: principally, the red portion (0.6–07 μm) is highly reflective in the near-infrared. NDVI is a direct indicator of the plant's photosynthetic activity. Therefore parameters such as water stress in vegetation (Shuaibu et al. 2016; Dinku et al. 2007) can be monitored successfully by analysing NDVI values. NDVI values usually indicate the vigour of vegetative activity when the chlorophyll is indirectly observed (Shuaibu et al. 2016). Grace et al. (2014) further expressed that, NDVI low values are usually attributed to lack of vegetation, dormant states of existing vegetation, over-irrigation, pest and diseases, water stress caused by agricultural droughts, etc. Already, outcomes of past satellite NDVI analyses have been used by governments and non-governmental organizations in responding to food crises (Funk and Brown 2005) and in developing livelihood strategies for drought affected communities (Grace et al. 2014; Andrea et al. 2011; Rojas et al. 2011).

In view of the above, development of a model to show the impacts of agricultural droughts on maize prices is imperative for improving food security in Kenya using satellite remote sensing techniques. Consequently, several time series of satellite derived vegetation indices at the global scale and with a high temporal frequency are freely available. These include: AVHRR (Advanced Very High Resolution Radiometer); MODIS (Moderate Resolution Imaging Spectroradiometer) and SPOT (*Satellite Pour l'Observation de la Terre*) VEGETATION. The SPOT VEGETATION (VGT) sensor was launched on board the SPOT 4 satellite in March, 1998, to monitor surface parameters on global basis at daily intervals with a 1 km resolution (Rijks et al. 2007; Rojas 2007; Zhou et al. 2009; Sawasawa 2003). The Normalized Difference Vegetation Index (NDVI) from SPOT VGT satellite has been used to assess droughts in semi-arid zones where rainfall fluctuates (Shuaibu et al. 2016; Pettorelli et al. 2005). A large number of the derived satellite products exist, and are repeatedly updated.

Therefore, the main objective of this study is to apply a novel remote sensing technique in modelling temporal fluctuations of maize production and prices in five (5) markets in Kenya. In doing so, we used NDVI values from SPOT-VEGETATION imagery to model the relationship between agricultural drought and maize wholesale prices. Therefore, monitoring agricultural production (especially maize production) in Kenya can provide important information about food security towards tracking significant changes over time. The outcome of this study could contribute to the actualization of number 2, SDG (Sustainable Development Goal, SDG) which is *zero hunger* by 2030.

2 Materials

2.1 Background on Study Area

The Republic of Kenya is located in East Africa, at latitudes 5°S to 5.5°N and longitudes 34°E to 42°E (FEWSNET 2013). The country borders Somalia to the East, Ethiopia to the North, Sudan to the Northwest, Uganda to the West and Tanzania to the Southwest and the Indian Ocean to the South. Its climate is tropical, but moderated by diverse topography (FEWSNET 2013). There is a steady rise of

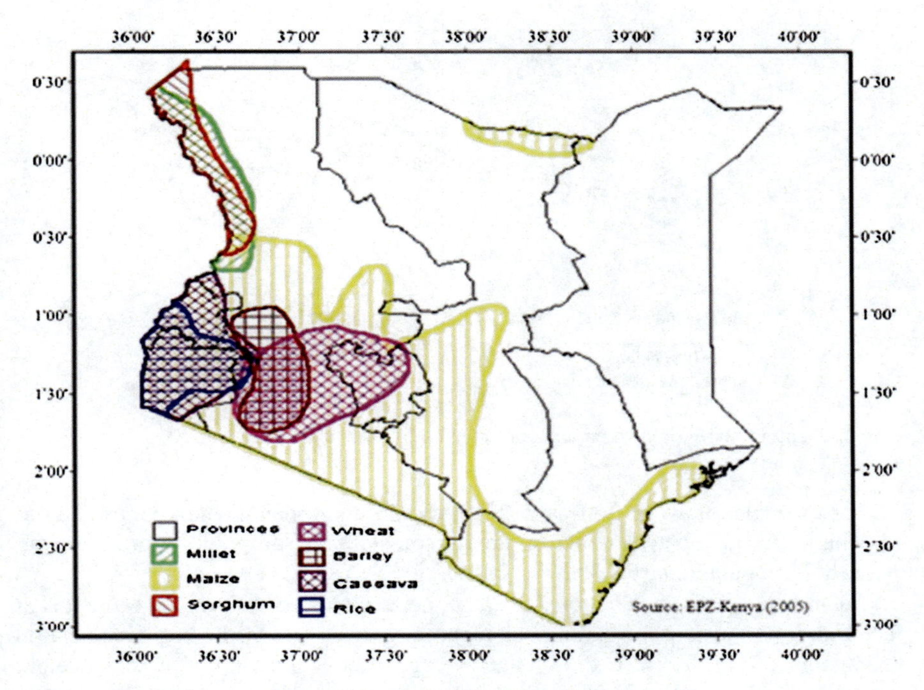

Fig. 1 Map of main crop zones of Kenya (FEWSNET 2013)

the land surface from the coastal plains to altitudes of over 5000 m. Nyanza, Western and Rift Valley provinces have a mono-modal rainfall distribution while Central, Eastern and Coast provinces exhibit an average bimodal distribution of rainfall with the possibility of having two rainfed cropping seasons per year (Ifejika et al. 2008; EPZ-Kenya 2005). The bimodal raining seasons commence from March to May (long rains period) and October to December (short rains period) on a yearly basis (Williams and Funk 2010; FEWSNET 2013). The average annual rainfall in Kenya ranges between 250 and 2500 mm (FEWSNET 2013). Given the geographic variation in intensity of seasonal rainfall, agricultural production differs regionally (FEWSNET 2013). One major small-scale agricultural crop (Fig. 1) that is widely cultivated and consumed in Kenya is maize.

History has it that, Kenya was hit with severe droughts in the periods: 1983/1984, 1991/1992, 1995/1996, 1999/2001, 2004/2005 and 2009/2010 (Andrea et al. 2011; Boken et al. 2005). The areas most affected by the droughts were the marginal agricultural lands at the north-eastern, north-western southern and south-eastern parts of Kenya (FEWSNET 2013; Andrea et al. 2011; Boken et al. 2005). These regions were also faced with various levels of emergency food insecurity, basically because, the drought occurrences fluctuated the average maize production outputs (FEWSNET 2013; Ifejika et al. 2008).

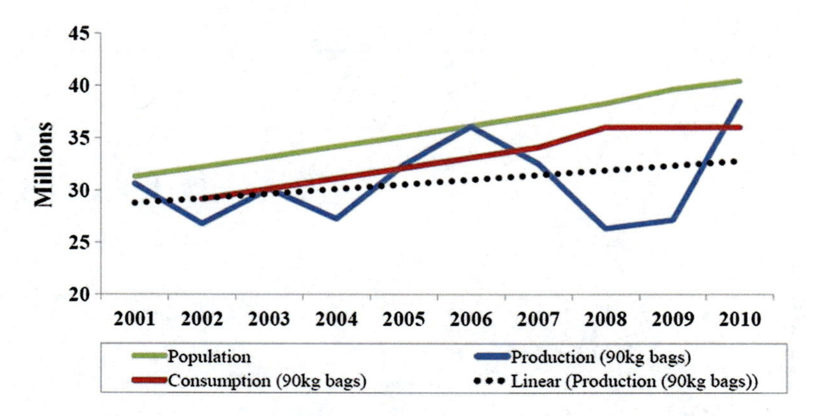

Fig. 2 Trend in maize production and consumption by Kenya's population, 2001–2010

For example, between 2001 and 2010, the drought phenomenon contributed to widening the gap between maize production and consumption (Fig. 2) among household populations (FEWSNET 2013).

Meanwhile, the Kenya national average maize yields per hectare are estimated at 1.8 tonnes per hectare; being an equivalent to 20 bags of a 90 kg bag specification (Kangethe 2011). Table 1 shows the average production of maize in the eight provinces of Kenya from 2005–2009.

The specific sampled market-sites are: Nairobi, Mombasa, Kisumu, Eldoret and Nakuru markets (Fig. 3).

Table 1 Average maize production from 2005–2009 (Copied from Kangethe 2011)

Provinces	Area cultivated (ha)	Production (90 kg/bag)	Yield (bags/ha)	Population
Rift Valley	644,895	13,225,039	20.50	10,066,805
Nyanza	262,453	3711,215	14.10	5442,711
Eastern	462,401	3903,141	8.40	5668,123
Western	225,302	4163,878	18.50	4334,282
Coast	129,379	1079,383	8.30	3325,307
Central	157,063	1047,879	2.20	2310,757
North Eastern	2525	5520	2.20	2310,757
Nairobi	1053	6420	14.4	3138,369

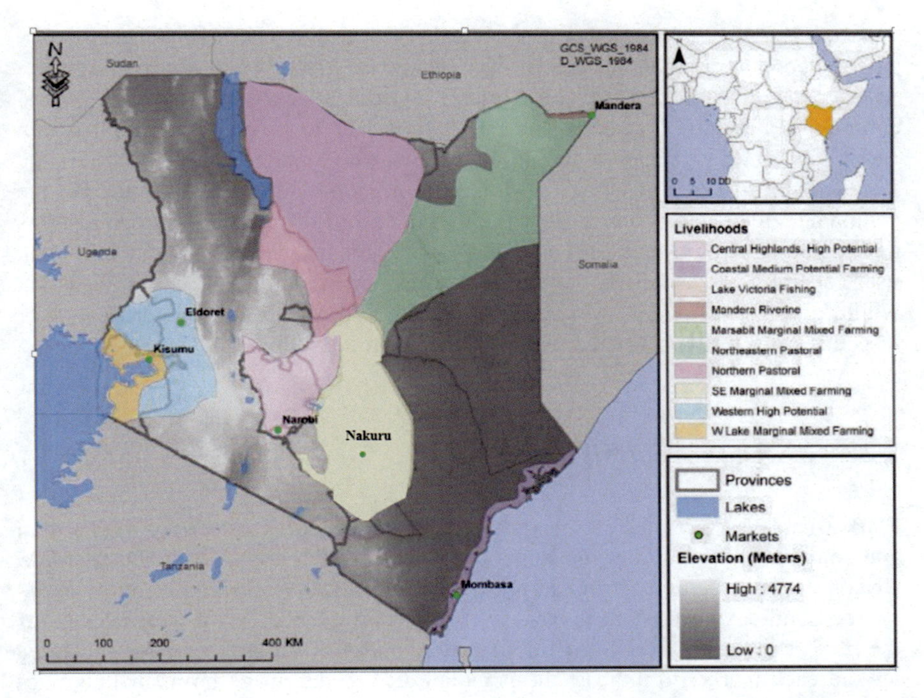

Fig. 3 Livelihoods map of Kenya showing the distribution of sampled markets (Modified from FEWSNET 2013)

2.2 Remote-Sensing Data

The products of SPOT VEGETATION acquired, are 10-day NDVI (Normalized Difference Vegetation Index) synthesis (S10) images, obtained through Maximum Value Compositing (MVC). The images are corrected for radiometry, geometry and atmospheric effects. The NDVI values were spatially averaged for each area comprised in the maize crop mask. Vegetation Index was derived from SPOT VEGETATION NDVI and a time series of the two (2) wet seasons during September, 1998 to August, 2011; were analyzed for maize crops.

3 Methods

3.1 Field Data Collection

Four criteria were adopted to select the markets used in this research: (a) The sites were selected because, the scenes of remote sensing image data that could be obtained over their locations were almost cloud-free; (b) sampled markets were located within the grains basket zone (Fig. 1) of Kenya; (c) For the Nairobi district, where there is more than one market, an average of three (3) markets was sampled. Selected Nairobi markets (for urban maize consumers) were in proximity to rural communities that are located within the grain basket zones of Kenya. Information about grain basket zonation in the study area was obtained from both: (i) the Kenyan Ministry of Agriculture's Market Information Bureau (MIB); and (ii) the National Cereals and Produce Board (NCPB); and (d) The other sampled districts have a particular day, set aside, as a market day. This means that, one day in a week was generally set aside for buying and selling commodities at a central market within a particular district. As such, maize price data were only recorded on the respective market days for the districts other than Nairobi.

3.2 GIS Data Analysis

Market analyses and maize prices were done in GIS software. Every market that was sampled is named after the Kenyan district in which it is situated. District GIS shapefiles were obtained from http://www.un-spider.org/links-and-resources/data-sources/land-cover-kenya-africover-fao. The center of each town was selected as the market place in ArcGIS 10.1 software. A buffer of 25 km was later generated around each market to indicate the potential area of the maize farms that supplies farm produce to each market. The 25 km buffered areas were intersected with the predicted maize fields. These maize cultivated lands were extracted around each market. Monthly SPOT-VGT NDVI was aggregated within the buffered maize areas to extract the time series of NDVI values for the selected maize areas. In this present study, the maize map used, was an unsupervised classification output of 1998–2011 SPOT-VGT NDVI time series.

3.3 Remote Sensing Data Analysis

SPOT VEGETATION (*Satellite Pour l'Observation de la Terre*) image data is freely available for vegetation studies. SPOT VEGETATION has four spectral imaging bands ranging from about 0.45 μm (blue light) to 1.67 μm (mid-infrared radiation) (Shuaibu et al. 2016). SPOT satellite has a field of view of 0–55° on both sides of the satellite tracking path. The SPOT satellite has a pixel size of 1.15×1.15 km at nadir. The major dataset used in this study is the high temporal frequency SPOT VEGETATION (SPOT VGT) 1 S10 imagery. This SPOT time series data consist of 10-day maximum-value composites at 1 km spatial resolution for the period September 1998 to August 2011. The advantages of using maximum-valued composite satellite imagery are two-fold: (a composite image is less influenced by cloud effects, sun-angle/shadow effects, aerosols, and water vapour effects (Rijks et al. 2007); and (b) reduces directional reflectance and off-nadir viewing effects (Rojas 2007). To further reduce the remaining atmospheric effects, an iterative *Savitzky-Golay* filter, as described by Huete et al. (2006), was applied to the time series of each pixel for temporal smoothing. The maize map used in this present study was an unsupervised classification output of 1998–2011 SPOT-VGT NDVI time series with a legend of 36 NDVI classes. These 36 legend classes output was the afore-mentioned product of a separability modelling on an initial 100 legend classes of the above-mentioned SPOT-VGT NDVI maize map. The highest average separability values from the 36 NDVI classified product, were chosen for further analysis.

3.4 Maize Price Data Analysis

The Kenyan wholesale maize prices used in this study were sampled from five markets: Nairobi, Mombasa, Kisumu, Eldoret and Nakuru. Maize wholesale price data collected in Kenya were further compared with Global maize price data, obtained from http://www.indexmundi.com/commodities/?commodity=corn&months=300. Maize prices were expressed in Kenyan shillings per 90 kg bag. Global wholesale prices (in US dollars per tonne) were converted to Kenyan shillings using the official exchange rate. Details on the above approaches, can further be found in this earlier publication: Shuaibu et al. (2016), been a segment of this present research.

3.5 Computation of NDVI Values from SPOT VEGETATION Satellite Image

Vegetation indices (NDVI) were derived using *NDVI* = (*refl.*NIR − *refl.*Red)/(*refl.*NIR + *refl.*Red) in ArcGIS 10.1 software. *refl.NIR* and *refl.Red* is each, the reflectance of green vegetation within the *NIR* and *Red* regions of the electromagnetic spectrum. NDVI is commonly used to indicate the vigour of vegetation in a satellite image and to also differentiate vegetated and non-vegetated areas. The NDVI parameter value ranges from '1' (for very high green vegetation cover) through '0' (reflectance from non-green vegetation) to '−1' (for bare-land/non-vegetation). Principal Components Analysis (PCA) was applied on the stacked NDVI images (Celik 2009). PCA application is useful in minimizing error that arise from overlaying different temporal satellite images and removes redundancy (Celik 2009; Li and Yeh 1998). Hence, it allows redundant data to be compacted because, the data is reduced. The PCA output image was further classified in 36 classes with a minimum class size of 100 pixels, and based on 10 iterations as sample interval per class. The output of thematic SPOT VGT image was reclassified using visual interpretation. To assess which class best correspond to maize area, a stepwise regression analysis was performed with the 36 NDVI classes (independent variable) and the average area statistics (dependent variable). Hence, the 36 classes (that were obtained from the PCA approach) were grouped into: maize farmland or none-maize class. The "none-maize" class was assigned NoData value, so as to enable masking of the result output (Li and Yeh 1998).

3.6 Analysis of LagNDVI with Maize Residual Price Series

NDVI values were aggregated on a temporal scale. Two (2) variables were created: (a) maize residual price series (dependent variable) and (b) lagged NDVI (lagNDVI) values (independent variable). Twelve (12) lags of the NDVI seasonal anomalies were adopted. Here, lag0 indicated minimum NDVI (otherwise known as lagNDVI) and the last value of the successive increased lags (e.g. 12), as maximum lag (maxNDVI). Lag0 was kept constant throughout the model. The twelve lagNDVI values were used because, the time difference between when maize was produced and when maize market price are raised; can be up to twelve months (Tachiiri 2003). More-so, that maize that was harvested in a year can be stored for sale in the following year. The distributed lag model (Eq. 1) was used to examine the relationship between maize residual price series and NDVI seasonal anomalies.

$$\mu_t = a + b_1 NDVIA_t + b_3 NDVIA_{t-1} + \cdots b_n NDVIA_{t-12} + \mu \tag{1}$$

where *NDVI* = monthly NDVI anomalies, b = the impulse response weight vectors describing the weight assigned to the current and past monthly NDVI anomalies series, a = the constant term and μ = model residual and μ_t = residual from optimal lag identified from model of wholesale price and global price.

3.7 Analysis of maxNDVI with Maize Wholesale Price Correlation Statistics

Maximum NDVI (maxNDVI) values were obtained using parameters for maize start-of-season (planting date) to end-of-season date in the maize cropping cycle (Shuaibu et al. 2016). Deriving maxNDVI during the maize cropping cycle is valuable for forecasting and verifying maize production through hind-casting (Lewis et al. 1998). There are two maize growing seasons in Kenya. The first crop season starts from February to August yearly and the last cropping season is from September to January (EPZ-Kenya 2005). Hence, maxNDVI values for the growing seasons, during September, 1998–August, 2011(a period of thirteen years) was chosen from the aggregated pixel of NDVI of each market. The maxNDVI values (independent variable) were correlated with the wholesale price series (dependent variable) of maize, one month after harvest of each of the thirteen years, to evaluate the direction and strength of the correlation between the two (2) variables.

To further analyse the relationship between monthly wholesale price series and NDVI Seasonal anomalies, an NDVI seasonal anomaly for the growing season of March to August, 2011, for each market was computed in excel (Pettorelli et al. 2005) using Eq. 2:

$$\mu_1 = a + b_1 NDVIA_t + b_2 NDVIA_{t-1} + \cdots b_2 NDVIA_{t-4} + NDVI_{max} + \mu_t \quad (2)$$

where *NDVIA* = seasonal NDVI anomalies b = the impulse response weight vectors describing the weight assigned to the current and past monthly NDVI anomalies series, a = the constant initial lag value and μ = regression model and t = wholesale price identified per successive increasing lagged value.

Afterwards, the wholesale price series after the growing season were regressed against the lagged NDVI anomalies during the growing season. In this case only four lags of NDVI seasonal anomalies were considered due to the length of the growing season (which is about six months). This growing season regression analysis was useful because, vegetation production anomaly of the growing season, has been found to be an indicator for predicting price after harvest (Brown et al. 2010). Meanwhile, wholesale maize price series of one month after harvest was used. This was done as to account for time delay in transportation of maize to the market after harvest. This analysis was repeated for every sampled market.

4 Results and Discussion

Shuaibu et al. (2016) present detailed results of the analyses illustrated in Sects. 3.2–3.4 above.

4.1 Computation of NDVI Values from SPOT VEGETATION Satellite Image

Figure 4 shows a maize farmland that is situated at 01° 21′ 59″ S and 36° 44′ 17″ E (WGS84) in Nairobi, Kenya. This output was obtained based on a stepwise regression analysis that was performed based on the PCA output, in which 36 NDVI classes were grouped into: maize crop (NDVI value = 0.95) or none-vegetated class (NDVI = −1). The ‘0.05’ NDVI value in Fig. 4c, indicates a reflectance from either grass, maize cobs, or maize flowers (yellow colour). NDVI is more sensitive to chlorophyll content in plant leaves (Huete et al. 2006; Tucker et al. 2005).

Studies have asserted that, fine spatial resolution imagery captures fine structures of maize plant such as the leaves while coarse resolution data cover canopy level. In their assessment of effect of spatial resolution on maize yield relationship, Xue and Su (2017) found out that by varying spatial resolution from very fine (0.02 m) to intermediate (0.04 m) and fine resolution (0.06 m); the inter-intermediate or fine

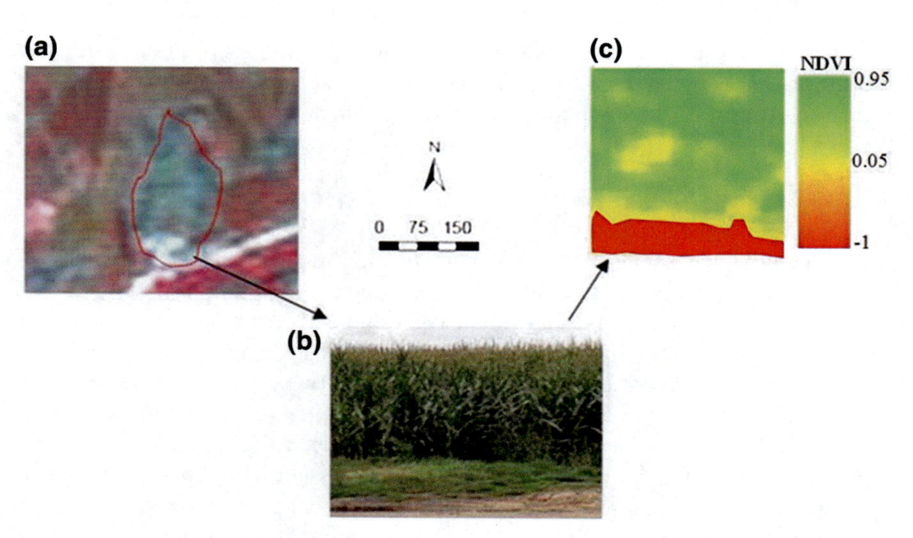

Fig. 4 An output of a PCA SPOT VGT NDVI spectral reflectance of a matured maize farm: **a** 1 km SPOT VGT. Satellite image (RGB = 421); **b** mono-cropped maize with cobs and flowers; and **c** pixel-based NDVI output

resolution index relationship with maize yield shows a better R^2 than very fine spatial resolution data in which they attributed to high noise from soil and non-maize vegetation. However, for coarse resolution data, the R^2 is degraded by the mis-match between maize fields and the pixel sizes especially for the 1 km resolution SPOT VEGETATION imagery. Therefore, in deriving maize yield with NDVI, it should always be pertinent to consider the stage of maize growth and the type of vegetation cover that can be found within the maize farm.

Furthermore, a number of previous studies have expressed that NDVI saturates with dense vegetation cover and maintains high greenness value (Wardlow et al. 2006; Tucker 1979). Maize phenology is divided majorly into vegetative (emergence to tasselling according to a number of leaves) and reproductive (silking to physiological maturity according to the degree of kernel development). Thus, the spectral information captured by a single image will measure spectral information from different stages of development. Such maize stages of development could influence spectral variations of NDVI models. This is obviously the reason for the spectral output that is illustrated in Fig. 4c.

In conclusion, the NDVI models as applied on SPOT VGT image is capable of spatially discriminating either, between photosynthetic and non-photosynthetic maize leaf elements and/or non-vegetated landcover.

4.2 Analysis of Lag NDVI with Maize Residual Price Series

A regression method is commonly used to generate a cause-effect relationship among variables. Different methods had been employed in showing a relationship between NDVI and price data; including correspondence analysis and probabilistic *Markov* price model (Brown and de Beurs 2008). The regression model used in this paper (i.e. distributed lag model) was able to analyse the current and past effects of NDVI anomalies on current maize prices, rather than predicting its future prices (Brown et al. 2010). Although, disadvantage of using a distributed lag model is multicollinearity of independent variable. However, the multicollinearity test applied in this present study of lagged NDVI anomalies per market indicate that, the variables are weakly correlated (Table 2); hence, the justification for using this model.

A number of studies have express that, maize yield, which is dependent of maize development in its growing season, do have serious effects on residual maize prices (Mkhabela et al. 2011). Generally, during maize development, the unfavourable conditions in the crop growing period (anthesis and physical maturity) has been found to likely impair pollination and reduce the fertilized kernels that are destined to be filled (Meier 2001). During maize development stage, a maximum yield could be achieved (Basso et al. 2013), if there is sufficient supply of nutrients under favourable conditions such as soil moisture, sunlight and favourable environmental conditions. Unfavourable conditions at the beginning of the reproductive cycle (i.e. *anthesis and tasselling*) are likely to impair pollination of maize crop; thus, reducing the number of fertilized kernels that are to be filled (Mkhabela et al. 2011). Any adverse condition during the grain filling period (*between anthesis and fruit development*) are likely to impair pollination. Therefore, detecting an early onset of *senescence* is necessary, because; it can facilitate bomber yield. The flowering and

Table 2 Results of correlation analysis of residual maize price series and NDVI seasonal anomalies illustrating the strength of correlation between these variables

Residual maize price versus lagNDVI	Pearson correlation coefficient (r)
Residual price series versus NDVI anomalies Lag0	−0.281
Residual price series versus NDVI anomalies Lag1	−0.390
Residual price series versus NDVI anomalies Lag2	−0.468
Residual e price series versus NDVI anomalies Lag3	−0.479
Residual price series versus NDVI anomalies Lag4	−0.463
Residual e price series versus NDVI anomalies Lag5	−0.422
Residual price series versus NDVI anomalies Lag6	−0.334
Residual e price series versus NDVI anomalies Lag7	−0.249
Residual price series versus NDVI anomalies Lag8	−0.188
Residual e price series versus NDVI anomalies Lag9	−0.169
Residual price series versus NDVI anomalies Lag10	−0.143
Residual price series versus NDVI anomalies Lag11	−0.091
Residual price series versus NDVI anomalies Lag12	−0.180

grain filling periods are the most critical stages for most crops; any water stress encountered during these crop growth stages, could result in reduced grain yields (Basso et al. 2013; Mkhabela et al. 2011). Therefore, variation in date of planting is always important, in that maize planted early will experience the different stages of development as compared to those planted late; thereby, contribute to keeping residual maize prices, at its lowest level.

In a nutshell, maize yield is an end product; whereas, maize crop undergoes a number of stages to produce a yield. Therefore, understanding when yield components can be determined is important in interpreting management and environmental factors that influence maize production (Sacks et al. 2010; Salami et al. 2010). It is therefore, concluded in this paper that, remotely sensed vegetation indices can assess the optimal stages of maize growth and estimate maize yield. This study further infers that, late planting of maize has a significant negative effect on maize yield, its availability and sales.

4.3 Analysis of maxNDVI with Maize Wholesale Price Correlation Statistics

Figure 5 visualises the result of the maximum NDVI (maxNDVI) for the maize growing season's verses wholesale maize price, a month after harvest, illustrating the strength and direction of the correlation.

A weak correlation coefficient exists for all the study sites as follows: Nairobi ($r^2 = 0.29$), Mombasa ($r^2 = 0.27$), Kisumu ($r^2 = -0.11$), Nakuru ($r^2 = 0.44$) and Eldoret ($r^2 = 0.05$). Kisumu market is the only site with a negative correlation value. Hence, positive and negative correlations values can be interpreted as thus: a *plus* r^2 ($+r^2$) value implies a high NDVI value at a low maize price (after harvest) while *minus* r^2 ($-r^2$) correlation coefficient represents a high NDVI together with a high maize price (after harvest) (Chatfield 1995). In other words, maize prices were high during drought periods (i.e. negative anomalies) and low during wet seasons (i.e. positive anomalies).

Aside drought, the rising dependence on maize imports in Kenya also contributes to wholesale maize price fluctuations (Short et al. 2012; Jayne et al. 2008; Groote et al. 2002). For example, in 2008, the Kenya National Cereal Pricing Board (NCPB) imported nearly 150,000 tonnes of maize from South Africa through Kenya's main port in Mombasa at a price of over USD 400/tonne (Short et al. 2012). This high import price was as a result of the surge in world prices that hit global markets in 2007–2008.

In this study, Kisumu market ($r^2 = -0.11$) is the only site with a negative correlation value. Positive and negative correlations values can be interpreted as: $+r^2$ value (implies high NDVI with a corresponding low maize price, after harvest) and $-r^2$ (a high NDVI with a high maize price, after harvest). This statistics output might not necessary be due to effect of drought; but could possibly be attributed to the *size* of Kisumu or the *subsistence type* of Kisumu market (with only a small

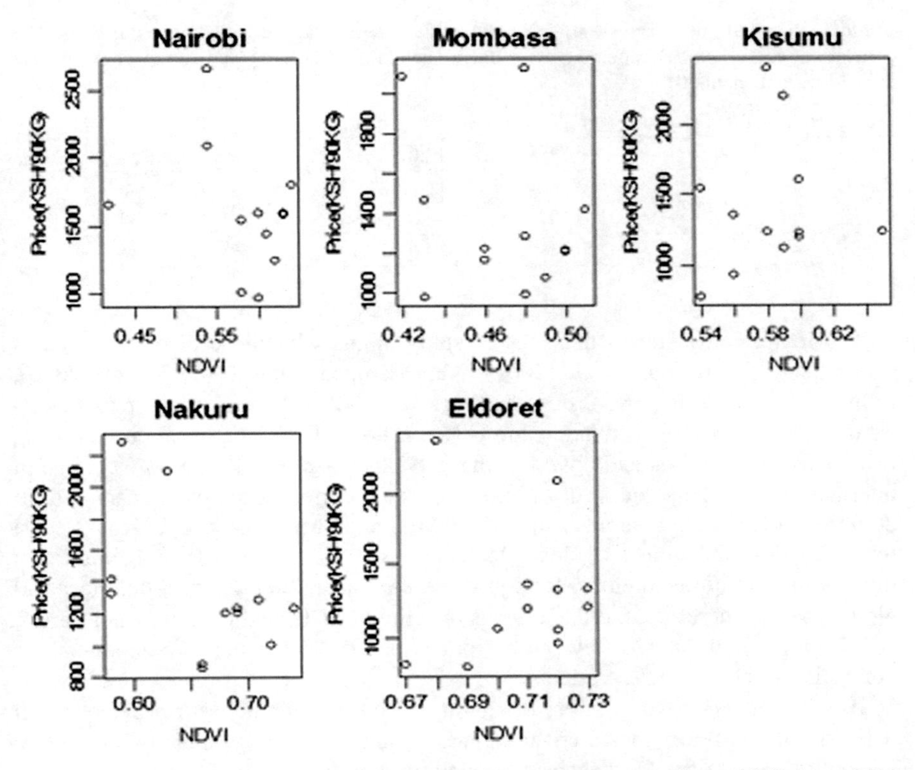

Fig. 5 Scatter plot of maximum NDVI during the growing season (Sept., 1998–Aug., 2011) versus maize price; one month after harvest (October, 2011) showing the direction and strength of the correlation

amount of maize production or consumption traded). This finding agrees with Short et al. (2012), which asserted that maize prices in Kisumu (capital city of the Nyanza Province in southwest Kenya) were often higher than maize prices in other parts of Kenya, by over USD 35/Tonnes.

Table 3 illustrates the result of the simple linear regression model in which wholesale prices were regressed against lagged global price. The direction of the correlation is positive at all lags. This result implies that, increasing global prices lead to a corresponding increase in maize wholesale price in Kenya. From Table 3, lag3 has a correlation coefficient (r) of 82%. Optimal lag3 implies that the dependence of wholesale prices on global prices was up to three months. In other words, the time it takes for trader in the wholesale market to notice and respond to the effect of global price was up to three months. The coefficient of determination (R-squared) is 0.52 for lag0, 0.58 for lag1, 0.63 for lag2 and 0.67 for lag3. These R-squared results show that, the amount of variations in maize wholesale price in Kenya, as influenced by global price is 52% for lag0, 58% for lag1, 63% for lag2 and 67% for lag3.

Table 3 Result of linear regression verses wholesale price and lagged global price illustrating the correlation coefficient (r), coefficient of determination (R-Squared) and significance of the regression coefficients (P)

Model (in months)	r	R-square	P
Lag0	0.73	0.52	1.65E−26
Lag1	0.76	0.58	1.22E−30
Lag2	0.79	0.63	1.11E−34
Lag3	0.82	0.67	2.05E−38

The regression result in this present study aligns with those of other previous studies such as Kuiper et al. (2003), Van Campenhout (2007), Minot (2010), Mundlak and Larson (1992). For example, Minot (2010) used autoregressive distributed model to show a relationship between domestic prices of 62 commodities in 11 African countries and observed that 54% increase based on a 54% change in international market prices of a produce, was correspondingly transmitted to domestic prices of the same commodity. Similarly, Mundlak and Larson (1992) regressed domestic prices against global prices in various forms and found that most of the variations in global food prices, were transmitted to domestic prices and global prices. Therefore, these analogies by Minot (2010) and Mundlak and Larson (1992), have both shown how global prices can drive the major components of domestic prices of maize products.

In summary, this study infers the ability of NDVI technique to provide useful data for maize price characterization and drought mapping, thereby, providing valuable inputs to the food security modelling community.

5 Conclusion and Recommendations

This study evaluated the relationship between remotely sensed vegetation indices and wholesale maize prices in five (5) markets in Kenya, using maximum Normalized Difference Vegetation Index (maxNDVI) values from SPOT-VEGETATION satellite imagery of 1998–2011. The correlation statistics as used in the study, indicates a weak relationship between the maximum NDVI (maxNDVI) and the Kenyan maize wholesale price: Nairobi ($r^2 = 0.29$), Mombasa ($r^2 = 0.27$), Kisumu ($r^2 = -0.11$), Nakuru ($r^2 = 0.44$) and Eldoret ($r^2 = 0.05$). Literally, maize prices were high during drought periods (i.e. negative anomalies) and low during wet seasons (i.e. positive anomalies). The positive correlation could be due to high prices of agricultural inputs which has also translated into making maximum NDVI to be low on maize cultivated area, thus resulting in high price after harvest. The negative correlation could imply correspondence of a good maize production with a low price after harvest and poor maize production to a high price after harvest.

Therefore, this paper concludes as thus: (a) maize prices in the major urban centres of Kenya are directly influenced by global maize prices due to high reliance on maize imports; (b) wholesale price of maize showed a lagged response to monthly NDVI anomalies in Nakuru and Mombasa; and, (c) NDVI remote sensing applications such as NDVI approach, has a general ability to provide useful data for maize price characterization and drought mapping, thereby, suggesting a potential for NDVI-based maize price monitoring, towards providing valuable inputs to the food security modelling community.

Meanwhile, the following two (2) salient recommendations are proffered towards improving future research: (i) since, the basic temporal NDVI profiles show some degree of spectral heterogeneity of a masked SPOT VGT image; a cumulative vegetation index (CVI) method is suggested as a way forward. CVI could be useful in reducing the differences that exist during maize growing stage; (ii) the use of a limited 25 km buffer as an area from which maize was supplied to the sampled markets, is possibly a limitation in this present study. Hence, to correct such, there is need to collect an on-spot (in situ) maize ground control points (GCP) at the farm level, so as to obtain a more accurate information on maize yield.

References

Andrea MB, Prakash NKD, Zhuohua T (2011) Strengthening institutional capacity for integrated climate change adaptation and comprehensive national development planning in Kenya. Millenium Institute, Washington DC, USA. Retrieved from www.millenium-institue.org. Accessed on 16 May 2016

Bassi A (2010) Review of the impacts of climate change in Africa: Kenya. Millennium Institute, Washington

Basso B, Cammarano D, Carfagna E (2013) Review of crop yield forecasting methods and early warning systems. In: The first meeting of the scientific advisory committee of the global strategy to improve agricultural and rural statistics, pp 1–56

Boken VK, Cracknell AP, Heathcote RL (eds) (2005) Monitoring and predicting agricultural drought: a global study. Oxford University Press, Great Britain

Bolton Douglas K, Friedl M (2013) Forecasting crop yield using remotely sensed vegetation indices and crop phenology metrics. Agric For Meteorol 173:74–84. Retrieved from http://www.fews.net/docs/Publications/Kenya%20Food%20Security%20Outlook_April%202011_final.pdf

Brown ME, de Beurs KM (2008) Evaluation of multi-sensor semi-arid crop season parameters based on NDVI and rainfall. Remote Sens Environ 112(5):2261–2271. https://doi.org/10.1016/j.rse.2007.10.008

Brown ME, de Beurs K, Vrieling A (2010) The response of African land surface phenology to large scale climate oscillations. Remote Sens Environ 114(10):2286–2296. https://doi.org/10.1016/j.rse.2010.05.005, https://research.utwente.nl/en/publications/the-response-of-african-land-surface-phenology-to-large-scale-cli

Celik T (2009) Unsupervised change detection in satellite images using principal component analysis and-means clustering. IEEE Geosci Remote Sens Lett 6:772–776

Chatfield C (1995) Model uncertainty, data mining and statistical inference. J R Stat Soc A 158:419–466

Deenapanray K, Tan Z (2011) Strengthening institutional capacity for integrated climate change adaptation, Final report. Millennium Institutem Washington DC, USA, p 157

Dinku T, Ceccato P, Grover-Kopec E, Lemma M, Conner SJ, Ropelewski CF (2007) Validation of satellite rainfall products over East Africa's complex topography. Int J Remote Sens 28 (7):1503–1524

EPZ-Kenya—Export Processing Zones Authority—Kenya (2005) Grain production in Kenya. Retrieved from http://www.epzakenya.com/UserFiles/files/GrainReport.pdf. Accessed on 26 Oct 2016

De Groote H, Doss C, Lyimo SD, Mwangi W (2002) Adoption of maize technologies in East Africa—what happened to Africa's emerging maize revolution? A paper presented at the FASID forum V. In: Green revolution in Asia and its transferability to Africa. Tokyo

Erenstein O, Kassie GT, Langyintuo A, Mwangi W (2011) Characterization of maize producing households in drought prone regions of Eastern Mexico

FAO—Food and Agricultural Organisation (2011) Situation analysis: improving food safety in the maize value chain in Kenya

FAO Food and Agricultural Organisation (2014) The state of food insecurity in the world 2014. Strengthening the enabling environment for food security and nutrition. FAO, Rome, Italy

FAO-WFP—Food and Agriculture Organization—World Food Summit (1996) Food security defination. In: Towards a food insecurity multidimensional index (FIMI) (2011). Retrieved from http://www.fao.org/fileadmin/templates/ERP/uni/FIMI.pdf. Assessed on 25 Oct 2016

FEWSNET—Famine Early Warning Systems Network (2009) Kenya food security outlook update. Retrieved from http://www.fews.net/docs/Publications/Kenya_FSU_2009_03_final.pdf . Accessed on 29 May 2011

FEWSNET—Famine Early Warning Systems Network (2010) Food security framework. Retrieved from http://www.fews.net/ml/en/info/Pages/fmwkfactors.aspx?gb=ke&l=en&loc=2

FEWSNET—Famine Early Warning Systems Network (2011) Kenya food security outlook. Retrieved from http://www.fews.net/docs/Publications/Kenya%20Food%20Security%20Outlook_April%202011_final.pdf

FEWSNET—Famine Early Warning Systems Network (2013) Kenya food security brief. Retrieved from http://www.fews.net/sites/default/files/documents/reports/Kenya_Food%20Security_In_Brief_2013_final_0.pdf

Funk CC, Brown ME (2005) Intra-seasonal NDVI change projections in semi-arid Africa. Remote Sens Environ 101(2006):249–256

Gitelson AA, Kaufman YJ, Stark R, Rundquist D (2002) Novel algorithms for remote estimation of vegetation fraction. Remote Sens Environ 80(1):76–87. https://doi.org/10.1016/S0034-4257(01)00289-9

Grace K, Brown M, McNally A (2014) Examining the link between food prices and food insecurity: a multi-level analysis of maize price and birthweight in Kenya. Elsevier J Food Policy 46(2014):56–65. https://doi.org/10.1016/j.foodpol.2014.01.010

Huete AR, Karl FH, Tomoaki M, Xiangming X, Didan K, Willem L (2006) Vegetation index greenness global data set. National Aeronautics and Space Administration

Ifejika S, Chinwe Kiteme B, Wiesmann U (2008) Droughts and famines: the underlying factors and the causal links among agro-pastoral households in semi-arid Makueni district, Kenya. Glob Environ Change 18(1):220–233

Jayne TS, Myers Robert J, Nyoro James (2008) The effects of NCPB marketing policies on maize market prices in Kenya. J Int Assoc Agric Econ 38(2008):313–325

Kangethe E (2011) Situational analysis: improving food safety in the maize value chain in Kenya. FAO technical report. Retrieved from http://www.fao.org/fileadmin/user_upload/agns/pdf/working_paper_aflatoxin_reportdj10thoctober.pdf. Accessed on 22 Oct 2016

Kuiper EW, Lutz C, van Tilburg A (2003) Vertical price leadership on local maize markets in Benin. J Dev Econ 71(2):417–433

Lewis JE, Rowland J, Nadeau A (1998) Estimating maize production in Kenya using NDVI: some statistical considerations. Int J Remote Sens 19(13):2609–2617

Li X, Yeh AGO (1998) Principal component analysis of stacked multi-temporal images for the monitoring of rapid urban expansion in the Pearl River Delta. Int J Remote Sens 19:1501–1518

Livingston G, Schonberger S, Delaney S (2011) Sub-Saharan Africa: the state of smallholders in agriculture. In: Conference on new directions for smallholder agriculture 24–25 Jan, Rome IFAD HQ, pp 1–31

Mariara J, Karanja F (2007) The economic impact of climate change: a Ricardian approach. World Bank Policy Research Working Paper, Washington

Meier U (2001) Growth stages of mono-and dicotyledonous plants. BBCH Monographie, p 166

Minot N (2010) Transmission of world food price changes to markets in sub-Saharan Africa. International Food Policy Research Institute, Washington

Mishra AK, Singh VP (2010) A review of drought concepts. J Hydrol 391(1–2):202–216

Mkhabela MS, Bullock P, Raj S, Wang S, Yang Y (2011) Crop yield forecasting on the Canadian prairies using MODIS NDVI data. Agric For Meteorol 151(3):385–393. https://doi.org/10.1016/j.agrformet.2010.11.012

Mundlak Y, Larson D (1992) On the transmission of world agricultural prices. World Bank Econ Rev 6(3):399–422

Mutanga O, Skidmore AK (2004) Narrow band vegetation indices overcome the saturation problem in biomass estimation. Int J Remote Sens 57:263–272

Nyoro JK (2002) Kenya's competitiveness in domestic maize production: implications for food security. Tegemeo Institute, Egerton University, Nairobi, Kenya

Nyoro JK, Kirimi L, Jayne TS (2004) Competitiveness of Kenyan and Ugandan maize production: challenges for the future. Tegemeo Institute of Agricultural Policy and Development, Nairobi

Ogbodo JA, Anarah SE, Sani Mashi A (2018) GIS-Based assessment of smallholder farmers' perception of climate change impacts and their adaptation strategies for maize production in Anambra State, Nigeria. In: Corn - production and human health in changing climate. https://www.researchgate.net/publication/328201491_GIS-Based_Assessment_of_Smallholder_Farmers'_Perception_of_Climate_Change_Impacts_and_Their_Adaptation_Strategies_for_Maize_Production_in_Anambra_State_Nigeria

Pettorelli N, Vik JO, Mysterud A, Gaillard J-M, Tucker CJ, Stenseth NC (2005) Using the satellite-derived NDVI to assess ecological responses to environmental change. Trends Ecol Evol 20(9):503–510

Rijks O, Massart M, Rembold F, Gommes R, Leo O (2007) Crop and rangeland monitoring in Eastern Africa. In: Proceedings of the 2nd international workshop, Nairobi, 28–30 Jan 2007, pp 95–104

Rojas O (2007) Operational maize yield model development and validation based on remote sensing and agrometeorological data in Kenya. Int J Remote Sens 28:3775–3793. https://doi.org/10.1080/01431160601075608

Rojas O, Vrieling A, Rembold F (2011) Assessing drought probability for agricultural areas in Africa with coarse resolution remote sensing imagery. Remote Sens Environ 115(2):343–352

Sacks WJ, Deryng D, Foley JA, Ramankutty N (2010) Crop planting dates: an analysis of global patterns. Glob Ecol Biogeogr 19(5):607–620. https://doi.org/10.1111/j.1466-8238.2010.00551.x

Salami A, Kamara AB, Brixiova Z (2010) Smallholder agriculture in East Africa: trends, constraints and opportunities. Working paper No. 105

Sawasawa HLA (2003) Crop yield estimation: integrating remote sensing, GIS and management factors: a case study of Birkoor and hortgiri mandals-Nizambad District, India. M.Sc. thesis, ITC, Enschede

Short C, Mulinge W, Witwer M (2012) Analysis of incentives and disincentives for maize in Kenya. Technical notes series. MAFAP, FAO, Rome

Shuaibu Sakirat M, Ogbodo JA, Wasige EJ, Sani Mashi A (2016) Assessing the impact of agricultural drought on maize prices in Kenya with spot-vegetation NDVI remote sensing approach. Future Food J Food Agric Soc 4(3). University of Kassel, Germany and the Federation of German Scientists (VDW). Retrieved from https://www.researchgate.net/publication/312328545_Assessing_the_impact_of_agricultural_drought_on_maize_prices_in_Kenya_with_the_approach_of_the_SPOT-VEGETATION_NDVI_remote_sensing

Smale M, Byerlee D, Jayne TS (2011) Maize revolutions in Sub-Saharan Africa. https://ageconsearch.umn.edu/bitstream/202592/2/Wp40-Maize-Revolutions-in-sub-Saharan-Africa.pdf. Accessed 3 June 2018

Tachiiri K (2003) Time lag between seasonal change in precipitation, NDVI and cereal prices in Burkina Faso. Paper presented at the geoscience and remote sensing symposium, 2003. IGARSS proceedings. IEEE International

Tucker CJ, Pinzón JE, Brown ME, Slayback DA, Pak EW, Mahoney R (2005) An extended AVHRR 8 km NDVI dataset compatible with MODIS and SPOT vegetation NDVI data. Int J Remote Sens 26(20):4485–4498

Tucker CJ (1979) Red and photographic infrared linear combinations for monitoring vegetation. Remote Sens Environ 8(2):127–150

Van Campenhout B (2007) Modelling trends in food market integration: method and an application to Tanzanian maize markets. Food Policy 32(1):112–127

Vrieling A, de Beurs KM, Brown ME (2011) Variability of African farming systems from phenological analysis of NDVI time series. Clim Change 109(3–4):455–477. https://doi.org/10.1007/s10584-011-0049-1

Wardlow BD, Kastens JH, Egbert SL (2006) Using USDA crop progress data for the evaluation of greenup onset date calculated from MODIS 250-Meter Data. Photogramm Eng Rem S 72:1225–1234. https://doi.org/10.14358/PERS.72.11.1225

Williams A, Funk C (2010) A westward extension of the tropical pacific warm pool leads to March through June drying in Kenya and Ethiopia. US Geological Survey Open file Report

Xue J, Su B (2017) Significant remote sensing vegetation indices: a review of developments and applications. J Sens 2017(Article ID 1353691):1–17. https://doi.org/10.1155/2017/1353691

Zhou L-M, Li F-M, Jin S-L, Song, Y (2009) How two ridges and the furrow mulched with plastic film affect soil water, soil temperature and yield of maize on the semiarid Loess Plateau of China. https://www.sciencedirect.com/science/article/pii/S0378429009000951. Accessed 30 June 2018

John A. Ogbodo, M.Sc. is the Founder and Chief Executive Director (CED) of STEi Foundation—Sustainable TransEnvironment International Foundation. He is a United Nations alumnus-Fellow on Partnership for Actions on Green Economy (PAGE) and a Ford International alumnus-Fellow/Social Justice Advocate. The author is currently working with Nnamdi Azikiwe University Awka–Nigeria, where he doubles as a remote sensing/GIS lecturer in the Department of Forestry and Wildlife Management, Faculty of Agriculture; and, an Associate Researcher of the University's Center for Sustainable Development (CSD). Ogbodo's research interests bother on remote sensing/GIS for natural resources management. He holds a M.Sc. degree in Geo-Information Science and Remote Sensing for Natural Resources Management from the University of Twente, Enschede, Netherlands; and a Bachelor of Forestry degree from the Federal University of Agriculture, Makurdi–Nigeria.

Ejiet John Wasige, Ph.D. I am a Geoscientist and Ecosystem modeler trained in remote sensing and Geographic Information Science (GIS), agricultural sciences, soil science, hydrology and Ecosystem modeling. I am interested in examining spatial and temporal environmental changes, and their implications for landscape-level management. I have profound experience in systems analysis to explore future scenarios for land use with a focus on food production and sustainable environmental management. I analyze big volume satellite imagery and crop-hydrological modeling to examine landscape and climatic changes over space and time. I have excellent skills in processing and managing multi-sensor remote sensing data from wide array of sensors such as Landsat, MODIS, AVHRR, SPOT and other systems.

Sakirat M. Shuaibu, M.Sc. I am a natural resources economist and a holder of M.Sc. in Earth Observation and Remote Sensing for natural resources management from the Faculty ITC, University of Twente in Enschede, Netherlands.

Timothy Dube, Ph.D. I hold a Ph.D. in GIScience and Earth Observation for water resources. My research interests include: forest biomass and carbon stocks mapping for climate change, landuse and landcover change; crop modelling for food security; ecosystems productivity and degradation mapping for biodiversity and climate change monitoring; water use and stress accounting in semi-arid regions.

Samuel Emeka Anarah, M.Sc. is an academic technologist in the Department of Agricultural Economics and Extension, Nnamdi Azikiwe University Awka, Anambra State, Nigeria. He obtained his B.Tech Agricultural Economics and Extension from Federal University of Technology Minna, Niger State and MSc. in Resource and Environmental Economics from Nnamdi Azikiwe University Awka. His research interests include: farm management, agricultural credit and finance, resource and environmental economics.

Anthropic Action Effects Caused by Soybean Farmers in a Watershed of Tocantins—Brazil and Its Connections with Climate Change

José Jamil Fernandes Martins, Amadeu M. V. M. Soares, Ulisses M. Azeiteiro and Mauro Lucio Torres Correia

Abstract This chapter presents the design and development of an Environmental Sustainability Index for the Evaluation of Environmental Agricultural Sustainability (impacts from anthropic action) applied to the Brazilian agricultural production Soybean monoculture area. The Agricultural Sustainability Index (*ISAGRI*) aims to express the degree of environmental sustainability of any agricultural production system, regardless of the system being of organic or conventional production. As such, it was designed to contain a minimum set of representative indicators of the quality of soil and water, of degradation and of measures compatible with sustainability. These indicators are useful both to compare the degree of sustainability of production systems, watersheds and regions, and to evaluate in any of these dimensions, its evolution over time. The *ISAGRI* employs mainly quantitative data, being a result of the aggregation of sub-indexes and indicators and it consists of a numerical value. This is because a high degree of comparability, as it proposes, is necessarily linked to the use of numerical data. The indicators part of these indexes were grouped into three dimensions named: Pressure—characterization of the environmental degradation caused by its use; State—identification of environmental quality and of natural resources and Response—measures taken for preservation of environmental quality. For a better comprehension of the circumstances under review, four alternative scenarios were proposed, preceded by some hypotheses, where the parameters that characterize each of them are: changing the type of planting, land use and used area or a combination of the above (planting and area). From these numbers, which are significant, it remains very clear the need for strict

Electronic supplementary material The online version of this chapter (https://doi.org/10.1007/978-3-319-75004-0_15) contains supplementary material, which is available to authorized users.

J. J. F. Martins (✉) · A. M. V. M. Soares · U. M. Azeiteiro
Department of Biology & CESAM, Aveiro University, Aveiro, Portugal
e-mail: jamil@ecologica.org.br

M. L. T. Correia
IFTO Tocantins Federal Institute of Education, Palmas, TO, Brazil

P. Castro et al. (eds.), *Climate Change-Resilient Agriculture and Agroforestry*,
Climate Change Management, https://doi.org/10.1007/978-3-319-75004-0_15

compliance with the environmental law that deals with forest reserve, including the restrictions to soil slopes, as well the use of no-tillage.

Keywords Agro-ecosystems · *Cerrado* · Soybean · Brazil · Soil health
Agricultural sustainability index

1 Introduction

Agro-Ecosystems are subjected to various agricultural pressures and interconnected environmental driving forces at the regional and global scales. Monoculture exploitation and expansion may have important consequences for the sustainability in a changing climate scenario.

Soybean production in Brazil has grown rapidly in recent years and according to the Brazilian Agricultural Research Corporation (EMBRAPA) that is under the aegis of the Ministry of Agriculture, Livestock, and Food Supply, Brazil is the leading producer of soybeans in South America being the world's second largest producer of grain.

Surveys conducted by RADAM-BRASIL project and by EMBRAPA projects at the state level (Previero 2006), indicate that 60% of the surface of the State of Tocantins soil is tillable and 25% more, approximately, is likely to be improved using available technology (Tocantins is the newest state in the Brazilian Federation that has about 200,000 km^2 of flatlands, fit for mechanized farming, easily irrigable and with a stable tropical climate appropriate to agropastoral activities) (Martins 2014, p. 18). To accomplish this, the studies carried out by EMBRAPA since the seventies that developed the technologies for grain crops, especially soy, in the areas of the *cerrado* biome were fundamental (Contini et al. 2010).

From the standpoint of the technology adopted by local soybean farmers, at Pedro Afonso (TO), where this study was carried, it was noted in loco that they all employ the no-tillage system. According to the observations and information obtained from the Pedro Afonso Agricultural Cooperative (COAPA 2009), provided by technicians and soybean farmers, throughout the cultivation process around 350 kg/ha^{-1} of fertilizer and 4.85 kg/ha^{-1} of pesticides are used. Apropos, Bernardi et al. (2003) recommend the use of the same level of pesticides and fertilizers in the case of no-tillage soybean crops in the *cerrado*, with an average yield of 2.67 t ha^{-1} (COAPA 2009). The average obtained in Brazil, 2015/2016 crop, was 2882 kg/ha (with losses) (EMBRAPA). In general, the farming process under consideration is carried out in the conventional manner.

As a result of the aforementioned farming process, an important aspect to be considered refers to the real possibility of the presence of negative externalities in terms of anthropic activities, practiced by soybean farmers. Can be found severe environmental problems like those caused by industrial pollution in terms of contamination and pollution of the water, erosion of cultivated areas, with loss of fertility and desertification (due to inappropriate land use), degradation, habitat destruction and reduction of biodiversity by deforestation and waste discarded in the soil, silting of streams and rivers and the influence of deforestation in the rainfall cycle and in desertification, as well as the impacts on biogeochemical cycles and regional climate change caused by pasture and soybean cropland expansion (Sampaio et al. 2007).

It's noticeable that in the short term there was a significant expansion of soybean monoculture in Tocantins *cerrado*. In the 1990/1991 crop, its acreage was 4500 ha (IBGE 2009). In the 2015/2016 crop this figure surpassed 870,000 ha (SEAGRO 2016). Until the 2003/2004 harvest Pedro Afonso was the Tocantins municipality with the highest growth rate of the planted area and of soybean production in the state (IBGE 2009).

Evaluating the agricultural anthropic impacts is a guarantee of sustainable production and agricultural sustainability together with forest and ecosystem preservation with increased food production. It is proposed to analyze the effects of the anthropic actions caused by soybean farmers established in the Lajeado Stream watershed of Pedro Afonso—TO, Brazil, and to discuss the impacts caused by climate change observed in the area. The aim is to assess the degree of sustainability of this farming method and also their interrelations with the ongoing climate change.

For a better comprehension of the circumstances under review, four alternative scenarios were proposed, preceded by some hypotheses, where the parameters that characterize each of them are: changing the type of planting, land use and used area or a combination of the above (planting and area). This study is based on Martins (2014), which employs the model developed by Silva (2007). It adopts the concept of watershed expressed in CATI (2000), establishing it be "a geographical area bounded by watersheds (spikes), drained by a river or stream, which drains rainwater."

2 Evaluation of Environmental Agricultural Sustainability due to Anthropic Action

Pimentel et al. (1995), apud Silva (2007) state that approximately ten million hectares of arable land become unproductive on the planet annually. Therefore, assessing the impacts caused by agricultural activity, using the environmental indicators and in the form of sustainability indexes, is fundamental for proper management and mitigation of the loss of arable land. Hence, it comes to the object of the study, using the Agricultural Sustainability Index—*ISAGRI* created by Silva (2007).

2.1 Model of Pressure-State-Response of Environmental Indicators

The following are important aspects regarding ISAGRI preparation.

Nowadays, the biggest source of environmental indicators were developed by the “Organization for Economic Cooperation and Development” (OECD 1993), apud Silva (2007), which provided the first tool for environmental monitoring. Its group of indicators covers a wide range of environmental issues and additionally incorporates indicators obtained from some industry groups and from environmental accounting systems. It must be emphasized that this system employs the Pressure-State-Response model (PSR), proposed by the OECD (1997), apud Silva (2007), which has been accepted and adopted worldwide (Lira and Cândido 2008). This model is based on the concept that human activities exert pressures on the environment, changing the quality and the amount of natural resources, namely, changing its state. Society reacts to these changes through environmental, economic and sectoral policies. Although that model may suggest a linear interaction between activities and the environment, it should be considered that such relationships are complex and therefore the externalities do not always manifest themselves in a linear fashion. From it three types of environmental indicators are specified (Fig. 1): (a) Environmental Pressure Indicators; (b) Status and Environmental Conditions Indicators and (c) Indicators of Social Responses.

In this model the pressures on the environment are reduced to those caused by man’s actions, excluding those from the action of nature. According to Lira and Cândido (2008), in order to better integrate environmental aspects to sectoral policies, the OECD (1998) sought to group the indicators by themes and sectors. Classification by themes was divided into: climate change, depletion of the ozone layer, eutrophication, acidification, toxic contamination, environmental urban quality, biodiversity, cultural landscapes, waste, water resources, forest resources, fisheries, land degradation (desertification and erosion) and general indicators. The sectors were classified mainly in industry, energy, transport and agriculture.

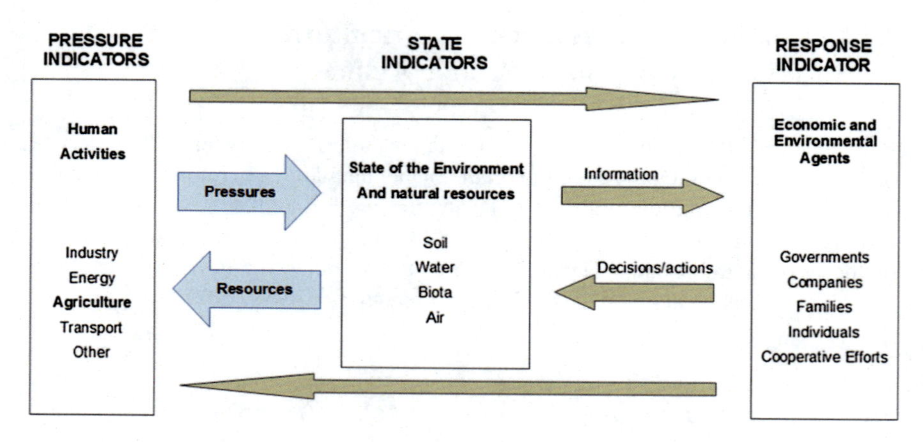

Fig. 1 PSR model for environmental monitoring. *Source* Tomasoni (2006)

2.2 Preliminary Information About ISAGRI

Theoretically, the preparation of *ISAGRI* was based on the concept of environmental agricultural sustainability seen before. Conceptually, the index is an instrument that was designed to inform decision makers, the media and the general public about the conditions of environmental sustainability of a particular farming system. In *ISAGRI* soil quality is considered a major factor in maintaining environmentally sustainable agricultural systems. Water quality plays a more modest contribution, contributing only one (1) indicator of the eight (8) in which *ISAGRI* is based. For Larson and Pierce (1994), Karlen et al. (1997), apud Silva (2007), a sustainable agricultural system is one that maintains the quality of the soil in the long term, through management practices considered conservationists such as no-tillage; crop rotation; use of green and organic fertilizers; employing cultivation methods that maximize the biological activity and maintain the productivity and fertility of the soil; as well as natural and biological control of pests, diseases and weeds. The soil quality is determined by its ability to perform its function in nature, keeping its ability to function as a medium for plant growth, regulating and compartmentalizing water flow in the environment, stocking and promoting the cycling of elements in the biosphere and functioning as an environmental "buffer" in the formation, mitigation and decomposition of harmful compounds in the environment.

The *ISAGRI* aims to express the degree of environmental sustainability of any agricultural production system, regardless of the system being of organic or conventional production. As such, it was designed to contain a minimum set of representative indicators of the quality of soil and water, of degradation and of measures compatible with sustainability. These indicators are useful both to compare the degree of sustainability of production systems, watersheds and regions, and to evaluate in any of these dimensions, its evolution over time. The *ISAGRI* employs mainly quantitative data, being a result of the aggregation of sub-indexes and indicators and it consists of a numerical value. This is because a high degree of comparability, as it proposes, is necessarily linked to the use of numerical data.

The indicators part of these indexes were grouped into three dimensions named: Pressure—characterization of the environmental degradation caused by its use; State—identification of environmental quality and of natural resources and Response—measures taken for preservation of environmental quality (Table 1). These partial indexes were named as: (a) *IDEG*—degradation vector index; (b) *IEA*—Index of agricultural ecosystem status; (c) *ICOR*—Index of prevention and correction measures, respectively (Table 1).

The *ISAGRI* is therefore a synthetic index composed of 8 indicators, which were converted into indexes, and then aggregated to the dimensions to which they belong. The indicators proposed for the composition of the sub-indexes were separately analyzed according to their function, coverage degree and unit of measure (Table 1).

Table 1 Partial indexes, sub-indexes and components ISAGRI indicators

Environmental agricultural sustainability indexes (ISAGRI)		
Partial indexes/dimension	Sub-indexes	Indicators
Agricultural ecosystem state index (IEA)/State	Water quality index (IQA)	1—Water physical-chemical parameters
	Soil physical quality index (IQF)	2—Soil penetration resistance (MPa) 3—Porosity (%)
Degradation vectors index (IDEG)/Pressure	Erosion index (IERO)	4—Soil Loss (kg ha^{-1})
	Defensive use contamination potential index (IDEF)	5—Quantity of defensives used (kg ha^{-1})
	Fertilizer use contamination potential index (IFERT)	6—Quantity of de P_2O_5 and N used (kg ha^{-1})
Prevention and correction measures index (ICOR)/ Response	Soil management index (IMANEJ)	7—Soil use and management 8—Conservationist practices

3 Study Area

The Lajeado Stream watershed is located in the Pedro Afonso—TO municipality (Fig. 2); the Stream is about 30 km long, on average 15 m wide and has a flow of 32.70 cubic meters per second in its mouth, comprising a total area of 45,411.03 ha, distributed as follows: 15,154.41 ha with soybeans crops; 26,069.32 ha of land with typical *cerrado* cover; 2273.16 ha in fallow and 1914.14 ha with sugarcane crops (COAPA 2009). An important part of this area consists of flat lands, easily mechanized, offering favorable conditions for growing the crop in study.

4 Methodology

4.1 Methodology for Construction of the Agricultural Environmental Sustainability Index (ISAGRI)

The preparation of *ISAGRI* started by the indicators of agricultural environmental sustainability (Table 1) that were transformed into indexes whose values range between zero and one, so that those higher indicate greater sustainability. To quantify these indexes the worst and the best possible values of the indicator were chosen.

The best value (bv) is the reference established by the literature or the one found in the list of events that occur in the studied watershed, as long as this value is better

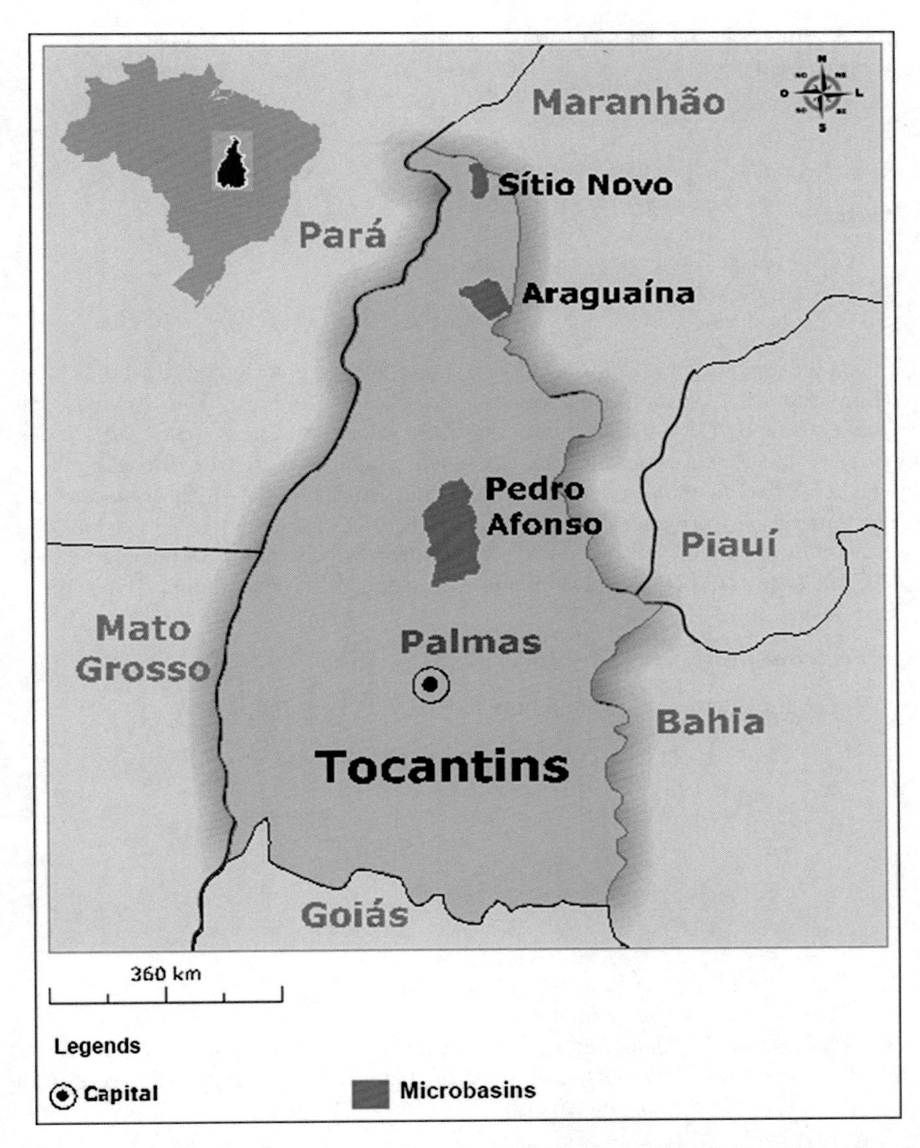

Fig. 2 Location of the Lajeado Stream watershed of Pedro Afonso municipality, in the State of Tocantins, Brazil

than the one recorded in the literature. In such way, a higher degree of rigidity of the index is employed, taking into account the provisions of the mathematical expression (1).

The worst value, chosen from the perspective of sustainability, is the most aggressive to the environment, within the universe that includes all observations collected for that variable in all watersheds considered (supplementary information 1).

$$Index = \frac{\overline{vo} - pv}{mv - pv} \tag{1}$$

where:

(a) $\overline{vo}$ = average value observed for the indicator;
(b) pv = worst value;
(c) mv = best value.

For the preparation of *ISAGRI*, different weights were attributed to all indicators of environmental sustainability of each dimension (Table 1). The Soil Physical Quality Index (*IQF*), which composes the Agricultural Ecosystem Status Index (*IEA*) of the status dimension was assigned weight two. It was also assigned a weight of two to the Conservationist Practices Index (*IP*), which composes the Prevention and Correction Measures Index (ICOR) of the answer dimension. The Erosion Index (*IERO*) was assigned weight three, which composes the degradation Vectors Index (*IDEG*) of the pressure dimension. The other indexes were given weight one.

Agricultural Ecosystem Status Index (IEA)

The Agricultural Ecosystem Status Index (*IEA*) is as follows:

$$\mathrm{IEA} = \frac{\mathrm{IQA} + 2\mathrm{IQF}}{3} \tag{2}$$

$$\mathrm{IQA} = \frac{\sum_{i=1}^{7} \mathrm{IPFi}}{8} \tag{3}$$

where

(a) (IEA) = Agricultural Ecosystem Status Index;
(b) (IQA) = Water Quality Index;
(c) (IQF) = Soil Physical Quality Index (supplementary information 2);
(d) IPFi = Average index of water physical-chemical parameters in each collection point (Supplementary information 3).

To compose the *IQF* the following soil physical variables were used: porosity and penetration resistance. Therefore, the *IQF* was calculated by the mathematical expression:

$$\mathrm{IQF} = \frac{\mathrm{IPOR} + \mathrm{IRP}}{2} \tag{4}$$

where:
IPOR = soil porosity index

$$IPOR = \frac{\sum_{i=1}^{n} IMACi}{n} \quad (5)$$

where:
IMACi = Soil Macroporosity Index, average, in each collection point (supplementary information 2)

$$IRP \frac{\sum_{i=1}^{n} IRPi}{n} \quad (6)$$

where:

(a) IRP = Soil Penetration Resistance Index;
(b) IRPI = Soil Penetration Resistance Index, average, in each collection point (supplementary information 4).

Degradation Vectors Index (IDEG)

The Degradation Vectors Index (*IDEG*) is made-up of the Index Agricultural Pesticides Use Contamination Potential (*IDEF*), together with the Fertilizer Use Contamination Potential.

Index (*IFERT*) and the Erosion Index (*IERO*) as below:

$$IDEG = \frac{IDEF + IFERT + 3IERO}{5} \quad (7)$$

$$IDEF \frac{\sum_{i=1}^{n} IQDEFi}{n} \quad (8)$$

where:
IQDEFi = Quantity of Agricultural Pesticides Used Index (supplementary information 5):

$$IFERT \frac{\sum_{i=1}^{n} IQFERTi}{n} \quad (9)$$

where:
IQFERTi = Quantity of Used Fertilizer Index (supplementary information 6)

$$IERO = \frac{\sum_{i=1}^{n} IPSi}{n} \quad (10)$$

where:

(a) IPSI = Soil Loss Index (supplementary information 7)

Soil Management Index (IMANEJ)

The Soil Management Index (*IMANEJ* = ICOR) is obtained from the mathematical equation:

$$\mathrm{IMANEJ} = \frac{\mathrm{IC} + 2\mathrm{IP}}{3} \quad (11)$$

$$\mathrm{IC} = \frac{\sum_{i=1}^{n} \mathrm{ICi}}{n} \quad (12)$$

where:

(a) IC = Soil Use and Management Index;
(b) ICi = Soil Use and Management Index of the considered crops. For this case, we adopt the weighted average applied to each type of crop within the studied watershed.

For the municipality of Pedro Afonso, i = 1 Þ soybean; i = 2 Þ sugarcane; i = 3 Þ *cerrado*; and i = 4 Þ fallow.

$$\mathrm{IP} = \frac{\sum_{i=1}^{n} \mathrm{IPCi}}{n} \quad (13)$$

where:

(a) IPCI = Conservation practices Index in the considered crops.

For this case, the weighted averages applied to each type of crop within the watershed are adopted.

For the municipality of Pedro Afonso, i = 1 soybean; i = 2 sugarcane; i = 3 *cerrado*; and i = 4 fallow.

The Agricultural Environmental Sustainability Index (ISAGRI) is therefore made by:

$$\mathrm{ISAGRI} = \frac{\mathrm{IEA} + \mathrm{IDEG} + \mathrm{ICOR}}{3} \quad (14)$$

Those interested in more details on the methods of water and soil analysis, whose results were used to determine the Agricultural Ecosystem Index Status (*IEA*) and the Soil Management Index (*IMANEJ*) may get them in Martins (2014).

4.2 *Sustainability Indication Thermometer*

Aiming to evaluate the sustainability level of anthropogenic activity, Silva (2007) established the "sustainability indicator thermometer", based on key indicators of agricultural environmental sustainability (Fig. 3).

This thermometer classifies a given anthropogenic activity in levels, and the closer the value of the index is to 1.00, the higher the level of agricultural environmental sustainability.

5 Results and Discussion

5.1 *Agricultural Environmental Sustainability Index and Its Sub-indexes and Partial Indexes*

Agricultural Ecosystem Status Index (IEA)

We begin the analysis of the results by looking at the Water Quality Index (*IQA*), the first component of the *IEA* (Table 2), in the Lajeado Stream micro watershed.

The average *IQA* value for the three seasons and for the samples collected at the source, located at the Xerente reservation was 0.829 (Table 2), the best value observed in that micro watershed. This is justified, evidently, due to its location,

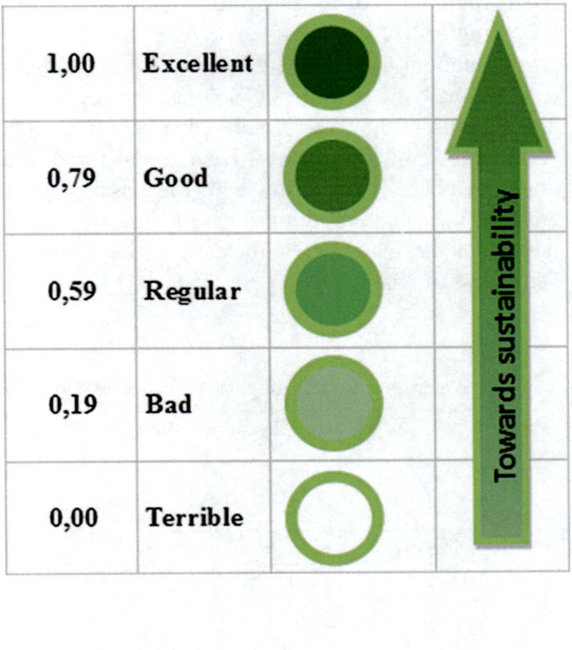

1,00	**Excellent**		Towards sustainability
0,79	**Good**		
0,59	**Regular**		
0,19	**Bad**		
0,00	**Terrible**		

Fig. 3 Sustainability indication levels *Source* Silva (2007)

Table 2 Water quality indexes in the Lajeado stream watershed

Microbasins	Identification of the water resource	Water quality index (IQA)			
		Spring average[a]	Middle average[a]	Mouth average[a]	Total average
Pedro Afonso—TO	Lajeado stream	0.829	0.646	0.707	0.727

[a]Average of three samples collected at three different periods

outside the farmers' action area. The average value of 0.646, recorded in the intermediate region, was the lowest, since it concentrates the largest volume of soybean crops. This result reflects the direct effect of anthropogenic activities in the area and the consequences of pesticides and fertilizers use, which impact the water springs. The second worst rate, 0.707, recorded at the mouth, reflects the improvement caused by the greater distance of the core of the plantations and the resilience effect. For a better understanding of the fact, it should be noted that the flow of the Lajeado Stream at its mouth is 32.7 cubic m/s (Table 3), while in its spring it is only 0.217 cubic m/s and in the intermediate region is 19.312 m^3/s. It reinforces the hypothesis of the most obvious resilience action in the mouth region.

Therefore, from the *IQA* viewpoint, with a final value of 0.727 (Table 2) under current culture conditions, it was found that the soybean planting is not causing significant environmental damage to the Pedro Afonso—TO region. However, looking deeper into the variables of the *IQA*, one realizes that ammonia nitrogen and phosphorus, present discreet alterations between the samples collected at the mouth of Lajeado Stream compared to those taken at its spring. The observed levels for ammonia nitrogen and phosphorus in the spring are respectively 0.029 and 0.070 mg/L^{-1}. However, in its mouth, the values are increase to 0.172 and 0.228 mg/L^{-1} (Table 4).

Throughout the considered period the ammonia nitrogen was within the limit set by the Brazilian authorities (Table 5). However, due to phosphorus levels observed at the mouth of the river, which only in the last collection held (Table 6) reaches the acceptable limit, rings the proviso signal for the risk of eutrophication, even if it does not occur in loco, but with the possibility of it happening in accumulation areas of these waters, downstream of the studied watershed (with the recent completion of the Estreito Hydroelectric Power Plant, located north of Tocantins and southwest of Maranhão states).

Table 3 Water flow at the spring, at the midpoint and at the mouth of the studied watershed

Microbasins	Identification of the water resource	Flow of water resources at collection points (m^3 s^{-1})		
		Spring	Middle	Mouth
		m^3 s^{-1}		
Pedro Afonso—TO	Lajeado stream	0.217	19.312	32.700

Table 4 Data for IQA calculation in the Lajeado stream

Scenarios description		IQA	pH	Turbidity	Dissolved oxygen	Total ammoniac nitrogen	Nitrate	Nitrite	Phosphor total	Chl	IPOR	IRP	IQF	IEA
Pedro Afonso original		**0.727**	5.503	6.556	7.722	0.204	0.001	0.140	0.249	0.306	**0.572**	**0.836**	**0.704**	**0.712**
Spring (cerrado)	**Indexes**	**0.829**	0.311	0.986	0.527	0.969	0.953	0.946	0.961	0.976	**0.888**	**0.979**	**0.933**	**0.898**
	Raw Data					*0.029*			*0.070*					
Middle (soy)	**Indexes**	**0.646**	0.315	0.870	0.415	0.410	0.728	0.819	0.639	0.976	**0.277**	**0.688**	**0.482**	**0.537**
	Raw Data					*0.412*			*0.449*					
Mouth (soy)	**Indexes**	**0.707**	0.301	0.903	0.492	0.760	0.720	0.659	0.827	0.992				
	Raw Data					*0.172*			*0.228*					

Table 5 Water quality indicators determined in water samples collected at three different points in the Lajeado Stream micro basin in Pedro Afonso—TO

Parameter	CONAMA index (Class II waters)	Best value	Worst value	Observed value	Parameter index
				Lajeado stream	
pH	6.00 < pH < 9.00	7.50	4.61	5.50	0.309
Turbidity (UNT)	<100.000	1.000	70.000	6.556	0.919
Dissolved oxygen (mg L^{-1})	>5.000	8.400	7.100	7.722	0.478
Total ammoniac N (mg L^{-1})	<3.700	0.008	0.693	0.204	0.713
Nitrite (mg L^{-1})	<1.000	0.000	0.005	0.001	0.800
Nitrate (mg L^{-1})	<10.000	0.068	0.443	0.140	0.808
Total phosphor (mg L^{-1})	<0.030	0.024	1.200	0.249	0.809
Chlorophyll a (μg L^{-1})	<30.000	0.000	16.430	0.306	0.981

Table 6 Values of total phosphorus (PO_4^- mg/L) measured in the Lajeado stream micro basin in Pedro Afonso

Collection date	CONAMA index	Collection point (Mouth)
04/08/08	<0.03	0.030
23/01/2009		0.056
08/06/2009		0.6
16/07/2009		0.074
27/08/2009		0.012

Eutrophication will increase in a climate change scenario (Moss et al. 2011). Ongoing and future monitoring programs to address the effects of eutrophication and climate change (see Fragoso et al. 2011, for the use of modeling and simulation effects of nutrient enrichment and climate change) together with research on the impacts of climate change factors on eutrophication (see Xia et al. 2016) are necessary.

Therefore, the growth of agriculture in the region, from the *IQA* viewpoint, based on available technology, should already be preceded by more accurate studies regarding environmental protection. It should also be noted that in the vicinity of the area now planted with soybeans a sugarcane cultivation project is being implemented, with an estimated 33,000 ha area. Thus, the deleterious effects to the local environment may worsen if appropriate preventive measures are not taken. It's noteworthy to mention it is being considered the implementation of this project in part of the area presently being used for soy crops.

Soil Physical Quality Index (IQF)

The result of applying the Soil Physical Quality Index (*IQF*) in the Lajeado Stream micro basin show some interesting aspects (Table 7). In the area where the Xerente Indians are settled, also bathed by the Lajeado Stream, upstream of the area with soybean crops, as expected, the results for the Porosity Index—IPOR; of Soil Penetration Resistance—*IRP* and, consequently, the soil physical quality index—*IQF* were excellent, respectively, 0.888; 0.979 and 0.933, very close to the theoretically ideal physical quality (due to it being a long preserved environment). However, when the contiguous area occupied by soy crops is analyzed, which is the central object of our study, we are facing a worrying *IPOR* of only 0.277, even in a context where direct tilling is used.

It should be noted that the porosity is of great importance to the physical-chemical and biological processes such as infiltration, conductivity, drainage, water retention, nutrients diffusion, microorganisms and root growth (Moreira and Siqueira 2002), and these in turn are affected by the drivers of global change.

Soil properties, functioning and quality are affected by climate change and, as stated by Allen et al. (2011), understanding how changes in climate relate to soil health is vital (see Allen et al. 2011 for a review of "soil health indicator under climate change").

The worst Soil Quality Index, out of all the areas assessed in the state—*IQF* of 0.482, was observed in soybean crops (Table 7), showing that even in one of the least aggressive forms of cultivation, the direct tillage practiced in the region, there are serious environmental risks in terms of soil compaction. The Soil Porosity Index becomes significantly reduced with soybean monoculture crops.

Regarding the penetration resistance index: *IRP* = 0.688 (Table 7), recorded at the mouth of the Lajeado Stream, notwithstanding that it's in an area still considered good, it also inspires care. As previously seen, the trend is that as the level of soil porosity lowers, the penetration resistance increases and reduces the IPOR in the planted area, which is already considered very low.

In brief, despite the severe restrictions above commented, in its aggregate, the Agricultural Ecosystem Status Index (*IEA*), 0.712, found in the Lajeado Stream watershed, presented results considered good (Table 8).

Table 7 Porosity, resistance to penetration and soil physical quality indexes in the study watershed

Microbasins	Soil occupancy	Porosity index (IPOR)	Soil penetration resistance index (IRP)	Soil physical quality index (IQF)
Pedro Afonso	Xerente reservation	0.888	0.979	0.933
Pedro Afonso	Soybean crop	0.277	0.688	0.482

Table 8 Agricultural ecosystem status index in the studied micro basins

Description	Pedro Afonso	
	Value	Weight
Water quality index—IQA	0.727	2.0
Soil physical quality index—IQF	0.704	1.0
Agricultural ecosystem state index—IEA	0.712	

Soil porosity and pore size distribution needs attention in future studies, namely on the relationship of soil health indicators and climate change (see Allen et al. 2011), and the impacts of extreme weather events (droughts and/or erosion events) (Salvador Sanchis et al. 2008; Mullan 2013) and the necessary dedicated management strategies to prevent water erosion and the climate effect on soil erodibility (Salvador Sanchis et al. 2008; Mullan 2013).

Degradation Vectors Index (IDEG)

The results obtained by applying the Erosion Index—*IERO*, a component of *IDEG*, in the Lajeado Stream micro basin are presented in Table 9. At first it can be seen the minimal loss of soil per hectare/year, in the not yet explored *cerrado* region, where the Xerente Indians village lies. Only 65 kg, which yields an excellent 0.999 IERO. However, in the area effectively planted with soybean, the value of the loss rises to 34,726 ton. which generated a very low *IERO*, 0.428. This result really causes concern, given the known adverse effects of erosion in the production process. Two points are important to explain the difference in results, considering that the primitive coverage and pattern of soil from two areas were the same. The first point is that the area for soybeans crops was cleared and the second the production process, where the soil is subjected to mechanical actions of plowing, ploughing and chemical and biological processes seeking its correction, fertilization, pest control (remembering that in this case there was a reduction in the sequelae presented due to the practice of no-tillage).

In this sense, according to Dos Anjos and Van Raij (2004), the no-till system on straw without soil disturbance by plowing and ploughing, constitutes a management system that has been effective in erosion control. This procedure, in some way, at least minimizes the effects of deforestation and erosion. In a more systematic way, it is possible to say that its adoption results in a decrease in the loss of soil and nutrients and increased levels of organic matter and stored water. Consequently, there is improvement in soil quality with emphasis on its physical, chemical and biological qualities, in terms of greater availability of nutrients and microbial biomass volume, and the consequent increase in carbon and nitrogen content.

According to Table 9, regardless of the above restrictions, in general, the *IERO* found for the Lajeado Stream micro basin, 0.761, can be classified as regular. This positive outcome is the result of the high percentage of lands not arable or under permanent crops and the practice of no-tillage.

Soil erosion and its impacts (nutrient-holding, organic matter content and soil depth for the biota), under a climate change scenario, is a growing environmental

Table 9 Erosion index, IERO, by category of use in the Lajeado stream watershed

Microbasin	Soil occupancy	Area (ha)	Annual soil loss (A)	Best value (mv)	Worst value (pv)	R (MJ mm ha^{-1} h^{-1} $year^{-1}$)	K (t h MJ^{-1} mm^{-1})	LS	C	P	IERO
Pedro Afonso (TO)	Cerrado	26,069.32	0.065	0.00	60.70	11,788.80	0.03	1.33	0.0004	0.30	0.999
	Soybean	15,154.41	34.726						0.1069	0.60	0.428
	Sugarcane	1,914.14	23.677						0.0754	0.58	0.610
	Fallow	2273.16	37.620						0.1069	0.65	0.380
	Original	45,411.03[a]	14.507[b]						–	–	0.761[c]

[a]Total Microbasin Area evaluated at Pedro Afonso—TO, as a function of different forms of soil occupancy
[b]Weighed average of soil losses, as a function of different forms of soil occupancy of the evaluated microbasin in Pedro Afonso—TO
[c]IERO weighed average in Pedro Afonso—TO, considering the area total and forms of soil occupancy

problem in the future (see Salvador Sanchis et al. 2008 and Mullan 2013). Once again, mitigation measures and management of soil erosion and its impacts should be put in place for future sustainability in the area.

Contamination Potential Index from Agricultural Pesticides Use (IDEF)

Due to the volume of pesticides used in soybean crops, at the Lajeado Stream micro basin in Pedro Afonso, it appears that its *IDEF* is relatively low, 0.671 (Table 10).

According to some perspectives, from the special circumstances surrounding the market along the past five decades there had been an evolution of pesticides with its chemical characteristics presenting significant changes, especially in regard to the agronomic efficiency, ecotoxicology and toxicology (Menten 2009). With regard to the question of agronomic efficiency, in the 1960s, on average, it was consumed in the Country 1097.50 g a.i. ha^{-1} of insecticides and that in the 1990s this value fell steeply to 69.75 g ha^{-1} which shows the agronomical efficiency achieved, whereas smaller doses or smaller amounts of active ingredients are required to achieve the same goals in pest control (Seron 2010). According yet to Seron, op. cit., comparing the averages applied in the decades of 1960–1970 with the decades of 1990–2000, there was a percentage reduction in herbicide doses estimated at 88.4; insecticides by 93.6 and fungicides 86.7. In the same period the toxicity of pesticides was reduced by 160 times.

Regarding ecotoxicology, there's also been an evolution of products for situations involving more specific actions for the target species, reaching a large degree of specificity, targeting only specific species. Furthermore, another noteworthy aspect is the reduction of the waste of active molecules applied to the crops. In this sense, the legislation has barred, with increasing rigidity, products with residual effect in crops and agricultural products, according to Menten (2009).

Analyzing the issue from another viewpoint, Spadotto and Gomes (2004, p.117) present the following view to analyze pesticide consumption in Brazil:

> In 1964 the consumption of active ingredient (a.i.) was of 16,000 tons, but in 1960 the area planted with crops was 28 million hectares. However, in 1998 were consumed more than 128 thousand tons while the planted area was only 50 million hectares, i.e., the area of crops grew 78% while the consumption of pesticides rose by 700%, although these are average figures, because there are great differences in consumption between regions.

Table 10 Potential contamination index by the use of pesticides, at the studied watershed

Microbasin	Soil coverage	Area (ha)	Volume applied	Average	Worst value	Best value	IDEF
			L ha^{-1}				
Pedro Afonso (TO)	Soybean	15.154,41	5.17	1.79	5.43	0	0.671
	Cerrado	26.069,32	0.00				
	Sugarcane	1.914,14	1.50				
	Fallow	2.273,16	0.00				

Contamination Potential Index by Fertilizer Use (IFERT)

The *IFERT* (last *IDEG* component presented in the Lajeado Stream micro basin) of 0.659 (Table 11), represents an important negative value in the composition of *IDEG* (Table 12) and therefore of the *ISAGRI*, considering the relatively large volume of fertilizers used in soybean crops. This fact is easily explained, taking into account the adverse effects of the environmental contamination caused by their use.

Notwithstanding the provisos above outlined, by the classification parameter adopted, the *IFERT* found in the studied watershed was classified as good.

Analyzing the relevant values of *IERO*, *IDEF* and IFERT, it is possible to finally comment on the aggregate of the three, which makes-up the *IDEG* (Table 12), which was calculated at 0.723, a result also considered good within the parameters adopted by the survey. Note that there was a small increase with respect to the *IDEF* and *IFERT* due to the *IERO* found, somewhat higher than both.

In summary, under the proviso of index aggregation, the *IDEG* result suggests that there is no significant environmental impairment to environmental degradation vectors to date.

Soil Management Index (IMANEJ)

It is known that there is a direct, positive, relationship between the size of the area with forest cover plus the planted area using the no-till system and the result of applying the IMANEJ. If this case the two variables are present. According to Brazilian law, in plantations in the State of Tocantins 35% of the property area must be preserved. In the Pedro Afonso—TO watershed the area for forest preservation is 57.41%. On the other hand, no-till is practiced throughout the planted area. This has made a significant influence, respectively, in the factors P, conservation practices and C, use and management of soil and, therefore, the *IMANEJ* as a whole (Table 13). Nevertheless, opening up the index, it is clear that in the unexplored *cerrado* region, it was very good, 0.805. But when considering the cultivated soil, it drops alarmingly to 0.156, which means that, from the point of view of sustainability, in relation to the currently existing management and conservation practices, the level reached by the activity is still well below the desirable. On average the *IMANEJ* was 0.533 (Table 13), considered regular.

Table 11 Contamination potential index by fertilizers use in the Lajeado Stream micro basin

Microbasin	Use	Area (ha)	Volume applied	Average	Worst value	Best value	IFERT
			kg/ha				
Pedro Afonso (TO)	Soybean	15,154.41	350.00	125.23	367.50	0	0.659
	Cerrado	26,069.32	0.00				
	Sugarcane	1914.14	200.00				
	Fallow	2273.16	0.00				

Table 12 Degradation vector index in the Lajeado Stream micro basin—IDEG

Description	Pedro Afonso (TO)	
	Value	Weight
Erosion index—IERO	0.761	3.0
Index of potential contamination by the use of pesticides—IDEF	0.671	1.0
Index of potential contamination by the use of fertilizers—IFERT	0.659	1.0
IDEG	0.723	

Table 13 Soil management index in the Lajeado Stream micro basin

Microbasin	Use	C	P	IC	IP	IMANEJ
Pedro Afonso	Cerrado	0.0004	0.30	1.000	0.610	0.805
	Soybean	0.1069	0.60	0.091	0.221	0.156
	Sugarcane	0.0754	0.58	0.360	0.247	0.303
	Fallow	0.1069	0.65	0.091	0.156	0.124
Average						0.533

Table 14 Agricultural environmental sustainability index (ISAGRI) in the Lajeado Stream micro basin

Description	Pedro Afonso (TO)
ISAGRI = (IEA + IDEG + IMANEJ)/3	0.656

Agricultural Environmental Sustainability Index (ISAGRI)

Finally, the *ISAGRI* of 0.656 (Table 14), weighted average of the previously seen subindexes, applied to soybean farmers established in the Lajeado Stream micro basin, in the Pedro Afonso—TO municipality, representing, in the aggregate, the degree of their activities sustainability. A result considered good, calculated in accordance with the criteria adopted in this work. It should be noted, however, it cannot be considered reassuring, taking into account the provisos presented in the analysis of the various partial indexes and sub-indexes that compose it.

As stated by Sabiha et al. (2016) measuring the extent of environmental degradation in agriculture is essential (namely for countries dependent on agriculture). In this study the environmental factors and measurements allowed accommodate different types of environmental impacts arising from various environmental sources trying to cover the environmental complexity in the study area. This first study constitutes a baseline allowing future research to follow the environmental situation and to accommodate in the future other parameters like the ones superimposed by climate change.

5.2 Hypothetical Scenarios for the Pedro Afonso Watershed

Table 15 presents the *ISAGRI*, *IQA*, *IPOR*, *IRP*, *IQF*, *IEA*, *IERO*, *IDEF*, *IFERT*, *IDEG* and *IMANEJ* values and soil loss for each of the four scenarios.

The first scenario analyzed was the one in which the conventional planting and the planted area is that established by law, with regards to the forest reserve, occur. It must be remembered that the legal requirement—RL for the region is to maintain 35% of the native vegetation, plus the Permanent Preservation Areas (APP's). In this case the loss of soil is 26.357 t ha^{-1} 1 $year^{-1}$, observing that the *ISAGRI* is only 0.466, which makes the possible embodiment of said scenario unsustainable (Table 15).

Keeping the new scenario, were compared the results of the previous hypothesis with, in the same area and instead of the conventional tillage, it was applied the technique of direct planting. Then, the loss would be of 22.595 t ha^{-1} 1 $year^{-1}$ (Table 15), causing a decrease in soil loss of 3.76 t ha^{-1} 1 $year^{-1}$ when opting for direct tillage rather than conventional, which, in percentage terms, equals to 14.27%. There would be a slight increase in *ISAGRI*, which would rise to 0.492 (Table 15). However, the new scenario would still be in the range of unsustainability.

Next there is the scenario where the environmental legislation is not complied with, doing away with the legal reserve, i.e. happens cultivation throughout the existing area. In this case, if the planting occurs in the conventional manner, the loss of soil is 40.514 t ha^{-1} 1 $year^{-1}$ (Table 15), where the *ISAGRI* reaches the lowest value observed among all projected scenarios, namely 0.291, which makes the worst sustainability index. On the other hand, in the case of tillage, the loss fell to 34.726 t ha^{-1}1 $year^{-1}$ and ISAGRI was 0.332, slightly higher than the previous, and yet, still within the unsustainability range (Table 15). In this situation there was a reduction of 5788 t ha^{-1} 1 $year^{-1}$ from one system to the other, which is equivalent in percentage terms to a decline of 14.29% in soil loss by t ha^{-1} 1 $year^{-1}$.

What interests to highlight here is the existing situation in Pedro Afonso, in the current conditions, where the weighted average loss found in the region was only 14,507 t ha^{-1} 1 $year^{-1}$ considering the occurrence of tillage associated to the existence of a forest reserve covering 57.4% of the area, i.e. 22.4% higher than required by law, compared to the situation exemplified in this scenario. Therefore, there was a smaller difference 26.007 t ha^{-1} 1 $year^{-1}$ the lowest, when the scenario is conventionally tilled and 20.219 t ha^{-1} 1 $year^{-1}$ as deduced from the data presented in Table 15, in the case for a direct planting situation. From these numbers, which are significant, it remains very clear the need for strict compliance with the environmental law that deals with forest reserve, including the restrictions to soil slopes, as well as the use of no-tillage.

Table 15 Hypothetical scenarios of land use for the studied watershed

Scenery Description	ISAGRI	IQA	IPOR	IRP	IQF	IEA	IERO	IDEF	IFERT	IDEG	IMANEJ	Soil loss (t ha^{-1} $year^{-1}$)
Pedro Afonso original	0.656	0.727	0.572	0.836	0.704	0.712	0.761	0.671	0.659	0.723	0.533	14.507
Pedro Afonso Soybean conventional tillage obeying the law	0.466	0.727	0.277	0.688	0.482	0.564	0.566	0.381	0.381	0.492	0.341	26.357
Pedro Afonso Soybean no till obeying the law	0.492	0.727	0.277	0.688	0.482	0.564	0.628	0.381	0.381	0.529	0.383	22.595
Pedro Afonso Soybean conventional tillage total area	0.291	0.727	0.277	0.688	0.482	0.564	0.333	0.048	0.048	0.219	0.091	40.514
Pedro Afonso Soybean no till total area	0.332	0.727	0.277	0.688	0.482	0.564	0.428	0.048	0.048	0.276	0.156	34.726

6 Conclusions

In a changing climate scenario the sectors of agriculture, agroforestry, sustainable agriculture, water management and soil conservation are of great concern at present and for the future. The definition/conceptualization and application of these indexes allow to establish a baseline and therefore to structure long-term monitoring programs towards sustainable management of the studied agro-systems.

The 0.656 value of ISAGRI found to measure the sustainability of the production process is relatively low, despite still being rated as good sustainability index, according to the criteria adopted here. On the other hand, there is the concern about the very low rate of porosity found in the planted area, with an IPOR equivalent to 0.277. It is a worrying value, especially considering that local producers have adopted certain soil protection measures. The practice of no-till and nonetheless registers a low IPOR. The fear is that in the future the resistance index to soil penetration might increase, together with lower water absorption; that there is an increase in the volume of the rain water flow, more soil entrainment, impairing production and agricultural productivity. The future monitoring of the environmental situation is essential to assess the impact of production and possible mitigation measures such as good management and sustainable farming practices.

Another very serious concern is the presence of phosphorus in the waters of the Lajeado Stream, with contents above the allowed by Brazilian authorities, which was aggravated by the construction of a dam downstream of the growing region, where, as already said, it can cause eutrophication. It is important to remember that this stream crosses the soybean planting area considered in this work.

The situation described here is also conditioned by ongoing climate change scenario. In this sense, the analysis of results and future adequacy of the contents of this reality is also fundamental. All parameters studied for this work relate to the ongoing climate change and its assessment must also be made in that framework. In addition to the effects of human activities (nutrients and eutrophication, pesticides and toxicology and health of the soil) its own agricultural production is going to be affected by climate change.

In the context herein analyzed, there is a very serious problem related to sustainability in its relations with environmental and public health. An important part of the aggressions caused to soil, water and living things in general, especially to people, comes from the heavy use of pesticides, acidity correctives and fertilizers throughout the rural production process.

Finally, the projected scenarios emphasize the need for strict compliance with the environmental law that deals with forest reserve, including the restrictions to soil slopes, as well as the use of no-till system.

References

Allen DE, Singh BP, Dalal RC (2011) Soil health indicators under climate change: a review of current knowledge. In: Singh BP et al (ed) soil health and climate change, soil biology 29. https://doi.org/10.1007/978-3-642-20256-8_2

Bernardi ACC et al (2003) Sistema PAQLF 1.0 para a administração do programa de análise de qualidade de laboratórios de fertilidade da Embrapa solos. EMBRAPA, Rio de Janeiro

CATI (2000) Programa estadual de microbacias hidrográficas. São Paulo

Contini E, Gasques JG, Alves E, Bastos ET (2010) Dinamismo da agricultura brasileira. In: Revista de Política Agrícola, publicação da Secretaria de Política Agrícola do Ministério da Agricultura, Pecuária e Abastecimento. Ano XIX. Edição Especial de Aniversário do MAPA – 150 anos. Brasília, July 2010. p. 42/64. Available at: https://ainfo.cnptia.embrapa.br/digital/bitstream/item/80771/1/Dinamismo-da-agricultura-brasileira.pdf. Accessed on 25 Sept 2016

COOPERATIVA AGROPECUÁRIA DE PEDRO AFONSO-COAPA (2009) Pedro Afonso. Tocantins

Dos Anjos LHC, Van Raij B (2004) Indicadores de processos de degradação de solos. In: Romeiro AR (ed) Avaliação e contabilização de impactos ambientais, Romeiro AR (Org). Unicamp, Campinas; Imprensa Oficial do Estado de São Paulo, São Paulo, pp 87–111

EMBRAPA—Empresa Brasileira de Pesquisa Agropecuária. Soja—Portal Embrapa, safra 2015/2016. Available at: https://www.embrapa.br/soja/cultivos/soja1. Accessed on 27 Sept 2016

Fragoso CR, Marques DMLM, Ferreira TF, Janse JH, van Nes EH (2011) Potential effects of climate change and eutrophication on a large subtropical shallow lake. Environ Model Softw 26:1337–1348. https://doi.org/10.1016/j.envsoft.2011.05.004

IBGE (2009) Levantamento Sistemático da Produção Agrícola. Estado do Tocantins: quadro comparativo da produção de grãos. Available at: www.ibge.gov.br/home/estatistica/indicadores/agropecuaria/lspa/. Accessed on 16 Mar 2016

Karlen DL, Mausbach MJ, Doran JW, Cline RG, Harris RF, Schuman GE (1997) Soil quality: a concept, soil quality: a concept, definition, and framework for evaluation. Soil Sci Soc Am J 61:4–10

Larson WE, Pierce FJ (1994) The dynamics of soil quality as a measure of sustainable management. In: Doran JW et al (ed) Defining soil quality for a sustainable environment. Madison Soil Science Society of America Special Publication, pp 37–51

Lira WS, Candido GA (2008) Análise dos modelos de indicadores no contexto do desenvolvimento sustentável. Perspectivas Contemporâneas, Campo Mourão 3(1):31–45

Martins JJF (2014) Padrões Econômico-Ambientais da Agropecuária no Estado do Tocantins: Estudo Comparativo de Microbacias correspondentes a três Sistemas Agrários relevantes. NAEA, Belém, p 274. ISBN 978-85-7143-121-8

Menten JOM (2009) Importância de novos defensivos agrícolas para a sustentabilidade da produção vegetal 31. Semana da Citricultura, ANDEF, São Paulo

Moreira FMS, Siqueira JO (2002) Microbiologia e Bioquímica do Solo. Editora UFLA, Lavras, pp 85, 97–98. Available at: http://www.prpg.ufla.br/solos/wp-content/uploads/2012/09/MoreiraSiqueira2006.pdf. Accessed on 30 Sept 2016

Moss B, Kosten S, Meerhoff M, Battarbee RW, Jeppesen E, Mazzeo N, Havens K, Lacerot G, Liu Z, De Meester L, Paerl H, Scheffer M (2011) Allied attack: climate change and eutrophication. Inland Waters 1:101–105. https://doi.org/10.5268/IW-1.2.359

Mullan D (2013) Soil erosion under the impacts of future climate change: Assessing the statistical significance of future changes and the potential on-site and off-site problems. Catena 109:234–246. http://dx.doi.org/10.1016/j.catena.2013.03.007

OECD (1993) Organization for Economic Cooperation and Development: core set of indicators for environmental performance reviews; a synthesis report by the group on the state of the environment. OECD, Paris

OECD (1997) Environmental indicators for agriculture. Paris: Organization for Economic Cooperation and Development. Issue and policies. The Helsinki Seminar, OECD, Paris

OECD (1998) Agriculture and the environment: issues and policies. The Helsinki Seminar, OECD, Paris

Pimentel D, Harvey C, Resosudarmo P, Sinclair K, Kurz D, McNair M, Crist S, Shpritz L, Fitton L, Saffouri R, Blair R (1995) Environmental and economic costs of soil erosion and conservation benefits. Science 267:1117–1123. https://doi.org/10.1126/science.267.5201.1117

Previero CA (2006) Caracterização e análise da Oepa UNITINS Agro/To. Relatório Final. Centro de Gestão e Estudos Estratégicos. Ciência, Tecnologia e Inovação. Palmas – To. ago. 2006. p 6. Available at: www.cgee.org.br/atividades/redirect/3403. Accessed on 25 Sept 2016

Sabiha N-E, Salim R, Rahman S, Rola-Rubzen MF (2016) Measuring environmental sustainability in agriculture: a composite environmental impact index approach. J Environ Manage 166:84–93. https://doi.org/10.1016/j.jenvman.2015.10.003

Salvador Sanchis MP, Torro D, Borselli L, Poesen J (2008) Climate effects on soil erodibility. Earth Surf Proc Land 33:1082–1097

Sampaio G, Nobre C, Costa MH, Satyamurty P, Soares-Filho BT, Cardoso M (2007) Regional climate change over eastern Amazonia caused by pasture and soybean cropland expansion. Geophys Res Lett 34:L17709. https://doi.org/10.1029/2007GL030612

SEAGRO (2016) Secretaria do Desenvolvimento da Agricultura e Pecuária. Governo do Tocantins. Available at: http://seagro.to.gov.br/noticia/2016/3/15/tocantins-deve-colher-mais-de-dois-milhoes-de-toneladas-de-soja-safra-201516. Accessed on 28 Sept 2016

Seron E (2010) Inovação e contribuições para o agronegócio brasileiro. VIII Reunião da Câmara Setorial de Oleaginosas e biodiesel—MAPA. Associação Nacional de Defesa Vegetal. ANDEF (2004). Brasília—DF.11/07/2010. Available at https://pt.slideshare.net/LeomachadoFAEG/andef. Accessed on 8 Oct 2016

Silva LF (2007) A construção de um índice de sustentabilidade ambiental agrícola (ISA): uma proposta metodológica. Campinas, SP. 2007. 232 f. Tese (Doutorado em Economia Aplicada)—Universidade Estadual de Campinas, Campinas

Spadotto CA, Gomes MAF (2004) Impactos ambientais de agrotóxicos: monitoramento e avaliação In: Romeiro AR (Org). Avaliação e contabilização de impactos ambientais. Editora da UNICAMP, Campinas, SP. Imprensa Oficial do Estado de São Paulo, São Paulo, SP, pp 112–122

Tomasoni MA (2006) Contribuição ao estudo de indicadores ambientais. Geonordeste 15(2): 90–118

Xia R, Zhang Y, Critto A, Wu J, Fan J, Zheng Z, Zhang Y (2016) The potential impacts of climate change factors on freshwater eutrophication: implications for research and countermeasures of water management in China. Sustainability 8:229. https://doi.org/10.3390/su8030229

Enhancing Food Security and Climate Change Resilience in Degraded Land Areas by Resilient Crops and Agroforestry

Muhammad Saqib, Javaid Akhtar, Ghulam Abbas and Ghulam Murtaza

Abstract Land degradation is threatening food security, life quality and climate change resilience of rural communities in many parts of the world. Salinity and sodicity of soil and water, and drought are considered major causes of increasing land degradation. The soil and water problems of these areas vary in intensity and type hence need site specific solutions. These solutions depend on the needs of the farmers and the capability of the farmers to adapt a specific solution. These solutions are the use of stress resistant genotypes of crops, grasses and trees along with different amendments including organic and inorganic. The crops may be grown in slightly degraded areas whereas grasses and trees may be used for moderately to severely degraded areas. A win-win situation can be created in these areas by reversing land degradation through integrating use of crops, grasses, trees and, organic and inorganic amendments. The cultivation of barren lands not only ensures food security but also contributes to the environmental conservation through carbon sequestration and ecological rehabilitation. This chapter discusses the salt-induced land degradation, its causes and potential solutions to profitably utilize these degraded areas for enhancing food security and climate change resilience in these areas on sustainable basis.

Keywords Carbon sequestration · Climate change resilience · Food security Plants · Land degradation · Salinity

M. Saqib (✉) · J. Akhtar · G. Murtaza
Institute of Soil and Environmental Sciences, University of Agriculture, Faisalabad, Pakistan
e-mail: Muhammad.Saqib@ernaehrung.uni-giessen.de; m.saqib@uaf.edu.pk

M. Saqib
Institute of Plant Nutrition, Justus-Liebig University, Giessen, Germany

G. Abbas
Department of Environmental Science, COMSATS University Islamabad, Vehari Campus, Vehari, Pakistan

P. Castro et al. (eds.), *Climate Change-Resilient Agriculture and Agroforestry*, Climate Change Management, https://doi.org/10.1007/978-3-319-75004-0_16

1 Introduction

Land degradation is considered a major threat to food security, life quality and climate change resilience of rural communities in many parts of the world. Salinity, sodicity and drought are considered the main reasons of land degradation around the globe and large tracts of land have been rendered unproductive by these stresses. The estimated salt-affected area in the world is over 800 mha (Munns 2005). It occurs in 75 countries of the world and according to an estimate about 2000 ha of cropland in irrigated areas is affected by salinity and sodicity per day (Qadir et al. 2015). The San Joaquin Valley in the United States, the Murray Darling Basin in Australia, the Yellow River Basin in China, the Indus Basin in Pakistan, Indo-Gangetic Basin in India, Aral Sea Basin (Amu-Darya and Syr-Darya River Basins) in Central Asia, and the Euphrates Basin in Syria and Iraq are commonly known hot spots of land degradation induced by salt (Qadir et al. 2014). Climate change induced increase in temperature is aggravating the situation. The main fresh water resources are insufficient to fulfil the agricultural needs. To overcome the water shortage for agriculture, tubewells are extensively installed which mostly pump poor quality water. In Pakistan for example, about 70–80% of the pumped water from such tubewells is marginal to hazardous quality (Ghafoor et al. 2008; Latif and Beg 2004) which is increasing the soil salinity and sodicity, and leading to land degradation. However, it looks inevitable to use salt-affected soils and water in agriculture (Beltran and Manzur 2005; Bouwer 2002; Qadir and Oster 2004). It is essential for feeding and supporting the livelihood of the people living in the affected areas in particular and for the other people around the world in general. The use of poor quality water in agriculture is also important as it decreases the problem of disposal of the poor quality drainage water. This may be the only water available for the utilization and remediation of the degraded and barren lands as the availability of good quality water for such lands does not seem possible in future.

The land and environmental degradation endangers food security and climate change resilience of rural people of the salt-affected degraded areas in particular and of the society as a whole. The common staple food crops including rice, maize and wheat are the major focus of modern day agriculture as they provide about half of the carbohydrates, proteins and calories required by the human beings (Lopez-Noriega et al. 2012). About 95% of the dietary energy obtained by the human beings comes from 20 of the 100–120 plant species used in production of food today (Parry et al. 2009). According to another estimate, only 6 species including wheat, soybean, maize, rice, barley and rapeseed cover about 50% of world arable land (FAO 2010). The agricultural and developmental research have also been focused on these crop species. About one billion world people suffer from starvation and deficiencies of one or more micronutrients particularly iodine and iron whereas vitamin A deficiency affects about two billion people around the world (Alnwick 1996).

The situation of food security, malnutrition and climate change resilience can be improved in degraded land areas through crop diversification and use of salinity tolerant crops and tree species. The Earth's plant life evolved in salty water of oceans about 3 billion years ago but this high salt adaptation of most of the plants was lost overtime (Rozema 1996). However, still about 1% of the land plant species have the ability to grow and reproduce under saline conditions (Rozema and Flowers 2008). These plants include crops, grasses, trees and different ornamental species. The purpose of this chapter is to discuss the causes of land degradation and potential solutions for profitable utilization of degraded land areas. These solutions are important for enhancing food security and climate change resilience in degraded land areas on sustainable basis.

2 Salt Induced Land Degradation

The insufficient precipitation to wash out salts from the soil profile along with irrigation without a proper drainage facility leads to land degradation in arid and semi-arid parts of the world. The salts which accumulate in the root zone badly affect different physical and chemical soil properties resulting in a loss of soil productivity. Salt-affected lands may be generally grouped as saline, sodic and saline sodic. The sodic and saline sodic soils are also known as alkali and saline-alkali soils, respectively. Saline soils have large amount of soluble salts which can impair growth of most of the crop plants. The sodic soils are characterized with high sodium contents, deterioration of soil structure and surface crusting. The saline sodic soils have both high soluble salts and sodium contents therefore share the properties of saline as well as sodic soils. In addition to these soils, there are also soils having high magnesium content (Vyshpolsky et al. 2008) which also decreases plant growth due to poor water flow and poor distribution of water as a result of large clod formation when plowed.

Salinization is an important cause of land degradation and loss of soil productivity. Salt-affected lands are scattered in different regions of the world including Middle East, North and South America, North Africa, and Central and South Asia with an extent of about one billion hectares and a loss of productivity of about two million hectares per year around the world associated with secondary salinization (Menzel and Lieth 1999; Yensen 2006). These arid and semi-arid areas lack good quality water and enough precipitation to leach down surface salts leading to soil salinization and desertification. An improper use of good as well as poor quality water also adds to the problem. As good quality water is decreasing with time, large volumes of brackish water as well as sea water are available. Although these poor quality water resources are considered a problem yet these may be turned into an opportunity if used properly. The frequency of drought and climate extreme events, and soil salinity is subject to increase as a result of climate change (Pachauri and Reisinger 2007).

3 Salt-Affected Degraded Lands, Plant Growth and Environment

Plant growth is usually reduced due to salinity and sodicity however with a differential affect based on the plant species, salt concentration and the environmental conditions. Salinity reduces plant growth by osmotic stress, ion toxicity and nutritional imbalance (Grattan and Grieve 1999; Saqib et al. 2008; Schubert et al. 2009). Sodium being the dominant cation in the salt-affected soils is taken up in abundance and mostly becomes a problem for the plants. Small quantitates of sodium are considered essential as a mineral nutrient in C4 and Crassulacean Acid Metabolism (CAM) plants (Brownell and Crossland 1972). Most of our crop plants are categorised as salt sensitive with a large yeild reduction under salt-affected conditions. About 50% reduction in dry matter production is recorded for rice (*Oryza sativa*) when salinity is 80 mM NaCl, for wheat (*Triticum turgidum* ssp. durum) when salinity is 100 mM NaCl and for barley when salinity is 120 mM NaCl (Munns and Tester 2008). A lower salinity of less than 2 dS m^{-1} reduced yield (per dS m^{-1}) of beans (*Phaseolus vulgaris*) by 19%, pepper (*Capsicum annuum*) by 14%, corn (*Zea mays*) by 12% and potatos (*Solanum tuberosum*) by 12% (Maas 1990). On the other hand, the salt-loving plants, the halophytes are able to withstand a wide range of salinities and therefore offer good opportunity as crops (Yensen 2006) for salt induced degraded lands. There is large diversity among the halophytic plants that leads to their multifaceted applications including their use as food, feed, ornamentals and biofuels or bioenergy. The possibility of crop cultivation in seawater has been demonstrated in the earlier years of 1960s (Boyko 1964).

Soil and water salinity/sodicity decreases crop growth and farm income which limit the ability of farmers of the affected areas to sustain a better livelihood. It also reduces vegetation cover and biodiversity which result in increased environmental degradation and human health risks. Improper disposal of poor quality water causes environmental pollution and several health hazards. The poor agricultural production, poor vegetation and poor diversity in the degraded land areas lead to decreased food security and climate change resilience. Land degradation induced by salinity and sodicity mainly results from the inability of these soils to sustain vegetation cover. The salt-affected lands usually lack any vegetation cover or have very thin vegetation cover resulting in their poor or un-healthy contribution to the environement. Therefore, salinity and sodicity results in increased desertification in many parts of the world.

4 Reclamation of Salt-Affected Soils

Salt-affected soils may be utilized for agricultural crops either through soil reclamation and/or through enhancing the salt resistance of different crop plants and/or through the use of naturally salt resistant plants. The situation of salt-affected soils may be improved through different surface and subsurface drainage techniques. This is termed as an engineering approach. Saline soils are usually reclaimed by heavy irrigations with good quality water whereas the sodic and saline sodic soils need a calcium source like gypsum followed by heavy irrigation. However, the cost of reclamation is increased as the degree of the salinization (Murtaza 2013) or sodication (Vyshpolsky et al. 2008) increases. Good quality water is an important pre-requisite for reclamation of salt-affected soils. Only 2.5% of the global total water is fresh water and about two-thirds of this fresh water is present in the form of glaciers (Gleick 2009). About 65% of the global water use is in agriculture particularly for irrigation (Postel et al. 1996). An increasing shortage of good quality water makes its availability for reclamation impossible in future.

5 Food Security and Climate Change Resilience in the Degraded Land Areas

Food security is defined as access of all people to sufficient, safe and nutritious food physically and economically according to their requirements and likings for a healthy and active living (IDRC 2010). The communities in the salt-affected areas are usually smallholder farmers and this small scale farming is important in food security and reduction of poverty in these areas. The biophysical dimensions of the lands degraded due to salts are more known and studied as compared to the economic and social aspects (Zekri et al. 2010). The people of salt-affected degraded land areas are mostly very vulnerable to changing climatic conditions. The resilience to climate change and food security of the people of the salt-affected degraded land areas can be improved by profitable and sustainable utilization of these soils. Saline agriculture provides a sustainable solution for such areas through the use of resilient crops and agroforestry.

6 Saline Agriculture

Saline agriculture is an integrated use of genetic resources including plants, animals, fish and insect, and better agronomic practices for utilization of salt-affected land as well as water resources profitably and sustainably (Qureshi and Barrett-Lennard 1998). It is the need of the day to increase the productivity of irrigated salt-affected areas to fulfill the food, feed and fiber needs of the growing population as not sufficient productive land may be available for this purpose (Qadir and Oster 2004). The reclamation of the salt-affected areas is an established technology however it cannot be practiced in many saline sodic areas due to high cost, shortage of good quality water and poor soil permeability (Qadir et al. 2008; Qureshi and Barrett-Lennard 1998). The biological approach of growing salt-resistant crops, trees and forage species provides an alternate option. The utilization of organic and inorganic amendments as well as better agronomic practices improves the establishment and performance of different plant species in degraded land areas.

There are differences among the plant species and genotypes for salt tolerance and to grow on salt-affected lands (Maas and Hoffman 1977). The crop species usually grow at low to moderate salt levels and are called glycophytes (Lauchli and Epstein 1990). On the other hand salt-resistant plants mostly comprise of shrubs and trees and are called halophytes. Historically, there have been research efforts on two pathways including the breeding for improving the salt resistance of the existing crops and secondly to domesticate the salt resistant plants. However, it has been learned that breeding for salt resistance is not easy because of the multigenic nature of the salt resistance trait (Flowers and Yeo 1995; Flowers et al. 2010). The second approach has also not resulted in fast progress but the researchers have kept up both the approaches. It may be easier and more economical to develop and maintain sustainable agro-ecosystems in highly salt-affected lands using the salt resistant crop genotypes as well as halophytic plant speicies.

Salt-affected degraded lands and water may provide a good opportunity for raising forests as the good land and water resources are becoming unavailable for this purpose. A tree based approach directly helps the rural people by supplying fodder and fuel wood and indirectly by enhancing the ecological rehabilitation and environmental sustainability. Many countries of the world are using similar strategies. The salient benefits of growing trees in agriculture include biodiversity conservation, goods and services for the people, enhanced soil fertility and improved social and economic conditions of the people (Pandey 2007; Singh and Dagar 2009). A study from Pakistan revealed a contribution of 63 and 37% respectively by crops and trees in an agro-forestry system. This study also showed that the trees increased the overall income of farm and decreased the dependency of household on natural forest for fuel wood and timber which has a positive effect on conservation of natural forest (Essa 2004). The revegetation of the barren lands increases carbon sequestration (Lal 2001), reduces erosion and improves environmental conservation and ecological rehabilitation. The carbon sequestration potential through reclamation and revegetation of salt affected lands may be 0.4–1 Pg C $year^{-1}$ (Lal 2009).

Quinoa (*Chenopodium quinoa*) is a good example, it has good adaptation to different abiotic stresses and can survive and produce under a range of agroecosystems. In Incan Empire it was called as the ‘mother grain’ (Risi et al. 1984). Its seeds have balanced amount of essential minerals, vitamins, carbohydrates, lipids, amino acids and fibre (Vega-Gálvez et al. 2010) along with low glycaemic index and are glutin free (Gordillo-Bastidas et al. 2016). In recognition of its importance and potential, United Nations declared the year 2013 as the International Year of Quinoa. In spite of its importance and potential it is still an understudied and underutilized plant which needs further improvement in different agronomic characteristics.

Salt-affected wastelands are of poor economic value but may provide an opportunity of producing timber, different biomaterials and biomass for energy production. With growing population, increasing living standards and pressures on fossil-fuel supplies, the demand for other resources is growing, providing new opportunities for the salinized marginal land and water resources (Corbishley and Pearce 2007). This presents an approach which is low cost and environment friendly, and can be used for the utilization of the salt-affected soil and waters on sustainable basis in many parts of the world. Plants growing on salt-affected lands also contribute to phytoremediation and help in mitigating the climate change impact. The H^+ ions released by roots helps in dissolution of $CaCO_3$ and removal of Na^+ from calcareous saline-sodic/sodic soils (Qadir et al. 2003). Release of organic acids as root exudates acidifies rhizosphere (Marschner 2003) and may also help in dissolving $CaCO_3$ and Na^+ removal from a calcareous saline sodic soil hence improving soil conditions for plant growth. The halophytic salt-tolerant plants have been successful under saline conditions of India (Tomar et al. 2003), Pakistan (Qureshi et al. 1993) and Australia (Barrett-Lennard 2002). Growing salt-tolerant forage and tree species provides good opportunity for the profitable and sustainable use of salt-affected degraded soils and water (Masters et al. 2007; Qadir et al. 2008).

The living conditions of the people of the salt-affected areas are very poor because of no/very low production from these lands. Studies have shown poor health and housing conditions, poor literacy rate and purchasing power of the people of the salt-affected areas (Qureshi and Barrett-Lennard 1998). The poor and non-skilled people of the salt-affected areas usually have no other option except to relly on the lands to earn their livelihood. The poor people are more vulnerable to the impacts of air and water pollution and bear more effect of environmental degradation when crop yields and agricultural productivity are reduced (ADB 2002). Therefore, reduced impact of salinity and other environmental degradation factors will reduce rural poverty and improve living conditions of the people.

As described earlier, saline agriculture advocates an integrated use of the available genetic resources including plants, animals, fish and insect, and better agricultural practices for profitable and sustainable use of salt-affected land and water resources (Qureshi and Barrett-Lennard 1998). As discussed by Jacobsen et al. (2015), Qadir et al. (2008), Qureshi and Barrett-Lennard (1998) and Rozema and Schat (2013), Saline Agriculture may comprise of a general strategy, site specific practices and, short and long term programs of Saline Agriculture as described here.

6.1 General Strategy of Saline Agriculture

General strategy of Saline Agriculture includes an evaluation and assessment of the prevailing situation and aspiration of the farmers. The available water and soil is analysed and evaluated for different quality parameters. The local needs of the farmers and of the community, the aspiration of the farmers concerned and, the market trends are assessed. Based on this information, a general strategy is devised for the use of different plant species and agronomic practices as shown below:

- Slightly to moderately salt-affected soils
  - salt and drought tolerant varieties of crops and grass species can be grown on such soils with improved agronomic practices.
- Highly salt-affected soils
  - These soils may be used for tree species of high salt tolerance (*Eucalyptus, Acacias, Tamarix* etc).
  - Salt tolerant grass species and forage shrubs like *Atriplex* species etc. can be planted in between the tree rows.
  - In these areas, small animals may be introduced after about 2-years of intervention and large animals may be allowed after about 3 years of intervention for controlled grazing.
- On the basis of the farmer's interest and requirements of the local market, salt and drought tolerant fruit trees, apiculture and fishery can be included in the program.
- The crops and forage species can provide economic returns to the farmer in 1–2 years however trees grown in the waste lands may provide economic return in 5–7 years. A list of common salt tolerant plant species with particular benefits is given in Table 1.

6.2 Site Specific Practices for Saline Agriculture

Salt-affected degraded land areas are diverse in nature and characteristics and requires site specific practices for successful utilization of the land. Depending on the site characteristics, these site specific practices may include the followings:

- Poor germination and seedling establishment may be tackled by seed priming, higher seed rate and by increasing number of plants.
- In soils with a hard pan at a certain depth, trees may be planted after breaking the hard pan. Hard pan usually restricts the root growth.
- To decrease salt load in the root zone while new tree plantation, the tree seedlings may be planted at some height on the side/shoulder of the irrigation trenches.

Table 1 Some salt tolerant plant species for saline agriculture

S. no.	Plant species	Potential use and salt tolerance	Country of study/References
1	*Acacia ampliceps*	• As fuel and fodder • As windbreak and for soil conservation • EC (electrical conductivity): 15–20 dS m^{-1}	Australia/Marcar et al. (1995)
2	*Acacia nilotica*	• As fuel, fodder, timber, gum and lac • EC:15–21 dS m^{-1}	Pakistan/Qureshi et al. (1993)
3	*Albizzia lebbeck*	• As fuel, fodder and ornamental plant • As windbreak and for soil conservation • EC:15–21 dS m^{-1}; SAR (sodium adsorption ratio): 38	Pakistan/Qureshi et al. (1993)
4	*Sesbania sesban*	• As fuel, fodder and ornamental plant • As windbreak and for soil conservation • EC:15–21 dS m^{-1}; SAR: 38	Australia, Pakistan/Evans and Macklin (1990), Qureshi et al. (1993)
5	*Leucaena leucocephala*	• As fuel and fodder • As a nitrogen fixing agent • EC:14 dS m^{-1}; SAR: 30	Pakistan/Qureshi et al. (1993)
7	*Atriplex amnicola*	• As fodder and fuel • EC: 30–40 dS m^{-1}	Australia, Pakistan/ Barrett-Lennard and Malcolm (1995), Perveen (2002)
8	*Atriplex lentiformis*	• As fodder and fuel • EC: 30–40 dS m^{-1}	Australia, Pakistan/ Barrett-Lennard and Malcolm (1995), Perveen (2002)
9	*Grewia asiatica*	• As fruit • Sticks used for making baskets etc • EC:16 dS m^{-1}; SAR: 28	Pakistan/Qureshi et al. (1993)
10	*Zizyphus mauritiana*	• As a fruit • As fuel, fodder and lac • High salinity and sodicity tolerant	Pakistan/Qureshi et al. (1993)
11	*Sorghum sudanese*	• As fodder • EC: 14 dS m^{-1}	Pakistan/Malik et al. (1986)
12	*Leptochloa fusca*	• As fodder and as soil cover • To produce mushrooms and biogas • EC: 22.3 dS m^{-1}; SAR: 150	Pakistan, India/Malik et al. (1986), Gupta and Abrol (1990)
13	*Chloris gayana*	• As fodder • As a soil cover • EC: 23 dS m^{-1}	Saudi Arabia, Pakistan/Fisher and Skerman (1986), Qureshi et al. (1993)

(continued)

Table 1 (continued)

14	*Azadirachta indica*	• As a medicinal and shade plant • Moderately salt tolerant	India/Ahmad and Chang (2002)
15	*Cymbopogon flexuosus*	• As a medicinal plant • EC: 10 dS m^{-1}; SAR: 50	India/Patra and Singh (1998)

- Use of gypsum as a source of calcium is very effective in soils having sodicity problem due to high exchangeable sodium. Gypsum is a cheap and easy to use source of calcium for such soils.
- The use of Ca containing fertilizers such as single super phosphate is preferable.
- Appropriate use of Zn, B and other nutrients is helpful for plant establishment as degraded lands are usually poor in these essential plant nutrients.

6.3 *Short and Medium-Long Term Programs of Saline Agriculture*

Saline agriculture is a comprehensive approach to tackle the problem of salt-induced land degradation and it comprises of short as well as medium to long term solutions based on the local needs and future planning of the farmers. The short and medium to long term solutions include the followings:

6.3.1 Short Term Programs

The immediate start of the utilization of degraded land areas contributes to farmer income and environmental betterment. Therefore, it is very important to start cultivation of the degraded land areas at the earliest. The short term solutions available for these areas include the followings:

- Site specific land use and management practices:
  - Salt resistant crops and fish farming may be used in slightly salt-affected areas
  - Salt resistant crops, grasses and livestock farming may be used in moderately salt-affected areas
  - Salt tolerant trees and shrubs may be grown on highly salt-affected lands
  - The establishment of different crops and plants may be improved through better irrigation, seed priming and placement, bed sowing, fertilizer management, pit planting and transplantation of seedlings etc.
- Creation of seed banks of native species may improve the provision of seeds to the farmers at the time of need.

- Diversification and mix farming is better than using a single crop or crop variety as diversification may improve the success of cropping on such soils.
- Flooding removes the salts from the root zone temporarily and plants can get established easily in this situation. These salts again come to the surface soil with time and may cause problem of plant establishment. Therefore, the flooded degraded lands should be planted with trees and other suitable plants as soon as possible.

6.3.2 Medium-Long Term Programs

The rehabilitation of degraded land areas on sustainable basis is the need of the day and requires medium to long term solutions. These solutions should be considered and started as soon as possible alongside the short term solutions. These solutions include the followings:

- Selection and development of stress tolerant crop varieties and domestication of wild plants. There should be continuous programs at national and international levels for this purpose as these are time consuming processes.
- GIS (Geographic Information System) and remote sensing based monitoring of the existing and periodic changes in the soil salinization and land degradation should be carried out for proper documentation and planning.
- Development and improvement of irrigation and drainage system for sustainable utilization of degraded lands and water.
- Development of Science and Technology Parks for demonstration of research to the stake holders. Field demonstrations in the degraded land areas showing the successful utilization of these lands for different purposes may give a confidence to the farmers and stakeholders to use the same or similar technology for the rehabilitation and profitable utilization of their degraded lands on sustainable basis.

7 Conclusions and future outlook

Land degradation can be controlled and a sustainable situation (i.e. land use and income) can be created on the degraded wastelands through a site specific integrated approach involving stress tolerant crops, grasses and trees along with organic and inorganic amendments. The degraded lands may become a resource as carbon sequestration is an emerging opportunity for these land resources and these may be a source of energy and carbon credits. The utilization of the salt-affected lands and water will have a direct and long term impact to the improvement of the socio-economic conditions and the local and regional environment in the area. Therefore, the degraded lands and water should be treated as valuable resources and not as wastes. These areas may be developed into opportunities in future by the consistant efforts of the scientific community, all the stakeholders and the policy makers. They all should work for improved solutions and technologies for the profitable and sustainable utilization of degraded land resources.

References

ADB (2002) Poverty in Pakistan: issues, causes and institutional responses. Asian Development Bank, Manila, Philippine

Ahmad R, Chang MH (2002) Salinity control and environmental protection through halophytes. J Drain Water Manag 6:17–25

Alnwick D (1996) Significance of micronutrient deficiencies in developing and industrialized countries. In: Combs GF, Welch RM, Duxbury JM, Uphoff NT, Nesheim MC (eds) Food-based approaches to preventing micronutrient malnutrition. An International Research Agenda, Cornell University, Ithaca, p 68

Barrett-Lennard EG (2002) Restoration of saline land through revegetation. Agric Water Manag 53:213–226

Barrett-Lennard EG, Malcolm CV (1995) Saltland pastures in Australia: a practical guide. Bulletin 4312, Western Australian Department of Agriculture, South Perth, p 112

Beltran JM, Manzur CL (2005) Overview of salinity problems in the world and FAO strategies to address the problem. In: Proceedings of the international salinity forum. Riverside, California, 25–27 Apr 2005, pp 311–313

Bouwer H (2002) Integrated water management for the 21st century: problems and solutions. J Irrig Drain Eng 28:193–202

Boyko H (1964) Principles and experiments regarding irrigation with highly saline and seawater without desalinization. Trans New York Acad Sci Ser 2:1087–1102

Brownell PF, Crossland CJ (1972) The requirement for sodium as a micronutrient by species having the C4 Dicarboxylic photosynthetic pathway. Plant Physiol 49:794–797

Corbishley J, Pearce D (2007) Growing trees on salt-affected land. ACIAR impact assessment series report No. 51. ACIAR, Centre for International Economics, Canberra, Australia

Essa M (2004) Household income and natural forest conservation by agroforestry: an analysis based on two agro-ecological zones: Bagrot and Jalalabad in Northern Pakistan. M. Sc. thesis. Department of International Environment and Development Studies (Noragric), Norwegian University of Life Sciences, Norway

Evans DO, Macklin B (1990) Perennial sesbania production and use: a manual of practical information for extension agents and development workers. Nitrogen Fixing Tree Association, Waimanalo, Hawaii, USA, 41 p

FAO (2010) The state of food insecurity in the world—addressing food insecurity in protracted crises. FAO, Rome, p 62. http://www.fao.org/docrep/013/i1683e/i1683e.pdf

Fisher MJ, Skerman PJ (1986) Salt tolerant forage plants for summer rainfall areas. Reclam Reveg Res 5:263–284

Flowers TJ, Yeo AR (1995) Breeding for salinity resistance in crop plants: where next? Aust J Plant Physiol 22:875–884

Flowers TJ, Galal HK, Bromham L (2010) Evolution of halophytes: multiple origins of salt tolerance in land plants. Funct Plant Biol 37:604–612

Ghafoor A, Murtaza G, Ahmad B, Boers THM (2008) Evaluation of amelioration treatments and economic aspects of using saline-sodic water for rice and wheat production on salt-affected soils under arid land conditions. Irrig Drain 57:424–434

Gleick PH (2009) Peak water. In: The world's water 2008–2009. The biennial report on freshwater resources. Island Press, Washington, Covelo, London, pp 1–16

Gordillo-Bastidas E, Díaz-Rizzolo DA, Roura E et al (2016) Quinoa (Chenopodium quinoa willd). From nutritional value to potential health benefits: an integrative review. J Nutr Food Sci 6:497

Grattan SR, Grieve CM (1999) Salinity–mineral nutrient relations in horticultural crops. Scientia Horti 78:127–157
Gupta RK, Abrol IP (1990) Salt-affected soils: their reclamation and management for crop production. Adv Soil Sci 11:223–288
IDRC (2010) Agriculture and food security program prospectus for 2010–2015. International Development Research Centre, Canada
Jacobsen SE, Sørensen M, Pedersen SM, Weiner J (2015) Using our agrobiodiversity: plant-based solutions to feed the world. Agron Sustain Dev 35:1217–1235
Lal R (2001) Potential of desertification control to sequester carbon and mitigate the greenhouse effect. Clim Change 51:35–72
Lal R (2009) Carbon sequestration in saline soils. J Soil Saline Water Qual 1(1–2):30–40
Latif M, Beg A (2004) Hydrosalinity issues, challenges and options in OIC member states. In: Latif M, Mahmood S, Saeed MM (eds) Proceedings of the international training workshop on hydrosalinity abatement and advance techniques for sustainable irrigated agriculture, Lahore, Pakistan, 20–25 Sep, pp 1–14
Lauchli A, Epstein E (1990) Plant responses to saline and sodic conditions. In: Tanji KK (ed) Agricultural salinity assessment and management, Manuals and reports on engineering practices No. 71. American Society of Civil Engineers, New York, pp 112–137
Lopez-Noriega I, Galluzzi G, Halewood M et al (2012) Flows understress: availability of plant genetic resources in times of climate and policy change. CGIAR research program on climate change, Agriculture and food security (CCAFS) working paper 18. Copenhagen
Maas EV (1990) Crop salt tolerance. In: Tanji KK (ed) Agricultural salinity assessment and management, Manuals and reports on engineering practices No. 71. American Society of Civil Engineers, New York, pp 262–304
Maas EV, Hoffman GJ (1977) Crop salt tolerance-current assessment. J Irrig Drain Div Amer Soc Civil Eng 103:115–134
Malik KA, Aslam Z, Naqvi M (1986) Kallar grass: a plant for saline land. Nuclear Institute for Agriculture and Biology, Faisalabad, Pakistan, p 93
Marcar N, Crawford D, Leppert P et al (1995) Trees for saltland: a guide for selecting native species for Australia. CSIRO, Australia, 72 p
Marschner H (2003) Mineral nutrition of higher plants. Academic Press, London
Masters DG, Benes SE, Norman HC (2007) Biosaline agriculture for forage and livestock production. Agri Ecosys Environ 119:234–248
Menzel U, Lieth H (1999) Halophyte database vers. 2.0. In: Lieth H, Moschenko M, Lohman M, Koyro HW, Hamdy A (eds) Halophyte uses in different climates I: ecological and ecophysiological studies, vol 13. Progress in biometeriology. Backhuys Publishers, The Netherlands, pp 159–258
Munns R (2005) Genes and salt tolerance: bringing them together. New Phyt 167:645–663
Munns R, Tester M (2008) Mechanisms of salinity tolerance. Ann Rev Plant Biol 59:651–681
Murtaza G (2013) Economic aspects of growing rice and wheat crops on salt-affected soils in the Indus Basin of Pakistan (unpublished data). Institute of Soil and Environmental Sciences, University of Agriculture, Faisalabad, Pakistan
Pachauri RK, Reisinger A (eds) (2007) Climate change 2007: synthesis report, contribution of working groups I, II and III to the fourth assessment report of the intergovernmental panel on climate change. IPCC, Geneva, Switzerland, p 104
Pandey DN (2007) Multifunctional agroforestry systems in India. Current Sci 92:455–463
Parry M, Evans PM, Rosegrant W, Wheeler T (2009) Climate change and hunger: responding to the challenge. World Food Programme, Rome, p 108
Patra DD, Singh DV (1998) Medicinal and aromatic crops. In: Tyagi NK, Minhas PS (eds) Agricultural salinity management in india. Central Soil Salinity Research Institute, Karnal, India, pp 499–506
Perveen S (2002) Growth, water and ionic relations in atriplex species under saline and hypoxic conditions. Ph.D. thesis, Department of Soil Science, University of Agriculture, Faisalabad, Pakistan

Postel SL, Daily GC, Ehrlich PR (1996) Human appropriation of renewable fresh water. Science 271:785–788

Qadir M, Oster JD (2004) Crop and irrigation management strategies for saline-sodic soils and waters aimed at environmentally sustainable agriculture. Sci Total Environ 323:1–19

Qadir M, Steffens D, Yan F (2003) Proton release by N_2-fixing plant roots: a possible contribution to phytoremediation of calcareous sodic soils. J Plant Nutr Soil Sci 166:14–22

Qadir M, Tubeileh A, Akhtar J, Larbi A, Minhas PS, Khan MA (2008) Productivity enhancement of salt-affected environments through crop diversification. Land Degrad Develop 19:429–453

Qadir M, Quillérou E, Nangia V, Murtaza G, Singh M, Thomas RJ, Drechsel P, Noble AD (2014) Economics of salt-induced land degradation and restoration. Nat Res Forum 38:282–295

Qadir M, Noble AD, Karajeh F, George B (2015) Potential business opportunities from saline water and salt-affected land resources. In: International Water Management Institute (IWMI). Resource recovery and reuse series, vol 5. CGIAR Research Program on Water, Land and Ecosystems (WLE), Colombo, Sri Lanka, 29 p. https://doi.org/10.5337/2015.206

Qureshi RH, Barrett-Lennard EG (1998) Saline agriculture for irrigated land in Pakistan: a handbook. ACIAR, Canberra

Qureshi RH, Nawaz S, Mahmood T (1993) Performance of selected tree species under saline-sodic field conditions in Pakistan. In: Lieth H, Masoom AA (eds) Towards the rational use of high salinity tolerant plants, vol 2. Kluwer Academic Publishers, The Netherlands, pp 259–269

Risi C, Galwey NW (1984) The Chenopodium grains of the Andes: Inca crops for modern agriculture. Adv Appl Biol 10:145–216

Rozema J (1996) Biology of halophytes. In: Malcolm CV, Hamdy A, Choukr-Allah R (eds) Halophytes in biosaline agriculture. Marcel Dekker Inc, New York, pp 17–30

Rozema J, Flowers T (2008) Crops for a salinized world. Science 322:1478–1480

Rozema J, Schat H (2013) Salt tolerance of halophytes, research questions reviewed in the perspective of saline agriculture. Environ Exp Bot 92:83–95

Saqib M, Zörb C, Schubert S (2008) Silicon-mediated improvement in the salt-resistance of wheat (*Triticum aestivum*) results from increased sodium exclusion and resistance to oxidative stress. Funct Plant Biol 35:633–639

Schubert S, Neubert A, Schierholt A, Sümer A, Zörb C (2009) Development of salt-resistant maize hybrids: the combination of physiological strategies using conventional breeding methods. Plant Sci 177:196–202

Singh G, Dagar JC (2009) Biosaline agriculture: perspective and opportunities. J Soil Saline Water Qual 1(1–2):41–49

Tomar OS, Minhas PS, Sharma VK, Singh YP, Gupta RK (2003) Performance of 31 tree species and soil conditions in a plantation established with saline irrigation. Forest Ecol Manag 177:333–346

Vega-Gálvez A, Miranda M, Vergara J, Uribe E, Puente L, Martinez EA (2010) Nutrition facts and functional potential of quinoa (*Chenopodium quinoa* willd.), an ancient Andean grain: a review. J Sci Food Agric 90:2541–2547

Vyshpolsky F, Qadir M, Karimov A, Mukhamedjanov K, Bekbaev U, Paroda R, Aw-Hassan A, Karajeh F (2008) Enhancing the productivity of high–magnesium soil and water resources through the application of phosphogypsum in Central Asia. Land Deg Develop 19:45–56

Yensen NP (2006) Halophyte uses for the twenty-first century and a new hypothesis the role of sodium in C4 physiology. In: Khan MA, Weber DJ (eds) Ecophysiology of high salinity tolerant plants. Springer, The Netherlands, pp 367–396

Zekri S, Al-Rawahy SA, Naifer A (2010) Socio-economic considerations of salinity: descriptive statistics of the Batinah sampled farms. In: Mushtaque A, Al-Rawahi SA, Hussain N (eds) Monograph on management of salt-affected soils and water for sustainable agriculture. Sultan Qaboos University, Oman, pp 99–113

Muhammad Saqib is a Professor at the Institute of Plant Nutrition, Justus Liebig University, Giessen, Germany as a Fellow of the Alexander von Humboldt Foundation (Germany). He is also an Associate Professor at the Institute of Soil and Environmental Sciences, University of Agriculture, Faisalabad, Pakistan. He is a fellow of German Academic Exchange Service (DAAD), Germany and Japan Society for the Promotion of Science (JSPS), Japan. He works on land use and plant eco-physiology in the context of soil and water stresses, environmental pollution and climate change. His research work aims to understand and exploit soil-plant-water system under changing climatic conditions and to devise solutions for the efficient utilization of these resources for the betterment of growing human population.

Javaid Akhtar is Professor and Director of the Institute of Soil and Environmental Sciences, University of Agriculture, Faisalabad, Pakistan. He is a well-known scientist and expert in the field of Soil and Environmental Sciences. He has a vast experience as a teacher, researcher and administrator. He has played a leading role in a number of national and international projects. The main focus of his research has been the rehabilitation of salt-affected lands using brackish drainage water through saline agriculture technology.

Ghulam Abbas is an Assistant Professor at the Department of Environmental Sciences, COMSATS Inst. of Information Technology, Vehari, Pakistan. He works on the utilization and rehabilitation of salt-affected soils using different tree species. His research interest also includes the use of salt tolerant plant species for the remediation of metal contaminated saline and non-saline soils.

Ghulam Murtaza is a Professor at the Institute of Soil and Environmental Sciences, University of Agriculture, Faisalabad, Pakistan. He is a well-known scientist and expert in the field of Soil, Water and Environmental Chemistry. He works on the reclamation of salt-affected soils and on the development of technologies for the safe use of poor quality water.

Community Participation in Climate Change Mitigation Research in the Arid and Semi-Arid Areas of Sudan

N. A. Mutwali, M. E. Ballal and A. M. Farah

Abstract Three studies were conducted in three states within the arid and semi-arid zones of Sudan namely; Northern State, White Nile and North Kordafan States. The Northern State is located between latitudes 16° 53′ and 16° 56′ N and longitudes 31° 36′ and 31° 40′ E. The soil is sandy with inherent deficiency in nitrogen and organic matter. The mean annual rainfall is 40 mm only, and the mean minimum and maximum temperatures are 39 and 23 °C, respectively. The objectives of this study were to develop methods for controlling wind erosion using different methods of establishment for different tree species with the involvement of local communities in field work and protection of farms. The study site in the White Nile State is located in Central Sudan (32° 15′ N and 14° 45′ E). The average rainfall is less than 200 mm. The soil of the study area is classified as White Nile clays. The third study was conducted in North Kordofan State (11° 15′ and 16° 45′ N; 27° 05′ to 32° E) where the soil is sandy and the annual rainfall is about 318 mm and where the accumulation of sands affected up to 50% of the agricultural land. The Agricultural Research Corporation and the Forests National Corporation organized extension work and training for the local communities in controlling desertification and stabilizing sands using seeds and seedlings in home nurseries. The studies recommended the establishment of drought tolerant trees by integrating mechanical protection means with seedlings planting in the arid and semi-arid lands of Sudan for stabilizing the highly moving sand dunes.

N. A. Mutwali · M. E. Ballal (✉)
Forestry and Gum Arabic Research Centre, Agricultural Research Corporation, University of Khartoum, Khartoum, Sudan
e-mail: mohamedballal@yahoo.com

A. M. Farah
Soba Research Station for Reclaiming Saline and Sodic Soils, Khartoum, Sudan

P. Castro et al. (eds.), *Climate Change-Resilient Agriculture and Agroforestry*, Climate Change Management, https://doi.org/10.1007/978-3-319-75004-0_17

1 Introduction

Sudan, the third largest country in Africa, covers an area of about 1.86 million km^2, much of which is comprised of desert and arid lands. It lies within the tropical zone between 10° and 23° N and 21° 45″ and 38° 30″ E and bordering seven countries and shares surface and ground water with 12 countries. Its topography can be broadly characterized as vast plains interspersed by widely separated ranges of hills and mountains. The country is 29% desert, 19% semi-desert, 27% low rainfall savanna, 14% high rainfall savanna, 10% flood regions (swamps and areas affected by floods) and less than 1% mountain vegetation. The country's population is 40 million people and is divided administratively into 16 States. The economy has long depended on agricultural and livestock exports but in recent years gold and oil have been exploited.

An examination of Sudan's ecological zones indicates that the majority of its land is highly vulnerable to changes in temperature and precipitation. The mean annual temperature lies between 26° and 32° but in some areas it reaches 47 °C causing a lot of stresses and heat related diseases. The country's inherent vulnerability can be explained by the fact that food security is dependent on rainfall, particularly in the rural areas where about two thirds (65%) of the population lives. Rainfall is erratic and varies significantly from the North to the South. The unreliable nature of rainfall together with its concentration during the short growing season increases the vulnerability to failure of the rain-fed agricultural crops. A trend of decreasing annual rainfall in the last 60 years (0.5%) and increased rainfall variability is contributing to drought conditions in many parts of the country. This pattern has led to serious and prolonged drought episodes. For example, Sudan experienced a succession of dry years e.g. 1974, 1984 and 1991, which have resulted in severe social and economic impacts including many human and livestock fatalities, migration and displacement of several million people. Drought problems such as these will increase if trends continue (Fadl El Moula and Elgizouli 2008). Sudan experienced some severe floods of two specific types during the past several decades. The first type occurs during heavy rain when high levels of water overflow the River Nile and its tributaries, usually due to above normal rainy seasons in the Ethiopian Plateau. Severe floods were reported in 1946, 1988, 1994, 1998 and 2001. The other type of flood occurs as a result of heavy localized rainfall reported in 1952, 1962, 1965, 1978–1979, 1988 and 1997 (HCENR 2005). In addition to drought and floods there are other climate extreme events such as dust storms, thunder storms and heat waves.

In the arid lands of Sudan, soil erosion is also one of the major problems that have occurred as part of geological processes and climate change. Erosion is becoming more severe and it is accelerated by adverse human activities.

Wind erosion is an important land surface process which has received attention in many countries. The transport of the finest and most valuable soil particles by wind and water has led to degradation processes affecting the agriculturally used areas and pollutes the atmosphere (Roger and Michel 2013). Scientists have long been interested in the direct and indirect effects of wind erosion The earliest publication relating to aeolian processes was written by astronomer Godefroy Wendelin in 1646 (Stout et al. 2009). Wind erosion is a soil degrading process that affects more than 500 million hectares of land worldwide and creates between 500 and 5000 kg of fugitive dust annually (Grini et al. 2003).

The climate change and the associated effects on human society in general and agriculture in particular have set the problem of wind erosion in a broader perspective (Funk and Reuter 2006). Wind erosion has received more attention as a process responsible for decrease in soil fertility and as source of atmospheric pollution (Oldemann et al. 1990; Gobin et al. 2003; Warren and Bärring 2003). The removed soil particles may be deposited downwind to become part of the new landscape or may be transported to oceans where the rich nutrient dust enhances aquatic life (Morales 1977). The average annual soil loss resulting from raindrop splash and runoff from field slopes, is still most frequently used at large spatial scales (Renschler and Harbor 2002 and Panagos et al. 2014). In addition to soil fertility degradation, there is a disproportionate loss of soil organic carbon (Van Pelt et al. 2007) and soil fines may affect soil water infiltration and holding capacity, further affecting soil productivity in semiarid regions. However, estimating soil erosion rates and amounts began in the USA since the 1920s and rapidly advanced in the 1930s following the devastating impact of the “Dust Bowl” on the American Great Plains (Römkens 2010).

For effective biological control against wind erosion, some literature suggests that shelterbelts should be designed to contain 10 or more rows of trees and shrubs to provide maximum benefit to wildlife and land/home owners (IOWA 2010). Biological control through field investigations indicated the effectiveness of trees and shrubs to suppress moving sand compared to other mechanical measures (Al-Amin et al. 2006; Al-Amin et al. 2010; Shulin et al. 2008; Mohammed et al. 1999). However, in Sudan the main degraded zones are the arid and semi—arid zones where 76% of the human population lives (Ayoub 1998). Dregne et al. (1991) estimated that nearly 70 million hectares of Sudan land are very severely degraded. In North Kordofan, sand movement is a serious problem facing the land and people of Namla village. To alleviate this problem, wind breaks and shelterbelt are expected to reduce the severity of wind erosion. In this respect, experience derived from various international and national projects has demonstrated that people can devise their own alternative actions to deal with local problems. Therefore, community participation is a proven approach to addressing a number of agricultural and environmental problems. Hence, a number of experiments were conducted in Northern Sudan, White Nile and North Kordofan States so as to measure soil erosion and sand accumulation, to investigate the limitations for natural regeneration of vegetation, and to seek effective methods for controlling wind erosion with the involvement of local communities.

2 Materials and Methods

2.1 Mechanical and Biological Methods for Controlling Wind Erosion at Um Jawaseer Area in the Northern Sudan

This study was carried out at Um Jawaseer village in Merawe locality in the Northern State. Um Jawaseer area is located within the arid zone of Sudan. It lies between latitudes 16° 53′ and 16° 56′ N; and longitudes 31° 36′ and 31° 40′ E. The rainfall is generally low and erratic. The mean annual rainfall is 40 mm that falls between July and September. Average evaporation is 16.4 mm which is high enough to reduce effectiveness of precipitation. The relative humidity is 30.2%. Extreme high temperature characterizes the area during summer months reaching an average maximum of 39 °C and a minimum of 23 °C. The main wind direction constitutes the northeast strong dry winds blowing from October to June with sand storms resulting from the seasonal movement of the Inter Tropical Convergence Zone. However, during July to September the dominant wind direction is the southern wind but the direction is sometimes interchanged between west and east. Wind speed is 12.5 (km/h). The soil is sandy with low nitrogen and organic matter contents, low available P, very low CEC that can be considered as infertile but non-saline, non-sodic, with somewhat high bulk density, medium porosity and hydraulic conductivity and low available water capacity (AWC). Some patches of compacted soil and salinity are generally found in this area.

2.2 Effect of Mechanical Protection of Wind Erosion on Growth and Survival of Different Tree Species

A 3-factor randomized complete block experiment with three replications was conducted to study the effect of mechanical methods of protection against wind erosion as reflected on growth and survival of four indigenous tree species under rain fed conditions. The first factor comprised three methods of mechanical protection namely checkerboard, surface lying and a flat area with no mechanical protection as control. Checkerboards were made by tying palm leaves together to make a one meter high barrier which is constructed around the planted area for protection from wind. Surface lying was made by collecting dry bushes tied as a bundle so that the trees can be grown between two bundles. The control consisted of planting the trees without protection. The second factor comprised four tree species, namely *Salvadora persica*, *Moringa pregrenia*, *Acacia tortilis* and *Ziziphus spina christi*. The third factor consisted of the method of establishment by direct seeding or seedlings planting. Each treatment combination was represented by 4 trees making 96 trees per replication and a total of 288 trees. Survival and growth of the tree species were measured after one and two years from planting. Analysis of variance and Duncan's Multiple Rage Test at 0.05 probability level was used according to MSTAT-C software package.

2.3 Assessment of Soil Erodibility

Estimation of soil erodibility of individual grains is dependent upon their diameter, density and shape. Soil erodibility index was estimated using the procedure described by Chepil (1945). Soil erodibility is best estimated by carrying out direct measurements on field plots (Kinnell 2010). Panagos et al. (2012a) estimated soil erodibility based on soil texture and organic carbon data obtained from the Land Use/Cover Area Frame Survey (Toth et al. 2013).

To determine erodibility, a spade was pushed under the surface soil layer to a depth of 3 cm. five soil samples 0.5 kg each were taken and air-dried. The percentages of soil aggregates greater than 0.84 mm in diameter were determined using a sieve of the same size. Soil erodibility was determined using indices developed by Woodruff and Siddways (1965); and Skidmore and Siddways (1978). Erodibility is calculated by the following equation:

$$\text{I (erodibility)} = \frac{\text{weight of grains} > 0.84\ \text{mm} \times 100}{\text{Total weight of sample}}$$

2.4 Assessment of Wind Erosion Using Sand Traps

Vertical sand traps were fixed in the north and south directions of the physical barriers (checkerboard and surface lying) for measuring wind erosion and sand movement. The vertical trap is a metal pipe with three levels 15, 30 and 45 cm from the surface of the ground. It has two windows in opposite directions. Each window is 1×1 cm and one of the windows is open and the other is covered with mesh to control the sand grains entering though the open one. The traps were discharged every three months and wind erosion (ton ha^{-1} day^{-1}) was calculated using the following equation:

$$\text{Wind erosion} = \frac{\text{weight of soil sample (g)} \times 100}{\text{number of days} \times \text{area of trap}}$$

However, in the newly established young tree belt, plastic cans ($21 \times 21 \times 40$ cm dimensions) were erected as horizontal traps in the windward and leeward directions of the shelterbelt. Four traps were constructed per each replication in each side of the young trees' belt. These traps were discharged every three months and the wind erosion (ton ha^{-1} day^{-1}) is calculated using the above equation.

2.5 Effect of Shelterbelts on Sand Accumulation and Seedlings Establishment of Indigenous Tree Species

This experiment was conducted in the leeward side of Um Jawaseer shelterbelt which is composed of six rows of six-years-old tree species namely; *Eucalyptus microtheca, Prosopis chilenses, Ziziphus spina-christi, Azadirachta indica, Conocarpus lancifolius* and *Acacia amplicips*. A two-factor randomized complete block design experiment with three replications was used. Four tree species namely *Acacia tortilis, Acacia oerfota, Leptadenia pyrotechnica* and *Zizipus spina christi* as factor one, were planted using 2 × 2 m between trees and 3 × 3 m between rows. Two methods of sowing viz. direct seeding and seedlings planting constituted the second factor. Each treatment combination was represented by ten trees making a total of 240 trees. Tree height, diameter, survival rate and number of branches were measured every 3 months. Soil erodibility was determined using Woodruff and Siddways (1965); Skidmore and Siddways (1978) indices.

2.6 Degradation of Vegetation at Elgetaina Area in the White Nile State

This study was carried out in central Sudan, east of Elgetaina town in the White Nile State (32° 15′ North and 14° 45′ East) during 2004–2006.The climate is atypical tropical continental, characterized by warm dry winter and hot rainy summer. The mean temperatures are 37 and 21 °C for summer and winter, respectively. Relative humidity is lowest in April (10%) and highest in August (67%). The mean daily evaporation is highest in April (20 mm) and lowest in August (10.8 mm). The wind blows in the dry season from the south-east direction at about 4.5 m/s. Dust storms (Haboobs) are common during the summer season. The soils of study area are classified as White Nile clays (Brawn et al.1991) with 60–70% clay content and are rather uniform in texture and profile features.

2.7 Sampling

Four sample plots of 500 × 500 m each were selected randomly in the study area and composition of the vegetation cover was assessed. A total of 40 soil samples were taken from two depths (0–30 cm)-(30–60 cm) during April 2004. Soil analysis was carried out at the Land and Water Research Centre of The Agricultural Research Corporation (ARC) Sudan.

2.8 Measuring Wind Erosion and Sand Accumulation at Namla in Western Sudan

This study was carried out at Namla village in North Kordofan State which lies in the Northern part of western Sudan at 71 km North East of Elobeid town. About 50% of the agricultural land (approximately 3000 ha) were covered by sand dunes. The annual rainfall is about 318 mm. The vegetation cover is composed of *Acacia senegal*, *Balanites aegyptiaca*, *Acacia tortilis*, *Ziziphus spina christi*, and *Leptadenia pyrotechnica*. But in the study area the soil is devoid of vegetation because of tree cutting, over grazing, and agricultural expansion. However, the understory vegetation is compound of *Cenchrus ciliaris*. *Cenchrus biflorus* and *Cassia tora*. The total population is 1645 mainly from Gawamaa tribe. The agricultural system is traditional, where sesame, sorghum and gum arabic constitute the main food and cash crops.

The Agricultural Research Corporation and the Forests National Corporation organized extension work and training for the local communities in order to control desertification thought the stabilizing of moving sands using seeds and seedlings grown in home nurseries.

2.9 Assessment of Wind Erosion Using Sand Traps

Vertical and horizontal sand traps were established for measuring wind erosion and sand movement in the study area. The horizontal traps were namely plastic cans with 21 × 21 cm dimension erected in the windward and leeward directions of the shelterbelt and arranged in three replicates used in the northern and southern wind directions. The traps were discharged every month and the wind erosion (ton ha^{-1} day^{-1}) was calculated using the previous equation. Vertical traps were erected in the northern and southern wind directions. The vertical trap is a metal pipe with three levels: 15, 30 and 45 cm from the surface of ground. It has two windows in opposite directions. Each window is 1 × 1 cm and one of the windows is open and the other is covered with a piece of mesh to control the sand grains entering though the open one. These traps were discharged every month and wind erosion (ton ha^{-1} day^{-1}) was calculated using previous equation. These vertical traps were constructed in front of and behind the shelterbelt in the northern and southern directions.

2.10 Assessment of Soil Erodibility

Estimation of soil erodibility of individual grains is dependent upon their diameter, density and shape. Soil erodibility index was estimated using the procedure described by Chepil (1945). Erodibility of soils with different percentages of non-erodible fractions was determined by standard dry sieving for a fully crusted soil surface regardless of soil texture. However, good soil structure and high aggregate stability are important parameters for improving soil fertility, enhancing porosity and decreasing erodibility (Bronick and Lal 2005). In practice, erodibility represents an integrated average annual value of the total soil and soil profile reaction to a large number of erosion and hydrological processes (Bonilla and Johnson 2012).

In measuring erodibility, a spade was pushed under the surface soil layer to a depth of 3 cm. Five soil samples 0.5 kg each were taken and air-dried. The percentages of soil aggregates greater than 0.84 mm in diameter were determined using a sieve of the same size. The soil erodibility index was estimated for each field sample using previous erodibility indices.

2.11 Effect of Shelterbelt on Sand Accumulation and Seedling Establishment

This experiment was conducted in the leeward side of Namla village. The tree species which were selected to control sand movement were *Leptadeniapyrotechnica, Acacia tortilis, Acacia senegal, Balanitesaegyptiaca, Faidherbiaalbida,* and *Panicumturgidum*in the under story. Two methods of sowing viz. direct seeding and seedlings planting were used. Tree height, diameter, survival rate and number of branches were measured every three months. Soil erodibility was determined using Woodruff and Siddways (1965); Skidmore and Siddways (1978) indices.

2.12 Home Nurseries (Individuals)

The Agricultural Research Corporation and the Forests National Corporation organized an extension work for the local community by training them on controlling desertification and sand dune stabilization using seedlings and seeds in home nurseries. The main species produced by the local people in home nurseries were *Acacia senegal*, *Acacia mellifera,* *Grewiatenex*, *Azdirachtaindica*, *Balanitesaegyptiaca* and *Ziziphus spina christi.*

3 Results and Discussion

3.1 Mechanical Control of Wind Erosion

The average erodibility of the soil at Um Jawaseer in the Northern Stateof Sudan was 16.8 (Table 1). Erodibility was calculated on the basis of indices developed by Woodruff and Siddways (1965) or 37.2 according to Skidmore and Siddway (1978). The former erodibility values falls within the values given in Skidmore and Siddoways (1978) table of erodibility. This result evidenced moderate wind erosion in the study area, where some materials of the soil were removed and some deposited. Breshears et al. (2008) measured sediment transport rates ranging from 0.17 to 27.4 g m^{-2} d^{-1}, respectively.

Table 2 shows soil accumulation (ton ha^{-1} day^{-1}) according to trap height, the highest (76.8 ton ha^{-1} day^{-1}) sand accumulation was obtained at the 15 cm height of trap followed by the 30 cm and the 45 cm heights, respectively. This result proved that sand movement in this area is in the form of surface creep. Seedlings planting under the three methods of mechanical protection showed that checkerboards has the highest survival (61.4%), followed by surface lying (57.8%) and the control (52.1%) as shown in Table 3. In this respect, studies conducted by Chepil (1945), Chepil and Woodruff (1963), Woodruff and Siddways (1965); and Skidmore and Siddoways (1978) showed that the quantities of various crop residues needed to protect soils from wind erosion have been determined and compared to an equivalent amount of flat small grain. The relationship between physical properties of the residues and erosion was modeled by Lyles and Allison (1976, 1980, 1981). Where the sand is fixed, soil formation can begin; fine particles are accumulated and a hard soil crust is formed on the dune surface improving both the micro-environment and the stability of the dune surface. The straw checkerboard can significantly increase the content of organic matter of the surface soil. The technique has several advantages such as: remarkable effect on dune fixation, ease of construction, rapid results, and no pollution of the environment. However, its labor cost is high and the replacement of the straw after 3–5 years can be a problem. The straw checkerboard decreases wind velocity near the ground surface and can prevent wind erosion of the soil. After establishment, the straw gradually rots and

Table 1 Estimation of soil erodibility at Um Jawaseer in the Northern State, Sudan

Sample	Aggregates greater than 0.84 mm	Erodibility (Woodruf and Siddways)	Erodibility (Skidmore and Siddways)
1	65	16.5	37
2	60	21.0	47
3	61	20.0	45
4	66	15.6	35
5	71	11.0	22
Average	64.6	16.8	37.2

Table 2 Sand accumulation (ton ha^{-1} day^{-1}) according to trap height at Um Jawaseer in the Northern State, Sudan

Trap height (cm)	Soil accumulation (ton ha^{-1} day^{-1})
15	76.79 a
30	56.69 b
45	15.96 c
P	P = 0.0001
SE±	3.1782

Means with the same letters are not significantly different according to DMRT at $P \leq 0.05$

Table 3 Effects of establishment and protection methods on tree survival (%) at Um Jawaseer in the Northern State, Sudan

Method of protection	Method of establishment	
	Seeds	Seedlings
Checkerboard	46.9 d	61.4 b
Surface lying	44.3 e	57.8 b
Control	43.3 e	52.1 c
P	0.0001	
SE±	0.690	

Means with the same letters are not significantly different according to DMRT at $P \leq 0.05$

converted to organic matter. Checkerboards are widely used for dune fixation in arid and semi-arid regions of China (Maki 1999). Research has shown that the straw checkerboard significantly increases the roughness length and decreases the sand content in the sand flux by as much as 99.5% (SDRS 1986; Liu 1987; Liling 1991; Wang 1991; Xu et al. 1998). The extent of dune fixation depends on the dimensions of the checkerboard. In this respect, a size of 1 × 1 m, as in the present study has remarkable wind break and dune fixation effects (SDRS 1986; Wang 1991).

There was also a highly significant interaction of species and method of protection ($P \leq 0.01$) on height growth (cm). *Ziziphus spina Christi se*edlings have the highest growth (>23.0 cm) in height under surface lying and control and also under checkerboard method.

The lowest growth in height for all the selected species was in the control compared with the other methods of protection. *Moringapregrina and Acacia tortilis*have reached about the same height using surface lying. Similarly *Ziziphus spina christi and Salvadorapersica* has the same height using checkerboard method. On the other hand, *Moringapregrina and Acacia tortilis* gave the lowest height growth (17.0 cm) using checkerboard method. *Salvadorapersica*has the lowest height growth (17.9 cm) as control. Surface lying gave the highest height growth in all the selected species followed by checkerboard and control (Table 4).

Table 4 Effect of species and method of protection on height growth (cm) at Um Jawaseer in the Northern State, Sudan

Species	Methods of protection		
	Checkerboard	Surface lying	Control
Salvadora persica	19.68 bc	19.11 bcd	17.93 cd
Moringa pregrina	16.96 d	19.88 bc	18.95 bcd
Acacia tortilis	17.90 cd	20.96 b	19.93 bc
Ziziphus spina christi	19.76 bc	23.65 a	23.21 a
P	0.01		
SE±	0.6803		

Means with the same letters are not significantly different according to DMRT at $P \leq 0.05$

3.2 *Biological Control of Wind Erosion Using Trees*

The combined analysis of two-years on the effect of biological protection against wind erosion using the four species revealed highly significant ($P<0.01$) effect of the interaction of species and years on height growth (cm). *Acacia oerfota* has the highest growth (65.7 and 108.7 cm) compared with the other species in both years, respectively. *Ziziphus spina christi*has the lowest growth of 27.1 and 46.0 cm in both years, respectively. *Lyptadenia pyrotechnica* and *Acacia tortilis* showed no significant difference in their height growth in the first year and also in the second year of establishment (Table 5). The effect of species on diameter growth (mm) was found to be significant ($P<0.05$). The four species, i.e. *Ziziphus spina christi, Acacia oerfota, Leptadenia pyrotechnica and Acacia tortilis* have similar diameter growth in the first year of establishment with no significant difference among them. In the second year of establishment, *Acacia oerfota* has the highest growth (2.6 mm) in diameter which was significantly different from *Ziziphus spina Christi* but similar to the other three species (Table 6). Field investigations indicated the effectiveness of trees and shrubs to suppress moving sand compared to other mechanical measures (Al-Amin et al. 2006; Al-Amin et al. 2010; Shulin et al. 2008; Mohammed et al. 1999). However, the effectiveness of a barrier depends on its porosity. On the other hand, low porosity creates turbulence on the leeward side of the belt (Ki-Pyo and Young-Moon 2009). According to Kaul (1969) a windbreak strip is to be established on sand to prevent them from being blown away by winds for better established of field crops. Above authors suggested a five row shelterbelt in a pyramidal shape in a similar work to this study in dry areas. However, some literature suggests 10 or more rows of trees and shrubs to provide maximum benefit to wildlife and human dwellings (IOWA 2010). In this respect, shelterbelts in Sudan were established to stop drifting sand damage in Gezira scheme (Stigter et al. 1989).

Table 5 Effect of species and year on height growth (cm) of young trees at Um Jawaseer in the Northern State, Sudan

Species	2008	2009
Ziziphus spina christi	27.1 e	46.0 d
Acacia oerfota	65.7 c	108.7 a
Leptadenia pyrotechnica	42.7 d	67.9 c
Acacia tortilis	46.9 d	76.2 b
P	0.002	
SE±	2.372	
CV%	6.8	

Means with the same letters are not significantly different according to DMRT at $P \leq 0.05$

Table 6 Effect of species and year on diameter growth (mm) of young trees at Um Jawaseer in the Northern State, Sudan

Species	2008	2009
Ziziphus spina christi	1.2 c	2.1 b
Acacia oerfota	1.3 c	2.6 a
Leptadenia pyrótechnica	1.1 c	2.3 ab
Acacia tortilis	1.2 c	2.3 ab
P	0.05	
SE±	0.036	
CV%	10.8	

Means with the same letters are not significantly different according to DMRT at $P \leq 0.05$

3.3 Sand Accumulation

The 15 cm trap height accumulated significantly ($P<0.05$) the highest (73.25 ton ha^{-1} day^{-1}) amount of sand compared with traps of heights above the surface creep level (Table 7). Accumulation of sand in the horizontal traps during the two years using tree belts showed that autumn has significantly ($P<0.05$) the highest accumulation of sand compared with accumulation in the summer or winter season (Table 8). In addition, the high accumulation of soil in the horizontal traps supported the previous findings that most of the erosion in northern Sudan is by surface creep. Farah (2003) in his study of wind erosion in Khartoum State, found increased erosion from winter to summer and that Khartoum state is suffering from wind erosion in its four directions. He also found that the intensity of wind erosion in the soil surface decreased with the height of trap, which means that the major wind erosion is surface creep. Similarly, in this study, the trap height of 15 cm also gave significantly ($P<0.05$) the highest accumulation of sand in vertical traps. In the present study the mean daily wind load or accumulation was estimated at 73.25 ton ha^{-1} day^{-1} as compared with 76.79 ton ha^{-1} day^{-1} in the mechanical control method, indicating better wind erosion protection resulting from tree belts as compared to mechanical methods.

Table 7 Accumulation of sand (ton ha^{-1} day^{-1}) in vertical traps at Um Jawaseer in the Northern State, Sudan

Trap height (cm)	Sand accumulation (ton ha^{-1} $day^{-1)}$
15	73.25 a
30	48.64 b
45	41.84 b
P	0.0001
SE±	3.1038

Means with the same letters are not significantly different according to DMRT at $P \leq 0.05$

Table 8 Effect of season and direction of sand traps on accumulation of sand in horizontal traps at Um Jawaseer in the Northern State, Sudan

Seasons	Accumulation (mean)	Direction	Accumulation (mean)
Summer	335.3 b	North	405.2
Autumn	524.7 a	South	430.2
Winter	393.1 b		
P	0.0002		0.559
SE±	28.8		29.9

Means with the same letters are not significantly different according to DMRT at $P \leq 0.05$

3.4 Degradation of Vegetation in the White Nile

The study area has a low woody vegetation cover. Only 10 trees in the whole study area were found and these are 5 shrubs of *Capparis decidua,* 3 trees of *Acacia tortilis* and 2 shrubs of *Acacia nubica* (Table 9). Most parts of the study area are bare land especially in plot 3. Vegetation consisted of the following species *Cassia senna* and *Panicum turgidum*. This is in addition to some sparse vegetation of *Chrozophora oblongifolia,* which is highly desired by camels. The ground vegetation was estimated at 20% in the study area. There were no viable seeds in the whole study area because there were no trees in this area and it contains some dead grass seeds.

Generally the soil at the study site is low in N and Organic carbon (Table 10). Cl content is high especially in the third sample. In all four samples we noticed that $CaCO_3$ was very high in all samples. Electric conductivity (EC) and exchangeable sodium percentage (ESP) increase with depth. The soil of sample 3 and 4 is saline and sodic (EC>4ds/m and ESP>15). In these two samples there were no trees but some grasses and bushes e.g. *Panicum turgidum* are present.

The high sodicity decreases the infiltration rate and inhibits the emergence of seeds. The high PH inhibits P availability. These soils need to be reclaimed by leaching the soluble salts by good quality water and addition of farm yard manure and gypsum to reduce the sodicity level and to improve the physical properties of the soil.

Table 9 Vegetation cover in sample plots at Elgetaina area in the White Nile, Sudan

Sample no.	Location	Trees	Shrubs	Species
1	14° 49′ 20 N 32° 26′ 35 E	2	3	*Trees: Acacia oerfota and, Acacia tortilis subsp. Radiana; Shrubs: Capparis decidua*
2	14° 49′ 17 N 32° 26′ 37 E	1	3	*Trees: Acacia oerfota and, Acacia tortilis subsp. Radiana; Shrubs: Capparis decidua*
3	14° 49′ 00.3 N 32° 26′ 54E	0	0	–
4	14° 48′ 43 N 32° 27′ 13E	0	1	*Only Capparis decidua* shrub.

Table 10 Soil chemical properties in 4 sample plots at Elgetaina area in the White Nile, Sudan

Sample	Depth (cm)	Exchangeable bases (Coml.+/kg)					Soluble cations and anions (meq/L)				
		Ca Co3	Na	O.C %	N %	PH paste	E.C (ds/m)	ESP %	Na	CL	HCO_3
1	0–30	1.28	1.93	0.06	0.02	7.76	1.40	6.6	8.55	5.7	2.25
	31–60	1.64	2.99	0.07	0.03	7.94	3.26	12.8	24.7	22.0	2.30
2	0–30	1.84	2.33	0.06	0.02	8.16	1.9	8.8	13.9	7.05	1.0
	31–60	1.52	2.24	0.07	0.02	8.28	1.55	8.6	11.4	2.3	1.15
3	0–30	5.08	12.06	0.10	0.04	8.36	8.45	38.4	74.5	61.0	4.2
	31–60	4.84	11.44	0.15	0.02	8.32	9.29	38.6	80.0	71.7	4.9
4	0–30	3.32	3.49	0.09	0.02	8.52	1.99	14.6	15.9	13.9	3.3
	31–60	3.36	7.08	0.05	0.02	8.56	4.41	24.8	38.4	35.0	3.2

The analysis indicates that the inherent fertility status of this soil is very low. It also has low water holding capacity and is susceptible to wind erosion. Due to the current droughts and misuse, the vegetation cover in this area is very low resulting in low carrying capacity of livestock. Hence, this study recommended reseeding of the area with drought and salt resistant trees for stabilizing the highly mobile sand dunes.

Table 11 Estimation of soil erodibility at Namla village in North Korrdofan State in western Sudan

Sample No.	Erodibility (Woodruf and Siddway)	Erodibility (Skidmore and Siddway)
1	80.1	179
2	83.2	187
3	90.1	206
4	75.6	172
5	80.6	182
6	77.5	175
7	75.3	169
8	79.5	178
9	86.3	193
10	92.8	208
Average	82.1	184.9

3.5 *Estimation of Soil Erodibility in Western Sudan*

Based on Skidmore and Siddway (1978) and Woodruf and Siddways (1965) the soil erodibility at Namla village was 82.1 and 184.9, respectively. The results showed that there is high erodibility in the study area (Table 11). However, the highest soil accumulation (106.1 ton ha^{-1} day^{-1}) was obtained at the 15 cm trap height. Accumulation by the 30 cm and the 45 cm height traps was also high reaching >30 ton ha^{-1} day^{-1}. This result proved that the wind in this area is blowing at different levels carrying sand accumulation at varying heights

Table 12 Accumulation of sand (ton ha^{-1} day^{-1}) in vertical traps at Namla village in North Kordofan State in western Sudan

Trap height (cm)	Soil accumulation (ton ha^{-1} day^{-1}) North of belt	Soil accumulation (ton ha^{-1} day^{-1}) South of belt	Reduction %
15	106.11	88.37	17.0
30	93.32	50.71	45.7
45	68.54	35.2	48.7

Table 13 Soil accumulation (ton ha^{-1} day^{-1}) in horizontal traps at Namla village in North Kordofan State in western Sudan

Trap distance from belt (m)	Soil Accumulation (ton ha^{-1} day^{-1}) North of belt	Soil accumulation (ton ha^{-1} day^{-1}) South of belt	Reduction %
01	536.6	339.33	36.8
10	254.42	156.62	38.4
20	159.63	109.73	31.3

Table 14 Threats faced by the community at Namla village in western Sudan before establishment of shelterbelts

Threats to environment	Respondents' views (%)
Desertification	86.1
Wind erosion	72.3
Overcutting	76.6
Environmental pollution	97.9
Land degradation	100

(Table 12). However, the soil accumulation is decreasing in the southern direction of the belt by more than >30%. Accumulation of sand is decreasing according to the distance from the belt (Table 13). However, Rubio and Recatalá (2006) proposed stoniness to be included in the soil erodibility index qualitative estimation. Erodibility is best obtained from direct measurements on natural plots (Kinnell 2010). Measured erodibility values have also been related to soil properties.

3.6 Environmental and Socioeconomic Impact of Shelterbelts at Namla Village

According to the respondent's views, land degradation, environmental pollution and desertification are the main environmental hazards faced by the community before establishment of the shelterbelts. However, wind erosion and overcutting of trees for fuel, are also negatively affecting the area and will also increase land degradation (Table 14). All respondents believe that land degradation is the major threat to agriculture and livestock production.

However, after the establishment of the shelterbelts, the respondent's views depicted high to moderate reductions in the previous hazards. Sand movement and wind erosion were reduced by more than 40%. The community mentioned that negative effects of drought and desertification were reduced by more than 60%. Over seventy percent of the respondents believed that the shelterbelts offered protection to their farms from sand blasts. In addition, the shelterbelts provided high protection or shelter to their animals and birds (Table 15). Therefore, as a result of above environmental enhancement, the shelterbelts are considered the main recreational site within the village.

Table 15 Effects of shelterbelts on the farms at Namla village in western Sudan

Benefits to environment	%
Decrease in sand movement	44.4
Decrease of wind erosion	48.9
Decrease of drought and desertification	62.2
Decrease of environmental degradation	66.7
Protection of the village and farms	73.3
Protection of animals and birds	97.6

References

Al-Amin NKN, Stigter CJ, Mohammed AE (2006) Establishment of trees for sand settlement in a completely desertified environment. J. Arid Land Res Manage 20(4):309–327

Al-Amin NKN, Stigter CJ, Mohammed AE (2010) Wind reduction pattern around isolated biomass for wind-erosion control in desertified area of central Sudan. J Environ Earth Sci 2 (4):226–234

Ayoub AT (1998) Extent, severity and causative factors of land degradation in the Sudan. J Arid Environ 38:397–409

Bonilla CA, Johnson OI (2012) Soil erodibility mapping and its correlation with soil properties in central Chile. Geoderma 189–190:116–123

Brawn M, Burgstaller H, Hamdoun AM, Walter H (1991) Common weeds of central Sudan. GTZ, Eschborn

Bronick CJ, Lal R (2005) Soil structure and management: a review. Geoderma 124(1–2):3–22

Chepil WS (1945) Dynamics of wind erosion. The nature of movement of soil by wind. Soil Sci 60:305–320

Chepil WS, Woodruff NP (1963) The physics of wind erosion and its control. Adv Agron 15:211–302

Dregne H, Kassas M, Rozanov B (1991) A new assessment of the world status of desertification. Desertification Control Bull 20:6–18

Fadl El Moula I, Elgizouli I (2008) Climate change and impacts in Sudan and the future prospective to mitigate climate change

Farah AM (2003) Wind erosion in Khartoum State. Soil and environmental sciences. Department of soil sciences, PhD. University of Khartoum, Sudan

Funk R, Reuter HI (2006) Wind erosion in Europe. In: Poesen J, Boardman J (eds) Soil erosion in Europe, pp 563–582

Gobin A, Govers G, Jones R, Kirkby M, Kosmas C (2003) Assessment and reporting on soil erosion. In: European Environment Agency (EEA), Technical Report 94, 103 pp

Grini A, Myhre G, Zender CS, Sundet JK, Isaksen ISA (2003) Model simulations of dust source and transport in the global troposphere: effects of soil erodibility and wind speed variability. Institute Report Ser. 124. Dep. of Geosciences, Univ. of Oslo

HCENR (2005) Adaptation to climate change and related impacts, the Case of Sudan, prepared by UN commission on sustainable development and the ministry of environment and physical development, Sudan

IOWA (2010) http://www.iowapf.Org/page/1100/IowaShelterbelts-Windbreaks.jsp. Accessed 10 Oct 2011

Kaul RN (1969) Shelterbelt to stop creep of the desert. Sci Cult 24:406–409

Kinnell PIA (2010) Event soil loss, runoff and the universal soil loss equation family of models: a review. J Hydrol 385:384e397

Ki-Pyo Y, Young-Moon K (2009) Effect of protection against wind according to the variation porosity of wind fence. Environ Geol 56:1193–1203

Liling Y (1991) Physical principles of blown sand and application to the design of a railway protection system, case study of the Baotou-Lanzhou railway line. Res Desert Control 2:297–308

Liu Y (1987) Establishment and effect of a protective system along the Bautou-Lanzhou railway in the Shapotou sandy area. J Desert Res 7(4):1–11

Lyles L, Allison BE (1976) Possible effects of wind erosion on soil productivity. J Soil Water Conserv 30:279–283

Lyles L, Allison BE (1980) Range grasses and their small grain equivalents for wind erosion control. J Range Mgmt 33:143–146

Lyles L, Allison BE (1981) Equivalent wind-erosion protection from selected crop residues. Trans. ASAE. (in press)

Maki T (1999) Pictures of food, environmental, and agricultural issues in China. Tsukuba Press, Tsukuba, p 174

Mohammed AE, Stigter CJ, Adam HS (1999) Wind regimes windward of a shelterbelt protecting gravity irrigated crop land from moving sand in the Gezira scheme (Sudan). Theor Appl Climat 62:221–231

Morales DW (1977) Saharan dust. Mobilization, transport, deposition. Scope 14, pp 233–242. Wiley, Uk

Oldemann LR, Hakkeling RTA, Sombroek WG (1990) World map of human-induced soil degradation: an explanatory note, global assessment of soil degradation (GLASOD), ISRIC and united nations environment program (UNEP), FAO-ITC, 27 pp

Panagos P, Ballabio C, Yigini Y, Dunbar M (2012) Estimating the soil organic carbon content for European NUTS2 regions based on LUCAS data collection. Sci Total Environ 442:235–246

Panagos P, Meusburger K, van Liedekerke M, Alewell C, Hiederer R, Montanarella L (2014) Assessing soil erosion in Europe based on data collected through a European network. Soil Sci Plant Nutr. (in press)

Renschler CS, Harbor J (2002) Soil erosion assessment tools from point to regional scales—the role of geomorphologists in land management research and implementation. Geomorphology 47:189–209

Roger F, Michel R (2013) Leibniz-centre for agricultural landscape research, ZALF Müncheberg, Eberswalder Str. 84, 15374 Müncheberg, Wageningen University, Erosion and Soil and Water Conservation Group, Droevendaalsesteeg 4, P. Box 47, 6671 AA Wageningen, The Netherlands

Römkens MJM (2010) Erosion and sedimentation research in agricultural watershedsin the USA: from past to present and beyond. In: Banasik K, Horowitz AJ, Owens PN, Stone M, Walling DE (eds) Sediment dynamics for a changing future, vol 337. IAHS Publication, pp 17–26

Rubio José L, Recatalá L(2006) The relevance and consequences of Mediterranean desertification including security aspects. Desertification in the Mediterranean Region. A security issue NATO security through science series. vol. 3, pp 133–165

SDRS (Shapotou Desert Research Station, Institute of Desert Research, Academia Sinica, Lanzhou) (1986) The principles and measures taken to stabilize shifting sands along the railway line in the southeastern edge of the Tengger Desert. J Desert Res 6(3):1–19

Shulin L, Tao WC, Guangting G, Jian X, Shaoxiu M (2008) Field investigation of surface sand and dust movement over different sandy grass lands in the otindag sandy land China. Environ Geol 53:1225–1233

Skidmore EL, Siddoways FH (1978) Crop residue requirements to control wind erosion. In: Oschwald WR (ed) Crop residue management systems. ASA special pub no 31, pp 17–33

Stigter CJ, Coulson CL, El-Tayeb Mohamed A, Mungai DN, Kainkwa RMR (1989) Users' needs for quantification in tropical agrometeorology: some case studies. Instruments and observing methods report 35, WMO/TD 303, WMO. Geneva, pp 365–370

Stout JE, Warren A, Gill TE (2009) Publication trends in aeolian research: an analysis of the bibliography of aeolian research. Geomorphology 105:6–17

Toth G, Jones A, Montanarella L (2013) The LUCAS topsoil database and derived information on the regional variability of cropland topsoil properties in the European Union. Environ Monit Assess 185(9):7409–7425

Van Pelt RS, Zobeck TM, Ritchie JC, Gill TE (2007) Validating the use of 137Cs measurements to estimate rates of soil redistribution by wind. CATENA 70:455–464

Wang K (1991) Studies on sand dune stabilization in the shapotou area. Res Desert Control 2:13–26

Warren A, Bärring L (2003) Introduction. In: Warren A wind erosion on agricultural land in Europe. European Commission, EUR 20370, pp 7–12

Woodruff NP, Siddways FH (1965) A wind erosion equation. Soil Sci Soc Am Proc 29:602–608

Xu X, Hu Y, Pan B (1998) Analysis of the protective effect of various measures of combating drifting sand on the Tarim Desert Highway. Arid Zone Research 15(1):21–26

Using of Optimization Strategy for Reducing Water Scarcity in the Face of Climate Change

Mohammad Javad Zareian and Saeid Eslamian

Abstract Climate change is one of the most important factors that affect the availability of water resources in recent years. Many areas of the world are involved with droughts and water shortages due to climate change. Iran is one of the most notable areas in this field. The Zayandeh-Rud River Basin is of one of the most strategic areas in the central part of Iran in terms of water resources and consumption. In this study, the outputs of 15 GCMs related to the fourth assessment report of IPCC (AR4) are used for the analyzing the effects of climate change. A combination of the various GCM models is used through a weighting approach to generate the different climate change patterns including the ideal, medium and critical patterns. Then, the IHACRES model as a simple model designed to describe the dynamic response characteristics of catchments, was used to predict the natural inflow to the Zayandeh-Rud dam. Also, the Agro-Ecological Zones (AEZ) method that has been developed by the Food and Agriculture Organization (FAO) and International Institute for Applied Systems Analysis (IIASA), was used to determination of the effects of climate change on agricultural water demand. Based on the changes in the water resources and consumptions in the region, a non-linear optimization model was proposed to allocate the water between the different demands. The objective function was defined based on the minimizing of the water shortage during the next 30 years (2015–2044). The General Algebraic Modeling System (GAMS) software was used to solve this function. The results indicated that the annual water deficit of 610–1031 MCM will be occurred in different climate change patterns.

Keywords Zayandeh-Rud River · Climate change · IHACRES
GAMS · Water scarcity

M. J. Zareian
Department of Water Resources Research, Water Research Institute (WRI),
Ministry of Energy, Tehran, Iran

S. Eslamian (✉)
Department of Water Engineering, Collage of Agriculture,
Isfahan University of Technology, 8415683111 Isfahan, Iran
e-mail: saeid@cc.iut.ac.ir

P. Castro et al. (eds.), *Climate Change-Resilient Agriculture and Agroforestry*,
Climate Change Management, https://doi.org/10.1007/978-3-319-75004-0_18

1 Introduction

Since 1950, the number of heat cycles has increased and it has increased the warm nights. In addition, areas affected by drought have also risen since rainfall has declined, but evaporation has increased due to warming areas. In general, the number of days with heavy rainfall that leads to an explosion has increased. The frequency of storms and tidal gaps varies greatly from year to year, but evidence suggests that their intensity and duration have increased since the 1970s.

While there are numerous evidence of drought in recent years, in many studies, the combination of total rainfall and average temperature are used to calculate drought indices. These indices have been calculated from the mid-twentieth century, showing widespread dryland trends in many parts of the northern hemisphere since the 1950s, as well as widespread drylands in many parts of southern Eurasia, South Africa, Canada and Alaska. The number of 4th and 5th storm categories has risen to 75% since 1970. The highest increase was observed in the Pacific North, India and the Southwest Pacific (Gellens and Roulin 1998).

Other parts of human life are also affected by climate change. Studies have shown that climate change reduces the yield of agricultural products and increases water consumption in this sector (Osborne et al. 2013; Roudier et al. 2011; White et al. 2011; You et al. 2009). Increasing the concentration of carbon dioxide in the atmosphere leads to a change in agricultural potential and water utilization efficiency of various products, as well as the expansion of agricultural-ecological potential towards poles and geographic latitudes. These positive results can be reduced with changes in temperature, quantities and distribution of rainfall, evapotranspiration pattern, radiation regime and indirect effects on land production capacity, such as increased pests, diseases and weeds, resulting in negative results. Therefore, in the long run, changes in climate patterns and agricultural potential, land potential and the ability of future generations to provide food and agricultural products significantly change (Fischer and Van Velthuizen 1996; Fischer et al. 2001; Sembroek and Antonie 1994). Other effects of climate change on agriculture include changes in the flow and storage of materials, the ecology of pests and diseases, the dynamics of rainfall regimes, plant response to temperature, CO_2 accumulation and plant tolerance to salinity.

In 2002, heat and drought severely reduced agricultural production in the world. As a result, agricultural production in the world was 90 million tones or 5% less. In 2003, the intense heat of Europe led to a decline in grain yield and 90 million tones more grain yield than its consumption. In 2004, rainfall improved, but in 2005 the United States drought reduced the world's grain harvest by 34 million tons.

The occurrence of extreme and rare floods is another part of the effects of climate change that causes a lot of damage (Arnell and Lloyd-Hughes 2014; Arnell and Gosling 2014). Melting the ice is another side effect of the heat of the planet. As the melting of the ice, more water enters the ocean. So the volume of water will increase and the evaporation rate will also increase. The water that rises with evaporation should land somewhere. So the number of storms and their severity will increase. Due to the heat of the atmosphere, rain is more likely to occur in the form of rain and the amount of precipitation is reduced to snow. As a result, mountain glaciers, the world's main reservoirs of water, are getting weaker, and storms and floods will become more severe.

Thus, the effects of climate change on the Earth's dynamism status are very broad and extend to socioeconomic segments (Stern 2006). These influences are to a degree that prompts politicians to provide incentive policies to prevent greenhouse gas emissions (Brunner et al. 2012; Jeffrey et al. 2001).

Climate change is a process that, in the long run, affects the Earth's climate processes based on changes in human greenhouse gas emissions. As previously mentioned, the greatest effects of climate change on the temperature and precipitation of the planet. The most important environmental processes affecting the future of mankind are due to the phenomenon of climate change. These processes include a wide range of atmospheric and hydrologic interactions, including floods, droughts, storms, and so on (IPCC 2007). So, among researchers, there is the conclusion that water resource management, not taking into account the effects of climate change, would not be very useful (World Water Assessment Program 2009). An important part of the effects of climate change on the hydrological cycle of water occurs on Earth. In other words, a large part of the hydrological properties in different parts of the world, the northern runoff, the surface water and underground water reserves, as well as the degree of flow (floods and droughts), undergo a change in climate. Under the conditions of climate change, many parts of the water consumer sector, including agriculture, urban, industrial and environmental, will have to respond to this situation. This answer will in many cases lead to conflicts over water demand. This is due to the increase in the human population, and many parts of the world will suffer from a shortage of supply and an increase in demand for water. Therefore, regardless of the comprehensive management of water resources in the absence of adequate water supplies, one cannot expect sustained supply and demand in the field of water resources. Such management should, by examining different aspects of water allocation between different uses, determine the priorities of each of the consumer sectors and, on the other hand, with regard to the availability of available water resources and the ability to store and release these resources, provide management scenarios Sustain in the region. Without such management, water resources, both surface and underground, will be seriously damaged in the long run (Smit et al. 2000).

Preparing different parts of water resources and resources to face the hydrological, social and economic impacts of climate change should be taken into account with regard to the adaptive capacities of the set in terms of pessimistic and optimistic forecasts. In such a situation, it is very important to pay attention to the hydrological parameters due to climate change. Under such a situation, the response of different consumers and their vulnerability to water scarcity. Therefore, the assessment of the effects of climate change on different water use sectors should be made according to appropriate responses to these conditions (Hutchinson et al. 2010).

Adapting a water resource system to climate change is based on understanding the talents and capabilities of that system to change the components of water resource allocation. To do this, the presentation of different approaches in the event of the occurrence of extreme values of hydrological parameters is inevitable. In such a situation, the consumption system will be able to respond to climate change in terms of minimum damage and sustainable development (Cohen et al. 1998; Srinivasan and Philipose 1996; Stakhiv 1994). The collection of relevant social, economic, and environmental information in each water supply system will help to define the limits and take appropriate management measures (Rayner and Malone 1998).

The degree of consistency of a water resource management system relative to climate change is not explicitly explored. The reason for this is the creation of many changes that are forced into the system and may lead to different and heterogeneous responses from consumers. Therefore, first and foremost, the components of compatibility with water scarcity conditions should be classified based on the desired responses (Krankina et al. 1997). This classification is subject to multiple time and space priorities. It can be assumed initially that no special management is performed in dehydration conditions, and thus ensure the maximum response of the system to the existing situation. Subsequently, based on the observed responses, it is possible to define different management scenarios for adapting the system to the provided responses (Bryant et al. 2000). In other words, some compatibility methods are presented after understanding the maximum responses. These methods can be defined according to different scheduling (Stakhiv 1994).

Another aspect of climate change-compatible measures can be found in the type of use. Part of the water consumption is dependent on the natural conditions in each region. Environmental expenditures cover a large part of this area, which in the long run adapts to underwater conditions. But we should not expect that the natural adaptation of the system would be low in accordance with the wishes of relevant managers in the water resources sector. Another part of these expenditures is done by human societies, which should be adapted to climate change and dehydration based on managerial vision. Different urban, industrial, and socioeconomic sectors need to define suitable water patterns, which should be defined primarily by water resource planners (Krankina et al. 1997). Planning to adapt water consuming sectors is mainly done by government agencies. These plans are designed to increase consumer awareness in order to reduce losses and increase water use efficiency (Bryant et al. 2000). The researchers refer to such plans as independent adjustments and against the natural adaptations of water users (Smit et al. 2000). The prediction and estimation of probable future adaptations are essential components of impact assessment studies and vulnerability of climate change. The likelihood of future climate change in a system depends to a large extent on the compatibility efficiency. In studies where no system compatibility is considered, the net or residual effects are too realistic estimates, resulting in greater systemic vulnerability, while in studies that have an effective adaptation to the system there is an estimated amount of vulnerability and the remaining effects less than their actual value. Therefore, having a proper understanding of the adaptation process and obtaining more information about the expected situations in which types of adaptation occurs is essential. It is important to have information on how, why, and when to make consistency in order to make informed judgment about the vulnerability of sectors, geographic regions and various communities (Kane et al. 1992; Tol et al. 1998).

The study of the effects of climate change on the Fraser River basin in Canada by Morrison et al. (2002) showed that during the period 2099–2099, the average flow of the river would increase by 5% and the maximum average flow would decrease by 18% (Morrison et al. 2002).

Hughes (2003) explores the climate change of the last century, illustrating future climate change and its ecological effects in Australia. The study of climate variables in the last century has shown that the average continental temperature in Australia has increased to 0.8 °C since 1910. Rainfall also showed a slight increase in the last century that it was more than summer in summer. In the case of rainfall, this trend is increasing on a regional and seasonal scale with a wider, broader continental scale. Also, a significant decrease in the average coverage was estimated during this period. The results showed that by 2030, the average annual temperature in most parts of Australia would be an increase of 0.6–2 °C. These changes are expected to be around 1–6 °C in 2070. Models also showed that the increase in maximum and minimum annual temperatures in the future would be similar to the variations in average temperature. While in past observation periods, the temperature rise was at most and at least more than the average temperature (Hughes 2003).

The results of the Kamga study on the upstream of the Benno River in Cameroon showed that by 2100 this region could see a decrease in rainfall of 4–13%, an increase in temperature of 1–3 °C, and a moderate decrease of the river flow by 3–18% (Kamga 2001).

Bekele and Knapp (2010) used three climate change scenarios (A1B, B1 and A2) and the SWAT model to assess the potential impacts of climate change on water supply in the Canadian Fox River Basin. Using climatic models, they produced climate scenarios for the study area and, using the Clear Delta switch, prepared these data for entering the hydrologic model. The produced climate scenarios showed an increase in temperature range from 0 to 3.3 °C and an annual rainfall change of −127 to 127+. The analysis of the results showed that rainfall changes cause significant fluctuations in the flow in the late summer and autumn, while the temperature increase strongly affects winter currents due to snow melt (Bekele and Knapp 2010).

Maurer (2009), using 16 models of GCM and A2 and B1 emission scenarios, investigated the hydrological effects of climate change in the Rio Lempa basin in Central America. To evaluate the effects of climate change, the VIC model was used as a distributed physical base model. The average temperature variation in the study area was estimated to be between 1.9 and 3.4 °C by 2099. 11 of the 16 models used predict the amount of rainfall in the future less than the present, so that the average rainfall under the A2 and B1 emission scenarios was reduced by 10.4 and 5%, respectively. Also, the amount of inflow into the dams in the area decreased by 13–24%, which was reported by the A2 scenario to be higher than the scenario B1 (Maurer 2009).

Minville et al. (2008) examined the uncertainty of the effects of climate change on the hydrology of one of the Scandinavian watersheds. In this study, five general circulation models and two climate scenarios together with the HSAMI integrated concept model were used as a rainfall/runoff model. The results of climate change in the future indicate a rise in temperature for all models and scenarios. In the case of rainfall, the results for diffusion models and scenarios were different. Considering all sources of uncertainty in this research, the most effective factor on the final results was the choice of the GCM model (Minville et al. 2008).

Gosain et al. (2006) examined the hydrological effects of climate change in Indian watersheds. Using SWOT distributed hydrologic model, they investigated the potential impacts of climate change in 12 river catchment areas. The results showed that the severity of droughts and floods in different parts may be more severe in the future. Also, the reduction of runoff was confirmed under greenhouse gas emission scenarios (Gosain et al. 2006).

The impact of climate change on the flow of Tamsi River flow in England was studied by Wilby and Harris (2006). In this research, uncertainty sources related to GCM models, microscopic methods, greenhouse gas emission scenarios, different models of rainfall-runoff simulation and their uncertainty related to their parameters, taking into account different weights and the Monte Carlo method Simulated. The results showed that the uncertainty associated with GCM models has the highest contribution and climate scenarios have the lowest contribution in estimating the runoff probability function. However, the main weakness of this research is the lack of weighting the uncertainty associated with GCM models that have the greatest impact on the outcomes of the system (Wilby and Harris 2006).

1.1 Case Study

The Zayandeh-Rud River Basin is located at 26,917 km^2 in the longitude 50° 24′–53° 24′ and latitude 31° 11′–33° 42′. The lowest and highest point in the basin is 1450 and 4300 m above sea level, respectively (Strauss et al. 2013). The region, located in the central and semi-arid region of Iran, is one of the strategic and complex regions of Iran in terms of the status of resources and water resources (Eslamian et al. 2012; Madani and Marino 2009; Zareian et al. 2014, 2017; Zareian and Eslamian 2016). The rainfall in this basin varies from 1500 mm in the west of the basin to 50 mm in the eastern regions of the basin (Moghaddasi et al. 2013). In general, a large part of the basin has a rainfall of less than 150 mm throughout the year. The average annual rainy days in this area are 70 days (Salemi and Murray-Rust 2004). Figure 1 shows the overall position of the Zayandeh-Rud River Basin in the central part of Iran.

Zayandeh Rud River is the most abundant river in the semi-arid region of Iran, which supplies most of the water needed in Zayandeh-Rud River Basin.

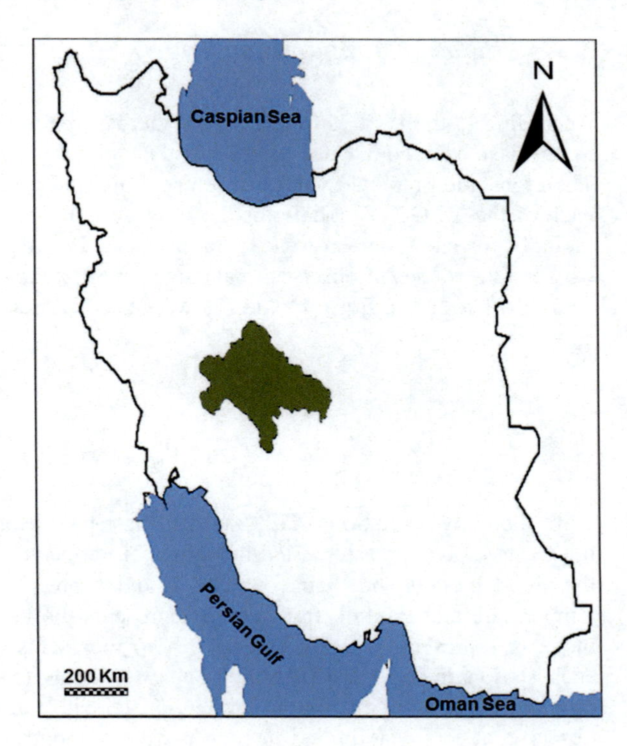

Fig. 1 The overall position of the Zayandeh Rud basin in the central part of Iran

The existence of this river in this basin is one of the main foundations of the formation of the ancient civilization of this region. Different sectors of water consumption in the Zayandeh-Rud River Basin, which include agricultural, urban and industrial areas, are highly dependent on the flow of water in the Zayandeh Rud River River. The expansion of urban areas, agricultural lands and industrial settlements along the river is quite clear.

The attraction that has been caused by the increasing development of this basin in various dimensions has encouraged the people of neighboring areas to migrate to this region, and along with that, the population of the basin has increased dramatically and created numerous industries and factories within the basin (Hashmi et al. 2011).

The synchronization of the conditions mentioned with the occurrence of climate change has made the situation more difficult for the catchment area. The increase in temperature and future rainfall in the Zayandeh-Rud River Basin has been reported by researchers (Fischer et al. 2001). As a result of future climate change, quantity and quality of water will also be affected, and the need for newer water resources will be felt (Zayandab Consulting Engineering Co. 2008; Zareian et al. 2014a, b, 2017; Zareian and Eslamian 2016).

1.2 Climate Change Modelling

According to the IPCC's fourth report, the base period of 1971–2000 has been set as the last base course for analyzing the output of GCM models (IPCC 2007). Therefore, the predicted values of temperature and precipitation were extracted by each of the 15 GCM models for the base period. This process was repeated individually for each meteorological station used. In order to calculate the difference between the values of observational data as well as the output data of GCM models from 1971 to 2000, Eqs. (1) and (2) were used (Zareian et al. 2014):

$$\mathrm{TE}_m^{G_i} = \left|\left(\bar{T}_m^B\right)_{G_i} - \left(\bar{T}_m^B\right)_O\right| \quad (1)$$

$$\mathrm{PE}_m^{G_i} = \left|\left(\bar{P}_m^B\right)_{G_i} - \left(\bar{P}_m^B\right)_O\right| \quad (2)$$

In the above equations, $\mathrm{TE}_m^{G_i}$ and $\mathrm{PE}_m^{G_i}$ are the absolute error values of each of the GCM models in estimating the values of temperature and rainfall. G represents the GCM models and their counters. $\bar{T}$ and $\bar{P}$ are average values of 30 years of temperature and rainfall. Indicator B represents the base year 1971–2000 and the index m represents the target month. Also represents the actual data observed at each station in 1971–2000. Simply put, Formulas (1) and (2) show how much difference between the output of GCM models and the actual values of data observed at each station and in each particular month of the year.

The period that was selected to predict the effects of climate change on the Zayandeh Rud River basin was the 30-year period from 2015 to 2044. Due to the use of two A2 and B1 emission scenarios, two general combinations for extracting temperature and rainfall parameters were defined using GCM models. The temperature and rainfall changes of each station are defined using relationships 3–6:

$$\Delta T_m = \sum_{G=1}^{15} \left\{ \mathrm{WT}_m^{G_i} \times TCF_{Gi} \right\} \quad (3)$$

$$TCF_{Gi} = \left(\left(\bar{T}_m^F\right)_{G_i} - \left(\bar{T}_m^B\right)_O \right) \quad (4)$$

$$\Delta P_m = \sum_{G=1}^{15} \left\{ \mathrm{WP}_m^{G_i} \times PCF_{Gi} \right\} \quad (5)$$

$$PCF_{Gi} = \left(\frac{\left(\bar{P}_m^F\right)_{G_i}}{\left(\bar{P}_m^B\right)_O} \right) \quad (6)$$

where ΔT_m is the difference between the temperature between the next period and the base period (°C), ΔP_m, the amount of precipitation difference between the next period and the base period (%) and $\left(\bar{T}_m^F\right)_{G_i}$ and $\left(\bar{P}_m^F\right)_{G_i}$ are the average 30 years of temperature and precipitation in the upcoming period, respectively (Zareian et al. 2014b, 2017).

In order to use the output of GCM models in meteorological studies and water resource issues, they should be converted to quantitative or monthly data values at any point by means of downscaling methods. This is done by weather generators (Hashmi et al. 2011). The LARS-WG model is one of the most well-known random weather generators capable of producing a daily weather data series.

1.3 Water Allocation Modelling

In order to allocate water in the future, at the downstream of the Zayandeh Rud dam, first all future imaginable water resources for this area were investigated. These resources include the surface water of the Zayandeh Rud dam, the transfer of water from other basins, and the underground water source.

The main source of surface water in the area was Zayandeh Rud Dam, fed by Zayandeh Rud River. The dam, which started its operation in 1970, is used as a means of adjusting the flow in cold and dry seasons.

This dam is located in the upper reaches of the selected land area, and its water, in addition to consumption in irrigation and drainage networks, also consumes industry and drink.

For the purpose of determining the lateral water consumption behind the Zayandeh Rud Dam, the average bill of 10 years of the dam (from 2004 to 2014) was considered average.

Therefore, other than usual uses in the fields of agriculture, drinking and industry, the share of evaporation from the Zayandeh Rud River Dam and unauthorized impressions should also be taken into account on the basis of the amounts obtained. Also, the volume as the base volume in the reservoir of the dam should be avoided in order to prevent the full drainage of the dam. This amount is between 150 and 200 million cubic meters, which is estimated at 200 million cubic meters. In other words, the total amount of water entering the dam can not be allocated to costs and the unwanted upstream costs (evaporation, remaining water volume in the tank and unauthorized harvest) should be deducted from it.

The main objective of optimization in this research is to reduce the difference between supply and demand of water based on a quadratic function. The objective function is intended to minimize water scarcity in the studied area as follows:

$$\min Z = \sum_{j=1}^{30} \sum_{i=1}^{12} (Supply - Demand)^2 = \sum_{j=1}^{30} \sum_{i=1}^{12} \left(Sh_{ij}\right)^2 \tag{7}$$

In this equation, Z is the objective function of minimizing the water deficit and Supply and Demand, respectively, the amount of water allocated to each of the expenditures. j represents the target year for optimization that varies from 2015 to 2044 for this research, or in other words, varies from 1 to 30. i represents each of the months of the year, which varies from 1 to 12.

1.4 Results

Figures 2 and 3 show respectively the annual water allocation for the baseline scenario in A2 and B1 scenarios. Table 1 also shows the summary of the annual water allocation values for both of the published scenarios. The results of the A2 emission scenario indicate that the average annual water requirement of different sectors in the ideal, moderate and critical climate change patterns will be 3505, 3572, and 3641 MCM, respectively. Of these, an average of 3051, 2939 and 2837 million cubic meters will be allocated. For the B1 emission scenario, the average annual total water demand in the three climate change models was calculated to be 3386, 3450, and 3498 million cubic meters, and the allocation would be equal to 3039, 2852 and 2746 million cubic meters. Therefore, on average, there will be an annual water shortage for climate change patterns in the A2 emission scenario of 454, 633, and 803 million cubic meters, respectively. These values in the B1 emission scenario will be 347, 576, and 752 MCM.

The annual amounts reported for allocations and water scarcity are based on a 30-year average over the period 2015–2044. In some years the amount of deficiency will be higher and in some years will be less than the average. For the A2 emission scenario, the highest amount of water requirement in all climate change patterns will occur in 2042, the smallest in 2029. The maximum water requirement for an ideal, moderate, and critical climate change pattern in the A2 emission scenario will be 4191, 4251 and 4333 MCM, respectively, which will account for 733, 852 and 1031 MCM of water deficit, respectively.

The highest amount of water requirement in the region is observed in the B1 emission scenario in 2041 and the lowest in 2017. The greatest deficit in the three climate change patterns will be 3876, 3450, and 3498 MCM respectively, with the lowest of 2924, 2980 and 3025 MCM, respectively. Therefore, the largest water shortage of the region will be 610, 802, and 906 MCM, respectively.

In Figs. 2 and 3, in addition to determining the total water consumption during the period from 2015 to 2044, the total amount of water allocation from the Zayandeh Rud Dam, groundwater resources, as well as the deficiency occurred on the basis of the baseline water allocation scenario. Climate change will affect all of the above mentioned parameters.

The results show that different patterns of climate change in two states will affect water scarcity. On the one hand, with increasing water requirements in the agricultural sector, they increase the total water requirement of the entire area and, on the other hand, by reducing the flow of the dam, reduces the possibility of the

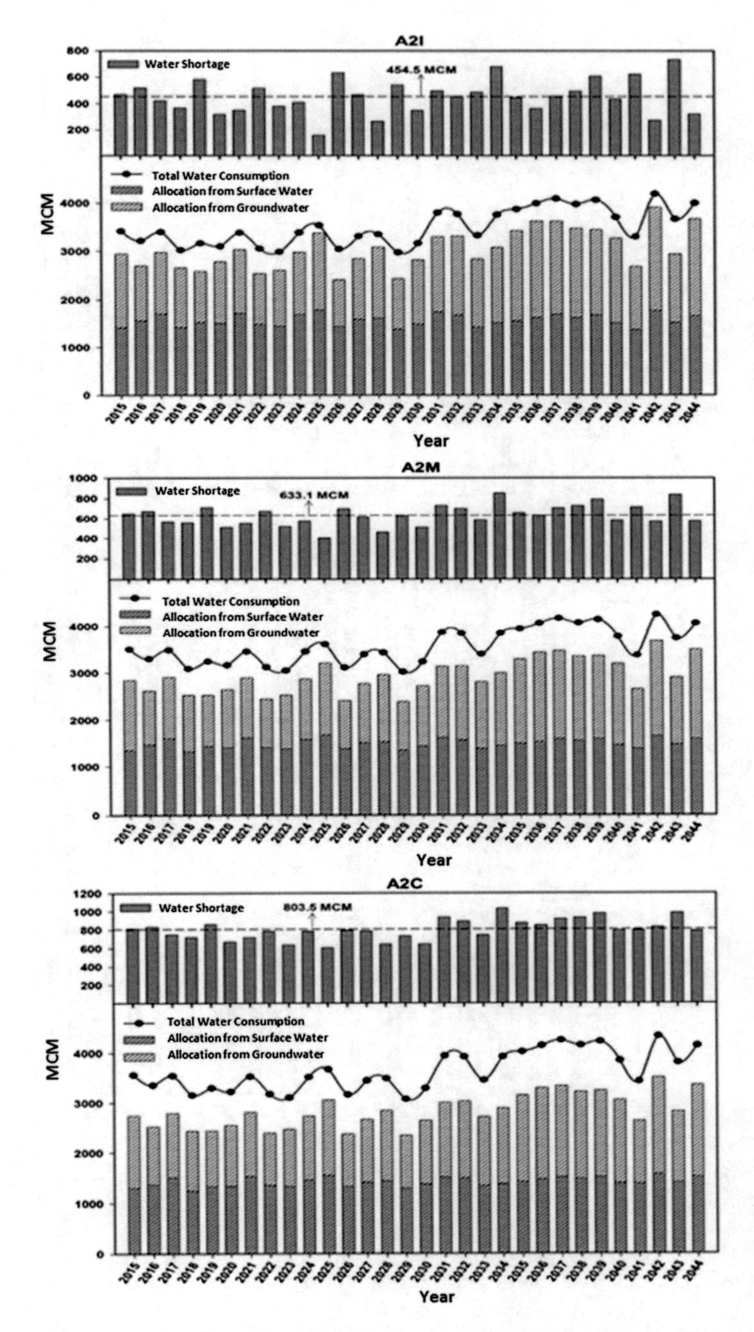

Fig. 2 Annual water allocation scheme in the basic management scenario under different patterns of climate change in the A2 scenario

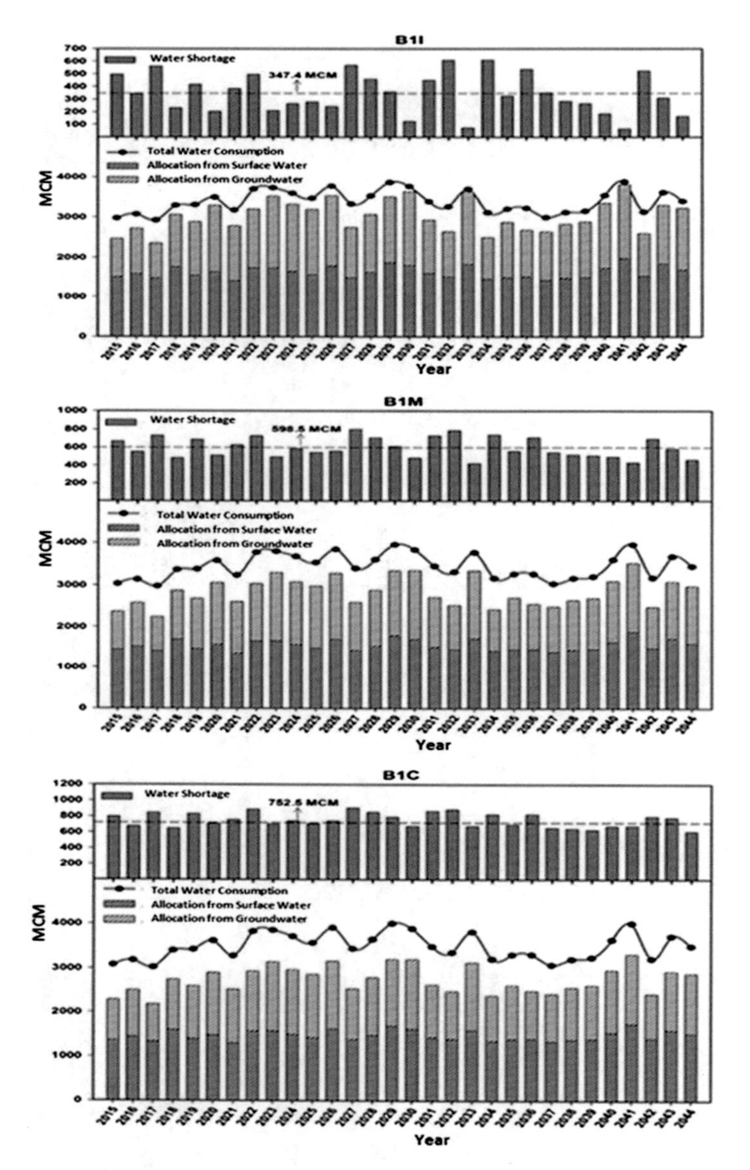

Fig. 3 Annual water allocation scheme in the basic management scenario under different patterns of climate change in the B1 scenario

allocation of surface water. In addition, the reduction of water consumption in the basin and the release of surface water will affect the return of water from different uses to groundwater resources. For example, in the case of moderate climate change, the amount of water allocation from the dam and underground water would be 1494 and 1445 MCM, respectively, in the A2 emission scenario and 1549 and 1303 MCM respectively in the B1 emission scenario. Because the amount of abandoned surface water in the B1 climate change pattern is more than A2, the amount of groundwater extraction in the B1 emission scenario can be reduced to a lesser extent than the A2 emission scenario.

References

Arnell NW, Gosling SN (2014) The impacts of climate change on river flood risk at the global scale. Clim Chang 134(3):387–401

Arnell NW, Lloyd-Hughes B (2014) The global-scale impacts of climate change on water resources and flooding under new climate and socio-economic scenarios. Clim Chang 122:127–140

Bekele EG, Knapp HV (2010) Watershed modeling to assessing impacts of potential climate change on water supply availability. Water Resour Manage 24:3299–3320

Brunner S, Flachsland C, Marschinski R (2012) Credible commitment in carbon policy. Clim Policy 12:255–271

Bryant CR, Smit B, Brklacich M et al (2000) Adaptation in Canadian agriculture to climatic variability and change. Clim Chang 45:181–201

Cohen S, Demeritt D, Robinson J, Rothman D (1998) Climate change and sustainable development: towards dialogue. Glob Environ Chang 8:341–371

Eslamian SS, Gohari SA, Zareian MJ, Firoozfar A (2012) Estimating Penman-Monteith reference evapotranspiration using artificial neural networks and genetic algorithm: a case study. Arab J Sci Eng 37:935–944

Fischer G, Van Velthuizen HT (1996) Climate change and global agriculture potential project: a case study of Kenya. International Institute for Applied Systems Analysis, Laxenburg, Australia

Fischer G, Tubiello FN, Van Velthuizen H, Wiberg DA (2001) Climate change impacts on irrigation water requirements: effects of mitigation, 1990–2080. Technol Forecast Soc Chang 74:1083–1107

Gellens D, Roulin E (1998) Stream flow response of Belgian to IPCC climate change scenarios. J Hydrol 210:242–258

Gosain AK, Rao S, Basuray D (2006) Climate change impact assessment on hydrology of Indian river basins. Curr Sci 90:346–353

Hashmi MZ, Shamseldin AY, Melville BW (2011) Comparison of SDSM and LARS-WG for simulation and downscaling of extreme precipitation events in a watershed. Stoch Environ Res Risk Asses 25:475–484

Hughes L (2003) Climate change and Australia: trends, projections and impacts. Austral Ecol 28:423–443

Hutchinson CF, Varady RG, Drake S (2010) Old and new: changing paradigms in arid lands water management. In: Schneier-Madanes G, Courel MF (eds) Water and sustainability in arid regions, vol 3. Springer, Berlin, pp 311–332

IPCC (2007) Summary for policymakers in climate change: the physical science basis. Contribution of working group I to the fourth assessment report of the intergovernmental panel on climate change. Cambridge University Press, Cambridge, pp 1–18

Jeffrey SJ, Carter JO, Moodie KB, Beswick AR (2001) Using spatial interpolation to construct a comprehensive archive of Australian climate data. Environ Model Softw 16:309–330

Kamga FM (2001) Impact of greenhouse gas induced climate change on the runoff of the Upper Benue River (Cameroon). J Hydrol 252:145–156

Kane SM, Reilly J, Tobey J (1992) An empirical study of the economic effects of climate change on world agriculture. Clim Chang 21:17–35

Krankina ON, Dixon RK, Kirilenko AP, Kobak K (1997) Global climate change adaptation: examples from Russian boreal forests. Clim Chang 36:197–215

Madani K, Marino MA (2009) System dynamics analysis for managing Iran's Zayandeh-Rud River Basin. Water Res Manage 23:2163–2187

Maurer EP (2009) Climate model based consensus on the hydrologic impacts of climate change to the Rio Lempa Basin of Central America. Hydrol Earth Syst Sci 13:183–194

Minville M, Brissette F, Leconte R (2008) Uncertainty of the impact of climate change on the hydrology of a nordic watershed. J Hydrol 358:70–83

Moghaddasi M, Araghinejad S, Morid S (2013) Water management of irrigation dams considering climate variation: case study of Zayandeh-Rud reservoir. Iran. Water Resour Manage 27:1651–1660

Morrison J, Quick MC, Foreman MGG (2002) Climate change in the Fraser River watershed: flow and temperature projections. J Hydrol 263:30–244

Osborne T, Rose G, Wheeler T (2013) Variation in the global-scale impacts of climate change on crop productivity due to climate model uncertainty and adaptation. Agric For Meteorol 170:183–194

Rayner S, Malone EL (1998) Human choice and climate change. The tools for policy analysis. Battelle Press, Columbus, OH, USA

Roudier P, Sultan B, Quirion P, Berg A (2011) The impact of future climate change on West African crop yields: what does the recent literature say? Global Environ Chang 21:1073–1083

Salemi HR, Murray-Rust H (2004) An overview of the hydrology of the Zayandeh Rud Basin, Iran. Water Waste 1:2–13

Sembroek WG, Antonie J (1994) The use of geographic information systems in land resources appraisal, land and environment information systems. FAO, Rome, Italy, pp 543–556

Smit B, Burton I, Klein RJT, Wandel J (2000) An anatomy of adaptation to climate change and variability. Clim Chang 45:223–251

Srinivasan K, Philipose MC (1996) Evaluation and selection of hedging policies using stochastic reservoir simulation. Water Resour Manage 10:163–188

Stakhiv EZ (1994) Managing water resources for adaptation to climate change. In: Engineering risk in natural resources management, pp 379–393

Stern N (2006) Stern review on the economics of climate change. Her Majesty's Treasury, London

Strauss F, Formayer H, Schmid E (2013) High resolution climate data for Austria in the period 2008–2040 from a statistical climate change model. Int J Climatol 33:430–443

Tol R, Fankhauser SJS, Smith JB (1998) The scope for adaptation to climate change: what can we learn from the impact literature? Glob Environ Chang 8:109–123

White JW, Hoogenboom G, Kimball BA, Wall GW (2011) Methodologies for simulating impacts of climate change on crop production. Field Crop Res 124:357–368

Wilby RL, Harris I (2006) A framework for assessing uncertainties in climate change impacts: low-flow scenarios for the River Thames. Water Resour Res, UK. https://doi.org/10.1029/2005WR004065

World Water Assessment Program (2009) The United Nations world water development. Report 3: water in a changing world. UNESCO, Paris and Earthscan, London

You L, Rosegrant MW, Wood S, Sun D (2009) Impact of growing season temperature on wheat productivity in China. Agric For Meteorol 149:1009–1014

Zareian MJ, Eslamian S, Hosseinipour EZ (2014a) Climate Change impacts on reservoir inflow using various weighting approaches. World Environmental and Water Resources Congress, Portland, USA

Zareian MJ, Eslamian S, Safavi HR (2014b) A modified regionalization weighting approach for climate change impact assessment at watershed scale. Theor Appl Climatol 122:497–516

Zareian MJ, Eslamian S, Gohari A, Adamowski J (2017) The effect of climate change on watershed water balance. In: Furze JN, Swing K, Gupta AK, McClatchey R, Reynolds D (eds) Mathematical advances towards sustainable environmental systems. Springer International Publishing, Switzerland, pp 215–238

Zareian MJ, Eslamian S (2016) Variation of water resources indices in a changing climate. International J Hydrol Sci Technol 6(2):173–187

Zayandab Consulting Engineering Co. (2008) Determination of resources and consumptions of water in the Zayandeh-Rud River Basin, Iran

Climate Change Impact on Agriculture and Irrigation Network

Zohreh Dehghan, Farshad Fathian and Saeid Eslamian

Abstract In the last century, impacts of climate change phenomenon and its consequences such as changes in temperature, precipitation, evapotranspiration, and amount of available water have been observed in various aspects of human life, especially in agricultural sector. In the agricultural sector, performance of irrigation networks is known as one of the important and effective factors on crop yield and development of sustainable agriculture. These networks can be affected by climate change because the irrigation networks are designed based on the climatic variables and crop pattern of considered region. In this chapter, sprinkler irrigation network of Bilesavar, located in the north of Iran, is considered as study area in order to investigate the effects of climate change on climatic variables. Moreover, according to existing agro-industry in the study area, the impacts of climate change are investigated on available crop pattern in order to specify conditions in the region in terms of maintaining sustainable agriculture. Since the most irrigation systems are designed based on the water requirement of crop pattern and water delivery based on farmer demand in the required pressure and discharge, the performance of irrigation network are evaluated in terms of equity and adequacy indices. Eventually, the adaptation strategies to climate change will be discussed for maintaining performance, stability of pressurized irrigation system, productivity improvement, and an increase in efficiency.

Keywords Agriculture · Climate change · Irrigation network · Performance evaluation · Iran

Z. Dehghan · S. Eslamian (✉)
Department of Water Engineering, Faculty of Agriculture,
Isfahan University of Technology, P.O.Box 8415683111, Isfahan, Iran
e-mail: saeid@cc.iut.ac.ir

F. Fathian
Department of Water Engineering, Faculty of Agriculture, Vali-E-Asr
University of Rafsanjan, P.O.Box 7718897111, Rafsanjan, Iran

P. Castro et al. (eds.), *Climate Change-Resilient Agriculture and Agroforestry*,
Climate Change Management, https://doi.org/10.1007/978-3-319-75004-0_19

1 Introduction

According to the fifth IPCC report, increasing concentrations of CO_2 and other greenhouse gases have led to fundamental changes due to anthropogenic activities in the global climate over the course of the last century. The future global climate will be characterized by uncertainty and change, and this will affect water resources and agricultural activities (IPCC 2013; Adamowski et al. 2010). In the agricultural sector, there are multiple variables which will be affected due to the impacts of climate change; so that, in the longer term, the potential significant effect of climate change is nowadays recognized worldwide. Furthermore, the impact of climate change will depend on hydrometeorological and environmental conditions as well as the capacity to adapt to change. For instance, evapotranspiration is controlled by climatic variables (such as temperature, precipitation, net radiation, wind and relative humidity), and changes in climatic regimes can affect hydrological processes, yield production, and the development of agricultural activities (Allen et al. 1991; Yang et al. 2007; Perez Urrestarazu et al. 2010). Given the importance of agriculture, there is a significant need to understand the implications of climate change on agriculture, and to explore different adaptation options (Gohari et al. 2013).

Performance and vulnerability assessments of irrigation networks are an increasingly significant area of research with respect to impacts and adaptation to climate change (Fathian et al. 2016). From the viewpoint of agricultural water management, the phenomenon of climate change will have unfavorable effects on evapotranspiration and crop water requirements (CWRs), and there will be impacts on the performance of irrigation networks (Perez Urrestarazu et al. 2010). Irrigation networks will have to be designed for longer and higher peaks in irrigation water demand, which may cause problems in some of the present networks (Rodríguez Díaz et al. 2007). These changes will influence the planning and operational management, flow regime, and performance of irrigation networks. As such, the analysis of climate change impacts and adaptation strategies to maintain the performance and stability of irrigation systems, as well as improve their efficiency, are necessary (Calejo et al. 2008; Huang et al. 2011).

To date, many studies have been completed on the topic of climate change impacts on agriculture (e.g., Rosenzweig et al. 1995; Thomson et al. 2006; Brunsell et al. 2010; Vaze et al. 2011). Some of these studies are related to climate change impacts on agricultural products (e.g., Tatsumi et al. 2011; Dalezios et al. 2017); while others are related to modeling of climate change impacts on crop water requirements, irrigation systems and ect (e.g., De Silva et al. 2007; Fischer et al. 2007; Rodríguez Díaz et al. 2007). Since most studies have evaluated the performance of surface and pressurized irrigation networks in the current climate status. In recent years, the performance of irrigation networks, especially under climate change scenarios, has become a growing concern of the researchers, irrigation policy makers and donor agencies. For this purpose, a few studies evaluated the effects of future changes in the climate on irrigation networks. Perez Urrestarazu et al. (2010) used climate data in three periods (2050, 2079 and a base period), and evaluated an irrigation network using vulnerability indices such as equity and adequacy. The results of their study showed that climate change would have a major impact on network performance with existing cropping patterns, but that changes in cropping patterns could reduce this impact. Daccache et al. (2010) considered the impacts of climate change and the performance of pressurized irrigation water distribution networks under Mediterranean conditions. The results showed that under the current design, the irrigation system can tolerate a peak demand discharge, which is below the 2050s average. Accordingly, the performance of the system will fall significantly as the number of unreliable hydrants will increase so that, assuming the same cropping pattern, the threshold discharge will fail during peak demand periods. In another study, Karamouz et al. (2010) evaluated the improvement of urban drainage networks in Iran under climate change. However, to date, the effect of climate change on irrigation distribution networks has not been explored in Iran. On the other hand, introducing of water resources supply scenarios, as another contribution, has not been tackled yet in any study. Therefore, the main objective of this study was to explore and evaluate the performance of irrigation networks in terms of the vulnerability of systems under climate change and water supply scenarios. To accomplish this, a methodology is proposed in this study to assess the possible effects of future changes on irrigation networks, and it was applied in the Bilesavar irrigation network district in Northern Iran.

2 Materials and Methods

2.1 *Study Area, Cropping Patterns and Data*

The Bilesavar irrigation network is located in the Moghan plain in Ardebil province in Northern Iran (48° 15′ to 48° 20′ East longitude and 39° 21′ to 39° 28′ North latitude). In this area, increases of around 5 and 13% in the CWRs are predicted by 2010–39 and 2050–79, respectively, due to climate change (Dehghan 2012). The Bilesavar irrigation district is 3200 ha, with 21 field units or sectors and cropping patterns that are devoted to a wide range of crops such as wheat (40%), barley (15%), alfalfa (20%), cotton (20%) and lentil (5%). The area is covered by a well maintained pressurized network, which is designed to provide irrigation on demand with an irrigation efficiency of 80%, and a distribution efficiency of 100%. Table 1 shows the planting calendar and crop coefficients of the cropping pattern for the estimation of crop evapotranspiration and water requirements in this district. It can be seen that alfalfa is cultivated all months of the year, while wheat, barley, lentil, and cotton are cultivated during 8, 8, 4, and 7 months of the year, respectively. Water is taken from the Aras river and pumped using a system of 55 pumps from 21 pumping stations. The configuration of the system is not a looped network, in other words; the 21 pumping stations supply the water demand for 21 field units in

Table 1 Planting calendar and crop coefficients

Crop	Jan.	Feb.	Mar.	Apr.	May	June	Jul.	Aug.	Sep.	Oct.	Nov.	Dec.
Wheat	0.4	0.4	1.2	1.3	1	1.04					0.4	0.44
Barley	0.4	0.4	1.3	1.3	1	1.04					0.4	0.44
Alfalfa	1	1	1	1	1	1.1	1.1	1.1	1.06	1	1	1
Lentil				0.4	1.1	1.1	0.75					
Cotton					0.4	0.6	0.84	1.2	1.15	1.11	0.88	

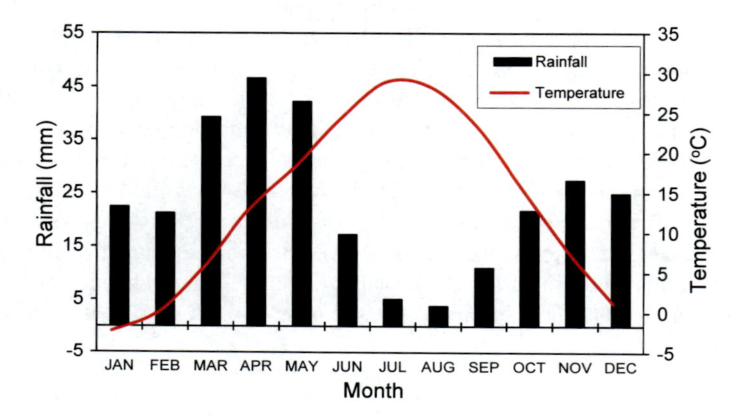

Fig. 1 Variations of monthly rainfall (mm) and temperature (°C) during the 1971–2000 period

independent networks. The main network is able to supply 3 L/s/ha on demand in each one of the 863 outlets, simultaneously. The irrigation period is 6 days and the irrigation time is 19 h. The assigned pressure in the outlet is 30 m. Data from the Pars Abad meteorological station, which was used as the data for the baseline period, was applied to predict climate variables as well. The record of the data is 30 years (1971–2000), and as shown in Fig. 1, the average, maximum, and minimum annual temperature and the average annual precipitation are 13.7, 30, and −2.3 °C, and 283 mm, respectively, in Pars Abad region (http://www.weather.ir/). The driest month is August, with less than 5 mm of rain. Most rainfall falls in April, with an average of 47 mm. July is the warmest month of the year. The temperature in July averages 28.9 °C. In January, the average temperature is −2.3 °C. It is the lowest average temperature of the whole year. There is a difference of 42.8 mm of precipitation between the driest and wettest months. The average temperatures vary during the year by 31.2 °C.

2.2 General Circulation and SDSM Downscaling Models

To estimate future climate change resulting from the continuous increase of greenhouse gas concentrations in the atmosphere, general circulation models (GCMs) are used. GCMs demonstrate significant skill at the continental and hemispherical scales, and incorporate a large proportion of the complexity of the global system, but are inherently unable to represent local subgrid scale features and dynamics (Wigley et al. 1990; Carter et al. 1994). The outputs of GCMs cannot be used directly for climate change studies, and they do not provide a direct estimation of the hydrological response to climate change (Dibike and Coulibaly 2005). This is due to the mismatch in the spatial resolution between the GCMs and hydrological models, and the large coarse resolution.

As a result, downscaling techniques are used to convert the coarse spatial resolution of the outputs of GCMs into finer resolutions, which may involve the generation of point/station data of a specific area using GCM climatic output variables (Dibike and Coulibaly 2005; Hashmi et al. 2009). Downscaling techniques can be classified into statistical and dynamical downscaling methods (Maraun et al. 2010). This study used a statistical downscaling method, SDSM, which has been widely applied to downscale GCMs, as can be seen by the over 170 publications that used this model (Wilby and Dawson 2012). In this study, the output from the Hadley GCM3 model (HadCM3) was utilized to downscale temperature and rainfall data for region study. The HadCM3 model is one of the major models used in the IPCC Third and Fourth Assessments, and it also contributed to the Fifth Assessment. It is widely considered to be quite useful, and as such was used in this study (Reichler and Kim 2008). Additional details regarding GCMs and downscaling models can be obtained in Hashmi et al. (2009), and Hassan and Harun (2012).

2.3 Climate Change and Water Supply Scenarios

The Intergovernmental Panel on Climate Change (IPCC) published a new set of scenarios in 2000 for use in the Third Assessment Report (Special Report on Emissions Scenarios—SRES). The SRES scenarios were constructed to explore future developments in the global environment with special reference to the production of greenhouse gases and aerosol precursor emissions. The SRES team defined four scenarios, labelled A1, A2, B1 and B2, which describe the way in which global population, economies, and non-climate policies may evolve over the coming decades. It has been useful to describe the scenarios with respect to two dimensions. The first dimension relates to a more economic (A scenarios) or a more environmental (B scenarios) orientation, while the second dimension relates to a more global (1) or a more regional (2) orientation. The scenarios are summarized as follows (Nakicenovic et al. 2000):

- A1: a future world of very rapid economic growth, global population that peaks in mid-century and declines thereafter, and rapid introduction of new and more efficient technologies;
- A2: a very heterogeneous world with a continuously increasing global population and regionally oriented economic growth that is more fragmented and slower than in other storylines;
- B1: a convergent world with the same global population as in A1 but with rapid changes in economic structures toward a service and information economy, with reductions in material intensity, and the introduction of clean and resource-efficient technologies;
- B2: a world in which the emphasis is on local solutions to economic, social, and environmental sustainability, with continually increasing population (lower than A2) and intermediate economic development.

In this study, A2 and B2 scenarios related to the temperature and precipitation variables were applied in order to include the most salient scenario variables of demographic and economic development on the high and intermediate side (scenarios A2 and B2 respectively), and their results were subsequently compared (Hassan and Harun 2012). Furthermore, in this study, two water supply scenarios (as new contributions) were also considered and combined with A2 and B2 emission scenarios for adaption strategies in the future. In the first water supply scenario, it was assumed that there would be no limitation of water resources (NLWR) to supply increasing water demands of crops in the irrigation district. In the second water supply scenario, it was assumed that there would be limitations of water resources (LWR), and that the available water resources would not be able to supply increasing water demands. Therefore, in order to prevent of the yield of crops, reduce the damage to farmers, supply the crop water requirement and respond to the appropriate planting area in LWR condition, it is necessary to optimize the planting area.

In this study, the planting area was optimized using the LINGO model (Moghaddasi et al. 2010; Xie and Xue 2005). The LINGO software solve the objective function based on linear optimization model. In this study, the objective function was to maximum benefit which is equal to the product ($Max\,Benefit = F \times P \times A$) of the product yield ($F$: kg per hectare), the price of each product (P: \$/kg), and the optimized planting area for each crop (A: ha). The constraints introduced in LINGO in order to determine the optimized planting area for each crop are: (i) changes in A along with $\pm 20\%$ of current A (ii) maximum consumed water in the future should not be more than the current consumed water, (iii) all variables are greater than zero. These constraints was adopted from Moghaddasi et al. (2010) and Paimozd et al. (2010) according to the different adaptation strategies to climate change and the amount of irrigation water available to achieve maximum benefit. Finally, in order to assess the irrigation network under the previously mentioned scenarios, the climate change scenarios were coupled with water supply scenarios; these are given as A2 + NLWR, B2 + NLWR, A2 + LWR and B2 + LWR for the 2010–39 and 2050–79 periods.

2.4 Methodology

Figure 2 shows the methodology that was used in this study. As it can be seen, the SRES scenarios have been used in addition with the water supply scenarios. Therefore, rainfall, temperature, and other climatic data were obtained for the baseline, A2 2010–39, B2 2010–39, A2 2050–79 and B2 2050–79 for the case study irrigation district in northern Iran.

The CROPWAT model (Clarke et al. 1998) was used to estimate ET_o, CWRs, gross irrigation water requirements (GIWR), and the maximum flow required using the rainfall, temperature, other climatic data and the crop patterns (as input data into the model) for each scenario. The FAO Penman-Monteith equation was used in this model to calculate ET_o (Smith 2000). The irrigation network was modeled in WaterGems (Fig. 3), one of the most widely used hydraulic simulation models. The inputs were the hydraulic characteristics of the network. After running the simulations in all situations (baseline, A2 2010–39, B2 2010–39, A2 2050–79 and B2 2050–79), the key variables were obtained for calculating selected indicators (Calejo et al. 2008; Perez Urrestarazu et al. 2009, 2010) as described below.

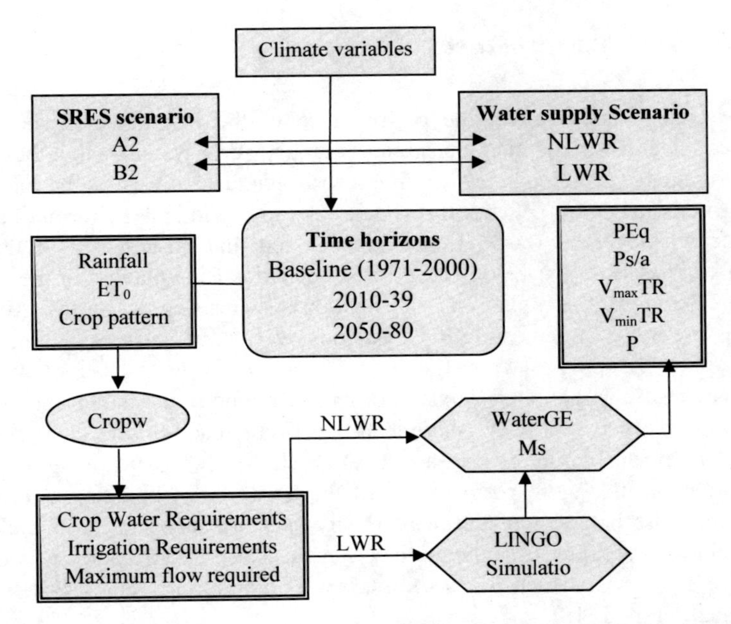

Fig. 2 Flow chart displaying methodology (*Note* PEq = Pressure Equity, Ps/a = Pressure ratio, $V_{max}TR$ = Maximum velocity ratio, $V_{min}TR$ = Minimum velocity ratio, P = Consumption energy of pumping)

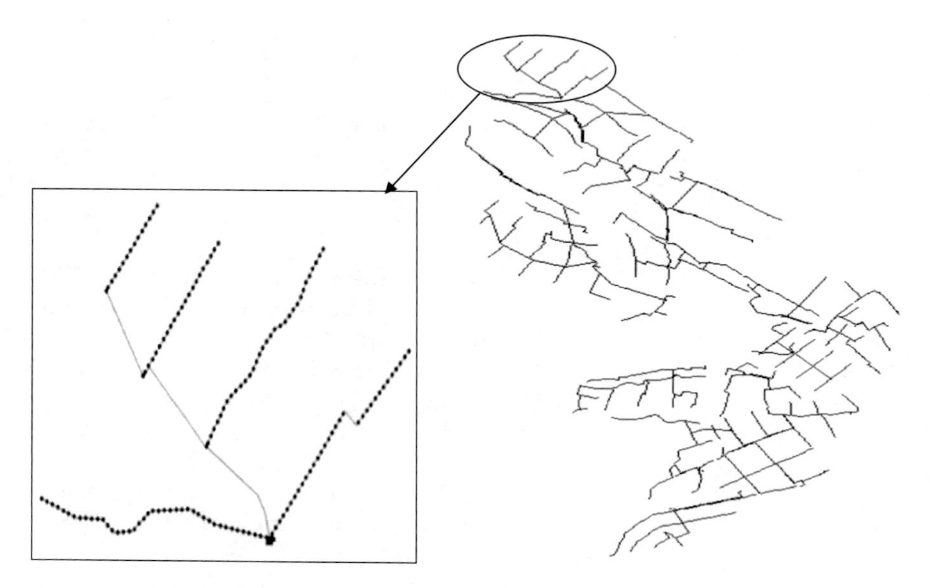

Fig. 3 Model of the network in WaterGems for the baseline

2.5 *Network Performance*

Climate change could affect the performance of the irrigation network due to increases in demand. The main constraint of an irrigation system is that the required flows should be supplied to water users with adequate pressure (Farmani et al. 2007). In this irrigation network the system is on-demand and the farmer has water available whenever he wants. It should be noted that the concept of the water available is based on the LWR and NLWR scenarios as explained in the previous section. Therefore, the limitations of irrigation systems depend on the given flow and pressure in the outlets, and if the system is overloaded, farmers may be obliged to cut off the supply and postpone irrigation (Rodríguez Díaz et al. 2007). Thereupon, these irrigation systems, in order to respond to the increasing demands, need to have a network with a higher distribution capacity, while it makes them much more expensive (Perez Urrestarazu et al. 2010). Hence, it is essential to have information on the future performance of the network. In this paper, in order to study the future behavior of the network, five performance indicators were used. Each indicator is expressed quantitatively, as an aspect of irrigation network performance standards, which helps to assess and monitor irrigation system (Perez Urrestarazu et al. 2010).

Pressure Equity (PEq)

This indicator expresses the uniform distribution of pressure between outlets in the network using the interquartile ratio:

$$PEq = \frac{\overline{P_{pq}}}{\overline{P_{bq}}} \quad (1)$$

where $\overline{P_{pq}}$ is the average pressure in the weakest quarter and $\overline{P_{bq}}$ is the average pressure in the best quarter, taking into account all the network's checkpoints. In this study, the pressure equity indicator refers to the pressure at the outlets which, for on-demand irrigation distribution systems, is not constant but depends on outlets simultaneously operating (Gorantiwar and Smout 2005; Perez Urrestarazu et al. 2010).

Simulated/Assigned Pressure Ratio ($P_{s/a}$)

This indicator represents the pressure obtained in the simulation with the one assigned in the outlets of the network:

$$(P_{s/a})_i = \left(\frac{P_s}{P_a}\right)_i \quad (2)$$

where P_s is the simulation pressure in the checkpoint and P_a is the design (or actual) pressure in the i checkpoint. If $P_{s/a} < 1$, it means that this outlet will be working below the required pressure (Perez Urrestarazu et al. 2009).

Maximum Velocity Ratio ($V_{max}TR$)

This indicator represents the ratio of the measured to the permissible maximum velocity in the outlets. If the value of the index is greater than zero, the occurrence probability of water hammer in pipes increases and it causes an unsteady flow conditions into network:

$$V_{\max}TR = \frac{V_{\max}T}{V_{\max}TP} - 1 \quad (3)$$

where, $V_{\max}TR$ is the indicator of maximum velocity of water transmission, $V_{\max}T$ is the measured maximum velocity and $V_{\max}TP$ is the permissible maximum velocity of water transmission in the pipes (Mahdavi 2008).

Minimum Velocity Ratio ($V_{min}TR$)

This indicator determines the ratio of the measured to the permissible minimum velocity in the outlets. If the value of the index is lower than 1, the probability of sedimentation due to the low velocity of water in pipes increases:

$$V_{\min}TR = \frac{V_{\min}T}{V_{\min}TP} \quad (4)$$

where, $V_{\min}TR$ is the indicator of minimum velocity of water transmission, $V_{\min}T$ is the measured minimum velocity and $V_{\min}TP$ is the permissible minimum velocity of water transmission in the pipes (Mahdavi 2008).

Consumption Energy of Pumping (P)

This indicator represents the energy consumption of the pumping station:

$$P = \frac{V \times H}{0.102 \times E_a} \tag{5}$$

where P is the value of energy consumption (Kw), and V, H and E_a are the water volume, height and efficiency of pumping, respectively (Rodrıguez Dıaz et al. 2009).

3 Results and Discussion

3.1 Comparison of Rainfall, Temperature and ET_o Values Among Scenarios

Climate variables were calculated for the Bilesavar irrigation district to give an overview of the general differences between the two time periods on a monthly basis. Figure 4a–c shows the monthly average of temperature, rainfall, and ET_o values for the baseline (1971–2000), 2010–39, and 2050–79 periods, respectively. In general, scenario A2 results in more increases in temperature than B2 in each time period, and according to these scenarios, higher temperatures will be seen in 2050–79 compared to 2010–39 (Fig. 4a). According to scenarios A2 and B2, the months of June to September will experience higher temperature increases. Temperature may increase around 10 and 23% on average by 2010–39 and 2050–79 in comparison with baseline period, respectively.

In the case of rainfall, the results for each time period are different (Fig. 4b). For 2050–79, scenario B2 results in a higher increase in rainfall than A2 in each time period, while for 2010–39, an increase can be observed from November to February. Compared to the baseline, both A2 and B2 scenarios show a decrease from April to October for the two time horizons. Rainfall is seen to decrease by 4% on average by 2010–39, and by 7 and 1% for A2 and B2, respectively, by 2050–79. The outcomes of the climate change scenarios in case of the changes in the rainfall and temperature values in this section are in accordance with the study of Ashofteh and Massah (2009). They showed that the temperature increases and precipitation decreases in the Aidoghmoush area in Iran (an area near Ardebil province with a similar climate).

For ET_o, in comparison with the baseline, both A2 and B2 scenarios show an increase for each time period (2010–39 and 2050–79). Generally, scenario A2 results in a higher increase in ET_o compared to B2 in each time period for 2050–79,

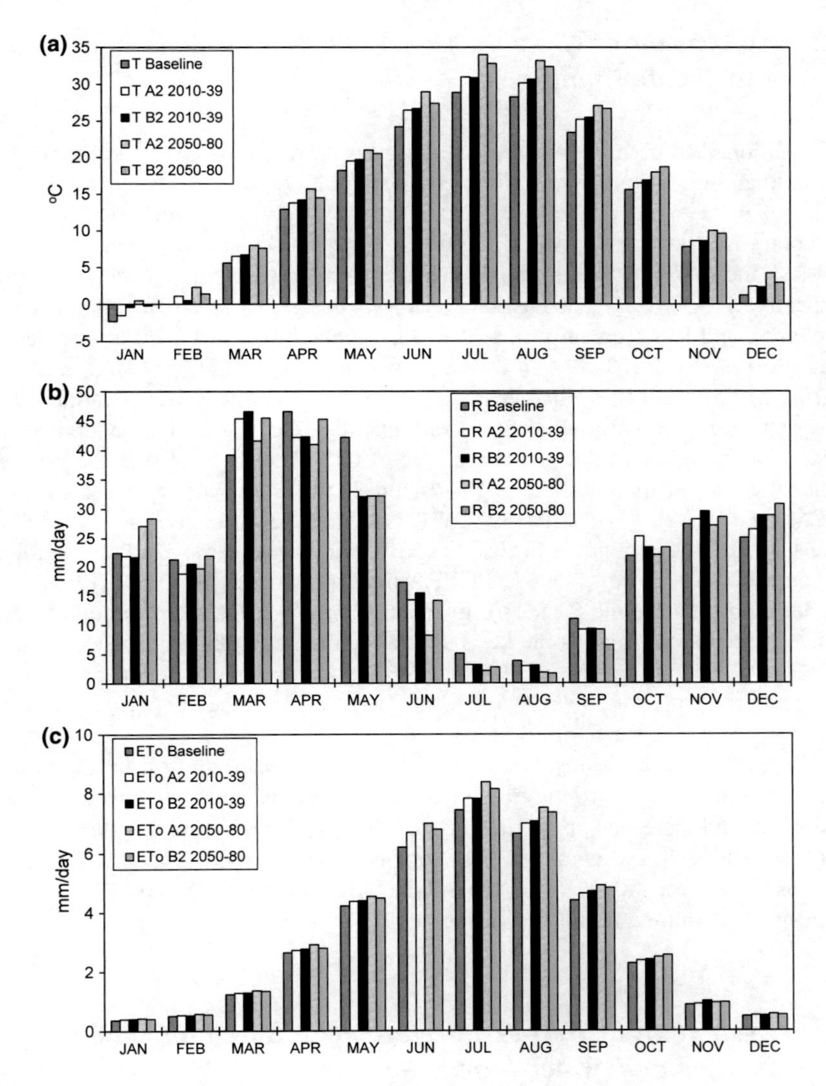

Fig. 4 Values of **a** temperature, **b** rainfall and **c** ET_o for all the scenarios

while for 2010–39, there is no difference between the values of ET_o in the A2 and B2 scenarios (Fig. 4c). ET_o is around 6% higher than the baseline for 2010–39 and 12% for 2050–79. Overall, the situation will be more unfavorable as in the months with peak demand, and the conditions will be more extreme (less rainfall, more temperature and ET_o).

3.2 Assessment of Cropwat Model Outputs Based on Climate Change

As mentioned in the previous section, because the value of ET_o in scenario A2 is higher than in B2 for both of future periods, the ET_o of scenario A2 was selected to estimate the GIWR, CWRs, and maximum flow of water supply (FWS) of the cropping patterns in the district. Figure 5a–c shows the annual values for GIWR, CWRs, and FWS for each crop in the baseline, 2010–39 and 2050–79 periods, respectively. GIWR is an effective factor in designing and managing irrigation networks, and is the quantity of water to be applied in reality, taking into account water losses. For 2010–39 (2050–79), GIWR may be 6 (13), 6 (13), 3 (12.3), 7.5 (10.7), and 5.8% (14.5%) higher than the baseline for wheat, barley, alfalfa, lentil and cotton, respectively (Fig. 5a). Consequently, the water volume of irrigation requirements increases due to the increase of GIWR. In light of this, the estimated water volume for the baseline, 2010–39 and 2050–79 periods are 16.5, 18 and 21 MCM, respectively. For 2010–39, CWR is higher than the baseline period for all crops, increasing around 5, 5, 2, 7, and 5% for wheat, barley, alfalfa, lentil and cotton by 2010–39. For 2050–79, CWR values are around 13, 13, 12, 12, and 12% higher than the baseline (Fig. 5b). In the case of FWS, there are more differences between scenarios which is the key factor affecting network performance (Fig. 5c). For 2010–39, FWS is 8, 8, 3, 10, and 6% higher than the baseline for all crops, respectively, while for 2050–79, the results show an increase of approximately 12% compare to the baseline for all crops.

The results of this study are in accordance with other studies. In an extensive study, Fischer et al. (2007) used climate change models and showed that irrigation water demand increased approximately 45% in the world during the 1990–2079. In another study, Knox et al. (2010) showed that irrigation water requirements increased between 20–22% for the A2 and B2 scenarios for the climatic period 2050 in the Mhlume region in Switzerland.

3.3 Evaluation of Network Performance Based on Climate Change and Water Supply Scenarios

Before interpreting the results of this section, it is important to briefly discuss the LWR scenario. The LWR scenario states that the network will not be able to supply water demands; therefore, there will be a water shortage because of the limitation of water resources supplies in the irrigation district (Banihabib et al. 2015). Assuming a constant volume of available water in all three time horizons, it will only be possible to supply 16.5 MCM of water. Given increasing water demand in the future periods, the outcomes of LINGO model showed that the optimum cultivated area is 159 and 155 ha (for the A2 and B2 scenarios, respectively) lower than the baseline in 2010–39 period, and 423 and 395 ha (for the A2 and B2 scenarios,

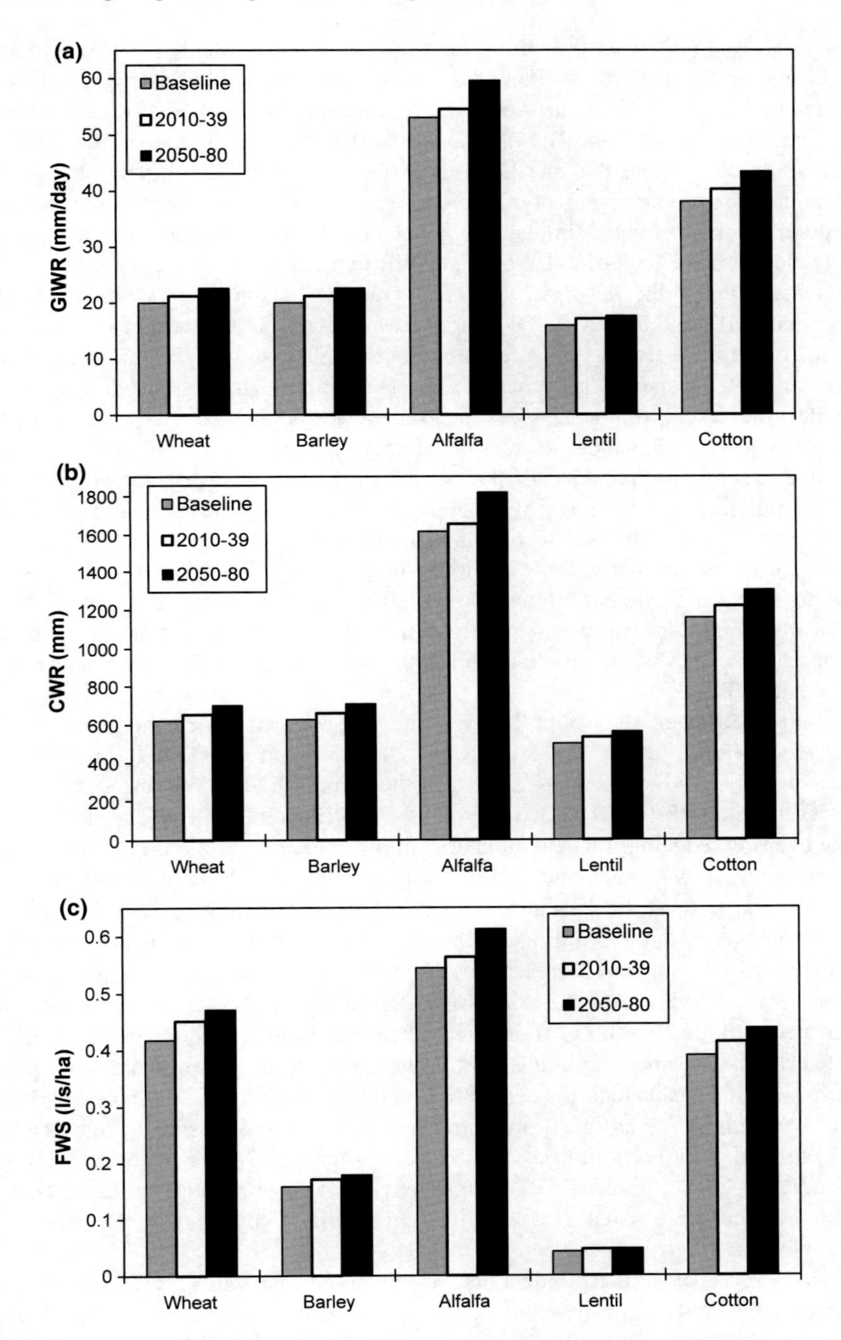

Fig. 5 Annual values of **a** GIWR, **b** CWR, and **c** FWS for all cultivated crops in district

respectively) in 2050–79 period. Furthermore, in order to calculate the performance indicators of the network, 9 different scenarios, as explained in Sect. 2.3, were used that were based on: 2 scenarios of water resources (LWR and NLWR), 2 scenarios of climate change (A2 and B2), and 2 time horizons (2010–39 and 2050–79) along with a baseline (note that the baseline is represented as no climate change and current water supply resources). The nine scenarios mentioned in this section were applied in the network simulation model created with WaterGems for a peak demand reference day using FWS and GWIR in the outlets. The results showed that the discharge of the network were 3.7, 4 and 4.3 cubic meters per second for baseline, 2010–39, and 2050–79 periods, respectively. In the case of the discharge of the outlets, the discharges were found to be 3.5, 3.7, and 4 L per second for these time horizons, respectively. When the necessary input data was introduced into the model, the results of the network pressure analysis showed that because of the increase in water demand, the pressures decreased in the network. It was found that for the baseline period, 61% of the outlets worked at a pressure between 40 and 50 m, but in future periods, more than of 55% of the outlets would work at a pressure between 30 and 40 m. This demonstrates the significant decrease of pressure in the network. Table 2 shows the average values of the indicators calculated for a peak demand reference day. Through the optimization of cultivated areas in conditions of water supply limitation, the water requirements of the cropping patterns was supplied and, indicators of network performance were evaluated (Table 2).

PEq are lower than the baseline in all scenarios, decreasing ~1.5% in B2 + NLWR to 22% in A2 + LWR for 2010–39, and ~2% in B2 + NLWR to 42% in A2 + LWR for 2050–79. On the other hand, PEq in LWR are ~20% (40%) lower than NLWR for 2010–39 (2050–79). The PEq values in LWR scenario shows that even by reducing the cultivated area in order to address the limitation of water supply, excess water demand still exists in the network. In case of the Ps/a indicator, compared to the baseline, both A2 + NLWR and B2 + NLWR scenarios show decreasing but minor differences (decreasing ~9%), while the other scenarios show major differences (decreasing ~17–47%). As a result, the number of outlets with Ps/a < 1 will increase ~63–146% (depending on the scenario type) in comparison with the baseline. Therefore, when the water demand increases in the network, the number of outlets with Ps/a < 1 in terms of pressure supply will increase. This means that, for example, in A2 + NLWR 2050–79, 450 more outlets will work under the assigned pressure. In addition, if LWR is taken into account, the number of outlets with Ps/a < 1 decreases ~6% and 7–13% in comparison with NLWR for 2010–39 and 2050–79, respectively. This is because in the LWR scenario, the cultivated area decreases in future periods, and part of the network is removed.

In the case of velocity indicators, results show that the $V_{min}TR$ and $V_{max}TR$ values compared to the baseline decrease (~7–25%, depending on the scenario type) and increase (~20–320%), respectively. In the Bilesavar district, by increasing the maximum velocity, changes in pressure between maximum and minimum discharge increases in the pipelines, and this may cause water hammer in

Table 2 Average values of the indicators calculated for a peak demand reference day

		Baseline	2010–39				2050–80			
			NLWR		LWR		NLWR		LWR	
			A2	B2	A2	B2	A2	B2	A2	B2
Pressure equity (PEq)		0.80	0.77	0.79	0.62	0.64	0.73	0.78	0.46	0.49
Simulated/assigned pressure ratio (Ps/a)	Max.	1.21	1.13	1.11	1.24	1.27	1.02	1.06	1.05	1.13
	Min.	0.83	0.73	0.72	0.34	0.35	0.60	0.62	0.23	0.25
	Av.	1.02	0.93	0.92	0.80	0.82	0.82	0.85	0.54	0.57
Outlets with Ps/a < 1		308	553	534	517	504	758	755	705	658
Velocity	Min.	0.27	0.22	0.24	0.22	0.25	0.20	0.20	0.23	0.23
	Min. TR	0.44	0.37	0.41	0.34	0.41	0.33	0.34	0.38	0.40
	Max.	2.24	2.79	2.81	2.95	2.98	3.01	3.05	3.50	3.74
	Max. TR	−0.10	0.12	0.13	0.18	0.21	0.20	0.22	0.40	0.42
Energy consumption (P)	Kw	1981.64	1990.01	1987.46	1890.51	1864.38	2272.83	2165.08	1963.71	1897.82

Table 3 Optimum temperature at planting date

Crop	Planting date	Optimum temperature (°C)	Germination (°C)
Wheat	Late Oct. –early Nov.	20–25	4
Barley	Late Oct. –early Nov.	20–25	4
Alfalfa	Apr. and early Sep.	25	1
Lentil	Late Mar.	18–30	15–25
Cotton	Early May.	34	15

the network. On the other hand, by decreasing the minimum velocity, the risk of deposits in pipes increases. With NLWR, energy consumption would be less than 0.5% higher than the baseline for 2010–39, but may increase ~15% and 9% in A2 and B2 in 2050–79, respectively. With LWR, because some pumping stations have been removed in the network, energy consumption is lower than the baseline (~1–6%). In general, the results clearly show how climate change would have a significant impact on network performance with existing cropping patterns. The results of this research confirm the study of Perez Urrestarazu et al. (2010), who used a combination of emissions and European agricultural policy scenarios to evaluate the performance of the Fuente Palmera irrigation network in terms of the equity and adequacy of pressure at the outlets.

3.4 Discussion of Adaptation Strategies to Climate Change

– *Changing the date of planting*

Temperature is one of the environmental factors influencing different stages of plant growth. The reproductive stage in plants compared to the growth stages is very sensitive to high temperature and reproductive organs are very vulnerable to increasing temperature before or during flowering stages. Climate change in a region causes a limitation in the date of planting of a plant, therefore; change in the date of planting can be one of the adaption strategies as reported in other researches (Shahkarami 2009). The optimal temperature for planting each of plants and at least temperature for germination, according to the crop pattern of study area, is presented in Table 3. Moreover, Table 4 presents the optimum temperature at planting date for 2050–79 in study area.

As shown in Tables 4 and 5, the optimum temperature for growth of wheat and barley is between 20 and 25 °C; therefore, November will be an appropriate month

Table 4 Optimum temperature at planting date for 2050–79

Scenario	Jan.	Feb.	Mar.	Apr.	May.	Jun.	Jul.	Aug.	Sep.	Oct.	Nov.	Dec.
T-A2	0.5	2.3	8	15.7	21	29	34	33.2	27.1	18	10	4.2

Table 5 Gross irrigation water requirement (mm) by changing the planting date of wheat and barley in 2050–79

State	Crop	Jan.	Feb.	Mar.	Apr.	May.	Jun.	Jul.	Aug.	Sep.	Oct.	Nov.	Dec.	Sum.
No adaption strategy	Wheat	0.1	0.2	1.2	4.6	5.5	9.9	–	–	–	–	0.2	0.1	21.8
	Barley	0.1	0.2	1.2	4.6	5.5	9.9	–	–	–	–	0.2	0.1	21.8
	Alfalfa	0.5	0.7	1.3	3.3	6	10.3	13.2	11.8	7.4	3.5	1	0.6	59.6
	Lentil	–	–	–	0.8	6.4	10.5	–	–	–	–	–	–	17.7
	Cotton	–	–	–	–	1.9	5.7	10	12.4	8.1	3.8	0.9	–	42.8
	Sum	0.7	1.1	3.7	13.3	25.3	46.3	23.2	24.2	15.5	7.3	2.4	0.8	163.7
Chang in the planting date of wheat and barely	Wheat	0.1	0.2	0.1	4.6	7.7	9.6	10	–	–	–	–	0.1	32.4
	Barley	0.1	0.2	0.1	4.6	7.7	9.6	10	–	–	–	–	0.1	32.4
	Alfalfa	0.5	0.7	1.3	3.3	6	10.3	13.2	11.8	7.4	3.5	1	0.6	59.5
	Lentil	–	–	–	0.8	6.4	10.5	–	–	–	–	–	–	17.7
	Cotton	–	–	–	–	1.9	5.7	10	12.4	8.1	3.8	0.9	–	42.8
	Sum	0.7	1.1	1.6	13.3	29.7	45.7	43.2	24.2	15.5	7.3	1.9	0.8	184.9

Table 6 Gross irrigation water requirement (mm) by changing the planting date of cotton in 2050–79

State	Apr.	May.	Jun.	Jul.	Aug.	Sep.	Sum.
No adaption strategy	–	1.8	5.7	14.3	11.9	6.2	39.9
Changing the planting date of cotton	1	3.1	8	14.4	12.4	–	38.9

for germination and growth of those plants. Furthermore, as shown in Table 5, by moving the planting date from early October to early November in 2050–79 period, irrigation water requirement for wheat and barley increases. One of the reasons for this increase can be due to increase in the crop coefficient of those plants and locating in the growth and development stages with high temperature. In the case of cotton, it will be possible to shift the planting date from May to early April in 2050–79. As seen from Table 6, with this change, annual gross irrigation water requirement is only 1 mm less than before adaption strategy, which is not remarkable for adaption to climate change.

- *Changing the hours of irrigation*

One of the other adaption strategies to reduce the network's vulnerability is changing the hours of irrigation. The results show that in 2050–79, due to the increase in evapotranspiration, irrigation interval reduces from 7 days to 5 days. Moreover, in order to supply new water requirement in the irrigation network, the hours of irrigation should be increased from 19 to 27.6 and 29 h for 2010–39 and 2050–79 periods, respectively. Since the maximum irrigation hours per day, including 2 h of rest for pumping stations, are 22 h; therefore, the maximum of 22 h is considered for the network and its performance is evaluated with the evaluation indicators of interest.

- *Pressure equity indicator*

The values of this indicator (Table 7) show that the equity of pressure distribution is notably improved by increasing the hours of irrigation in the network. The reduction of water abstraction at the outlets and improvement of the irrigation network performance can be as one of the reasons for increasing the pressure distribution equity; so that, when the hours of irrigation increase, the plant's water requirement is supplied for a longer time without reduction in the performance of irrigation system.

Table 7 Pressure equity indicator by changing the hours of irrigation

Status	Period	Ppq	Pbq	Peq
No change in the hours of irrigation	2010–39	24.8	40.4	0.62
	2050–79	14.2	31	0.46
Changing the hours of irrigation	2010–39	35.7	49	0.73
	2050–79	24.6	40.8	0.6

Table 8 Pressure adequacy indicator by changing the hours of irrigation

Status	Period	Ps/a ave	Ps/a max.	Ps/a min.	Ps/a < 1
No change in the hours of irrigation	2010–39	0.80	1.24	0.34	517
	2050–79	0.54	1.05	0.23	705
Changing the hours of irrigation	2010–39	1.04	1.40	0.67	56
	2050–79	0.80	1.24	0.34	202

Table 9 Maximum and minimum velocity ratio indicator by changing the hours of irrigation

Status	Indicator	2010–39	2050–79
Changing the hours of irrigation	V_{max}	2.70	3.10
	$V_{max}TR$	0.08	0.24
	V_{min}	0.24	0.26
	$V_{min}TR$	0.40	0.44

– *Pressure adequacy indicator*

According to the Table 8, the values of pressure adequacy indicator presents that the status of pressure has been improved and the number of outlets, which will be working at a pressure less than the required pressure, has been decreased by increasing the hours of irrigation and subsequently, reducing the discharge at the outlets for 2010–39 and 2050–79 periods.

– *Maximum and minimum velocity ratio indicator in water transmission pipes*

The values of the maximum and minimum velocity ratio indicator under the status of changing the hours of irrigation are presented in Table 9. The results indicate the reduction of maximum velocity in pipes so that this reduction in 2010–39 is more considerable than 2050–79. The results also show that by increasing the hours of irrigation, minimum velocity ratio indicator increases and the sedimentation of particles due to the low velocity of water in the pipes decreases.

4 Conclusions

This study was an attempt to investigate the potential effects of climate change on the irrigation network of the Bilesavar district, in Iran in terms of evaluation indicators. In the case of climate change scenarios, the results showed that temperature may increase by 10 and 23% on average by 2010–39 and 2050–79, respectively. Rainfall may decrease by 4% on average by 2010–39, and by 7 and 1% for A2 and B2 by 2050–79, respectively. ET_o is expected to be around 6% higher than the baseline for 2010–39, and 12% for 2050–79.

Due to changes in the climatic conditions in the Bilesavar district, irrigation requirements may be higher in 2010–39 and 2050–79 by around 6 and 13% on average (depending on crop type) respectively, which could lead to an increase in the maximum flow requirements in the pumping stations and at the outlets. Furthermore, the results of the model outputs indicated an increase in discharge of the network that will lead to a rise in water demands for cropping patterns, and a drop in pressure at the outlets.

Due to predicted changes in the climate, the irrigation network will have problems in terms of pressure and discharge supply. The total discharge of the irrigation system in its current condition will increase from 3.7 to 4 and 4.3 cubic meters per second for the two future periods. It was observed that pressure equity and pressure adequacy (Ps/a) decreased for 2050–79; the pressure equity dropped on average from 81 to 46%, and the adequacy dropped on average from 1.03 to 0.54. In addition, the minimum and maximum velocity in the pipes showed major differences with the permissible velocity for 2050–79. At present, the system would still work properly in these conditions, but according to the results of this study, the network will face problems. The significant changes in CWR or IR will have a major effect on the network performance in terms of pressure.

In order to adapt to the demand with water supply restrictions in the Bilesavar district, the planting area has to be optimally reduced. The results indicated a reduction of around 5% and 13% of agricultural lands for the 2010–39 and 2050–79 periods, respectively. In addition, the results of the study specified that the optimal reduction of the planting area, in conditions where sufficient amounts of water cannot be supplied, is an appropriate strategy. In the case of other adaption strategies, changing the hours of irrigation and changing the date of planting can be effective strategies to adopt the irrigation system to climate change for the future periods. The methodology presented in this paper can be used in other studies to help identify the network area that could potentially have problems in the future, and to assess other irrigation districts with different characteristics.

References

Adamowski J, Adamowski K et al (2010) Influence of trend on short duration design storms. Water Resour Manage 24:401–413

Allen RG, Gichuki FN et al (1991) CO_2-induced climatic changes in irrigation-water requirements. Water Resour Plann Manage 117:157–178

Ashofteh PS, Massah AR (2009) Impact of climate change uncertainty on temperature and precipitation of Aidoghmoush basin in 2040–2069 period. Water Soil Sci 19(12):85–98

Banihabib ME, Zahraei A, Eslamian S (2015) An integrated optimization model of reservoir and irrigation system applying uniform deficit irrigation. Int. J. Hydrol Sci Technol 5(4):372–385

Brunsell NA, Jones AR et al (2010) Seasonal trends in air temperature and precipitation in IPCC AR4 GCM output for Kansas, USA evaluation and implications. Int J Climatol 30(8):1178–1193

Calejo MJ, Lamaddalena N et al (2008) Performance analysis of pressurized irrigation systems operating on-demand using flow-driven simulation models. J Agric Water Manage 95:154–162

Carter TR, Parry ML et al (1994) IPCC technical guidelines for assessing climate change impacts and adaptations. London

Clarke D, Smith M et al (1998) Cropwat for windows: users' guide. FAO, Rome

Daccache A, Weatherhead K et al (2010) Climate change and the performance of pressurized irrigation water distribution networks under Mediterranean conditions: impacts and adaptations. Outlook Agric 39(4):277–283

De Silva CS, Weatherhead EK et al (2007) Predicting the impacts of climate change: a case study of paddy irrigation water requirements in Sri Lanka. Agric Water Manage 93(1–2):19–29

Dehghan Z (2012) Irrigation distribution networks' vulnerability to climate change: a case study on Bilesavar pressurized irrigation network. Dissertation, Tarbiat Modares University, Tehran, Iran

Dalezios NR, Gobin A, Tarquis Alfonso AM, Eslamian S (2017) Agricultural drought indices: combining crop, climate, and soil factors. In Eslamian S, Eslamian F (eds) Chapter 5 in Handbook of drought and water scarcity, vol 1: principles of drought and water scarcity. Francis and Taylor, CRC Press, USA, pp 73–90

Dibike YB, Coulibaly P (2005) Hydrologic impact of climate change in the Saguenay watershed: comparison of downscaling methods and hydrologic models. J Hydrol 307(1–4):145–163

Farmani R, Abadia R et al (2007) Optimum design and management of pressurized branched irrigation networks. J Irrig Drain Eng 133(6):528–537

Fathian F, Dehghan Z, Eslamian S, Adamowski J (2016) Assessing irrigation network performance based on different climate change and water supply scenarios: a case study in Northern Iran. Int J Water 3:191–208

Fischer G, Tubiello FN et al (2007) Climate change impacts on irrigation water requirements: effects of mitigation, 1990–2080. Technol Forecast Soc Change 74(7):1083–1107

Gohari A, Eslamian S, Abedi-Koupaei J, Massah-Bavani A, Wang D, Madani K (2013) Climate change impacts on crop production in Iran's Zayandeh-Rud River Basin. Sci Total Environ 442:405–419

Gorantiwar SD, Smout IK (2005) Performance assessment of irrigation water management of heterogeneous irritation schemes: a framework for evaluation. Irrigat Drain Syst 19(1):1–36

Hashmi MZ, Shamseldin AY et al (2009) Statistical downscaling of precipitation: state-of-the-art and application of Bayesian multimodel approach for uncertainty assessment. Hydrol Earth Syst Sci 6:6535–6579

Hassan Z, Harun S (2012) Application of statistical downscaling model for long lead rainfall prediction in Kurau River catchment of Malaysia. Malays J Civil Eng 24(1):1–12

Huang J, Zhang J et al (2011) Estimation of future precipitation change in the Yangtze River basin by using statistical downscaling method. Stoch Environ Res Risk Assess 25(6):781–792

IPCC (2013) Fourth assessment report of the intergovernmental panel on climate change. Cambridge University Press, Cambridge

Karamouz M, Hosseinpour A et al (2010) Improvement of urban drainage system performance under climate change impact. Hydrol Eng 16(5):395–412

Knox JW, Rodríguez Díaz J et al (2010) A preliminary assessment of climate change impacts on sugarcane in Swaziland. Agric Syst 103:63–72

Mahdavi P (2008) Development assessment model of transition and distribution utilities using with classic method. Dissertation, Tarbiat Modares University

Maraun D, Wetterhall F et al (2010) Precipitation downscaling under climate change: recent developments to bridge the gap between dynamical models and the end user. Rev Geophys 48 (3):RG3003. https://doi.org/10.1029/2009rg000314

Moghaddasi M, Morid S et al (2010) Assessment of irrigation water allocation based on optimization and equitable water reduction approaches to reduce agricultural drought losses: The 1999 drought in the Zayandeh Rud irrigation system (Iran). Irrig Drain 59(4):377–387

Nakicenovic N, Alcamo J et al (2000) Special report on emissions scenarios: a special report of working group III of the intergovernmental panel on climate change. Pacific Northwest National Laboratory, Environmental Molecular Sciences Laboratory (US), Richland, WA (US)

Paimozd S, Morid S et al (2010) Comparison of non-linear optimization and a system dynamics approaches for agricultural allocation (a case study: Zayande Rud Basin). Iran J Irrig Drain 4 (1):44–52

Perez Urrestarazu L, Diaz JAR et al (2009) Quality of service in irrigation distribution networks: case of Palos de la Frontera Irrigation District (Spain). Irrig Drain Eng 135(6):755–762

Perez Urrestarazu L, Smout IK et al (2010) Irrigation distribution networks' vulnerability to climate change. Irrig Drain Eng 136(7):486–493

Reichler T, Kim J (2008) How well do coupled models simulate today's climate? Bull Am Meteorol Soc 89:303–311

Rodríguez Díaz JA, Camacho Poyato E et al (2007) Model to forecast maximum flows in on-demand irrigation distribution networks. Irrig Drain Eng 133(3):222–231

Rodrıguez Dıaz JA, Lopez Luque R, Carrillo Cobo MT, Montesinos P, Camacho Poyato E (2009) Exploring energy saving scenarios for on-demand pressurized irrigation networks. Biosyst Eng 104(4):552–561

Rosenzweig C, Allen LH et al (1995) Climate change and agriculture: analysis of potential international impacts. ASA Special Publication 59, ASA, CSSA, and SSSA, Madison, Wis

Shahkarami N (2009) Climate change adaptation strategies involving risk analysis and comprehensive management of water resources in the Zayandeh Roud basin. Dissertation, Tarbiat Modarres University

Smith M (2000) The application of climatic data for planning and management of sustainable rain-fed and irrigated crop production. Agric Meteorol 103:99–108

Tatsumi K, Yamashiki Y et al (2011) Estimation of potential changes in cereals production under climate change scenarios. Hydrol Process 25(17):2715–2725

Thomson AM, Izaurralde RC et al (2006) Climate change impacts on agriculture and soil carbon sequestration potential in the Huang-Hai Plain of China. Agric Ecosyst Environ 114(2):195–209

Vaze J, Teng J, Chiew F (2011) Assessment of GCM simulations of annual and seasonal rainfall and daily rainfall distribution across south-east Australia. Hydrol Process 25(9):1486–1497

Wigley TML, Jones PD et al (1990) Obtaining subgrid scale information from coarse-resolution general circulation model output. J Geophys Res 95:1943–1953

Wilby RL, Dawson CW (2012) The statistical downscaling model: insights from one decade of application. Int J Climatol 33(7):1707–1719

Xie JX, Xue Y (2005) Optimization modeling and LINDO/LINGO software. Tsinghua University Press, Beijing

Yang X, Lin E et al (2007) Adaptation of agriculture to warming in Northeast China. Clim Change 84:45–58

Implication of Climate Change and Food Security Status on Rural Farmers in Kura Kano State North–Western Nigeria

Salisu Lawal Halliru

Abstract This paper explores the food security context and the socio-economic consequences of climate change on rural farmers in Kura local government, Kano state, Nigeria. The purpose of the study was to ascertain the food security status of the rural farmers in the study area. Socio- economic consequences were ascertained. Agro forestry will serve as a win-win solution to the difficult decision between reforestation and agricultural land use; hence it increases the storage of carbon and may also increase agricultural productivity. Lottery sampling procedure was used in the selection of local government, communities and farmers for the research study. Structured questionnaire were used to obtain the data for the study. Food security index was used to ascertain the level food insecurity among the rural farmers in the communities. Descriptive statistics as a tool for analysis was used to analyze the data obtained. 98.5% of the respondent was married with dependants and low annual income of # 80,000 and below. Most farmers experienced loss of investment on farm lands, lives and income respectively. This study reveals that rural farmers suffered serious hardship they cannot produce what to feeds their families for at least six month in a year (food insecure). They also suffered ill health, such as malaria, water born diseases and skin infections among others. The study recommends that policy makers should encourage more recognition of food security in the state, support for adaptation activities in rural areas, enhance the role of civil societies and adaptation and mitigation.

Keywords Climate change · Food security · Rural · Nigeria · Farmers

S. L. Halliru (✉)
Department of Geography, Federal College of Education Kano,
P.M.B 3043, Kano, Kano State, Nigeria
e-mail: lhsalisu09@gmail.com

P. Castro et al. (eds.), *Climate Change-Resilient Agriculture and Agroforestry*,
Climate Change Management, https://doi.org/10.1007/978-3-319-75004-0_20

1 Introduction

Climate change has effects on agriculture because an agricultural activity depends on climate condition. That effect threatens our ability to advance global food security. Agroforestry, is the growing of trees in agricultural landscapes, has the potential to achieve sustainable agriculture in smallholder farming. Various Agroforestry practices are suitable for adaptation of agroecosystem to climate change. In view of t that Agro forestry will serve as a win-win solution to the difficult decision between reforestation and agricultural land use, hence it increases the storage of carbon and may also increase agricultural productivity. Agro forestry will serve as a potential mitigation strategy.

It is important to point out that the global climate or the climate of any part of the world has never been static. Climate fluctuations and climate change impinge on human affairs in diverse and many ways, climate determines the ability of man to feed him through its influence on agricultural production.

Agriculture is an important occupation and operation that provide income, employment and food to Nigeria thereby enhancing food security in the country, but in a situation of climate change, Agriculture in northern Nigeria as in other northern parts of west Africa, would evidently be impacted (IPCC 2007). This is applicable to Kano state where most rural communities engaged in farming and crop production for their livelihoods and attainment of food security.

There is a growing consensus in the scientific literature that in the coming decades the world will witness higher temperatures and changing precipitation levels. The effects of this will lead to low/poor agricultural products. Evidence has shown that climate change has already affecting crop yields in many countries (IPCC 2007; Deressa et al. 2008; BNRCC 2008).

Food security, livelihoods, and poverty in rural communities in northern Nigeria are determined by the agricultural production of the individuals and communities. Also rainfall and temperature rate dictate the amount of agricultural production, annually. Agriculture everywhere in the country, being dependent on rainfall, will be adversely impacted by increased variability in timing and amount of rainfall.

Production of grain crops like maize, guinea corn, millet and rice can be depressed. The openness of the region to high temperature also affects the level of soil fertility (Adogi 2008). The state of agriculture in Nigeria in recent times shows a continuous decline in exportation and increase in importation of agricultural products into the country. The share of Nigeria's agricultural products in total exports plummeted from over 70% in the 1960s to less than 2% in 2010 (Adogi 2008).

The major contribution to the decline has been liked with the negative effects of climate change on crop production in sub-Saharan Africa (Okunnola and Ikuomola 2010). It is predicted that the majority of Nigerian and African countries will have novel climates over at least half of their current crop year by 2050 (IPCC 2007).

2 Agriculture and Climate Change

Though the extent and nature of the effect of climate change on agriculture has not yet been accurately forecast; its impact so far on diverse farming regions of the world has been profound (McClean, et al. 2005; FOA 2007; Revkin 2008). Water sources have become unpredictable; with excess, little or no rainfall and flooding and inundation in coastal areas (Brown 2006; Dore 2005; Hopkin 2005).

3 Food Security and Climate Change

According to the FAO, food insecurity exist when people are not able to secure access to an adequate and safe diet which constrains them from leading an active and health life today. In addition, those who are currently food secure may become vulnerable to food security in the future. Potential impact of climate change on food security includes both direct nutritional effects (changes in consumption quantities and composition) and lively livelihood effects (change in employment opportunities and cost of acquiring adequate nutrition). Climate change can affect each of these dimensions FAO(2002).

4 Food Security in Changing Climate

Food security (is) a situation that exists when all people at all times, have physical social and economic access to sufficient, safe and nutritious food that meets their dietary need and food preference for an active and healthy life FAO(2002). Climate change has already caused and will continue to cause change in global temperature and precipitation pattern as well as changes in soil processes and properties (Meehl et al. 2007). This has lead to considerable concern that climate change could compromise food security, which would lead to an overall decline in human health.

5 Implication of Climate Change and Food Security in Nigeria

Countries in sub-saharan Africa, including Nigeria are likely to suffer the most because of their geographical location, low incomes, low institutional capacity as well as their greater reliance on climate—sensitive renewal natural resources sector like agriculture (Ebon 2009). Climate change is threatening agricultural sector because food production is affected when there is a change on the climate.

Once effects occur on agriculture it will equally has an effects on people who depend on agriculture as a means of livelihood. According to IPCC 2007, BNRCC 2008. Rough estimates is of the view that for the next 50 years or so, climate change may likely have serious threat to meeting global food needs than other constrains on agricultural system.

The number of people without food to eat on regular basis is increasing in a geometric progression. Over 60% of the world's undernourished people live in Asia and a quarter in Africa (FAO 2002). Climate change phenomenon affects agriculture in many ways, such as unreliability in the onset of the farming season, due to changes in rainfall characteristics; this can result to an unusual sequence of planting and replanting which may lead to food shortage due to harvest failure (Okoh et al. 2011).

6 Significant of the Study

Most climate change impacts research studies have been focusing on human perception on climate change while implication of climate change and food security status remains a poorly investigated area in research. At the same time cultivation of indigenous foods found growing in the forests that are important locally but have, to date, been under-researched by the scientific community. Climate change is recognized as one of the major worldwide challenges facing men and his environment and has become one of the areas of urgent concern and focus.

Agro forestry is one of the potential options which can be us or practice to mitigate the impacts of climate change and therefore there is need to conduct a research on the implication of climate change and food security status in our communities.

7 Aim and Objectives of the Study

The main aim of this research is to explore the food security context and socio-economic consequences of climate change on rural farmers in Kura local government area, Kano state from which the following specific objectives were derived:

(a) To determine the relationship between climate change and food security in the study area.
(b) To examine the food security status of rural farmers in Kura emanating from climate change using food security index.
(c) To examine the impacts of climate change event on food security of the farmers in the local government.

8 Method and Material

This study was carried out in Kura local government area Kano state, Nigeria. The area was selected because agriculture is the major economic activities in the area. The study area is located at the southern part of Kano state with a population of 144,601 million people (NPC 2006) with a land mass of 206 km^2, is located between 11° 46′ 12.84″N and longitude 8° 35′ 29.02″E it is about 900 km from the edge of the Sahara desert and 1140 km away from the Atlantic ocean approximately.

The study area shares boundary north and east with Kumbotso local government and west–south it boarders with Madobi and Garun Malan local government area respectively, extreme south–east it boarders with Bunkure local government area. (Fig. 1) The area has three marked temperature regimes; warm, hot and cold with mean annual temperature of 26 and 21 °C main monthly range of maximum temperature in December/January and over 35 °C which is hottest (April/May)wet season start in May and ends October. While November to February is dry cool season with hamattan haze. Vegetation is savanna (grassland) of Sahel Sudan guinea type.

Lottery sampling procedure was used for the selection of the communities and farmers for the research. Six communities were selected to represent the local government area. The measures for selecting the communities include the following:

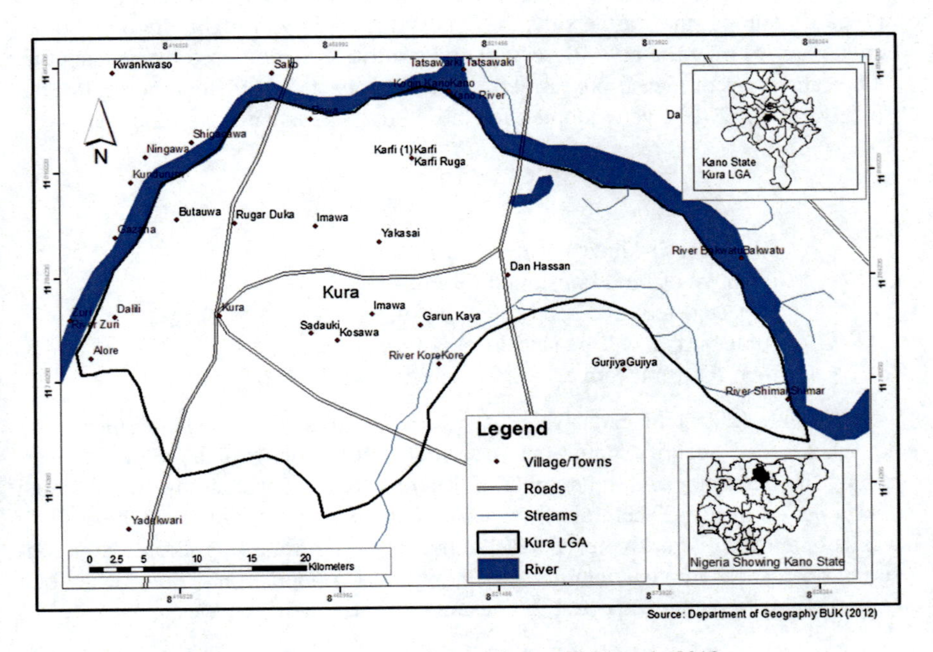

Fig. 1 Map of Kano state showing the study area *Source* Field study 2010

i. Community with a sizable number of farmers with at least ten (10) hectre of cultivable land.
ii. Community with all year round farmers i.e. engaged in farming activities during the dry and wet season.
iii. Agriculture- economy based community.

On view of the above, ***Imawa, Karfi, Gundutse, Bawa, Dukawa, Baure,*** were selected as the study location. A sample of 50 farmers were selected in each community based on snow ball method of one farmer directing the researcher to the next farmer this way a total of 300 sample were collected. The main source of data was through mean of annual temperature and rainfall from January 1991 to December 2013 were collected from Malan Aminu Kano International Airport Nigeria meteorological Agency (MAKIA-NIMET) for the study. The primary data were obtained through observation, discussion, interview and administration of structured questionnaire survey. Information sought was on personal information of respondents including their age, sex, marital status, educational level, additional information were sought based on their involvement in agricultural activities. Descriptive statistics were used to summarize the socio-economic characteristics and to determine the level of food insecurity as well as the food security status among the farmers. The researcher used period of six years in accessing farmers' agricultural activities in the study area from 2007–2012. The data gathered was analyzed using tables, percentages and other statistics techniques relevant for the data collected.

Food security index: food security index was used to determine the level of food in security among rural farmers that have been affected by drought, flood and other climate hazard over the past 40 years. Food security equation used by Felake et al. (2003) and ways of measuring farmers' food security status by Hoddinott (2001) in Emaziye et al. (2013) were adopted for this study. The equation is stated as:

$$C^x = Cj - Yi \tag{1}$$

C^x food security index of rural farmers
Cj quantity of food consumed (N = 1–5)
Yi expected required food to be consumed (N = 5)
If $C^{x=}0$ rural farmer will be said to be a food secure.
If $C^x < 0$ then the rural farmer will be said to be a food insecure

Hoddinott (2001) in Emaziye et al. (2013) outline four ways of measuring household food security status; such as dietary diversity which involves determining the frequency and the number of different foods consumed by an individual over a period of time. Therefore food security index of rural farmers in these study was adopted from Emaziye et al. (2013), based on the total household daily consumption (carbohydrate, vitamins, water, proteins, minerals and fat/oil). A food secured rural farmer is expected to consume all the time categories.

Food insecure category was further categorized in mild food insecure, moderately food insecure and severe food in secure.

$(C^x = C^j - Y^j) = 4 - 5 = -1$ (Mild food insecure)
$(C^x = C^j - Y^j) = 3 - 5 = -2$ (Moderately food insecure)
$(C^x = C^j - Y^j) = 2 - 5 = -3$ (severe food insecure)

9 Correlation

Correlation was used to determine the relationship between calculated climate change variables coefficient of variation and food security of rural farmers in Kura local government.

$$F_s = A_{tcv}T_{cv} + A_{rcv}R_{cv} + A_{ycv}Y_{cv} + e$$

where,

F_s	food security
T_{cv}	temperature coefficient of variation (%)
R_{cv}	rainfall coefficient of variation (%)
Y_{cv}	food production (yield) coefficient of variation (%)
E	error term
A_{tcv}, A_{rcv}, A_{ycv}	Model parameter

Adopted from: Emaziye et al. (2013).

10 Result and Findings

The data on the personal variable of respondents shows 65% of them were within the age 43–53 the remaining were aged 54 above. On the marital status of respondents a total of 200 which represents 66.67% were married, 65% which is 21.6 were divorced, and 35 which represent 11.8% were single. On the educational level of respondents, 120 respondents which is 40% had Qur'anic education, 100 respondents which is 33.33% had primary education, and 80 respondents which is 26.6% had adult education. The respondents mean annual income is N 65,642 (Naira) about $378USD which is less than $1(one dollar) a day which shows a poverty situation of the rural farmers. This might probably due to climate change impact in the state.

In Table 1 above reveal that food security has a significant relationship with climate change variables using Pearson correlation, as temperature and rainfall contributed toward food security it is a well known fact that every crop has rainfall and temperature requirement for it survival as equally observed by the rural farmers in the study area these coincided with the findings of Emaziye et al. (2013).

Table 1 Relationship between climate change variables (temperature and rainfall) and food security in Kura local government

Correlation	Food security (Fs)	Temperature (T_{cv})	Rainfall (R_{cv})
F_s	1.000	–	–
Pearson T_{CV}	–	1.000	1.000
Correlation R_{cv}	–	−000	−000
F_s	–	–	–
Sig. T_{cv}	−000	–	–
(1-tailed) R_{cv}	−000	–	–
F_s	5	5	5
T_{cv}	5	5	5
N R_{cv}	5	5	5

Source field study 2013

Table 2 Respondents food security index

Food security index	Kura local government (n = 300)	Percentage (%)
Food secure	10	3
Mild food insecure	15	5
Moderately food insecure	125	42
Severe food insecure	**150**	**50**
Mean	Severe food insecure	

Source field study 2013

Table 2 above shows how food security index was used to ascertain the level of food security in the study area, and it is reveals that the area falls under severe food insecure with 50% due to shortage of rainfall and 42% moderately food insecure respectively. Based on these findings only 3% were food secured and 5% mild food secure. This was attributed to climate (drought) that resulted to crop failure and loss of investment on farmlands in the study area.

Table 3 above indicated that farmers in the study area experienced a severe impact of losses with 55% because the farmers in the study area depend mostly on climate-sensitive resources for livelihood. This is as a result of climate change event. These findings coincided with statement of Ebon (2009) where he stated that countries in sub Saharan Africa, including Nigeria are likely to suffer the most because of their geographical location.

Limitation and constraint of the paper

- Some of the respondents were illiterate which pose a problem of language barrier. Though the researcher spent time with the respondents trying to translate and interpret the questionnaire for them. However, the research assistants were familiar with local language mainly *Hausa* in order to reduce the limitation.

Table 3 Impact of climate change event on food security

Impact of loss	Kura local government(n = 300)	Percentage (%)
No effect	25	8.33
Low	16	5.33
Moderate	46	15.33
Severe	**165**	**55**
Very severe	48	16

Source field study 2013

- The scope of the study is only limited to Kura local area which is not enough to make generalization about the investigated issue.
- Inadequate finance to cover wider affected area in the region.

11 Conclusion and Recommendation

Climate change poses a serious threat in Nigeria Kura local government inclusive (study area) especially in the area of agriculture. As a result of reduction in precipitation and high temperatures and evatranspiration during droughts period has negatively impacted staple food production in the study area. Thus, a negative impact from climate change in Kura brings about increased poverty, water scarcity and food in security. First we learned that there is a significant relationship between climate change and food security in the study area as rainfall and temperature were contributors. Secondly food security index revealed severe food insecure situation. Thirdly the study revealed severe impact of losses of crops and investment. The study therefore recommends the following for future prospect;

Additional research is needed to further develop local farmers' ability to understand and address issues related to climate change, Agriculture and Agroforestry.
Campaigns' to raise awareness on the role on indigenous species in climate change adaptation.
There is need for development of appropriate policies and institutional infrastructure to catalyze adoption of Agroforestry.

Acknowledgements I am very grateful to Dr (Mrs.) R.J Muhammad (provost F.C.E Kano) for her support and encouragements at all the time directly or indirectly, My Supervisor Dr Kassim B. Sekabira Kampala International University Uganda, Mohammed A.Mohammed, Sani Danladi, Musa Sule, Dr Sadi Sirajo, All academic staff Department of Geography F.C.E Kano and the entire Management and Staffs of Federal College of Education Kano, who have in one way or the other contributed to the success I have achieved National and International. I am greatly indebted to Tertiary Education Trust Fund (TETFund) for funding my Masters programmed in Uganda Thanks also to Professor A. I. Tanko Department of Geography Bayero University Kano - Nigeria, Professoor Abubakar Sani Mashi DG Nigerian Meteorological Agency(NIMET),

Professor S.U Abdullahi (former VC A.B.U Zaria), Prof. Tony Binns, University of Otago, Aisha Gidado Muhammad Maryam Salisu (Ilham), Maimuna Salisu (Iman), Fatima Salisu (Ihsan) and Alhaji Ibrahim Ahmad Gundutse.

All errors and omissions, and all views expressed, remain solely my responsibility.

References

Adogi MM (2008): Climate change and Northern Nigeria center for education and leadership development (CELDEV) Publication sponsored and published by Heinrich Boll foundation Nigeria

Brown P (2006) Global warming: the last chance for change. Published by Dakini books

Building Nigeria's Response to climate change (BNRCC) (2008) Annual workshop of Nigerian Environmental study team (NEST): The recent global and local action on climate change Held at Hotel millennium, Abuja, Nigeria

Deressa TR, Hassen T, Akmu M, Yesuf, Ringer C (2008) Analyzing the determinant of farmers choice of adaptation measures and perception of climate change in Nile basin of Ethiopia. International food policy Research institute (IFPRI) Discussion paper N0. 00798. Washington, DC: IFPRI

Dore HIM (2005) Climate change and changes in global precipitation patterns: what do we know? Environ Int J 31(8):1167–1181

Ebon E (2009) Implications of climate change for economic growth and sustainable development in Nigeria Enugu forum policy paper 10. African Institute for Applied Economics, Nigeria

Emaziye PO, Okoh RN, Ike PC (2013) An evaluation of effect of climate change on food security of rural households in cross river state, Nigeria. Asian J Agric Sci 5(4):56–61

FAO (2007) Adaptation to climate change in agriculture forestry and fisheries: perspective, frame work and properties Interdepartmental working group on climate change, Food and Agriculture Organization of the United Nations Rome ff://ftp.fao.org/docrep/fao/009/j927le.pdf. Retrieved on 1 July 2011

FAO (2002). Declaration of the World Food Summit. Retrieved from: www.fao.org. Online on 25 Aug 2012

Feleke ST, Kilmer RL, Gladwin CH (2003) Determinants of food security in Southern Ethiopia. A selected paper presented at the 2003, American Agricultural Economics Association Meetings in Montreal, Canada

Hoddinott J (2001) Targetting: principles and practice. In: Hoddinott J (ed) Food security practice: method for rural development projects, International Food policing Research Instituttion, Washington, D.C.

Hopkin M (2005) Amazon Hit by worst drought for 40 years: warming Atlantic linked to both US hurricanes and rainforest drought Nature News. http://www.bioedoline.org/news./news.cfm?art=2094. Retrieved on 1 July 2011

IPCC (2007) Summary for policymakers. In: Parry M, Parry ML, Canziani O, Palutikof J, Van der Linden P, Hanson C (eds) Climate Change 2007: impacts adaptation and vulnerability. Contribution of Working Group II to the Fourth Assessment Report of the Intergovernmental Panel on Climate Change, Cambridge University Press, Cambridge, UK, p 7–22

McCleanC J, Lovett JC, Kuper W, Hannah L, Sommer JH, Bartthlott W, Termansen M, Smith GF, Tokumine S, Taplin RDJ (2005) African plant diversity and climate change. Ann Mo Bot Gard 92(2):139–152

Meehl GA, Stocker TF, Collins WD, Friedlingstein P, Gaye AT, Gregory JM, Kitoh A, Knutti R, Murphy JM, Noda A et al (2007) Global Climate Projections. In: Solomon S, Qin D, Manning M, Chen Z, Marquis M, Averyt KB, Tignor M, Miller HL (eds) Climate change (2007): the physical science basis; contribution of working group i to the fourth assessment report of the intergovernmental panel on climate change. Cambridge University Press, Cambridge, UK, pp 747–845

National population commission Federal Republic of Nigeria (2006) Population census official Gazatte (FGP71/52007/2500 OL24) Published by National Population Commission, Abuja Nigeria

Nigeria meteorological Agency (MAKIA-NIMET) (2013) Malan Aminu Kano International Airport

OkohR N, Okoh PN, Ijioma M, Ajibefu AI, Ajieh PC, Ovherhe JO, Emegbo J (2011) Assessment of impacts, vulnerability. Adaptive capacity and Adaptation to climate change in Niger Delta Region, Nigeria

Okunnola RA, Ikuomola AD (2010) The socio-economic implication of climate change, desert encroachment and communal conflicts in northern Nigeria. Am J Soc Manage Sci, 1(2):88–101

Revkin CA (2008).New climate report foresees big changes The New York Times. http://www.nytimes.com/2008/05/28/sceince/earth/28climate.html. Retrieved on 1 July 2011

Decentralised, Off-Grid Solar Pump Irrigation Systems in Developing Countries—Are They Pro-poor, Pro-environment and Pro-women?

Sam Wong

Abstract This systematic, evidence-based literature review examines the effectiveness of localised solar-powered small-scale irrigation systems (PVPs) in poverty reduction, environmental conservation and gender empowerment in developing countries. It suggests that PVPs are able to enhance farmers' adaptive capacity by raising agricultural productivity and their incomes. They also help mitigate climate change by reducing CO_2 emissions. The distribution of the benefits and costs, brought by PVPs, is, however, so uncertain that requires further scrutiny. PVPs are successful in rising energy-water efficiency, but the environmental trade-offs with the underground water depletion and e-wastes requires solutions. Using PVPs to achieve gender equalities may only be materialised if the structural discrimination against women in land ownership and access to resources is challenged, along with the interventions of PVPs in rural communities. This book chapter recommends more in-depth and longitudinal studies to explore the complex and long-term implications of PVPs. More evidence is also needed to assess the effectiveness of governance reforms in access of PVPs in poor communities.

1 Introduction

Climate change brings erratic weather patterns, which poses threat to rain-fed farming practice in developing countries (Alemaw and Simalenga 2015; Mendelsohn 2008). Irrigation proves effective in mitigating the impact of climate change by providing more predictable water supplies (GGGI 2017; IRENA 2016; FAO 2011). Yet, the heavy reliance on diesel in pumping the irrigation systems produces so much CO_2 that, in turn, speeds up climate change.

S. Wong (✉)
University College Roosevelt, Middelburg, The Netherlands
e-mail: s.wong@ucr.nl

S. Wong
Department of Geography and Planning, University of Liverpool, Liverpool, UK

P. Castro et al. (eds.), *Climate Change-Resilient Agriculture and Agroforestry*, Climate Change Management, https://doi.org/10.1007/978-3-319-75004-0_21

The rising popularity of solar-powered irrigation systems (also known as photovoltaic water pumps—PVPs in short) to serve off-grid areas and to replace diesel-based irrigation systems seems to offer an answer to these questions. PVPs have been considered a 'game changer' in agricultural development in developing countries (Burney and Naylor 2012: 121). IRENA (2016) suggests that PVPs provide 'reliable, cost-effective and environmentally sustainable energy for decentralised irrigation services' (p. 50). Both the World Bank (2015a) and GGGI (2017) stress that PVPs, as a 'pro-poor' and 'pro-women' technology, deserve up-scaling.

This rosy picture is, however, not shared by all development scholars and agencies. EEW (2016), for instance, urges for caution. It argues that the one-size-fits-all intervention has yet seriously taken the contextual specificity of different developing countries into account. The barriers to the access to PVPs, such as affordability, are a few unresolved issues (Shinde and Wandre 2015).

Around the controversies of PVPs, this paper is intended to draw on evidence from existing literature to assess the actual impact of PVPs in developing countries in three aspects: poverty reduction, environmental conservation and women empowerment. Through a systematic, evidence-based literature review, it aims to differentiate claims from facts and to assess what PVPs have achieved and what have not. To achieve these goals, this paper focuses, not only on the irrigation technology, but also on the complex political and economic processes that shape human incentives in energy switch and influence the distribution of costs and benefits in PVP adoption.

This paper will first discuss the impact of irrigation around the 'water-energy-livelihood' nexus. After the theoretical and methodological section, it will evaluate the impact of PVPs in the 'pro-poor', 'pro-environmental' and 'pro-women' aspects with case studies, examples and evidence.

2 Evolution of Micro Irrigation Systems

Solar-powered irrigation systems are devices that use the 'solar cell from the sun's radiation to generate electricity for driving the pump' (Yu et al. 2011: 3176). They usually insist of an array of photovoltaic cells, a controller, a motor pump-set that pump water from a well or a reservoir for irrigation. PVPs are a generic term which touches on various solar-related interventions, such as drip, potted and sprinkler irrigation. Apart from irrigation, solar pumps are also used for providing drinking water for humans and livestock. There are diverse models in promoting PVPs in terms of ownership (individual vs collective/communal approach), payment (daily, weekly, monthly or annually) and organisation (individual vs. groups).

The current debate over the role of irrigation in social and agricultural development in developing countries has been around the 'energy-water-livelihood' nexus, amid issues, such as climate change, food security and renewable energy (Biggs et al. 2015). Agriculture absorbs most of the employment. For example, 45 and 50% of the workforces in Bangladesh and India are farmers respectively (World Bank 2015b). 40% of global population rely on farming for livelihoods

(IRENA 2016). 15% of total GDP in India comes from the farming sector. Apart from the economic importance, agriculture consumes 70% of global freshwater resources. Irrigation is particularly crucial to countries without sufficient surface water. For instance, Namibia relies on 50,000 boreholes for water supplies (SELF 2008).

While 20% of global cultivated land are under irrigation, only 5% is recorded in Sub-Saharan Africa (IRENA 2016). Most subsistence smallholders rely on rain-fed farming which leads to food insecurity and malnutrition. The promotion of electric- and diesel-based irrigation since the 1960s and 1970s has encountered numerous challenges. For instance, Bangladesh has installed 1.43 million diesel-based pumps and 320,557 electric-based pumps (Islam et al. 2017). However, electric-powered pumps are not always an option since many poor villages in developing countries are beyond the reach of national power grids. Only 5% of rural population in Kenya, for example, are connected to the electric grids.

In contrast, diesel-powered irrigation seems to offer a more realistic solution because the liquid-based innovation provides flexibility and convenience. According to the World Bank (2015b), diesel pumps consume 1 million tons of diesel globally, which is worth US$900 million. Diesel does not, however, come cheap. Bangladeshi government imported 2.9 million metric tons of diesel fuel in 2011–2012 and subsidised US$0.3 of each litre of diesel, which has exerted substantial fiscal pressure (Islam et al. 2017). Some farmers in India spend 40% of their annual revenues on diesel (IRENA 2016). They were also exploited by the middlemen who charged them higher diesel prices during peak irrigation and cropping seasons. Worse still, diesel pumps emit 6.73 million tons of CO_2 in Bangladesh alone, which has ameliorated the problems of climate change (Hossain et al. 2015).

In the light of these problems, the solar-powered irrigation technology has caught the attention over the past decade. The long-term costs of PVPs, in terms of the costs of fuel and maintenance, are claimed to be lower than that of their diesel-based counterparts. The use of solar energy, as a form of renewable energy, fits very well into the Clean Energy Mechanisms. As a result, the Moroccan government, for example, has promised to install 1 million sets of PVPs by 2022. The numbers of solar pumps in India are also expected to rise from 13,000 units in 2014 to 100,000 by 2020 (GGGI 2017).

3 Theoretical and Methodological Frameworks

How good are the PVPs? What impact have they made on the environment as well as on poor people's livelihoods? The mainstream approach in the literature tends to compare and contrast PVPs with electric- or diesel-powered irrigation systems over costs, performance and impact.

Development scholars and practitioners are keen to use the comparison to justify if, and how, the new technology is able to challenge the already operating conventional systems.

Yet, the comparative approach is not problem-free. Some researchers have been criticised for taking a binary thinking over 'clean solar versus dirty diesel' or 'efficient solar versus inefficient diesel' (Chandel et al. 2017). The explicit biases towards PVPs could sometimes affect the quality of the analyses. To address the limitations of the research design, there is a need to conduct a systematic literature review and use a set of objective indicators to measure the effectiveness of PVPs. This evidence-based work is particularly keen to examine scientific research and literature, based on three major criteria of comparison: (1) the short-term and long-term impact; (2) the quantitative and qualitative changes of people's lives and the environment before and after the interventions; (3) the differences between the control and treatment groups.

To analyse the impact on poverty reduction, the following criteria will be taken into account:

- farmers' incomes
- farmers' adaptive capacity
- farmers' social capital
- access and affordability
- opportunity of participation (in terms of decision-making power and ability to negotiate)
- access to resources, such as finance
- food security
- time-saving
- fuel-saving
- crop yields
- crop diversification
- cropping intensity
- impact on health
- government subsidies on fuel.

Over the environmental impact, the analytical criteria are:

- ground water level
- 'water-energy' efficiency
- carrying capacity of land
- noise and air pollution
- borehole contamination
- CO_2 emissions.

To evaluate the 'pro-women' claim, the following criteria will be adopted:

- women's incomes
- women's decision-making power
- access or control to resources
- capability of participation.

Theoretically, this paper pays attention, not simply to technological, but also to structural and political aspects around irrigation systems. The access and affordability issues reflect the decision-making and negotiation power of technology users. The inclusion and exclusion of certain technologies are related to the (un) even distribution of costs and benefits between groups.

In conducting the literature review, this paper has encountered several methodological challenges. As mentioned before, solar-powered irrigation systems are a generic term, which comprise different systems, such as drip and sprinkler irrigations. Their impact on farmers' livelihoods and the environment could be different. The diverse delivery models of PVPs have also made the comparison difficult. For example, the research by GGGI (2017) suggests that PVPs in East India tend to adopt the 'service-based delivery model with a capital subsidy scheme', whereas the 'grid-connected buy-back scheme and solar cooperative model' is more common in West India (p. 2). While the adoption of different management models is related to water availability of different regions, how each model is comparable with the diesel systems requires a more robust research design.

Another challenge is that the majority of comparative studies are conducted in a snapshot, rather than longitudinally. The longer-term impact is less well-documented. Additionally, a fair and scientific comparison between solar and liquid-based systems should be based on similar circumstances and contexts, such as crop types, planting size, environmental conditions and water quantity and quality (Maurya et al. 2015). Yet, this essential information is not often available in literature.

4 Pro-poor Analysis

The World Bank claims that PVPs are a 'lower-cost' option which enhances 'geographical equity in access to modern irrigation' (EEW 2016: 12). How true is this claim? This section will focus on four particular aspects: changes over productivity and incomes, the life-cycle cost analysis, farmers' adaptive capacity, and social capital building.

4.1 Changes Over Productivity and Incomes

The majority of the literature have suggested that farmers switching to PVPs have increased the crop production. Kishore et al. (2017), for example, stress that farmers in Bihar have witnessed a 9–10% increase in productivity after gaining access to irrigation from solar pumps. The rising yields, according to the literature, are the consequences of the expanding irrigated areas, multiple cropping and increasing cropping intensity. EEW (2016) finds out that PVPs have enabled farmers in India to increase the cropping intensity, from previously one cropping to three cropping a year.

The Ruti irrigation scheme in Gutu District, Zimbabwe, sponsored by Oxfam (2015), has replaced the old gravity-fed irrigation method with PVPs. The study suggests that, after the introduction of PVPs, 60 ha of land were irrigated and three harvests a year were materialised. The project has benefited 270 farmers and the production of maize has been improved to four to five tonnes per hectare. While the household incomes were all increased amongst the participant farmers, Oxfam has found out that the PVP project has benefitted the very poor group most by rising their household incomes by 286%, whereas the other two groups, the poor and the middle-income were recorded an increase by 173 and 47% respectively.

Research by Burney and Naylor (2012) also support the claim that PVPs help increase farmers' incomes. It found out that farmers in Northern Benin involved with PVPs earned US$0.69 more than their non-PVP counterparts on a daily basis. The rising productivity and higher incomes have significant implications for poverty alleviation. UNEP (2012) has suggested that a 10% increase in farm productivity would reduce poverty by 5 and 7% in Asia and Africa respectively.

An unintended, but positive, consequence of the adoption of PVPs is the rising crop diversity. Rather than focusing on staple food, solar-pumped irrigations have helped farmers increase both of the production and consumption of fruit and vegetables. Based on the solar market garden project in Northern Benin, Alaofe et al. (2016) find out that the PVP-involved households have increased their vegetable production. Similarly, Oxfam's Ruti irrigation scheme (2015) has recorded farmers' changing cultivation practices by making rotation between food crops and cash crops, such as potatoes, tomatoes and sugar beans, on the same fields. Growing and consuming more nutritious cash crops have significant gender and health implications since women play a key role in domestic food preparation. More fruit and vegetable intake would reduce the chance of micro-nutrient deficiency.

The sales of surplus vegetables and tomatoes to local markets, according to SELF (2008), have enabled poor families to buy staples and protein-rich food, especially in dry seasons. Burney and Naylor (2012) have made a similar conclusion on annual and seasonal food security. They have suggested that PVP-involved farmers, being able to purchase more food in dry seasons, have reduced the household food insecurity score by 17%.

Yet, not all research share the same conclusion about the links between the adoption of PVPs and the increase of crop diversification. Burney and Naylor (2012) has stressed that their research in Ghana and Zimbabwe could not find any evidence to support the claim that the introduction of PVPs automatically leads to more higher-value crop cultivation. They believe that changing cropping strategies and patterns are related to perceived risks amongst irrigation users.

The effectiveness of PVPs is affected by the seasonal change of radiation. In their research in Punjab, Pakistan, Naseem and Imran (2016) have found that the annual crop productivity was recorded a 5% increase when compared to the previous year after switching to PVPs. However, since the solar systems were less effective in winter, it caused dissatisfaction amongst some participants. This example has, therefore, shown that how to manage expectations and how well the participants are informed about the seasonal impact of PVPs are equally important.

Whether it is a good idea to increase crop intensity and farm productivity, induced by PVPs, also depends on local ecological contexts. Without paying sufficient attention to annual water recharge rates, groundwater levels and soil characteristics, Sarkar (2011) warns that short-term benefits could lead to long-term problems, such as waterlogging, salinization and land degradation. Citing a case study in China, EEW (2016) stresses that the long-term costs of land degradation and the depletion of water supplies are disproportionately borne by resource-poor farmers.

4.2 *Cost Comparison*

Another approach to the pro-poor analysis is to compare the costs and benefits of PVPs with their diesel- and electricity-based counterparts.[1] The majority of the literature seems to reach a consensus that PVPs are cheaper than the other irrigation systems in the long-run, and hence pro-poor in nature.

One of the most common comparative analyses is life-cycle cost analysis (LCC). LCC measures the net present values by comparing the total costs and total benefits, normally over a 20–25 years period (IRENA 2016). In terms of water pumping systems, the cost components include initial capital costs (such as costs of transport and installation costs), operational and recurring costs (such as fuel expenditure, operational costs, maintenance services and replacement costs and labour). The total benefit components include fuel savings and yield increases. The purposes of conducting the LCC analysis are to examine which system is the cheapest in the short- and long-term and to calculate the payback time of investment.

Our study has indicated that most of the literature have suggested that solar is the cheapest option when compared to other three types of fuels: diesel, electricity and gasoline. The LCC analysis by GGGI (2017) suggests that the total cost of PVPs is 64.2% of the 10-year life cycle cost of diesel water pumps. Similarly, Emcon (2006) finds that the LCC of PVPs are 20% lower than that of their diesel counterparts. The study by GIZ (2013), based in India, also indicates that the LCC of a 746 W diesel-based pumping system for a ten year period is nearly 36% higher than that of a PV-powered system of the same capacity. This means that the investment payment for PVPs is between 4 and 6 years (Chandel et al. 2015).

Based on the irrigation costs, in term of INR/m^3, through 5HP capacity pump over a 25-year period, EEW (2016) indicates that solar is much lower than diesel—0.18 and 0.32 respectively. Comparing solar and gasoline, a Guilan province-based study by Niajalili et al. (2017) shows that the LCC of gasoline-based pumping systems are 1.56 times of their PVP counterparts. In terms of the costs of solar and

[1]Some studies take a different perspective, comparing and contrasting the costs amongst solar-based irrigation systems. For example, Ali (2018) compares the initial cost and the levelised energy costs of PVPs, CDPs (concentrating disk pumps) and PTP (parabolic trough pumps) in Sudan and suggests that PVPs are the cheapest.

grid-based electricity, EEW (2016) show that solar is slightly cheaper than electricity in term of irrigation costs—0.18 and 0.19 INR/m^3 respectively.

All these studies agree that the up-front costs of PVPs are much higher than that of their diesel and electricity counterparts. Yet, the overall costs of solar panels have been decreasing. Solar does not require fuel, and the operational and maintenance costs of PVPs are neglectable. In contrast, EEW (2016) underlines three major problems of the diesel-based systems: volatile diesel prices, unreliable supply of diesel and complicated maintenance support, including replacement of parts and oil change and filter replacements (SELF 2008).

However, not all LCC studies support the superiority of PVPs. EEW (2016), for instance, shows that solar is actually less cost-effective than electricity-based pumps. Based on their LCC analysis in Namibia, SELF (2008) stresses that solar is cost-effective only in small- and medium-sized wells. The LCC of the big-sized wells is yet conclusive.

Another concern is about how LCC is actually conducted. To make reliable comparison, LCC researchers need to draw on similar criteria and conditions, such as system size and capacity, the depth of wells and the quantity of water being pumped up. However, Emcon (2006) highlights the methodological and practical challenges. For instance, he points out the significant performance differences between low and high quality diesel engines. The diesel maintenance costs between major and minor services also vary. Similarly, the efficiency of PVPs depends on how they are connected. Without batteries, PV modules power pumps directly, which save the maintenance costs of batteries. In calculating the actual and potential benefits, the PVP users may raise their revenues by increasing crop intensity and productivity, as indicated in the previous section. Yet, Hossain et al. (2015) caution that the benefits depend on the choice of crops and cultivation patterns. For example, in their case study of Bangladesh, PVPs are economically viable for growing tomatoes, brinjal and wheat crops, but not rice.

Scholars also question the rational choice assumptions underlying LCC—energy users will switch to cheaper technological options because of lower fuel costs and higher efficiency. Yet, costs is only one of many factors in energy and fuel switch. Social acceptance and cultural contexts, for example, play equally significant roles in shaping the transition.

Another criticism about the LCC analysis is about the confusion of cost and affordability. Many LCC scholars have suggested that PVPs are lower in LCC and they will bring long-term benefits to poor farmers. Yet, the initial costs of PVPs are much higher than that of their diesel counterparts. For example, the research by Niajalili et al. (2017) in Iran finds out that the up-front costs of PVPs are seven times of that of diesels. It will take nine years for both systems to be equal in total costs. The high initial costs deter poor farmers from using PVPs. Without addressing the access issues, poor farmers are not able to enjoy the long-term gains, suggested by the LCC analyses.

Heavy subsidies from governments on grid-connected electric and diesel-based pumps have been criticised for creating an unfair playing field to PVPs since the subsidies have distorted the actual costs of the technology. Yet, in many cases,

PVPs also receive subsidies from the governments and development agencies, but the figures are not often clearly documented. Another challenge for LCC analysts is that the reduction of CO_2 emissions by switching diesel irrigation to PVPs would bring additional incomes to governments because of the carbon credit markets. However, this additional benefit is rarely mentioned in the LCC analysis.

4.3 Adaptive Capability and Social Capital Building

PVPs are often praised in the literature for enhancing poor farmers' adaptive capacity to tackle climate change and for making social capital to build trust and mutual support. Adaptive capacity is intended to improve farmers' resilience against erratic climatic patterns brought by climate change (Colback 2015). Social capital making fosters mutual learning of farmers based on social network building (Badenoch 2009). These two goals and impact are usually achieved through collective- or community-based solar pump irrigation models. In the case studies of Dhundi village in Gujarat, India, for example, the International Water Management Institute pulls six individual solar-pumped irrigators together and helps them form the Solar Pump Irrigators' Cooperative Enterprise (IWMI 2015). Through the mini-grid solar system and selling surplus power to the national grids, IWMI stresses that all participants are better off.

This rosy picture has, however, been challenged for who the actual participants are. Burney and Naylor (2012) suggest that successful PV adapters are usually not the poorest agricultural households. In their project in treadle pumps in Ghana, they compare and contrast 108 individual farmers (52 adopters vs. 56 non-adopters). Their study suggests that, although the adopters of the treadle pumps receive higher incomes that the non-adopter counterparts, they are more likely to have 'lower dependence ratios in their households' and have 'greater access to extension services' (p. 114). Similarly, asking who are more likely to adopt PVPs for irrigation, Ali (2018) find out that it is educated, young and wealthier Pakistan farmers in their case study who are more likely to embrace PVPs. Research by EEW (2016) and Kishore et al. in Rajasthan, India (2014) also reach similar conclusions about medium- to large-farm owners are more likely to install state-subsidised PVPs because of their knowledge and networks.

Regarding social capital building, Wong (2012) warns that switching to solar-based systems could, unwittingly, destroy poor people's social capital. Using rural villages in Bangladesh and India as examples, he suggests that solar home lighting systems could only benefit those who are in the same room. In contrast, diesel, kept in liquid, could be shared with more villagers who have plastic containers to keep diesel. Furthermore, how to keep the solar systems safe could easily become a security issue (Emcon 2006). Unlikely diesel pumps which can be kept indoors, solar panels are installed on roof. Apart from theft, physical damage by strong wind could bring additional stress to villagers.

The PVP literature has also mentioned the positive impact of PVPs on education and time saving. For example, Burney and Naylor (2012) examine the impact of PVPs in school attainment in their Ghanaian research and find out that villages having the PVP installation have recorded higher rates in school attainment. Yet, they have not provided explanations for the changes. Similarly, PVPs have also been acclaimed to free people from hand watering and walking long distance to fetch water. Yet, not many ethnographic studies have been conducted to examine the impact of the gain of time-saving on people's actual livelihoods.

In a nutshell, this section has demonstrated that the evidence to support the pro-poor claims, associated with PVPs, is mixed, and sometimes with contradictory results.

5 Pro-environmental Analysis

In this section, we will focus on four particular aspects: CO_2 emissions, groundwater extraction and depletion, soil conservation and land degradation, and e-waste. The general consensus in literature is that replacing diesel-based pumps by PVPs helps minimise CO_2 emissions. Yet, PVPs may not necessarily have lower environmental footprints than their diesel counterparts because of a lack of incentives in reducing groundwater over-extraction and the potential of e-waste.

5.1 CO_2 Emissions

The agricultural sector has been considered a big CO_2 emitter. 50 to 70% of total emissions are generated from energy activities in the agricultural sector (IRENA 2016). Numerous research, such as GGGI (2017) and Burney et al. (2010), has been conducted to compare and contrast the CO_2 emissions of different irrigation pumps. For example, an irrigation pump with 3.73 kW capacity, powered by solar, grid-electricity and diesel and run for 1250 h annually, the study by Jain et al. (2013) finds out that their annual CO_2 emissions are 0, 4 and 5.2 tonnes respectively. Based on this research, these authors estimate that reducing 50% of existing 10 million sets of diesel pumps in India would help cut 26 million tonnes of CO_2 emissions annually. Similarly, installing 50,000 PVPs in Bangladesh would help save 450 million litres of diesel.

5.2 Groundwater Depletion and Over-Extraction

Irrigation can affect groundwater cycle, especially the groundwater recharge. Research by Casey (2013) underlines the global retreat of water table by 0.3 metres annually. Improving water-energy efficiency, defined as 'more crop per drop or per kilowatt-hour' (IRENA 2016: 9), by better performed PVPs, could worsen the problems. As EEW (2016) explains, since the operational costs of PVPs are nearly zero, once installed, there is little incentive for farmers to conserve water. Without adequate groundwater recharge, it would lead to unsustainable water consumption, especially in arid and semi-arid regions.

The over-withdrawal problem could be worsened by the guaranteed buy-back schemes promoted by some NGOs. To conserve energy, surplus energy generated by PVPs is sold to national grid. One case study by GGGI (2017) illustrates that farmers in Gujarat, India, make US$900 profit for a farm as big as one hectare with a 7.5 kW solar pump. The fee-in tariff, according to the World Bank (2015a), turns PVP farmers to 'micro-level independent power producers' (p. 16). However, the short-term monetary gains become pervasive incentives for them to maximise water withdrawal, without paying sufficient attention to the long-term environmental sustainability. Worse still, the tariff schemes may only benefit rich farmers owning more PVPs with higher capacity. To address these problems, using remote surveillance technology to monitor farmers' water usage and strengthening the power of groundwater regulators have both been proposed (El-kader and El-Basioni 2013).

Yet, the issues of over-extraction are context-specific. Comparing East and West India, GGGI (2017) discovers that a controlled underground water extraction could be beneficial, especially to East India. Being a flood-prone region, PVPs may help lower underground water table and produce porous alluvial aquifers. They both reduce surface water runoff and risk of flooding.

5.3 Land Degradation and Soil Conservation

Irrigation can reduce soil erosion and slow down land degradation by raising carrying capacity of grasslands. Xu et al. (2013), for example, have managed to use solar pump irrigation to cover a total area of 1.73 million hm^2 in Inner Mongolia. However, concerns have been expressed about the negative impact of the drop of groundwater level on the dynamic equilibrium of plateau permafrost. The thinning of permafrost in Qinghai could result in land subsidence (Yu et al. 2011).

5.4 E-waste Pollution

PVPs, unlike diesel pumps, minimise the danger of borehole contamination. However, an improper disposal of contaminating solar items, such as poly-silicon and cadmium telluride-based thin-film solar cells, could have significant negative impact on the environment. Dissimilar to developed countries, such as EU, which has got the Waste Electrical and Electronic Equipment Directive and the Restriction of Hazardous Substances Directive in place for PV recollection and decommissioning process, developing countries are weaker in regulation. How to enforce appropriate regulations of recycling in developing countries to avoid e-waste pollution is of some concern (Shamim et al. 2015).

6 Pro-women Analysis

Women play a significant role in agriculture in developing countries. The migration of men from rural to urban areas leaves women in charge of the farming operation. Yet, women-headed households are not only statistically poorer, but they are also often socially excluded from irrigation-related activities and community meetings because canal irrigation is considered a man's job (Nagrath et al. 2016). Additionally, women are crucial for domestic water supply, food preparation and family hygiene. Their decision-making power is, however, often constrained by rigid gendered division of labour and restrictive cultural norms.

In the light of all these cultural barriers, some development practitioners have been asking if the interventions of PVPs could offer an opportunity to challenge the gendered inequalities by reducing women's workload in water collection, and simultaneously, empowering women in participating in irrigation and other communal matters.

Literature provides strong evidence to support the claim that PVPs could substantially save women both time and energy in water collection for food production. For instance, the Oxfam-sponsored project in Zimbabwe suggests that, in order to irrigate their gardens, women walk 4 km on average on a daily basis to collect water from nearby dams. Research by SELF (2015) shows that the Solar Electric Fund helped 45 women form a co-operative in Benin in 2007. The installation of the PVPs have saved each women up to four hours per day.

In addressing women's poverty issues, quite a few PVP interventions target particularly women and set up women groups. As mentioned in the previous sections, the Solar Market Garden project in Northern Benin succeeds in raising the overall vegetable production amongst poor households. Evaluating the gendered impact of the project, Alaofe et al. (2016) discover that the project has made more positive impact on female participants. For instance, the women group increased their production of fruit and vegetable crops by 26%, whereas only 13.4% was achieved in non-women groups. Similarly, women involved with this project were three times more likely to increase their fruit and vegetable consumption when compared to non-women groups. The increased consumption of vegetables and fruit during the dry season amongst women participants is particularly crucial to the well-being of their families because the improved access to, and consumption of, nutrition-rich food help reduce under-nutrition of their children. IRENA (2016) therefore makes an optimistic remark, suggesting that: 'these experiences illustrate the disproportionately greater benefits of solar pumping solutions for women' (p. 15).

There is also evidence to show that the involvement of women in PVPs has not only raised women's incomes, but also increased their control of money. Drawing on the PVP project by the International Crops Research Institute for the Semi-Arid Tropics in Burkina Faso, which helped 20 poor pilot farmers, including five women, Belemvire (2007) found out that the women group earned US$1.16 per person on a daily basis, when compared to US$0.98 for the non-women groups. He explained the differences for the rise of vegetable production of the women group. The additional incomes have also helped women diversify their family diet by buying rice, sorghum and beans as well as slightly expensive food items, such as fish and cooking oil.

Being asked about the control of money, 52% of female participants in the Solar Market Garden project in Northern Benin, mentioned before, have reported that they had control over their garden incomes (Alaofe et al. 2016: 117). 57% of them have also suggested that they would spend more incomes on food, 54% on health care, 43% on gas, water and telephone, and 25% on education. The women in the project, who are poor and uneducated, have also expressed a sense of empowerment by being able to write something and make simple calculation after the engagement. Equally important, the project has also helped women secure civil documentation for their lands and gain access to numerous financial institutions.

Despite the positive impact, some academics are not certain whether the changes, brought by PVPs, really help challenge the unequal gendered structures. PVPs help reduce women's time and energy in water collection. However, there is a lack of in-depth research to examine whether it has liberated women from gender stereotypes, or simply shifted their workload to other domestic chores. While women's role in improving the well-being of their families is undeniable, there are concerns that PVPs could be used as an instrumental tool to reinforce, rather than challenge, women's gendered roles and positions. For instance, in their research in Sub-Saharan Africa, Burney and Naylor (2012) discovered that PVPs might have helped women increase their farming production, but the female members deliberately under-reported their actual production since they feared that the money they earned would be taken away or upset the family and communal harmony (p. 120).

The question about which groups of women actually benefit from PVPs is related to the distribution issues of benefits, costs and power. A study by Gugerty and Kremer (2008) in Kenya suggests that PVP-related projects tend to attract women who are younger, wealthier and better educated. Once the projects become well-established and financially sound, they warn that the older and less productive female members are likely to be expelled. The competition between different women groups has demonstrated that, without addressing the unequal social hierarchies and cultural practices, the interventions of PVPs might simply reinforce the subordinated positions of the weaker and less-educated women.

Similarly, increasing women participation in collective irrigation groups does not automatically increase their decision-making power. Using irrigation in rural India as an example, Girard (2014) suggests that institutionalising women in irrigation management could increase their visibility in formal meetings. Yet, without genuine empowerment, women are less likely to speak out in public arena because their involvement is simply considered 'an extension of their domestic duties' (p. 1).

7 Conclusions and Policy Implications

This paper has conducted a systematic and evidence-based literature review to examine if, and how, PVPs are effective in tackling three burning, and related, issues in developing countries, which are: poverty reduction, environmental conservation and gender empowerment. This review has shown a very mixed and complex picture. The PVP literature have reached two general consensus: PVPs help strengthen poor farmers' adaptive capacities by raising their agricultural productivity, improving their household incomes and building social capital. They also help mitigate climate change by replacing diesel with solar energy to reduce CO_2 emissions. There is also some evidence to support the positive gendered impact of PVPs on raising women's control of income and diversifying diets, and that improves the overall well-being of their families.

Some literature has, however, highlighted the problematic assumptions underlying the promotion of PVPs which focus on economic rationality and informed choice. Multiple and complex factors, affecting energy switch and the preferences for technology, have not been paid sufficient attention. The assumed even distribution of costs and benefits, particularly between men and women, has also been questioned.

The environmental debate around PVPs has been focused too much on CO_2 emissions. Yet, the rising energy-water efficiency, because of the technical improvement of PVPs, has resulted in undesirable environmental trade-offs, such as the depletion of underground water and e-wastes. From the gender perspective, the success in using PVPs to achieve gender empowerment hinges on how the structural power inequalities, between men and women, are understood and tackled. This literature review has demonstrated that many PVP projects may be successful in helping women groups raise their farming productivity and increase their. Nevertheless, some PVP interventions may, unwittingly, help reinforce gendered stereotypes. Women do not feel empowered in the process of participation in the PVP projects.

This book chapter suggests that more in-depth, longitudinal and mixed quantitative-qualitative research is needed in order to capture the long-term and complex impact of PVPs. More attention should be paid to the trade-offs between the impact on poverty reduction, environmental conservation and gender equality.

One of the major limitations of this book chapter is an inadequate analysis of the impact of governance in enabling, and constraining, PVP improvement. Governance is rules governing access to, and use of, water, energy and technology (Badenoch 2009). The World Bank (2016) has proposed several institutional changes to address the PVP-related issues, such as social acceptance, affordability and access. How effective the institutional strengthening is requires more robust research.

References

Alaofe H, Burney J, Naylor R, Taren D (2016) Solar-powered drip irrigation impacts on crops production diversity and dietary diversity in Northern Benin. Food Nutr Bull 37(2):164–175

Alemaw B, Simalenga T (2015) Climate change impacts and adaptation in rainfed farming systems: a modeling framework for scaling-out climate smart agriculture in Sub-Saharan Africa. Am J Clim Change 4:313–329

Ali B (2018) Comparative assessment of the feasibility for solar irrigation pumps in Sudan. Renew Sustain Energy Rev 81:413–420

Badenoch N (2009) The politics of PVC: technology and institutions in upper water management in Northern Thailand. Water Altern 2(2):269–288

Belemvire A (2007) Atelier de Restitution de l'E' valuation du Programme Pilote de De′ veloppement de l'Irrigation Goutte a' Goutte dans le Nord et de Refle′xion sur la Seconde Phase du Programme. International Crops Research Institute for the Semi-Arid Tropics, Ouahigouya, Burkina Faso

Biggs E, Bruce E, Boruff B, Duncan J, Horseley J, Pauli N, McNeill K, Neef A, van Ogtrop F, Curnow J, Haworth B, Duce S, Imanari Y (2015) Sustainable development and the water-energy-food nexus: a perspective on livelihoods. Environ Sci Policy 54:389–397

Burney J, Naylor R (2012) Smallholder irrigation as a poverty alleviation tool in Sub-Saharan Africa. World Dev 40(1):110–123

Burney J, Woltering L, Burke M, Naylor R, Pasternak D (2010) Solar-powered drip irrigation enhances food security in the Sudano-Sahel. Proc Nat Acad Sci 107(5):1848–1853

Casey A (2013) Reforming energy subsidies could curb India's water stress. Worldwatch Institute, Washington

Chandel S, Naik N, Chandel R (2017) Review of performance studies of direct coupled photovoltaic water pumping systems and case study. Renew Sustain Energy Rev 76:163–175

Chandel S, Naik N, Chandel R (2015) Review of solar photovoltaic water pumping system technology for irrigation and community drinking water supplies. Renew Sustain Energy Rev 49:1084–1099

Colback R (2015) The case for solar water pumps. The Water Blog, World Bank

EEW (Council on Energy, Environment and Water) (2016) Sustainability of solar-based irrigation in India. Key determinants, challenges and solutions. CEEW working paper, November (prepared by Shalu Agrawal and Abhishek Jain)

El-kader S, El-Basioni B (2013) Precision farming solution in Egypt using the wireless sensor network technology. Egypt Inf J 14(3):221–233

EMCON Consulting (2006) Feasibility assessment for the replacement of diesel water pumps with solar water pumps. Final report for Ministry of Mines and Energy, Namibia

FAO (Food and Agriculture Organisation) (2011) The state of the world's land and water resources for food and agriculture. Rome

Girard A (2014) Stepping into formal politics. Women's engagement in formal political processes in irrigation in rural India. World Dev 57:1–18

GIZ (Deutche Gasellschaft fur Internationale Zusammenarbbeit) (2013) Solar water pumping for irrigation: opportunities in Bihar

GGGI (Global Green Growth Institute) (2017) Solar-powered irrigation pumps in India—capital subsidy policies and the water-energy efficiency nexus. Republic of Korea

Gugerty MK, Kremer M (2008) Outside funding and the dynamics of participation in community associations. Am J Polit Sci 52(3):585–602

Hossain M, Hassan M, Mottalib M, Hossain M (2015) Feasibility of solar pump for sustainable irrigation in Bangladesh. Int J Energy Environ Eng 6(2):147–155

IRENA (International Renewable Energy Agency) (2016) Solar pumping for irrigation: improving livelihoods and sustainability. IRENA Policy Brief

Islam M, Sarker P, Ghosh S (2017) Prospect and advancement of solar irrigation in Bangladesh: a review. Renew Sustain Energy Rev 77:406–422

IWMI (International Water Management Institute) (2015) Payday for India's first ever sunshine farmer: could new scheme conserve water, boost incomes and combat climate change?
Jain R, Choudhury P, Palakshappa R, Ghosh A, Padigala B, Panwar TS, Worah S (2013) Renewables beyond electricity. New Delhi
Kishore A, Tewari N, Shah T (2014) Solar irrigation pumps: farmers' experience and state policy in Rajasthan. Econ Polit Wkly 49:55–62
Kishore A, Joshi P, Pandey D (2017) Harnessing the sun for an evergreen revolution: a study of solar-powered irrigation in Bihar, India. Water Int 42:291–307
Maurya V, Ogubazghi G, Misra B, Maurya A, Arora D (2015) Scope and review of photovoltaic solar water pumping system as a sustainable solution enhancing water use efficiency in irrigation. Am J Biol Environ Stat 1(1):1–8
Mendelsohn R (2008) The impact of climate change on agriculture in developing countries. J Nat Res Policy Res 1(1):5–19
Nagrath A, Chaudhry A, Giordano Mark (2016) Collective action in decentralized irrigation systems: evidence from Pakistan. World Dev 84:282–298
Naseem Z, Imran S (2016) Assessing the viability of solar water pumps economically, socially and environmentally in Soan Valley, Punjab. Int J Environ Chem Ecol Geol Geophys Eng 10 (6):711–716
Niajalili M, Mayeli P, Naghashzadegan M, Poshtiri A (2017) Techno-economic feasibility of off-grid solar irrigation for a rice paddy in Guilan province in Iran: a case study. Sol Energy 150:546–557
Oxfam (2015) Transforming lives in Zimbabwe. Rural sustainable energy development project. Prepared by John Magrath
Sarkar A (2011) Socio-economic implications of depleting groundwater resource in Punjab: a comparative analysis of different irrigation systems. Econ Polit Wkly 46(7):59–66
SELF (Solar Electric Light Fund) (2008) A cost and reliability comparison between solar and diesel powered pumps
SELF (Solar Electric Light Fund) (2015) Solar electric light fund annual report
Shamim A, Mursheda A, Rafiq I (2015) E-waste trading impact on public health and ecosystem services in developing countries. J Waste Resour 5(4):1–18
Shinde V, Wandre S (2015) Solar photovoltaic water pumping system for irrigation: a review. Afr J Agric Res 10(22):2267–2273
UNEP (United Nations Environment Programme) (2012) Sustainable consumption and production for poverty alleviation. Nairobi, Kenya
Wong S (2012) Overcoming obstacles against effective solar lighting interventions in South Asia. Energy Policy 40:110–120
World Bank (2015a) Solar water pumping. Ready for mainstreaming? Powerpoint presentation by Kris Welsien and Richard Hosier on 2nd December
World Bank (2015b) Solar-powered pumps reduce irrigation costs in Bangladesh. The World Bank, 8 Sept 2015
World Bank (2016) Solar program brings electricity to off-the-grid rural areas in Bangladesh
Xu H, Liu J, Qin D, Gao X, Yan J (2013) Feasibility analysis of solar irrigation system for pastures conservation in a demonstration area in Inner Mongolia. Appl Energy 112:697–702
Yu Y, Liu J, Wang H, Liu M (2011) Assess the potential of solar irrigation systems for sustaining pasture lands in arid regions—a case study in Northwestern China. Appl Energy 88:3176–3182

Livestock Technologies and Grazing Land Management Options for Climate Change Adaption and Mitigation as a Contribution for Food Security in Ethiopia: A Brief Overview

Shigdaf Mekuriaw, Alemayehu Mengistu and Firew Tegegne

Abstract African countries, like Ethiopia, are particularly vulnerable to climate change because their economies largely depend on climate-sensitive agricultural production. Growth and Transformation Plan (GTP) of Ethiopia recognized climate change as a huge threat and focusing on mitigation issues. The GTP stipulates the country's ambition to build a climate resilient green economy by 2030. This paper looks at the potential of livestock technologies and grazing land management for mitigation and adaption to a changing climate. Research findings in Ethiopia showed that livestock technologies and management of grazing lands such as improving the quality of forage, feeding highly digestible forages, processing and preservation of feeds, use of controlled grazing instead of continuous grazing and inclusion of legumes in forage mixes, have a great response to climate change. The choices for application of the technologies and potential mitigation strategies primarily depend on the adoption and cost associated with it. Grazing land management not only mitigates climate change but also reduces soil erosion, increases carbon sequestration and contributes to the resilience of crop-livestock farming systems in Ethiopia. In conclusion, management of grazing lands and implementation of livestock technologies have good implications in mitigating climate change on top of income generation and thereby improving the livelihood of farmers in Ethiopia.

S. Mekuriaw (✉)
United Graduate School of Agricultural Sciences (UGSAS), Tottori University, 1390 Hamasaka, Tottori, Tottori Prefecture 680-0001, Japan
e-mail: shigdaf@gmail.com

S. Mekuriaw
Amhara Region Agricultural Research Institute, Andassa Livestock Research Center, P.O.Box 27, Bahir Dar, Ethiopia

A. Mengistu
Pasture and Range Scientist, Urael Branch, P.O. Box, 62291 Addis Ababa, Ethiopia

F. Tegegne
College of Agriculture and Environmental Sciences, Bahir Dar University, P.O. Box 79, Bahir Dar, Ethiopia

P. Castro et al. (eds.), *Climate Change-Resilient Agriculture and Agroforestry*, Climate Change Management, https://doi.org/10.1007/978-3-319-75004-0_22

Keywords Livestock technologies · Climate change · Grazing land
Methane · Carbon sequestration · Drought and food security

1 Introduction

Climate change is widely considered to be one of the greatest challenges to modern human civilization that has profound socio-economic and environmental impacts (Ahmed et al. 2013; Khatri-Chhetria et al. 2017). Climate change is the common phenomenon worldwide. Ethiopia is among the most vulnerable countries in Africa as its economy heavily depends on subsistence rain-fed agriculture within a fragile highland ecosystem, which has been threatened by population pressure and land degradation (Somorin 2010). Historically it has been portrayed as a food deficit country with its people and animals suffering from recurrent droughts and floods. The famine that followed the 1984 droughts, which caused the death of up to a million persons and the 2006 catastrophic flood in Dire Dawa are crucial examples (Takele and Gebretsidik 2015). Climate change would affect particularly the economies of the rural areas where people are more dependent on livestock, fisheries and agriculture related activities for their livelihoods (Urquhart 2009).

Livestock in Ethiopia are extremely important as they serve a wide variety of functions in society from social to subsistence purposes (Kassahun et al. 2008). The economic importance of livestock provided more than 45% of agricultural Gross Domestic Product (GDP) in 2008–09 (Behnke 2010). With 60–70% of the population's livelihoods dependent on livestock in one way or another, livestock provide both food and income (Kimball 2011, unpublished). For many small-holder farmers livestock provide draught animal power, transportation and manure for fertilizing croplands. Livestock are also socially and culturally important in Africa for payment of dowry, celebrations and gifts to family members and also as a source of savings that is often safer, despite diseases and drought, than banking systems and easier to manage for farmers living in remote areas (Abela 2005). Throughout their long history, Ethiopians have constantly relied on livestock in order to survive. As the oldest form of assets in Ethiopia, cattle and other types of livestock have traditionally and still today serve as a significant indicator of wealth. Ethiopia is generally recognized to have the largest population of livestock of any other African nation (Halderman 2004). Ethiopia's dependency on livestock has in turn created a need to expand livestock production, to help feed and support the population that is growing at tremendous rate of 2.56% as of 2010 (CRGE 2011).

The livestock production system contributes to global climate change directly through the production of methane (CH_4) from enteric fermentation and both CH_4 and nitrous oxide from manure management (CRGE 2011). Among Ethiopian livestock species, the major contributors to Green House Gas (GHG) emission are cattle. Given the extensive livestock production practices, the cattle population is likely to increase from around 53 million (CSA 2013) to more than 90 million in

2030 (CRGE 2011), thereby almost reaching the cattle carrying capacity of the country and doubling emissions from the livestock sector.

Consequently, it is essential to develop a portfolio of strategies that includes adaptation, mitigation, technological development and research to combat climate change. It is imperative for countries to take a proactive role in planning national and regional programs on adaptation to climate variability (Ahmed et al. 2013). Therefore, this paper was initiated to identify the potential livestock technologies and grazing land management options for climate change mitigation and adaptation towards sustainable livestock agriculture.

2 Effects of Climate Change on Livestock Agriculture

It is predicted that climate change will double (by 25–50%) the frequency of droughts in the dry lands of Sub-Saharan Africa by the end of the century and drought periods are likely to last for longer (Umesh et al. 2015). Climate change is, therefore, a threat to the Ethiopian agrarian economy and livelihoods of millions of the poor. The key climate hazards are flooding, drought and rainfall variability (Nachmany et al. 2015). The effect of climate change hits the poorest people first as they are more dependent on climate-sensitive livelihoods such as livestock farming. Ethiopia is one of the first victims of climate change as evidenced by the catastrophic drought of 1984–1985 (Philander 2008). The occurrence of droughts and prolonged dry seasons are directly and indirectly affecting the livestock sector. Unplanned deforestation and drought bring unavailability of feeds and sustainable water resources for livestock use. If an animal experiences heat strain, it decreases feed intake and that causes the reduction of gross production (Rowlinson 2008). The present and future impacts of different climate events on livestock require the quick response during the flood and drought for the sustainable rehabilitation. Africa's livestock sector has been affected by climate changes through more frequent catastrophic events, reduced water availability, changes in the pattern and quantity of rainfall, an increase in temperature, changes in seasonality, a decrease in feed and fodder production, changing patterns and distribution of disease and altered markets and commodity prices (Tubiello 2012).

Various studies indicated that the trends in inter-annual and inter-seasonal rainfall variability like declining in amount, increasing in intensity, varying in the length of growing seasons with increasing temperature have negative implication on crop and livestock productivity (Kassie et al. 2013; Getachew 2015). It is obvious that the availability of pasture and water for livestock is determined by climate conditions and land use change. Increasing in temperature has a negative impact on livestock productivity as warming is expected to alter the feed intake, mortality, growth, reproduction, maintenance, and production of animals (Thornton et al. 2009). The direct effects include temperature and other climate factors such as shifts in rainfall amounts and patterns on animal growth, reproduction and milk production (Thornton and Herrero 2010). While the indirect effects of climate change include influence on availability of water, the quantity and quality of animal feed such as pasture, forage, crop yield and the severity and distribution of livestock diseases and parasites (Seré et al. 2008). In Ethiopia, the deterioration of rangelands and increases in woody browses can be expected to result in an increasing number of pastoralists maintaining mixed herds of browsing animals like camels and goats with smaller numbers of cattle and sheep (Kefyalew and Tegegne 2012).

It is expected that as temperature changes, optimal growth ranges for different forage species also change; species alter their competition dynamics, forage quality (like nutrient content) and quantity, and the species composition of mixed grasslands changes (Thornton et al. 2009). Hopkins and Del Prado (2007) noted that climate change can be expected to have several impacts on feed crops and grazing systems including changes in herbage growth brought about by changes in atmospheric CO_2 concentrations and temperatures; changes in the composition of pastures, such as changes in the ratio of grasses to legumes and changes in herbage quality. Under climate change (increase in temperature) the structural constituents of plant materials such as lignin, cellulose and hemicelluloses is reported to increase (Ford et al. 1979). Crop losses due to extremes in climate (temperature and rainfall) could result in less animal feed (crop residue) being available, especially in crop-livestock systems that predominates in the Ethiopian livestock production.

3 Climate Change Mitigation and Adaptation Strategies in Ethiopia

Ethiopia has identified different adaptation options in Climate Resilient Green Economy Strategy to reduce vulnerability of the people and economy to climate change impacts (EPA 2011). The strategy outlines four pillars focusing on Agriculture with emphasis on improving crop and livestock production practices for higher food security and farmer income while reducing emissions (ECRG 2011; Nachmany et al. 2015). The Government of Ethiopia also identified 37 potential adaptation options to address highly vulnerable sectors mainly agriculture, water and health (FDRE 2006). Mitigation potentials in Ethiopian pastoral systems using improved crop and grazing land management will increase soil carbon storage, and improved livestock management has great contribution to reduce methane emissions (Steinfeld and Gerber 2010; Abebe 2017). Moreover, improving the nutritive value of low-quality feeds in ruminant diets which are much available in Ethiopia, could increase animal and herd productivity, and consequently reduce methane emission (Hristov et al. 2013).

4 Livestock Technologies for Climate Change Adaption and Mitigation

The Government of Ethiopia, climate change national adaptation programme of action (NAPA 2007) report identified a range of adaption interventions that could potentially assist pastoralists to adapt to climate change. These include: additional care to avoid over-grazing and better manage stocking rates with pasture production; grazing with mixed herds of grazers and browsers; increased use of forage crops; water resource development; and the increased use of livestock feed supplementation. Khatri-Chhetria et al. (2017) also reported that technologies or interventions for reducing the emission of methane from grazing livestock include reducing livestock numbers by improving productivity, increasing the efficiency of animal production, genetic improvement, manipulation of the rumen microbial ecosystem, feed additives and improvement of farm management.

Livestock keepers have traditionally adapted to various environmental and climatic changes by building on their in-depth indigenous knowledge of the environment in which they live. However, the increasing human population, urbanization and land degradation have rendered some of those coping mechanisms ineffective (Sidahmed et al. 2008). Moreover, changes brought about by global warming are likely to happen at such a speed that they will exceed the capacity of spontaneous adaptation of both human communities and livestock species. These well-organized and knowledge-based national strategies are required to combat the effect of climate change on livestock production.

Several studies reported on how to adapt the livestock technologies to reduce the climate change impacts and action measures with the issue of sustainability of the developing countries (Demir and Bozukluhan 2012; Wang et al. 2012). Livestock keepers have traditionally been capable of adapting to threats to their livelihood. Indeed, one of the most widespread livestock systems in Ethiopia, pastoralism, has often been defined by its capacity to adapt to climatic uncertainty and other hazards. However, it is important to recognize that the outcomes of climate change are uncertain and that the way livestock keepers adapt will vary from location to location and household to household. The production and productivity among small holders is much lower than the potential that they can be increased through utilization of improved technology and practices such as; improved feeds and feeding, breed and breeding, health services, market efficiency.

4.1 Feed Technologies

Various feeding and grazing land management strategies such as rotational grazing, reseeding pasturelands, crop residues treatment with urea, forage development; efficient feed production techniques such as silage and urea molasses block preparation lead to a reduction of emissions for climate change adaptation (Bryan et al. 2011).

It can be seen that a significant factor affecting methane emissions is the animal's diet and this is subject to modification through feeding strategies particularly where the animal is fed a diet with a significant forage component (grazed or ensiled). Such approaches build on the considerable success that has been achieved in improving quality traits for animal production e.g. rye grasses with higher water soluble carbohydrate (WSC) content and increased digestibility (Abberton et al. 2007). Generally, diets of higher digestibility have these characteristics. Improving the nutritive value of the feed given to grazing animals by balancing the diet with concentrates, or by breeding improved pasture plants, should result in reduced methane emission (Ulyatt and Lassey 2005).

4.1.1 Improved Feed and Feeding Management

Literature results indicate that increasing feed utilization efficiency and improving the digestibility of feed are some of the options to reduce GHG emissions and maximize production and gross efficiency (Ominski et al. 2006). Enteric CH_4 emissions are highest when the animal is presented with poor-quality forage and has limited ability to select higher quality forage components as a result of reduced dry matter availability. On the contrary, methane production in ruminants tends to decrease with the quality of the forage fed. Boadi et al. (2004) demonstrated that forage quality has a significant impact on enteric methane emissions. Study conducted on steers indicated that CH_4 emissions of grazing steers that had access to high quality pastures declined by 50% compared to emissions from matured pastures (Oudshoorn 2009).

Feed resources in the mixed crop-livestock system in Ethiopia are mainly natural pasture grazing, crop residues and aftermath grazing, which all are characterized by poor quality (Tesfaye and Chairatanayuth 2007). So, improving the quality of these feeds is an important strategy to reduce methane emission from livestock. Changes in feeding system (e.g. roughage: concentrate ratio may reduce methane emission. Compared to forages, concentrates are usually lower in cell wall components, ferment faster than forage, giving rise to elevated levels of propionic acid. Sejian et al. (2015) suggested that CH_4 production can be lowered by almost 40% when a forage rich diet is replaced by a concentrate rich diet. Supplementation with concentrate feed especially for dairy cattle in potential dairy area is the best feeding strategy to reduce GHG emission from cattle (Bannink 2007). Feeding more concentrates to ruminants improves productivity and reduces enteric methane (even though volatile GHG in manure is increased) (Mekuriaw et al. 2014).

4.1.2 Ammonization of Roughage (Urea Treatment of Fibrous Feeds)

There is a huge amount of crop residue and native pasture with poor quality for animal feed in Ethiopia. Lack of degradable nitrogen in many fibrous residues can be corrected by supplementation with urea (ammonization). Straw ammonization technology reduces 25–75% of methane emissions per unit of animal produce-meat, milk and work; increases digestibility (by 8 to 12 points); increases nitrogen content (more than doubled); increases the intake (by 25–50%) and thus to the nutritive value (Klopfenstein et al. 1972; Pires et al. 2010; Liu et al. 2017). In addition, use of urea-molasses multi-nutrient block as supplement to fibrous feed enhance efficient rumen fermentation; improve the daily feed intake; improve body weight gain and body condition.

4.2 Genetic Improvement of Animals to Reduce Methane Emission

Ethiopia has the largest livestock population in Africa with different animal species (cattle, sheep, goat and camel). The cattle population in Ethiopia has historically grown in line with the expansion of the human population. The cattle population in mixed crop-livestock system is projected to double in 2030. Consequently, GHG emission from cattle is expected to increase. Production and productivity of Ethiopian livestock is poor compared to other countries. For example milk production potential of indigenous cattle breed is low (around 529–713 L per lactation) (EPCC 2015). However, a good first generation Holstein Friesian crossbred cow with moderate level of management can produce from 1726 to 2428 L of milk (EPCC 2015). This shows that if due emphasis is given to the effective functioning of key technological input and output markets, the cattle industry has huge potential to develop and contribute to mitigation of GHG emission. Artificial insemination (AI) and estrous synchronization are efficient technologies to deliver improved genotypes to a large number of dairy farmers in a short period of time. These techniques lead to replacing unproductive indigenous cow population with less number but more productive crossbred cows. This gives chance to farmers shifting to better management of fewer productive animals to mitigate climate change. Genetic improvement coupled with diet intensification could lead to substantial efficiency gains in livestock production and CH_4 output. This would result in fewer but more productive animals being kept, which could have positive consequences for CH_4 production.

In Ethiopian context, for methane emission mitigation option from cattle, 13% of indigenous cattle population in mixed crop livestock system will be replaced by crossbred dairy cattle which will also enhance milk yield production by 400%. This projection indicated that replacement of 13% of indigenous cattle population in dairy potential area by more productive cross-breeds will result in abatement potential equal to roughly 6 Mt CO_2e per year by 2030, assuming the indigenous and cross-breeds will emit around 1.08 and 1.5 tons CO_2e per year/head respectively. Increasing off-take rate is another strategy in decreasing GHG emission per animal; increasing socio-economic growth of the country in general and small holder farmers in particular.

4.3 Diversification Towards Lower Emitting Animal Species

Animal mix diversification has the potential to decrease the vulnerability of the herd to climate change, depending on the livestock species chosen. There are differences among livestock species in their ability to produce under given climate change. However, the economics of herd diversification, for example from cattle to small ruminant or poultry depends on the relative costs of different species and relative revenues, both under current and future climate condition. Beef is the primary meat consumed in Ethiopia, and the demand for beef is a major driver of the size of the cattle population in addition to the requirement for draught power. For this reason in Ethiopia, poultry specifically chicken meat offers a particularly attractive lower-carbon alternative to beef. Partial replacements of cattle with lower emitting species (poultry, sheep and goat, fish) would be an alternative option for mitigation of GHG emission from livestock. These low-emitting animals are high feed converters and low GHG emitters as compared to large ruminants such as cattle (Pachauri et al. 2014). Chickens are the most efficient in Ethiopia in terms of producing the most meat and protein per amount of GHG emitted (Lipson et al. 2011).

5 Sustainable Grazing Land Management Options and Carbon Sequestration

Sustainable grazing land management requires an understanding of how to use grazing to stimulate grasses to grow vigorously and develop healthy root systems, usage of the grazing process to feed livestock and soil biota, ideally maintaining 100% cover (plants and litter), and 100% of the time and provision of adequate rest from grazing without over resting areas of land. Improved grazing conditions will increase livestock productivity in rangelands, in turn increasing food security. Sustainable grazing management (often termed “holistic”) is already being used in Namibia, South Africa, the Northern Rangelands of Kenya and Ethiopia (Reed et al. 2015).

5.1 Proper Grazing Pasture Management

Proper pasture management through rotational grazing would be the most cost-effective way to mitigate GHG emissions from feed crop production. Grazing management techniques intended to increase forage production through increased perennial species have the potential to increase above and below ground soil carbon stocks, and to restore degraded dry lands (IPCC 2007). The recommended management measures and strategies for free grazing are changing grazing intensity (number of cattle per area) and grazing period (starting time and length) depending on seasonal climatic conditions. Managing grazing intensity (stocking rate, rotations and their timing), including deep-rooted fodder species and legumes in fodder crops and pastures reduce synthetic nitrogen fertilizer, optimizing nutrient allocation in spatial herd management and grazing patterns. Further improvements of the grazing pasture are possible with the introduction of indigenous and exotic grasses, legumes, and drought resistant trees that can increase the supply of forage material available to livestock. It will also become increasingly important to conserve hay and promote improved management that can enable livestock to be moved between pastures to avoid over-grazing.

5.2 Forage and Grassland Carbon Sequestration Technology

Carbon sequestration in rangelands may provide an option to capitalize on the existing environmental management practices of livestock keepers and capture additional incentives for more effective management. Carbon stocks have been found to reduce when dry lands are converted from pasture to either plantation or arable land, whilst in some cases increases in carbon stocks are seen when native forests or croplands are converted to pasture. Carbon capture is increased with improved grazing management of rangelands and that also contribute significantly to improving the local and household economy. Batjes (2004) estimated that improved management of 10% of the African grazing lands could increase soil carbon stocks by 13–28 Mt-C $year^{-1}$. The primary reason given for increased carbon emissions and loss of soil carbon sequestered on degraded rangelands is overgrazing and so eliminating or moderating grazing intensities is proposed to increase carbon sequestered on these rangelands (Conant and Paustian 2002). Improved grazing management (management that increases production), leads to an increase of soil carbon stocks by an average of 0.35 t C ha-1 $year^{-1}$ (Conant et al. 2015). In addition, introducing grass species and legumes into grazing lands can enhance carbon storage in soils (Calvosa et al. 2009).

6 The Way Forward for Policy Makers

As a country in the developing world with a substantial population growth rate, Ethiopia is struggling to feed itself. Understandably, agricultural expansion in Ethiopia is the government's top priority, according to the Federal Policy and Investment Framework from 2010–2020 documents. Paramount to Ethiopia's agricultural expansion is the livestock sector, which is estimated to account for 45% of agricultural GDP or more. Agriculture in Ethiopia is extremely vulnerable to climate change. A range of climate change scenarios and models suggest that many parts of Ethiopia are likely to experience such climate variability in the future. The farmers and local communities are the direct beneficiaries, and ultimately the enforcers, of the environmental policies seeking to mitigate the environmental impacts of livestock management in Ethiopia. Options and strategies that are cost effective and have no/minimum negative effects on livestock production hold a greater promise. This requires the government and policy makers direct manifestation and course of action. Consequently, community-based adaptation of livestock and participatory approach can be the greatest options for the policymakers to save the livestock sector. Moreover, carbon sequestration efforts help to reduce the impact of greenhouse gas emissions generated from livestock production. In order to effectively enact and monitor any potential livestock policy, there must be a prerequisite of full participation of relevant stakeholders to promote sustainable grazing land and livestock management practices. Successful livestock policy will require that all be involved with the policy making process and support the proposed measures to improve livestock productivity in order to reduce the negative externalities associated with livestock production suited for environment. Ethiopia will also need to develop a climate mitigation and verification capacity in order to benefit from the potential of livestock production for food security.

7 Conclusion and Recommendation

In terms of their contribution to GDP and household assets, livestock are crucial in Ethiopia's economy and social well-being. The damage of climate change is supposed to be more sever in the future. There are no alternatives except accessing high technology and advance knowledge for the sustainable development of the livestock sector. Climate smart grassland management systems sustainably increase productivity and resilience (adaptation), reduce greenhouse gas emissions (mitigation), and enhance food security. Reducing productivity gaps and increasing livestock production efficiency would contribute to mitigate climate change. Many different management practices can improve livestock production efficiency and reduce greenhouse gas emissions. Some of the most effective technologies include: improving grazing land management, supplementing cattle diets with needed nutrients, developing a preventive herd health program and improving genetics and reproductive efficiency. Moreover there are techniques that improve intake and digestibility of low quality feed resources which include urea treatment (ammonization), supplementing with urea-molasses-mineral blocks, and supplementing with high quality legume fodder or concentrate rations. Changes in livestock production practices such as intensification and/or integration of pasture management, introducing mixed livestock farming systems like feedlot fattening, and improved pasture grazing are some of production adjustment strategies recommended for climate change adaptation in mixed crop-livestock system. There are also technologies that have proven contributions in terms of both economic development and GHG reduction as learned from experiences of several developing countries. These technologies are compatible with the existing national development policies and strategies of Ethiopia. Therefore, the academia and policy makers need to understand applicable options as a measure for sustainable livestock management against climate change. Moreover, integration of mitigation and adaptation frameworks into sustainable free grazing management and development planning are an urgent need, especially in the developing countries like Ethiopia.

References

Abebe K (2017) Effect of climate change on nutritional supply to livestock production. Acad Res J Agri Sci Res 5(2):98–106

Abela LS (2005) The contribution of livestock production to drought vulnerability reduction in Mwingi district, Kenya. Doctoral dissertation, University of Nairobi, CEES, Kenya

Abberton MT, MacDuff JH, Marshall AH, Mike W (2007) The genetic improvement of forage grasses and legumes to reduce greenhouse gas emissions. A paper prepared by Humphreys from the plant breeding and genetics programme, Institute of Grassland and Environmental Research, Aberystwyth, United Kingdom in collaboration with the Plant Production and Protection Division, Crop and Grassland Service, of the Food and Agriculture Organization of the United Nations, Dec 2007

Adhikari U, Nejadhashemi AP, Woznicki SA (2015) Climate change and eastern Africa: a review of impact on major crops. Food and Energy Security 4 (2):110–132

Ahmed F, Alam GM, Al-Amin AQ, Hassan CHB (2013) The impact of climate changes on livestock sector: challenging experience from Bangladesh. Asian J Anim Vet Adv 8:29–40

Behnke R (2010) Trout and salmon of North America. Simon and Schuster, New York

Bannink A (2007) Feeding strategies to reduce methane loss in cattle. Rapport 34. 46 pagina's, 6 figuren, 4 tabellen. Internet http://www.asg.wur.nl/po

Batjes NH (2004) Soil carbon stocks and projected changes according to land use and management: a case study for Kenya. Soil Use Manag 20(3):350–356

Boadi D, Benchaar C, Chiquette J, Massé D (2004) Mitigation strategies to reduce enteric methane emissions from dairy cows: update review. Can J Anim Sci 84(3):319–335

Bryan E, Ringler C, Okoba B, Koo J, Herrero M, Silvestri S (2011) Agricultural management for climate change adaptation, greenhouse gas mitigation, and agricultural productivity. IFPRI Discussion Paper 01098, June 2011. Insights from Kenya Environment and Production Technology Division

Calvosa C, Chuluunbaatar D, Fara K (2009) Livestock and climate change. Livestock thematic papers tools for project design. International fund for agricultural development. Via Paolo di Dono, 44, 00142 Rome, Italy. website at www.ifad.org/lrkm/index.htm

Climate change national adaptation programme of action (NAPA) of Ethiopia (2007) National Meteorological Services Agency, Ministry of Water Resources, Federal Democratic Republic of Ethiopia, Addis Ababa

Climate resilient green economy (CRGE) (2011) Ethiopia's vision for a climate resilient green economy. Addis Ababa, Ethiopia, p 28

Central Statistical Authority (CSA) (2013) National population statistics. Federal Democratic Republic of Ethiopia, Addis Ababa

Conant RT, Paustian K (2002) Spatial variability of soil organic carbon in grasslands: implications for detecting change at different scales. Environ Pollut 116:S127–S135

Conant RT, Drijber RA, Haddix ML, Parton WJ, Paul EA, Plante AF, Six J, Steinweg JM (2015) Sensitivity of organic matter decomposition to warming varies with its quality. Glob Change Biol 14(4):868–877

Demir P, Bozukluhan K (2012) Economic losses resulting from respiratory diseases in cattle. J Anim Vet Adv 11:438–442

Environmental Protection Agency (EPA) (2011) Inventory of US greenhouse gas emissions and sinks: 1990–2009
Ethiopia's Climate Resilient Green (ECRG) (2011) Green economy strategy. Addis Ababa, Ethiopia
Ethiopian Panel on Climate Change (EPCC) (2015) First assessment report, Working Group II Agriculture and Food Security. Published by the Ethiopian Academy of Sciences
Federal Democratic Republic of Ethiopia (FDRE) (2006) Poverty reduction strategy paper annual progress. IMF Country Report No. 06/27. https://www.imf.org/external/pubs/ft/scr/2006/cr0627.pdf
Ford CW, Morrison IM, Wilson JR (1979) Temperature effects on lignin, hemicelluloses and cellulose in tropical and temperate grasses. Aust J Agri Res 30(4):621–633
Getachew T (2015) Genetic diversity and admixture analysis of Ethiopian Fat-tailed and Awassi sheep using SNP markers for designing crossbreeding schemes. Doctoral dissertation, University of Natural Resources and Life Sciences, Vienna
Halderman M (2004) The political economy of pro-poor livestock policy-making in Ethiopia. Pro-poor livestock policy initiative working paper, 19
Hristov AN, Oh J, Lee C, Meinen R, Montes F, Ott T, Firkins J, Rotz A, Dell C, Adesogan A, Yang W, Tricarico J, Kebreab E, Waghorn G, Dijkstra J, Oosting S (2013) Mitigation of greenhouse gas emissions in livestock production—a review of technical options for non-CO_2 emissions. In Gerber PJ, Henderson B, Makkar HPS (eds) FAO animal production and health paper no. 177. FAO, Rome, Italy
Hopkins A, Del Prado A (2007) Implications of climate change for grassland in Europe: impacts, adaptations and mitigation options: a review. Grass Forage Sci 62(2):118–126
Kassahun A, Snyman HA, Smit GN (2008) Impact of rangeland degradation on the pastoral production systems, livelihoods and perceptions of the Somali pastoralists in Eastern Ethiopia. J Arid Environ 72(7):1265–1281
Kassie M, Jaleta M, Shiferaw B, Mmbando F, Mekuria M (2013) Adoption of interrelated sustainable agricultural practices in smallholder systems: evidence from rural Tanzania. Technol Forecast Soc Chang 80(3):525–540
Kefyalew A, Tegegne F (2012) The effect of climate change on ruminant livestock population dynamics in Ethiopia. Bahir Dar University, college of agriculture and environmental sciences, department of animal production and technology, Bahir Dar, Ethiopia and Mizan-Tepi University, college of agriculture and natural resources, department of animal sciences, Ethiopia. Livestock Research for Rural Development
Khatri-Chhetria A, Aggarwal PK, Joshi PK, Vyasc S (2017) Farmers' prioritization of climate-smart agriculture (CSA) technologies. Agric Syst 151:184–191
Klopfenstein TJ, Krause VE, Jones MJ, Woods W (1972) Chemical treatment of low quality roughages. J Anim Sci 35(2):418–422
Lipson J et al (2011) Implications of livestock series. Evans School Policy Analysis and Research. EPAR Brief, pp 155–157
Liu Z, Liu Y, Murphy J, Maghirang R (2017) Ammonia and Methane emission factors from cattle operations expressed as losses of dietary nutrients or energy. Agriculture 7(3):16
Mekuriaw S, Tegegn F, Mengistu A (2014) A review on reduction of greenhouse gas emission from ruminants through nutritional strategies. Acad J Environ Sci 2(1):006–014. http://www.academiapublishing.org/ajes
Nachmany M, Fankhauser S, Davidová J, Kingsmill N, Landesman T, Roppongi H, Schleifer P, Setzer J, Sharman A, Singleton CS, Sundaresan J, Townshend T (2015) Climate change legislation in Ethiopia. An excerpt from the 2015 global climate legislation study a review of climate change legislation in 99 Countries. www.lse.ac.uk/GranthamInstitute/legislation/
Ominski KH, Boadi DA, Wittenberg KM (2006) Enteric methane emissions from backgrounded cattle consuming all-forage diets. Can J Anim Sci 86(3):393–400
Oudshoorn FW (2009) Innovative technology and sustainable development of organic dairy farming: the case of automatic milking systems in Denmark

Pachauri RK, Allen MR, Barros VR, Broome J, Cramer W, Christ R, Church JA, Clarke L, Dahe Q, Dasgupta P, Dubash NK (2014) Climate change 2014: synthesis report. Contribution of Working Groups I, II and III to the fifth assessment report of the Intergovernmental Panel on Climate Change, p. 151, IPCC

Philander SG (ed) (2008) Encyclopedia of global warming and climate change. Sage, Beverley Hills

Pires AJV, Carvalho GGPD, Ribeiro LSO (2010) Chemical treatment of roughage. Revista Brasileira de Zootecnia, 39:192–203

Reed MS, Stringer LC, Dougill AJ, Perkins JS, Atlhopheng JR, Mulale K, Favretto N (2015) Reorienting land degradation towards sustainable land management: linking sustainable livelihoods with ecosystem services in rangeland systems. J Environ Manage 151:472–485

Rowlinson P (2008) Adapting livestock production systems to climate change-temperate zones. In Proceedings of the livestock and global change conference, Tunisia, 25 May 2008

Seré C, van der Zijpp A, Persley G, Rege E (2008) Dynamics of livestock production systems, drivers of change and prospects for animal genetic resources. Anim Genet Resour Inf 42:3–24

Sejian V, Gaughan J, Baumgard L, Prasad C (eds) (2015) Climate change impact on livestock: adaptation and mitigation. Springer India, New Delhi

Sidahmed AE, Nefzaoui A, El-Mourid M (2008) Livestock and climate change: coping and risk management strategies for a sustainable future. In Livestock and global change, pp 27–28

Somorin OA (2010) Climate impacts, forest-dependent rural livelihoods and adaptation strategies in Africa: a review. Afr J Environ Sci Technol 4(13):903–912

Steinfeld H, Gerber P (2010) Livestock production and the global environment: consume less or produce better? Proc Natl Acad Sci 107(43):18237–18238

Takele R, Gebretsidik S (2015) Prediction of long-term pattern and its extreme event frequency of rainfall in Dire Dawa Region, Eastern Ethiopia. J Climatol Weather Forecast pp 1–15

Tesfaye A, Chairatanayuth P (2007) Management and feeding systems of crop residues: the experience of East Shoa Zone, Ethiopia. Livestock Res Rural Dev 19(3):6–12

Thornton PK, Herrero M (2010) Potential for reduced methane and carbon dioxide emissions from livestock and pasture management in the tropics. Proc Natl Acad Sci 107(46):19667–19672

Thornton PK, van de Steeg J, Notenbaert A, Herrero M (2009) The impacts of climate change on livestock and livestock systems in developing countries: a review of what we know and what we need to know. Agric Syst 101(3):113–127

Tubiello F (2012) Climate change adaptation and mitigation: challenges and opportunities in the food sector. Natural Resources Management and Environment Department, FAO, Rome. Prepared for the High-level conference on world food security: the challenges of climate change and bioenergy, Rome, 3–5 June 2008

Ulyatt MJ, Lassey KR (2005) Methane emissions from pastoral systems: the situation in New Zealand. Archivos Latinoamericanos de Producción Animal, 9(2)

Urquhart P (2009) IFAD's response to climate change through support to adaptation and related actions. Comprehensive report: final version, International Fund for Agricultural Development (IFAD), Rome, Italy. https://www.unccleam.org/sites/default/files/inventory/ifad71.pdf

Wang JY, Lan ZR, Zhang XM (2012) Advances in molecular breeding research of goat fecundity. J Anim Vet Adv 11:449–453

Shigdaf Mekuriaw is an Associate Researcher of Ruminant animal Nutrition at Andasa Livestock research Center, Bahir Dar, Ethiopia. He has been conducting livestock research activities of the center by identifying new areas of livestock thematic areas that are in demand; develop appropriate and relevant academic and research programs and managing their delivery for the regional Agricultural development office and small holder farmers as well as investors. He has authored/co-authored over 15 scientific and professional articles and chapters in the areas of livestock production and related issues. He was awarded UNESCO/KEIZO OBUCHI RESEARCH FELLOWSHIPS PROGRAMME (UNESCO/Japan Young Researchers' Fellowship Programme),

Cycle 2015. Japan Funds-in-Trust Project. Moreover, he also received VLIR Scholarship award for International Training Program of Dairy Nutrition, University of Ghent, Belgium in 2016. Currently, he received Japanese Government Monbukagakusho Scholarship (MEXT Scholarship) and he is a Ph.D. student at Tottori University, Japan.

Alemayehu Mengistu is highly specialized in pasture and rangeland science from university of Uppsala, Sweden. He worked in/for SIDA, ILRI/ILCA, World Bank, Africa bank, FINIDA, FAO, International Universities, national government, and NGO's as research, trainer and development advisor. He also served for many years the project coordinator of the Fourth Livestock in Ethiopia. Moreover, he has also worked on many research projects in collaboration with various institutes such as Ethiopia Institute of Agricultural Research (EIAR), Amhara Region Agricultural Research Institute (ARARI), etc. He published more than 10 books, 15 book chapters and over 72 journal articles and project documents. Now he is working as visiting professor and researcher, trainer and development advisor. He received Gold Madelia from Ethiopian Society of Animal Production (ESAP).

Firew Tegegne Amogne (Ph.D.) is an Associate Professor of Animal Nutrition at Bahir Dar University. He was a visiting scientist at Tottori University, Japan (2014–2015). He has been lecturing, conducting research and providing community development services for more than two decades. He has authored/co-authored over 50 scientific and professional articles and chapters in the areas of livestock production (mainly feeds and nutrition including nutritional strategies to mitigate GHG emissions from livestock), higher education and cross-cutting issues (gender) which are published in peer reviewed journals, conference proceedings and book chapters. He supervised and examined a number of postgraduate students from Ethiopian Universities (Bahir Dar, Gondar, Haramaya, Hawassa, Jimma and Mekelle Universities). Their theses researches dealt with characterization of livestock feeding systems in mixed crop-livestock systems, animal nutrition, forage production, rangeland management, animal production (fattening, dairy, poultry and apiculture) and phenotypic characterization of sheep and goats. Firew Tegegne was/is involved in the initiation, development, implementation and coordination of different capacity building, research and development projects. Now, he is the President of Bahir Dar University.

Impacts of Climate Change on Food Security in Ethiopia: Adaptation and Mitigation Options: A Review

Tadesse Alemu and Alemayehu Mengistu

Abstract Climate change is happening and already affecting food security in Africa. Ethiopia is vulnerable to climate change because our economies largely depend on climate-sensitive agricultural production. Environmental changes, such as changes in rainfall variability, drought, warmer or cooler temperature (lead to change in growing seasons) and land cover change have increased concerns about achieving food security. Growth and Transformation Plan (GTP) recognized Climate change as a threat and opportunity for Ethiopia. Both climate change adaptation and mitigation issues considered; GTP stipulates the country's ambition to build a climate resilient green economy by 2030. Climate change impacts on agriculture and livestock is depending on changes in temperature, precipitation and climate variability (such as erratic rainfall, floods and droughts). The complex interaction of these variables makes it difficult to predict how climate change will impact at the regional level. Despite the relatively high knowledge of the subject among policy-makers and the prominent role being played by Ethiopia in International Climate Change Negotiations many factors, such as El Nlno, are contributing to the deterioration of the local climate and making the population ever more vulnerable to global and regional climate change. The Policies and implementation Strategies should emphasized on an integrated, evidence-based and climate smart approach to addressing food security at all levels, from the National to local levels, from research to policies and investments, and across private, public and civil society sectors to achieve the scale and rate of change required.

Keywords Adaptation · Coping strategies · Climate change · Environmental challenges · Climate smart agriculture

---

T. Alemu (✉)
Biology Department, College of Natural Science, Assosa University, Assosa, Ethiopia
e-mail: tadese07@yahoo.com

A. Mengistu
Pasture and Range Scientist, Urael Branch, P.O. Box 62291, Addis Ababa, Ethiopia

P. Castro et al. (eds.), *Climate Change-Resilient Agriculture and Agroforestry*,
Climate Change Management, https://doi.org/10.1007/978-3-319-75004-0_23

1 Introduction

Growing consensus in the scientific community indicates that higher temperature and changing precipitation levels resulting from climate change will reduce crop yields in developing countries. Evidence from the Intergovernmental Panel on Climate Change (IPCC 2007) is now overwhelmingly convincing that Green House Gases (GHGs) induced climate change is a real and that the poorest and most vulnerable people will be the worst affected. IPCC (2014a) also predicts that by 2100 the increase in global average surface temperature may be between 1.8 and 4.0 °C. With increases of 1.5–2.5 °C, approximately 20–30% of plant and animal species are expected to be at risk of extinction (FAO 2007; IPCC 2014a, b) with severe consequences for food security in developing countries (IPCC 2007; Mekuriaw et al. 2014).

The links between climate change and food security have, to date, largely been explored in relation to impacts on crop productivity and hence, food production. For instance, Gregory et al. (2002) summarized experimental findings on wheat and rice that indicated decreased crop duration (and hence yield) of wheat as a consequence of warming and reductions in yields of rice of about 5% per °C rise above 32 °C (Gregory et al. 2008). Cline (2007) also estimates that global agricultural productivity will be reduced by 15.9% and developing country experiencing a disproportionally larger decline of 19.7%. Similarly, simulation of maize production in Africa and Latin America for 2055 predicted an overall reduction of 10% (Jones and Thornton 2003).

Warming of the climate system is unequivocal, as is now evident from observations of increases in global average air and ocean temperatures, widespread melting of snow and ice and rising global average sea level (IPCC 2007). The average temperature rose by about 0.3 °C during the first half of the 20th century, and by another 0.5 °C in the second half up to the beginning of the 21st century (IPCC 2007) very likely due to the observed increase in anthropogenic GHG concentrations. According to IPCC (2014a) report, there has been an increase in seasonal mean temperature in many areas of Ethiopia. The average annual temperature in Ethiopia increased by 1.1–3.1 °C by 2006, with an increase in the average number of 'hot' days and 'hot' nights per year (McSweeney et al. 2010). This has a severe impact on food production and animal health.

Climate change will act as a multiplier of existing threats to food security; it will make natural disasters more frequent and intense, land and water more scarce and difficult to access, and increases in productivity even harder to achieve. The implications for people who are poor and already food insecure and malnourished are immense (Gregory et al. 2008; UNFCCC 2009). Despite the uncertainty of climate impacts, it is clear that the magnitude and rate of projected changes will require adaptation. Actions towards adaptation fall into two broad overlapping areas: (1) better management of agricultural risk associated with increasing climate variability and extreme events, for example using climate smart agriculture, improve climate information services and safety nets, and (2) accelerated adaptation to progressive climate change over decadal time scales, for example integrated packages of technology, agronomy and policy options for smallholder farmer and food systems (Leslie et al. 2015). Maximization of agriculture's mitigation potential will require, among others, investments in technological innovation and agricultural intensification linked to increased efficiency of inputs, and creation of incentives and monitoring systems that are inclusive of smallholder farmers. More than 45 published articles, policy documents and international climate change reports were used and analyzed systematically. Therefore the objective of this paper was to assess the impacts of Climate change on food security and adaptation and mitigation options in Ethiopia.

2 Impacts of Climate Change in East Africa

The mean global combined land and ocean surface temperature appears to have risen 0.65 to 1.06 °C over the period of 1880–2014 (IPCC 2014a). As a result snow cover in the northern hemisphere decreases and the sea level rises. Africa is the continent that will be hited hardest by climate change. Unpredictable rains and floods, prolonged droughts, subsequent crop failures and rapid desertification, among other signs of global warming, have in fact already begun to change the face of Africa (Gregory et al. 2008; Thornton et al. 2008). Many climate scientists agree that climate change is very real, it is happening and it is happening now. We can no longer consider it a threat that is yet to hit us (Amsalu and Gebremichael 2009).

The impacts of climate change across Africa will vary: At mid- to high latitudes, crop productivity may increase slightly for local mean temperature increases of up to 1 to 3 °C, while at lower latitudes crop productivity is projected to decrease for even relatively small local temperature increases (1–2 °C) (IPCC 2007). In the tropics and subtropics in general, crop yields may fall by 10–20% by 2050 because of warming and drying, but there are places where yield losses may be much more severe (Thornton et al. 2008). The predictions showed that temperatures are expected to increase across the continent (IPCC 2014a). Seasonal average temperatures have risen in many part of eastern Africa, which will lead to increased plant stress and increased risks of drought.

In East Africa large water bodies and varied topography gives rise to a range of climatic conditions, from humid tropical climate along the coastal areas to arid low-laying inland elevated plateau regions across Ethiopia, Kenya, Somalia and Tanzania. The presence of Indian Ocean to the east, Regional lakes as well as high mountains induce localized climatic pattern in this region. In most of these countries, there are places where rainfall means are likely to decrease in the coming decades (Mario et al. 2010). Therefore rainfall in east Africa is very variable in time and space. Several physical processes, including El Niño Southern Oscillation, affect rainfall (IPCC 2014b). According to IPCC (2014b) warming of Indian Ocean is the cause of less rainfall and/or drought over eastern Africa in the last 30 years.

Nearly two thirds of Sub-Saharan Africans depend on livestock for some part of their livelihood. Climate change will affect the productivity of agricultural products as a result, major changes can be anticipated in livestock systems, related to livestock species mixes, crops grown, feed resources and feeding strategies (Anderson et al. 2010). The challenges for development are already considerable for Africans, and climate change will multiply the stresses. There are 300 million poor people in sub-Saharan Africa. Projections indicate an increase of arid and semiarid lands, and, in some countries, yield reductions in rain-fed agriculture of up to 50% by 2020 (Anderson et al. 2010). Therefore failure to manage agricultural climate change adaptation will cause a sharp decline in food production, famine and unprecedented setbacks in the fight against poverty in East Africa. Adapting agriculture to climate change is the key to food security in the 21st century in Africa (Anderson et al. 2010).

3 Climate Change in the Context of Ethiopia

Ethiopia is highly affected by climate change due to three main reasons; (i) about 80% of the population is largely depend on rain fed agriculture (ii) low income country (iii) varied geographical locations with different magnitude of climate impacts. Climate change induced El-Nino increase the average temperature and affect rainfall pattern in time and space leading to a recurrent drought which results in food insecurity particularly in dry and semi dry areas of the country. The country has experienced 16 major national droughts since the 1980s, along with dozens of local droughts. Recently in 2015/15 10 million peoples, in 2017 5 million peoples are food insecure, as a result of drought caused by climate change induced EL Nino.

In Ethiopia climate change is already taking place now, thus past and present changes helps to indicate possible future changes. Over the last decades, the temperature in Ethiopia increased at about 0.2–0.37 °C per decade (Kassahun 2008). The increase in minimum temperatures is more pronounced with roughly 0.4 °C per decade (Mengistu 2008; Kassahun 2008). The temperature will very likely continue to increase for the next few decades with the rate of change as observed (Kassahun 2008; Mengistu 2008; Mengistu and Mekuriaw 2014; IPCC 2014a).

The average annual volume of rainfall over the past 50 years (from 1951–2000) remained more or less constant for the whole country (NMSA 2001). Many authors agreed that mean annual rainfall showed a slight decreasing trend and higher year to year variation was observed in 1950–2010. However, rainfall distribution across the country shows a marked difference. There is a tendency for less rain to fall in the northern part of the country where there is already massive environmental degradation. The same trend can be observed in the south east and north east of the country which is both often affected by drought. However, in central Ethiopia where most of the population and the country's livestock are located, and where the soil is severely depleted and degraded, more rain is falling. The western and north-west parts of the country have also received more rain (Mengistu 2008;

McSweeney et al. 2010). Farmers and pastoralists are experiencing that the rain is becoming more unpredictable or is failing to appear at all. In some places the rain falls more heavily and the degraded soil is unable to absorb this ran which falls over a shorter period. According to Kassahun (2008), the farmers in the central part of the country have lost up to 150 tons of soil per hectare.

The rise in temperature and fluctuations in rainfall crate many problems for the pastoralists who live in the already drought stricken areas which are receiving less and less rain. They have already switched from cattle to goats and camels, as they are more able to endure the long periods of drought. In the central part of the country more rain will mean further erosion of the soil and lower crop yields for small holder farmers and lead to flooding in the more low lying areas. Climate change is affecting how long the farmers have to grow their crops. In addition, warmer weather provides better growing conditions for pests and other diseases that attack crops and destroy the farmers' harvests (Mengistu 2008; Kassahun 2008; Deressa et al. 2008). Therefore, it is possible to conclude that not only the rainfall distribution that has changed but it has also become warmer in the last 60 years. Hence, there is already a great demand for improved seed which is more drought and pest resistant, and for seeds which mature faster as the rains have become more unpredictable and shorter in some places.

Today the forest covers is very low (less than 10%), so the soil has become more vulnerable to erosion. People cut down the forest to create more farmland and to harvest firewood for cooking. Population growth will put pressure on the already degraded soil, and marginal plots will be brought into use which worsens the situation (Mengistu 2008; Deressa et al. 2008; Mengistu and Mekuriaw 2014) (Table 1).

4 Implications of Climate Change in Food Security

A large body of literature demonstrates negative impacts of climate change on the agricultural sector in East Africa. Climate change affects agriculture and food production in complex ways. It affects food production directly through changes in agro-ecological conditions (e.g. changes in rainfall leading to drought or flooding, or warmer or cooler temperatures leading to changes in the length of growing season), and indirectly by affecting growth and distribution of incomes, and thus demand for agricultural products (Gregory et al. 2008).

Climate change is likely intensified high temperature and low precipitation in semi dry and dry areas, it is the most dramatic effects that will be felt by small holder and subsistence farmers (Mendelson and Dinar 2009). According to IPCC 5th report Climate change impacts in East Africa will increase risk of food insecurity and the breakdown of food systems, increase risks of loss of rural livelihoods and income due to insufficient access to drinking and irrigation water and reduced agricultural productivity, particularly for farmers and pastoralists with minimal capital in semi-arid regions. Risks due to extreme weather events leading to

Table 1 Sectoral impacts of climate change in Ethiopia

Sector	Potential impacts
Agriculture	Shortening of maturity period, crop failure and expanding crop diseases
Livestock	• Change in livestock feed availability and quality • Effect on animal health, growth and reproduction • Impact on forage crops quality and quantity • Change in distribution of diseases, decomposition rate, income and price • Contracting pastoral zones in many parts of the country
Forests	• Expansion of tropical dry forests, desertification • loss of indigenous species/expansion of toxic weeds
Water resources	• Decrease in river run-off and energy production • Flood and drought impacts
Health	• Expansion of malaria to highland areas • Threat from expanding endemic diseases and newly emerging varieties of human, plant and livestock diseases
Wildlife	• Shift in physiological response of individual organisms • Shift in species distribution and Shift in biomass over decades/centuries • Shift in genetic make-up of populations • Loss of key wetland stopover and breeding sites for threatening birds species • Out migration, of endemic and threatened species
Environment	Reduced productive capacity from degradation of forests, range and water recourses

Adapted from Mengistu and Mekuriaw (2014)

breakdown of infrastructure networks and critical services such as electricity, water supply, and health and emergency services are also linked to these areas of concern (IPCC 2013).

The overall effect of climate change on yields of major cereal crops in the African region is very likely to be negative, with strong regional variation (Niang et al. 2014). At even relatively low levels of warming of 1–2 °C, many unique natural systems are threatened and food productivity, human health and water resources could be negatively impacted in some regions. "Worst-case" projections (5th percentile) indicate losses of 27–32% for maize, sorghum, millet and groundnut for a warming of about 2 °C above pre-industrial levels by mid-century (Schlenker and Lobell 2010). The IPCC concludes that large-scale warming, of around 4 °C or above, will increase the likelihood of severe, pervasive and irreversible impacts to which it will be difficult to adapt.

Achieving food security and reducing poverty in the Ethiopia has been a major challenge for both governments and development agencies due to the result of many factors, some of which are: (1) land degradation or poor in nutrients; (2) the rapid population growth (3) the low and inappropriate use of technologies such as improved varieties, fertilizers, mechanization and irrigation that have stimulated agricultural development elsewhere in the world (Mekuriaw et al. 2008; Kassahun 2008). The agricultural sector employs between 85% of the active population and contributes close to 40 $ of the Gross Domestic Product (GDP), generates about

88% of the export earnings; and supplies around 73% of the raw material requirement of agro-based domestic industries (Gebreegziabher et al. 2011). Agriculture is a major source of food and plays a key role in generating surplus capital to speed up the country's socio-economic development and hence the prime contributing sector to food security. Yet, agriculture in degraded and semi-arid regions is a highly risky enterprise due to unreliable and variable rainfall. According to Zenebe et al. (2011) as the effects of climate change on agriculture become negative, incomes drop off considerably. At the end of 2050, because of climate change, average incomes will be reduced. In the no-total factor productivity -growth scenario model showed that, climate change a leads to a loss of some 30% of income, compared with the no-climate-change baseline (Gebreegziabher et al. 2011).

According to World Bank (2006), droughts and floods are very common phenomena in Ethiopia with significant events occurring every three to five years. Climate change is expected to exacerbate the problem of rainfall variability and associated drought and flood disasters in Ethiopia (Mesfin 1984; NMA 2006; World Bank 2006; Amsalu and Adem 2009; UN-ISDR 2010).

5 National Response Towards Climate Change: Adaptation and Mitigation Policies and Strategies

Ethiopian's anthropogenic GHG emissions contribution is marginal (only 0.3% of global total) (USAID 2015). Ethiopian's GHG profile showed that the agriculture sector contribute the highest (61%) followed by land use change (18%), energy (17%); (waste 3%) and only 1% from industrial processes and product use (USAID 2015) (Fig. 1). In its climate resilient green economy (CRGE), Ethiopia plans to cut its 2030 GHG emissions at 145 $MtCO_2e$ by 64% (225 $MtCO_2e$) reduction from

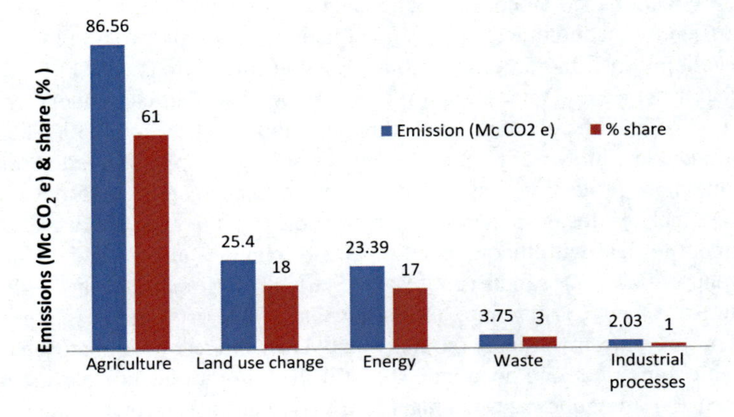

Fig. 1 Greenhouse gas emissions in Ethiopia; by sector. Adapted from (USAID 2015)

projected business as usual emission level by 2030. The reduction includes 130 $MtCO_2e$ from forestry, 90 $MtCO_2e$ from agriculture, 20, 10 and 5 $MtCO_2e$ from industry, transport and buildings, respectively(USAID 2015). Ethiopia has a potential to mitigate an estimated 2.76 billion tons of carbon through protection and sustainable management of forest resources (Moges et al. 2010).

To facilitate the country's response to climate change a comprehensive adaptation and mitigation mechanisms have been developed such as: CRGE strategy, GTP II, Sectoral GHGs Reduction Mechanism (SRM), National Disaster Risk Management and Strategy, National Adaptation Programme of Action (NAPA, now NAP), Sectoral/regional adaptation plans etc. Since 1992 many MEAs are signed and/or ratified. Ethiopia leads least developing countries group in the international climate negotiation agenda.

5.1 Current Policies on Environment and Climate Change

The Ethiopian government has recognized climate change as a threat to its national development. The country has signed most of the international environment conventions including those specifically focused on climate change: it ratified the UNFCCC in May 1994, UNCCD United Nations Convention to Combat Desertification in June 1997, and the Kyoto Protocol in February 2005. The country prepared a National Adaptation Programme of Action (NAPA) to fight the impacts of climate change and desertification (Amsalu and Gebremichael 2009). The program clearly states the urgency of taking practical adaptation and mitigation actions in many social and economic sectors (NMA 2006; Epsilon International 2011). CRGE (Climate Resilient Green Economy) strategy recommends the use of low carbon solutions to leapfrog other economic sectors while realizing the ambitions set out in the country's GTP. CRGE Present an overarching framework to marshal a coherent response to climate change, to generate both innovative thinking and a course of actions to meet the challenges associated with the transfer of climate-friendly technologies and finance for the construction of a climate resilient green economy in Ethiopia (NMA 2006; Epsilon International 2011).

GTP (Growth and Transformation Plan) recognized Climate change as a huge threat. It stipulates the country's ambitions to build a CRGE by 2030. The country has formulated a number of policies, strategies and action plans aimed at promoting Environmental protection, sustainable development and poverty reduction. However, lack of local specific focused policies and legislation were a serious impediment to deal with the adverse impacts of changes and variability in climate (Sintayehun 2008). Kassahun (2008) also stated, it is important and high time to take climate change issues into the country's policies, program and guidelines. However, the current policies, strategy and laws related to climate change and sustainable agriculture are adequate. But still they are not adequately incorporated into extension guidelines and manuals in a way that local farmers understand and participate in the implementation processes. Therefore, creating awareness about policies, strategies and implementation guidelines at all levels including agricultural extension workers and implementation of CSA (climate smart agriculture) is a key.

Despite the relatively high knowledge of the subject among policy-makers, and the prominent role being played by Ethiopia in international climate change negotiations, Ethiopia was still formulating its response in 2009 (Ayalew 2009). Ethiopia, on behalf of African continent plays a great role in climate change negotiations (COPs) which indicates that Africa as a whole and Ethiopians in particular are aware of the climate change impacts on global, regional and national scale. The efforts given to poverty alleviation and socio-economic development will be challenged by the impacts of a changing climate unless such issues are well integrated with adaptation plans (NMSA 2001). According to Deressa et al. (2008), since vulnerability to climate change in Ethiopia is highly related to poverty through loss of coping or adaptive capacity. Integrated rural development schemes can play a great role in reducing poverty and increasing adaptive capacity for dealing with climate change.

Yesuf et al. (2008) also suggested that farmers need timely information on predicted changes in climate in a readily accessible form to empower them to take appropriate steps to adjust their farming practices, such as adopting yield-enhancing adaptation strategies. The early warning system in the country is based on crop forecasts and assessments of food stocks, and deals mainly with preparedness for food emergency relief. In addition, efforts should also be made to reduce the risks of disasters, and extend access to credit markets and extension services in order to facilitate adaptation (Amsalu and Gebremichael 2009). However, the Government's response has been challenged by shortage of funds and lack of institutional capacity. Hence, the role of non-state actors and their contribution in enhancing local adaptive capacities is very crucial need to be encouraging (Amsalu and Adem 2009; Amsalu and Gebremichael 2009) and included in the plan.

5.2 Ethiopia's Program of Adaptation to Climate Change (EPACC)

EPACC (Ethiopian Program of Adaptation to Climate Change) strategy adequately understood climate change as a growing threat in Ethiopia and clearly elaborate the need to mainstream climate change in all spheres of development policy making and planning at all phases and stages of the planning and implementation process.

As a Party to the UNFCCC, Ethiopia is obliged by several articles of the convention to address climate change through the preparation of a national adaptation document and the integration of climate change into its sectoral development plans, policies and strategies. The NAPA, prepared in 2007, represented the first step in coordinating adaptation activities across government sectors, but was not intended to be a long-term strategy in itself. Ethiopia's NAPA projects are currently "on hold" whilst international adaptation funding mechanisms are under negotiation (Adem and Bewket 2011).

The former Federal Environmental Protection Authority and the present Ministry of Forest Environment and Climate Change (MFECC) of Ethiopia developed a separate work program for action on adaptation to climate change. The document interlinks climate change adaptation strongly with the economic development and physical survival of the country. The main objective of EPACC and CRGE is to create the foundation for a carbon-neutral and climate-resilient path towards sustainable development in the country. According to this programme, climate change will be implemented by inhabitants and farmers at local and district levels (NMA 2006 and Adem and Bewket 2011). The climate risks identified by EPACC are broadly in the areas of human, animal and crop diseases, land degradation, loss of biodiversity, decline in agricultural production, dwindling water supply, social inequality, urban waste accumulation, and displacement due to environmental stress and insecurity. The programme also identifies adaptation strategies and options in the various socioeconomic sectors including cloud seeding, crop and livestock insurance mechanisms, grain storage, societal reorganization, renewable energy, gender equality, factoring disability, climate change adaptation education, capacity building, research and development, and enhancing institutional capacity and the political momentum (NMA 2006). The program clearly explains the need to mainstream climate change in all spheres of development policy making and planning at all phases and stages of the planning and implementation process and the urgency of taking practical adaptation and mitigation actions in the various social and economic sectors.

5.3 *Climate Resilient Green Economy (CRGE) of Ethiopia*

Although international climate negotiations have made little progress, Ethiopia has started the race towards low-carbon development (LCD). LCD Plans have been developed and lay foundations for overall sustainable development planning of the country. In fact aggregate climate change mitigation commitments are still far apart from a level of ambition that effectively creates a realistic chance of limiting global warming to a maximum of 2 °C or possibly even lower. Many developing countries including Ethiopia seem to have already begun this process. In this regard, Ethiopia can be an example and tried to implement a new national strategic framework for a smooth transition to a climate resilient green economy by 2030. A climate resilient green economy is a long-term ambition of Ethiopia. The mission statement developed to facilitate the development of the Ethiopian CRGE strategy sets out a five step roadmap for moving towards a climate resilient low carbon economy. The roadmap identified the need for more work on Ethiopia's climate change institutions, monitoring and finance systems and sectorial and regional action plans. When combined, the work is expected to enable the EPA to draft a CRGE Strategy which will identify a clear path to the goal of a climate resilient green economy by 2030 (Adem and Bewket 2011). Building resilient means reducing the risk of becoming food insecure and increasing the adaptive capacity to cope with risks and respond to climate change (Gitz and Meybeck 2012).

Priority challenges and constraints for Addressing climate change impacts

Ethiopia faces a number of cross-cutting challenges and constraints in regard to climate change venerability assessment and adaptation and implementation. As noted in NAPA, these challenges include:

- Weak policy implementation and limited awareness
- Lack of research and development capacity to assess the impacts and consequences of climate change
- Lack of individuals with specialization in venerability and adaptation assessment in agriculture, water resources and health
- Limited skill capacity, facility, and technologies to provide accurate and timely weather and climate forecasts
- Weak institutional framework for dealing with climate change
- lack of coordination between research institutions and policy makers.

Addressing these capacity, institutional and coordination and needs will contribute to Ethiopians ability to continue to move forward effectively on implementation of adaptation that support long term climate-resilient green development.

5.4 *Improving Smallholder Livelihood and Resilience Through Climate-Smart Agriculture (CSA)*

Climate-smart Agriculture (CSA) is an approach that helps to guide actions needed to transform reorient agricultural system to effectively support development and ensure food security in the changing climate. CSA aims to tackle three main objectives: (1) increase agricultural production and income sustainability, (2) adapting and building resilience to climate change and (3) reduce and/or remove greenhouse gas emissions, where possible (IPCC 2013). CSA practices aimed at promoting efficient use of land, water and soil and other environmental resources. CSA promotes coordinated actions by farmers, researchers, private sectors, civil society and policy makers towards climate-resilient pathways through four main action areas: (1) Building evidence (2) increasing local institutional capacity and effectiveness (3) fostering coherence between climate and agricultural policies and (4) linking climate and agricultural financing (Fig. 2). CSA differs from 'business-as-usual' approaches by emphasizing the capacity to implement flexible, context-specific solution, supported by innovative policy and financing actions (Leslie et al. 2015; FAO 2016).

CSA emphasizes utilization of ecosystem service for agricultural systems to support productivity, adaptation and mitigation of climate change. CSA encourages integrated approaches (Leslie et al. 2015) for example:

- integrated crop, livestock, aquaculture and agroforestry systems;
- improved pest, water and nutrient management;

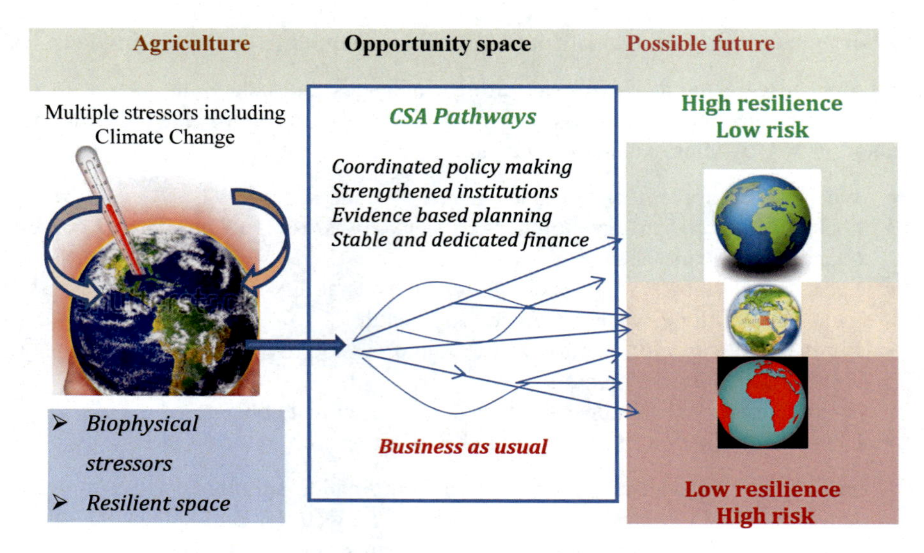

Fig. 2 Climate-resilient transformation pathways for agriculture. Adapted from ref. 4, © IPCC

- landscape approaches;
- improved grassland and forestry management;
- practices such as reduced tillage and use of adverse verities and breeds;
- integrating trees into agricultural systems;
- restoring degraded lands; improving the efficiency of water and nitrogen fertilizer use; and manure management, including the use of anaerobic bio-digesters.

All these integrated activities enhance soil quality and can generate high production. It also Enhances adaptation and mitigation benefits by regulating carbon oxygen and plant nutrient cycles leading to enhanced resilient to drought and flooding and to carbon sequestration (Leslie et al. 2015). Transformative change in agriculture can involve shifts in agricultural production (for example from crop to livestock) or source of livelihoods (increase resilient on non-farm income) (Leslie et al. 2015).

Transforming the current agricultural practice into CSA approach urgent actions from policy makers, public, private and civil society stakeholders at all levels is required in four areas: (i) building research based evidence and assessment tools; (ii) strengthening national, regional and local institutions (iii) developing coordinated and evidence- based policies and (iv) increasing finance institutional capacity and its effectiveness. The current evidence based research findings are inadequate, inaccessible to decision makers support effective decision making at the national, regional and local levels. Therefore the current research addressing climate change impacts on agriculture are not sufficient for national and local level planning. Research institutions should be coordinated to develop tools needed for evaluating the impact of climate change (both extreme events), adaptation and mitigation potential of different policies and technologies.

Challenges and opportunities for effective implementation of CSA in Ethiopia

The key challenges to implement CSA in Ethiopia are: weak implementing capacity on climate change adaptation and mitigation, lack of integration or coordination between federal and regional levels, public sector and civil society organizations, private sectors, and impacts of conventional agricultural practices such as open grazing and frequent ploughing. In spite of the above challenges the country has untapped opportunities to support the scale up of CSA in Ethiopia. These includes; Climate Resilient green economy (CRGE) strategy of Ethiopia, promotion of avoiding open and uncontrolled grazing by regional states, promotion of integrated watershed management to improve agricultural productivity, existence of extension and development agents to create climate related awareness and provide capacity building tanning at the local level. All these opportunities, in addition to NGOs, at the grass root level can promote climate smart agriculture activities in the country.

Improving smallholder livelihood and resilience, in the context of climate change, through climate-smart agriculture (CSA) includes improving farm level food security and productivity through the development of profitable and sustainable farming systems. This can be achieved through integrated and sustainable land and forest management program, integrated soil fertility management, small scale irrigation scams, integrating tree-food-crop livestock system, poultry, bee farming and animal fattening, soil and water conservation measures, rain and ground water harvesting practices. The existence of development agents and extension workers can start CSA practices if appropriate strategy, action plans and manuals are set in place.

6 Conclusion

Many studies convincing that climate change is real, that it will become worse, and that the poorest and most vulnerable people will be the worst affected. Agriculture completely dominates Ethiopia's economy and any climate-change impacts on agriculture will be considerable in the coming decades. Climate change affects agriculture and hence food security directly through changing agro-ecological conditions and indirectly by affecting growth and distribution of incomes. Environmental changes, such as changes in water availability and land cover, altered nitrogen availability and nutrient cycling, has increased concerns about achieving food security. These problems are further intensified by climate change. Shifts in rainfall and rise in temperature will bring major impacts in terms of crop and livestock feed yields, water availability, disease incidence and flood damage. CSA strategies for adaptation and mitigation options should be strengthen such as carbon-sequestration practices involving reduced tillage, increased crop cover, including agro-forestry, and use of improved rotation systems are needed. This transition to CSA will have to be improved by active adaptation policies on the part of the government and will surely need outside support. The countries Green Economy Policies and Strategies integrate the different sectors depending on water-rain fed and irrigated agriculture, livestock, fisheries, forestry, water and soil conservation and biodiversity protection activities. An integrated, evidence based and transformative approaches to addressing food and climate insecurity at all levels require coordination actions from national to local levels, from research to policies and investment and across private, public and civil society sectors to achieve the sale and rate of change required. With the right site specific practices, policies and investment's, the agriculture sector can move on to CSA pathways result in improved food security and decrease in poverty in the in the short term while contributing to reduce climate change as a treat to food security over a longer term.

References

Adem A, Bewket W (2011) Climate change country assessment. Epsilon International R&D, Ethiopia

Amsalu A, Adem A (2009) Assessment of climate change-induced hazards, impacts and responses in the southern lowlands of Ethiopia. Forum for Social Studies (FSS) Research Report No. 4. Addis Ababa, Ethiopia

Amsalu A, Gebremichael D (2009) An overview of climate change impacts and responses in Ethiopia

Anderson S, Gundel S, Vanni M (2010) The impacts of climate change on food security in Africa: a synthesis of policy issues for Europe

Ayalew (2009). Climate change in Ethiopia. BBC Research Briefing, Africa Talks

Cline WR (2007) Global warming and agriculture: impacts estimate by country. Centre for Global Development and Peterson Institute for International Economics, Washington, DC

Deressa T, Hassan R, Ringler C (2008) Measuring Ethiopian farmers' vulnerability to climate change across regional states. IFPRI (International Food Policy Research Institute), Washington, DC

Epsilon International (2011) Assessment of selected development policies and strategies of Ethiopia from a climate change perspective. Paper submitted to the Ethiopian civil society network on climate change (ECSNCC), Addis Ababa

FAO UN (2007) Viale delle Terme di Caracalla, 00153 Rome, Italy

FAO (2016) Ethiopia climate-smart agriculture scoping study. In Jirata M, Grey S, Kilawe E (eds) Addis Ababa, Ethiopia

Gebreegziabher Z, Stage J, Mekonnen A, Alemu A (2011) Climate change and the Ethiopian economy. A computable general equilibrium analysis

Gitz V, Meybeck A (2012) Risks, vulnerabilities and resilience in a context of climate change

Gregory PJ, Ingram JSI, Andreson R et al (2002) Environmental consequences of alternative practices for intensifying crop production. Agric Ecosyst Environ 88:279–290. https://doi.org/10.1016/S0167-8809(01)00263-8

Gregory PJ, Ingram JSI, Brklacich M (2008) The impacts of climate change on food security in Africa

IPCC (2007) Climate change: impacts, adaptation and vulnerability, Contribution of working group II to the fourth assessment report of the Intergovernmental panel on climate change, Cambridge University Press, Cambridge, UK, in press

IPCC (2013) Climate-smart agriculture sourcebook. Executive summary (Food and Agriculture Organization of the United Nations)

IPCC (2014a) Climate change; impacts, adaptation, and vulnerability. Part A: summary for policymakers: Global and Sectorial Aspects. In: Field CB et al. (ed) Cambridge University Press, Cambridge

IPCC (2014b) Climate change: mitigation of climate change. In: Edenhofer O et al. (eds) 29, note 4, Cambridge University Press, Cambridge

Jones PG, Thornton PK (2003) The potential impacts of climate change on maize production in Africa and Latin America in 2055. Global Environ Change 13:51–59. https://doi.org/10.1016/S0959-3780(02)00090-0

Kassahun D (2008) Impacts of climate change on Ethiopia: a review of the literature. In: Climate change—a burning issue for Ethiopia. Proceedings of the 22nd Green Forum, pp 9–35

Leslie L, Philip T, Bruce MC, Tobias B, Ademola B, Martin B et al (2015) Climate-smart agriculture for food security. Nat Clim Change Perspect https://doi.org/10.1038/nclimate2437

Mario H, Claudia R, Jeannette VDS (2010) Climate variability and climate change and their impacts on Kenya's agriculture center. International Livestock Research Institute (ILRI)

McSweeney C, New M, Lizcano G (2010) UNDP Country climate profile les. Ethiopia. http://country-profiles.geog.ox.ac.uk/

Mekuriaw S, Tegegn F, Mengistu A (2008) A review on reduction of greenhouse gas emission from ruminants through nutritional strategies. Acad J Environ Sci 2(1):006–014

Mekuriaw S, Tegegn F, Mengistu A (2014) A review on reduction of greenhouse gas emission from ruminants through nutritional strategies. Acadamia J Environ Sci 2(1):006–014

Mendelson R, Dinar A (2009) Climate change and agriculture. An economic analysis of global impact, adaptation and distribution effects. Elgar, Cheltenham

Mengistu A (2008) Climate variability and change. Ethop J Anim Prod 8(1):94–98

Mengistu A, Mekuriaw S (2014) Challenges and opportunities for carbon sequestration in grassland system: a review. Int J Environ Eng Nat Resour 1(1):1–12

Mesfin W (1984) Rural vulnerability to famine in Ethiopia 1958–1977. Vikas Publisher, New Delhi

MoFED (2010) Growth and transformation plan (2010/11–2014/15). Main Text: Part II, Addis Ababa

Moges Y, Eshetu Z, Nune S (2010) Ethiopian forest resources: current status state future management options in view of access to carbon finance. Prepared for the Ethiopian Climate research and Networking and UNDP

Niang I, Ruppel OC, Abdrabo MA, Essel A, Lennard C, Padgham J, Urquhart P (2014) Climate change in Africa: impacts, adaptation and vulnerability. Contribution of working group II to the fifth assessment report of the Intergovernmental panel on climate change. Cambridge University Press, Cambridge

NMA (2006) National adaptation programme of action of Ethiopia (NAPA). Addis Ababa

NMSA (National Meteorological Services Agency) (2001) Initial national communication of Ethiopia to the UNFCCC. Addis Ababa

Schlenker W, Lobell DB (2010) Robust negative impacts of climate change on African agriculture. Environ Res Lett 5(1), 014010. IOP Publishing Ltd

Sintayehu W (2008) Climate change: global and national response, pp 37–69. In Climate change—a burning issue for Ethiopia. Proceedings of the 22nd Green Forum

Thornton PK, Jones PG, Owiyo T, Kruska RL, Herrero M, Orindi V, Bhadwal S, Kristjanson P, Notenbaert A, Bekele N, Omolo A (2008) Climate change and poverty in Africa: mapping hotspots of vulnerability. Afr J Agric Res Extension 2(1):24–44

UNFCCC (2009) Climate change, food insecurity and hunger. Technical Paper of the IASC Task Force on Climate Change

UN-ISDR (2010) International strategy for disaster reduction (Africa). Country information Ethiopia. http://preventionweb.net/english/countries/africa/eth/

USAID (2015) Greenhouse gas emissions in Ethiopia. Available at: https://www.climatelinks.org/file/2347/

World Bank (2006) Ethiopia: Managing water resources to maximize sustainable growth. Country water resources assistance strategy, Washington, DC

Yesuf M, Falco DS, Deressa T, Ringler C, Kohlin G (2008) The impact of climate change and adaptation on food production in low-income countries: evidence from the Nile Basin, Ethiopia. IFPRI discussion paper no. 828, IFPRI, Washington, DC

Zenebe T et al (2011) Invasive bacterial pathogens and their antibiotic susceptibility patterns in Jimma University Specialized Hospital, Jimma, Southwest Ethiopia. Ethiop J Health Sci 21(1)

Comparative Study on Agriculture and Forestry Climate Change Adaptation Projects in Mongolia, the Philippines, and Timor Leste

Cynthia Juwita Ismail, Takeshi Takama, Ibnu Budiman and Michele Knight

Abstract The impacts of climate change, such as increasing temperature, erratic rainfall pattern, sea level rise, etc., are being increasingly reported. These impacts are destructive for human activities and thus the development and improvement of mitigation and adaptation strategies is a priority globally. In the least developed and developing countries, adequate adaptive capacities are required so to boost the resilience of communities towards the projected climate change projected. Moreover, activities of climate change adaptation not only provide solutions and strategies to deal with climate change, but also encourage sustainable development. This comparative study evaluates projects in three countries: Mongolia, The Philippines, and Timor Leste, by mapping and contrasting the factors that contribute to adaptive capacity and support sustainable development. A heuristic matrix was used to articulate the capacities that influenced the desired outcomes of each project. Some key components of adaptive capacity were identified in each context. The interaction of those components improved the generic and specific capacity at individual and system level then ultimately improved resilience towards climate change.

Keywords Adaptation · Climate change · Agriculture · Forestry
Water management · Sustainable development

C. J. Ismail (✉) · T. Takama · I. Budiman · M. Knight
Sustainability and Resilience Co (su-re.co), Bali, Indonesia
e-mail: cynthia.ismail@su-re.co

T. Takama
e-mail: ttak003@gmail.com

I. Budiman
e-mail: budimanibnu26@gmail.com

P. Castro et al. (eds.), *Climate Change-Resilient Agriculture and Agroforestry*,
Climate Change Management, https://doi.org/10.1007/978-3-319-75004-0_24

1 Introduction

The impacts and risks of climate change are already affecting many sectors crucial for human livelihoods including water resources and food security (UNFCCC 2014). Economically precarious communities, especially those in developing countries and rural localities, are considered the most vulnerable to negative climate change consequences (Hallegate et al. 2016). Adequate attention and proper measures should thus be given to climate change adaptation capacities to prepare communities for projected climate change impacts. As part of this, interventions that build capacity should ultimately support the sustainable development of the region.

Activities attempting to address the climate change impacts have been implemented worldwide, and are expected to be expanded and integrated to mutually promote both adaptation to climate change and sustainable development, yet the role of different capacity attributes promoted in these activities, not least their interaction in different contexts, is still poorly understood. UNFCCC (TEC-UNFCCC 2014) presented that agriculture represented the single most important sector in the economy of many low-income countries, and 75% of the world's population relies on the related activities. On the other hand, WorldBank (2013) highlighted that hundreds of millions of people around the world depend directly on forest resources for their income and livelihood, including many people living in extreme poverty. However, both sectors are under threats of climate variables (e.g. temperature, precipitation, radiation, and extreme weather events) which ultimately jeopardize the survival of people who rely on those sectors. IPCC (2007) added: "The inter-annual, monthly and daily distribution of climate variables affects a number of physical, chemical and biological processes that drive the productivity of agricultural, forestry and fisheries systems." IUFRO (2009) reported that human dimensions of adaptive capacity in subtropical and tropical forests are more variable due to constraints on access to capital, information and technology. Since the recent report of IPCC AR5 suggested the urgent efforts of adaptation, the establishment of adaptation strategies are crucial for both sectors (i.e. agriculture and forestry).

This study analysed adaptation projects as case studies at agriculture and forestry sector in three countries: Mongolia, Philippines, and Timor Leste. The idea of this study was to explore aspects that influence the success of increasing communities' adaptive capacities at the end of the projects by adopting a heuristic matrix developed by Eakin et al. (2014). This matrix can be an alternative way to evaluate the impacts of adaptation projects as a lesson learned in a more comprehensive way.

Monitoring and Evaluation (M&E) and Measuring, Reporting and Verification (MRV) for climate change adaptation receive increasing interest and attention at both political and operational levels. On the political side, the outcomes of Paris Agreement indicated an increasing focus on the national reporting of both future adaptation actions that have been implemented. On the other hand, the operational side tends to the scale of the financial resources flowing into climate adaptation, is likely to lead much stronger donor emphasis on documenting results and impacts in the future (Christiansen et al. 2016). Furthermore, M&E is crucial to capture the progress of the project whether sustainable development is achieved. In response to that, this study adopted qualitative analysis to evaluate the result of each case. The qualitative approach was selected to describe effectively the distinctive aspects and the contemporary phenomena of the project as they are unique cases and difficult to replicate. The qualitative approach is expected to reveal something about the case studies and contributes to a general understanding of the nature of this kind of activities.

2 Case Description

This study covered three adaptation projects in Mongolia, Timor Leste and the Philippines. These project activities aimed to improve the adaptation capacity and formulate the adaptation strategies at agriculture and forestry sector with the expectation of sustainable development stage is achieved. In Mongolia and Timor Leste case, the issues related to forestry management were highlighted, whereas the Philippines covered water management issues at agriculture sector. Improving adaptive capacity in the vulnerable areas are the focus of each project.

2.1 Mongolia Case

There were 7 vulnerable provinces (i.e. Tuv, Selenge, Khentii, Bulgan, Khuvsgul, Arkhangai and Uvurkhangai) selected as target areas aiming for improving livelihoods of rural communities through sustainable forest management and increasing apiculture. In those provinces, demographically, population densities are low and communities are usually dispersed and nomadic. Livelihood opportunities for local communities are limited by the short growing season and low yields for agricultural products. The communities consist mostly of herders with little access to non-herding income. The incidence of poverty generally increases with distance from municipal services and access to livelihood diversity. Thus, poverty in these areas is generally high—in 2012 (the baseline year of this project), according to World Bank assessments, more than 35% of the rural Mongolian population was considered impoverished following national standards. The 2006 Forest By-Laws, enacted in 2009, allow local communities to form community-based Forest User Groups (FUGs) to manage forest areas based on forest management plans approved by regional governments. To date, 1180 FUGs with about 26,000 members have been established, managing over 3 million ha of forest.

2.2 The Philippines

Lantapan, Bukidnon were selected as target areas to increase the adaptive capacity, especially capacity building of watershed management and up-land farming. Currently prolonged rains, impacts resulting from El Nino and La Nina, and early and delay onset of the rainy season have mainly negative impacts on crop yield, farm income, water and soil quality, and health of the farmers. Lantapan has an agricultural-based economy with 60% of the total labour force employed on bananas and pineapples plantations covering large tracks of land, and in commercial swine and poultry farms. Corn is the predominant crop, and is planted at higher elevations alongside coffee and other vegetables. Meanwhile, coffee is prevalent at the middle altitudes, with irrigated rice, and vegetables such as cabbage, tomatoes and potatoes being the other crops that are distributed within the watershed. Demographically, half of the population attended elementary school, a third entered high school, and slight less than 15% attained tertiary education. This limits their ability to seek non-farm employment in towns and cities. Many households live close to the poverty line. Thus, they seldom have the capital to start small, non-farm related scale business enterprises.

When the farmers are impacted by current climate hazards, they seek assistance from lending institutions for purchase of pesticides; local government units for provision of seeds, technical and financial assistance, and food subsidies/handouts; commercial plantations to seek additional employment; and local health centers and medicine men for the treatment of flue colds, coughs and fever. At times, the delay in the release of local budget can facilitate coping mechanisms, and other times constrain them, as there are challenges in arranging adjustments to budgets or to respond to climate variability and extreme events (e.g. to take on additional short-term staff), or results in a decrease in budgets.

2.3 Timor Leste Case

Climate data or forecasts is limited for Timor Leste, but climate change is predicted to cause hotter dry seasons, shorter and more unpredictable rainy seasons, more frequent extreme heavy rainfall and cyclone events and sea water intrusion. These natural disasters associated with droughts, floods, landslides and soil erosion, result in decreased capacity for agricultural production and damage to infrastructure. The country experiences a distinct 'hungry season' for up to four months of the year in many districts. In Timor Leste, a positive Indian Ocean Dipole equates to less rainfall, protracted dry seasons. During La Niña years above normal rainfall leads to increased flooding and landslides in Timor Leste, while El Niño years are associated with droughts. The most significant impact on the population during El Niño years is reduced ground water availability. Aileu is targeted as the project site. The communities are largely subsistence farmers who are very isolated, far from markets, little access to roads, have poor crop quality, low yields, poor food security, over reliance on maize and rice for income. The communities have a history of exploiting the natural resources of forest, and poor history of engagement with government. Education levels are extremely low across the project population, and lower for women than men. These baseline data put Aileu as the vulnerable area which requires proper adaptation measures.

3 Methodology

This study dealt with an evaluation on adaptation projects at agriculture and forestry sector in three countries: Mongolia, Philippines, and Timor Leste at agriculture and forestry sector. The data collection was done by qualitative approaches. The approach was conducted by site visit, observations and interview the respective stakeholders such as the project developer, the local government, and the impacted communities. Those was also complemented by reviewing the projects' final reports, and survey to the impacted communities (i.e. number of people, number of households). At the end of each project, Eakin et al. (2014) suggested a simple heuristic matrix to evaluate the impact of each project. This is a proper tool used to articulate the influential capacities that have led to the desired outcomes of a project. The capacities matrix consists of two dimensions: *generic capacity* and *specific capacity*, as illustrated in Table 1.

Table 1 The manifestation of different forms of capacity at different organizational levels

	Individual actor	System-level
Generic	• Income level and structure • Savings • Material assets • Health status • Education level • Population mobility • Participation in social organizations	• Economic productivity • Information infrastructure • Poverty levels • Economic and social inequality • Transparency in governance • Population-level education • Sanitation • Health care services • Built environment integrity
Specific	• Climatic information use • Protection of private property • Climate risk insurance • Adoption of technologies to reduce climate impacts • Cultural climate prediction • Traditional risk mitigation strategies	• Insurance provisioning systems • Early warning systems • Scenario development • Infrastructure investment • Disaster planning and compensation fund • Risk mitigation planning

Source Eakin et al. (2014)

Generic capacity is related to capacities which are basic human development needs, while specific capacity is defined as tools and knowledge required to anticipate and effectively respond to climatic threats. Thus, generic capacity includes education level, health, mobility, livelihood, and security whilst specific capacity refers to the knowledge and system concerning adaptive procedure. Although project indicators on one project are not analogous to those of other communities', it is possible to draw out the essential aspects of generic and specific capacity profiles. The matrix also furcates these capacities for examination at individual actor and system level.

To evaluate the interaction between specific capacity and generic capacity, Eakin et al. (2014) suggested 4 classifications as shown in Fig. 1. Firstly, when both generic and specific capacity are at a low level, the target community is classified to be in a "poverty-trap". In this state, the targeted community suffers from intense stress that erodes human welfare and social structure that would otherwise support effective risk management. Secondly, if generic capacity is low whilst specific capacity is high, the society will be considered as a "safety-first" population. The circumstances of this community lead them to prioritize present day safety and security over investments in generic capacities that might enable future welfare gains. There are typically weak safety nets at the level of governance ("system-level") of "safety-first" communities. Furthermore, capacity to invest in assistance for household risk management, or build generic capacity, is lacking.

A further classification, the "safe development paradox" is defined when the target society has high generic capacity but low levels of specific capacity. It describes a society with a good level of education or health, but limited ability to cope with the risk and impacts of climate change. At the system-level of the "safe development paradox" community, there may well be very strong safety nets and public investments in risk management and programs to ensure socioeconomic stability. Lastly, characterized by high generic and specific capacity, there is the community enacting "sustainable adaptation". Communities in this domain are characterized by conditions that would most likely lead to a sustainable outcome and potentially, transformative adaptation. In this condition, generic and specific

Fig. 1 Capacities matrix

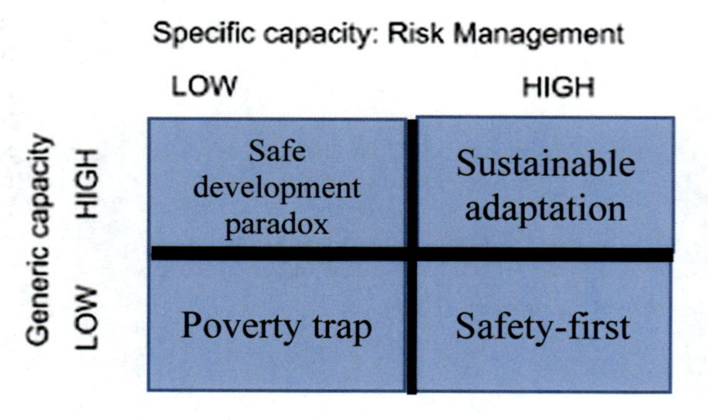

risk management is high at both individual and system-levels, and as such, development and adaptation policies are mutually reinforcing to the benefit of reduced overall vulnerability (Eakin et al. 2014).

3.1 Mongolia Case

The project has two approaches: the focus on protection and sustainable management of the forests and sustainable livelihoods for communities in proximity to forests. To execute the project, a field visit was conducted by World Vision Mongolia to hold focus group discussion and key informant interviews with respected community members and officials. Then, it was followed by a workshop to assess the priority environmental issues and the most important causes of those issues. The project supported 24 households (1.6% of the total number of beneficiaries) that had members with disability and some bee-keeping groups had active members who had disabilities. Furthermore, document report from the project were also conducted to evaluate the impact of the project.

3.2 Philippines Case

The overall goal of the project is to promote climate change adaptation by upland farmers and watershed management. To achieve this, the impacts of climate variability to crop yield, vulnerability and adaptation policies/strategies were assessed through a combination of one-on-one interviews with farmers and stakeholders, focus group discussions, workshops and review of literature. Results of the assessment were presented and validated during workshops. Then, capacity building activities were undertaken in the forms of magazine, video and other easy understanding media. Those are aimed for non-technical people to increase the level of awareness of farmers and stakeholders on climate variability, climate extreme and climate change. A pre-test of materials was performed before distributing to the project site. In addition, formal and informal training sessions for climate change 'champions', farmer groups, stakeholders and policy makers were held at the local level, and efforts were made to ensure that these were covered by the local media to attract more stakeholders involved.

3.3 Timor Leste Case

At the beginning of the project, meetings and visits created awareness and agreement for local leaders to implement and enforce "tara bandu" which is the local land law. In Timor Leste, tara bandu is enforced a strict local forest policy which prohibited cutting and burning of forests. Through the meetings (i.e. discussions, interviews), tara bandu was readopted and re-agreed to prohibit all burning, and limited and controlled wood harvesting. Stakeholders were able to advocate for more supportive arrangements for the management and utilization of natural forests. The project supported 46 people living with a disability (25 male and 21 female) to participate in training and project activities and were given priority access to agroforestry materials. Several assessments such as vulnerability assessment, participatory rural appraisal/PRA were performed.

4 Results

As result, Table 2 shows the overall outcome of projects in terms of adaptive capacity using the heuristic matrix suggested by Eakin et al. (2014). This result was based on the projects' final report, observation, interviews and survey. The logical view which lead to such results will be discussed in the next three sub-sections (i.e. Mongolia case, The Philippines case, and Timor Leste case).

Table 2 Project evaluation using a heuristic matrix suggested by Eakin et al. (2014)

No.	Location	Project name	Aim of the project	Status before project	Status after project
1.	Mongolia	Forest Protection and Enhanced Rural Livelihood Project (FPERLP)	To improve the livelihood of rural communities through sustainable forest management and increasing apiculture	Poverty-trap	Safety-first
2.	The Philippines	Mainstreaming climate change adaptation in watershed management and upland farming in the Philippines	To promote climate change adaptation by upland farmers and watershed management at the national and local levels in the Philippines	Safety-first	Safety-first
3.	Timor Leste	Building resilience to a changing climate and environment	Increased community and environmental resilience to climate change effects	Poverty-trap	Safety-first

4.1 Mongolia Case

4.1.1 Before the Project Implementation

Despite the potential to facilitate adaptation, most FUGs lack knowledge to establish forest inventories and management plans also capability on sustainably maintaining the forest- something the project attempted to address. On the other hand, the forest legislation constrains adaptation because they do not allow trees to be cut without a permit (permits being reserved for poorly regulated, private forest enterprises). Therefore, the climate vulnerability of local community members will be reduced by building the local community's capacity to protect, manage and create livelihoods from forests in an ecologically sustainable way. Given the situations mentioned before the project implementation, the community condition is categorized as "poverty-trap".

4.1.2 After the Project Implementation

The project increased community adaptive capacity by both increasing community's income by diversification of non-timber-forest-product livelihoods sources and improving awareness of environmental degradation and management. In terms of gender, there was no notable difference in vulnerability as it was not included in the scope of the project. Women's participation in the forest groups supported by the project was equal to men's, and women achieved slightly higher number of leadership roles than the men. The effort to diversify the livelihoods to non-timber-forest-production of the local community was assessed to have increased the specific capacity at both individual and system level shown in Table 4; as compared with the condition before the project implementation (Table 3). Specifically, diversifying the livelihood sources, at the generic level, reduced community reliance to exploit the primary resources and conditions required to profit solely from animal husbandry (livelihood diversification); while at specific capacity, it created new market opportunities, improved capacity to engage with market. Also, the mobility of community to urban areas as described in Table 3 was diminished. These outcomes may reduce the impact of climate change shocks and stressors. In addition, by explicitly educating community members

Table 3 The adaptive capacities at the beginning of the project in Mongolia case

	Individual actor	System-level
Generic	• Seasonal population mobility • Agrarian local knowledge and cultural appreciation for nature	• High population-level education rate
Specific	• Engaged in a livelihood activity (herding) • Some traditional use of forest resource	

Table 4 The adaptive capacities at the end of project of the project in Mongolia case

	Individual actor	System-level
Generic	• Social organization around resource management and markets • Increased awareness of environmental issues • Strengthened community and social cohesion through communal management and skills-transfer	• (Increased) economic productivity • Improved understanding of policies around sustainable forest management
Specific	• Engaged in an increased diversity of livelihood activities • Improved access/sustainable use of forest resources (across genders) • Improved capacity to engage with market • Improved community stewardship of forest resource	• New market opportunities and touchpoints • Improved DRR planning (specifically for bushfires)

around policy, environmental protection and livelihoods, as well as the links between these, the market economy activities related to forest resources have become molded. Overall, it was assessed that the project successfully improved the adaptive capacity from 'poverty trap" to "safety-first" community.

4.2 The Philippines Case

4.2.1 Before the Project Implementation

Considering the knowledge about climate change mitigation and access to funding, the state of the community is at "safety-first" where generic adaptation capacity is moderate at both individual and system levels, and specific adaptation capacity is relatively high. In the short-term, specific adaptation capacity is needed on the use of climate change information both at the individual actor and system level.

4.2.2 After the Project Implementation

The project found that the main factor related to vulnerability is low level of education that limits their ability to seek non-farm employment in towns and cities (Pulhin et al. 2016). Although the project has provided interactive education for non-technical people, it was claimed that there was no significant change in terms of adaptive capacity (Table 5) because the same table was produced at the end of project implementation. It was suggested that the development of insurance systems and the uptake on risk insurance should be introduced to alleviate food security

Table 5 The adaptive capacities before and after the implementation of the project in Philippines case

	Individual actor	System-level
Generic	• Population mobility • Participation in social organizations	• Economic productivity • Information infrastructure • Partnership strategies
Specific	• Climatic information use • Adoption of technologies • Traditional risk mitigation activities	• Risk assessment • Early warning system • Infrastructure investment • Risk mitigation/contingency plans • Disaster compensation/assistance funds • Recovery plans

issues during extreme climatic events in the short-term. In the long-term, climate scenario development and the analysis of positive and negative impact of climate change will help to inform development planning in the region. Additional investments in increasing education levels is required to improve incomes and ability of the community to find alternative employment that is less climate sensitive. Nevertheless, the project succeeded in identifying the impacts of current climate hazards on the local people, identifying their current coping mechanisms and raising their awareness of climate change. Overall, this project did not successfully improve the adaptive capacity of the respective communities, the safety-first condition.

4.3 Timor Leste Case

4.3.1 Before the Project Implementation

The condition of the community is categorized at "poverty trap" because of low capacity in generic and specific. For instance, low education level, lack of government support and social infrastructure are yet established in Aileu. On the other hand, climate change is hastening the decimation of the natural resource base, notably water, agricultural and forest resources.

4.3.2 After the Project Implementation

The project has significantly achieved its goal of increasing community and environmental resilience to climate change effects. The very significant reported decline in the incidence of burning is no doubt influenced by increased government messaging in this area, and in particular local level regulations promoted through tara bandu. A reduction in burning will greatly contribute to climate resilience. At the community level in the project area raising climate change awareness appears to have been effective. There are 76% of respondents claimed to know what climate

Table 6 The adaptive capacities at the beginning of the project in Timor Leste

	Individual actor	System-level
Generic	• Agrarian local knowledge • Tara bandu tradition	
Specific	• Some traditional use of forest resource in desperate times	

Table 7 The adaptive capacities at the end of the project in Timor Leste

	Individual actor	System-level
Generic	• Increased awareness of environmental issues • Improved understanding of prohibitions and rights around sustainable forest management • Decrease in destructive practices (specifically slash and burn)	
Specific	• Engaged in an increased diversity of livelihood activities • Improved community stewardship of forest resource	• Strengthened adherence to forest law (due to tara bandu governance)

change was but perhaps more notable was the rate of awareness on potential climate change impacts on agriculture with up to 84% of those aware of climate change able to describe at least one impact consistent with the general science. There was a reasonably high level of knowledge of suitable measures to mitigate climate change impacts. The improvement made by this project is shown in Table 7 when it is compared with the conditions before the project activity, shown in Table 6. Overall, the project successfully improved the adaptive capacity from "poverty trap" to "safety first" although mostly at individual level.

5 Discussion

A difference was ultimately observed for each project, and moreover differently affected the adaptive capacity of each target community. Herein, we will attempt to explain the interaction of adaptive capacities of each project that influence the outcome and subsequently provide comparative study especially in agriculture and forestry activity.

To further analyze factors that affect the success of each of the projects, this study highlighted the aspects to be evaluated including data availability, the importance of inception and participatory analysis, the education level of the target community, technology and knowledge transfer, policy and government support, and financial measures.

5.1 The Importance of Data Availability and Participatory Analysis

All cases suggest that inception and participatory analysis are significant to ascertain the success of a project (Pulhin et al. 2016; World Vision Mongolia 2015; World Vision Timor Leste 2016; Smit and Walden 2006). Likewise the development projects, data availability is significant in adaptation projects (Patwardhan 2003). For instance, the project proponent in the Philippines did assessment to picture the condition in the targeted areas by interviewing the respective stakeholders prior to assess climate vulnerability and impacts including the adaptation strategies (Pulhin et al. 2016). Understanding that the education of the community is low, then increasing the education level and awareness becomes crucial in the project.

5.2 Education Level, Technology and Knowledge Transfer

All cases imply that sufficient education level of local communities is crucial to assure the outcome of the project. The three cases described the education level of local communities are considerably low and various methods had been conducted to increase the awareness, knowledge and adaptive capacity (Pulhin et al. 2016; World Vision Mongolia 2015; World Vision Timor Leste 2016). However, the cases did not success in providing their adaptation strategies at the end of the projects because of the lack of knowledge and further need of capacity building. This illustrates the importance of education level in determining the pace of project implementation before proceeding to formulating the adaptation strategies. Therefore, an intensive and comprehensive capacity building is necessary as it was implemented intensively in the Philippines case and Mongolia. Focus Group Discussion, visual materials (e.g. book, video) and workshop could be options (Pulhin et al. 2016). In addition, high education level stimulates a good social organization that may contribute significantly on improvement at system-level (Williams et al. 2015). Having adequate education level also accelerates skill and knowledge transfer from one community to another.

According to the project report in the Philippines case, although the improvement is noticeable, it could have been better if the education level is not low (mentioned at the previous section). This implies capacity is necessary to ensure the enhancement of adaptive capacity occurs mutually at both individual and to an extent, at system level. Despite, the Philippines case study provides also a good example of a successful method to increase awareness and knowledge about climate change. As part of this project, the project team endeavored to use various media to increase the capacity at individual level based on familiar local means of communication, for example, all media used local dialect and easily-understood language for non-technical people to describe climate change and the importance of adaptive capacity (Pulhin et al. 2009).

Technology transfer is also considered plays important role in enhancing adaptive capacity of a community. Establishing improved pathways of technology transfer from the level of individual actor to system-level in the case of the Timor Leste project would improve specific capacity at both individual level and governance (system-level). The absence of this aspect, inhibits the enhancement of adaptive capacity as shown in Table 7.

5.3 Policy and Government Support

Policy is regarded as a notable attribute influencing the success of all projects in this study. This is in line with mainstream literature that attests institutional barriers are the most frequently reported barriers to climate change adaptation (Biesbroek et al. 2013; Brooks and Adger 2015). Governments have an important role in this context. They can help by creating an attractive environment for research, development and demonstration (RD&D) and safeguarding the drivers of innovation.

Well-designed targeted technology policies on both the supply and demand sides are a fundamental ingredient in a strategy to accelerate innovation. While the specific combination of policy measures will depend on country circumstances, it is important in all cases to construct the appropriate framework to allow breakthroughs to happen (IEA 2011). First and foremost, local and national responses to climate change need to be well coordinated. This ensures coherence of local and national action, while clearly acknowledging differences in the mandates of cities and national governments. City and sub-national regional leaders are generally best suited to design strategies to address their infrastructure needs, land use, geography, and economic profiles. Together they could work closer together to develop and exchange information about possible policy responses, to experiment with new solutions, to share experience and broaden and replicate successful initiatives especially in dealing with climate change. Mongolia is an illustrative case where inhibitory policy hindered the outcomes of the project. In Mongolia, the inhibitory forest legislation constrained adaptation so that adaptation was only realized at the individual level (Table 4).

Conversely, the case in Timor Leste is an example where amending policy to address the risk of climate change positively impacted adaptive capacity at the system level. This suggests that appropriate policy makes a significant contribution to adaptive capacity at system-level. Referring to Tables 6 and 7, at the beginning of the project, adaptive capacity at system level was absent. After readopting the local law of tara bandu, specific capacity at system-level was strengthened.

Furthermore, the Timor Leste case demonstrated government backing of favorable policy and/or governance contributes to development at both capacities, generic and specific, as well as at individual and system levels. Moreover, a supportive policy can facilitate the involvement of other stakeholders such as NGOs and local development agencies to contribute actively to accelerate project implementation to achieve its goals (Adger et al. 2011). Notably, such support expands the opportunities of partnership. These engagement principles are also suggested in

Timor Leste, where engagement is needed at the system-level to better support the individual actor, for example both in the improvement of forestry and agricultural industry to the development of markets that support sustainable forest management and ecological agriculture (World Vision 2016).

5.4 Financial Measures

Adaptation requires sufficient and sustained funding so that countries can plan for and implement adaptation activities. Indeed, the Intergovernmental Panel on Climate Change (IPCC) identifies economic wealth as a principal determinant of adaptive capacity (IPCC 2001). Central governments, in turn, can set out the broad goals and frameworks to encourage action in the right areas; they can also provide needed funding or other incentives for city initiatives. The costs of adaptation in cities will account for a significant proportion of this average, largely because of the expense required to adapt (or, in the case of many low- and middle-income countries, build new and resilient) infrastructure and services for densely populated areas. UNFCCC estimates that adapting infrastructure worldwide could require US$8–30 billion in 2030, one-third of which would be for low- and middle-income countries (UNFCCC 2007).

Most of the projects-those in Mongolia, Philippines, and Timor Leste, were precluded from being able to fully enhance community adaptive capacity to a sustainable state because of funding availability; e.g. lack of access to financial measures such as insurance, grants, markets or loans. Ultimately, the government plays a crucial role in mobilizing funds towards sustainable development at local, sub-national and national levels.

An example of an integrated financial measure is the Philippines case. When the farmers are impacted by current climate hazards, they are able to seek assistance from lending institutions for purchase of agricultural inputs; local government units for provision of technical and financial assistance, and food subsidies/handouts; commercial plantations to seek additional employment; and local health centers and medicine men for the treatment of flu, colds, coughs and fever. In addition, the development of insurance systems and the uptake of suitable risk insurance in the Philippines, may further alleviate food security issues during extreme climatic events in the short-term.

This measure does not effectively adapt the community to climate risk over the longer term however, nor provide the community with adequate and consistent financial reserve to support future-oriented adaptation activities (funding is aimed at recovering losses and responding to immediate needs). Additional investments to increase education levels would improve incomes and the ability of the community to find alternative employment options that are more advantageous, in accordance with the risk highlighted in climate scenarios. Unfortunately, the delayed release of the project budget hindered the project implementation. This implies that a sustainable and integrated funding is required to properly support projects building resilience to climate change.

In terms of progressing the requirements of sustainable development unmet at the project close, targeted monitoring data and strategies could promote follow-on projects. Funding is a need for all developing countries to develop and implement national adaptation plans and for these to exist at all levels: local, sub-national and national. This was found to be an important to support progress towards and perpetuation of sustainable development across the projects.

6 Conclusion

The analysis performed above identified the following attributes that contribute to improve adaptive capacity: data availability, the importance of inception and participatory analysis, the education level of community target, technology and knowledge transfer, policy and government support, and financial measures. How these attributes may change the generic and specific capacity at individual and system level was discussed.

We conclude from this analysis that the success of a project is contingent on meeting the following steps to ensure the formulation of effective adaptation strategies:

1. Strong scientific data basis for decision making
2. A pre-assessment on the vulnerability of climate change at local context including inception and participatory analysis
3. Education, training and public awareness on adaptation; including the establishment of pathways of technology transfer
4. Funding security
5. Project Evaluation and Monitoring to support follow-on projects in the future.

This study also reveals the complex interaction between project attributes. Notably, the presence of policy and government support is considered to give significant enhancement in generic and specific capacity at individual and system level. Likewise, both the analysis of positive and negative impacts of climate change through climate scenario development and the establishment of apprised pathways of technology transfer, will help informing and sustaining the development planning and innovation in the regions. It is these relationships and their interactions that determine the ultimate outcomes of adaptation activities. By doing so, it will be more likely to be implemented in a way that is effective, efficient and equitable. Furthermore, sustainable development stage is expected to achieve by implementing the suggested strategies above.

Acknowledgements We would like also to thank our colleague, Mariana Silaen who provided insight and expertise that greatly assisted the research.

References

Adger N, Brown K, Nelson DR, Berkes F, Eakin H, Folke C, Galvin K, Gunderson L, Goulden M, O'Brien K, Ruitenbeek J, Tompkins E (2011) Resilience implications of policy responses to climate change. Available via WIREs Climate Change 2. http://wires.wiley.com/WileyCDA/WiresArticle/articles.html?doi=10.1002%2Fwcc.133

Biesbroek GR, Klostermann J, Termeer C (2013) On the nature of barriers to climate change adaptation. Reg Environ Change 13(5):1119–1129

Brooks N, Adger WN (2015) Assessing and enhancing adaptive capacity: Technical Paper 7. http://www4.unfccc.int/nap/Country%20Documents/General/apf%20technical%20paper07.pdf

Christiansen L, Schaer C, Larsen C, Naswa P (2016) Monitoring & evaluation for climate change adaptation

Eakin HC, Lemos MC, Nelson DR (2014) Differentiating capacities as a means to sustainable climate change adaptation. Glob Environ Change 27:1–8

Hallegate S, Bangalore M, Bonzanigo L, Fay M, Kane T, Narloch U, Rozenberg J, Treguer D, Vogt-Schilb A (2016) Shock Waves: Managing the impacts of climate change on poverty. The World Bank, Washington DC

IEA (2011) Good practice policy framework for energy technology research, development and demonstration (RD&D). Available via IEA Data and Publications. https://www.iea.org/publications/freepublications/publication/good_practice_policy.pdf. Accessed 19 Nov 2018

IPCC (2001) Climate change 2001: determinants of adaptive capacity. Available via IPCC. http://www.ipcc.ch/ipccreports/tar/wg2/index.php?idp=651. Accessed 19 Nov 2018

IPCC (2007) IPCC fourth assessment report: current sensitivity, vulnerability and adaptive capacity to climate. Available via IPCC. https://www.ipcc.ch/publications_and_data/ar4/wg2/en/ch5s5-2.html. Accessed 19 Nov 2018

IUFRO (2009) Adaptation of forests and people to climate change—A global assessment report

Patwardhan A (2003) Climate risk and adaptation: importance of local coping strategies

Pulhin FB, Lasco RD, Victoria M, Espaldon, Gevana DT (2009) Mainstreaming climate change adaptation in watershed management and upland farming in the Philippines. Forestry Development Center, College of Forestry and Natural Resources, University of the Philippines Los Banos, Los Banos, pp 1–65

Pulhin JM, Peras RJJ, Pulhin FB, Gevana DT (2016) Farmers' adaptation to climate change variability: assessment of effectiveness and barriers based on local experience in southern Philippnes. J Environ Sci Manage 1:1–14

Smit B, Wandel J (2006) Adaptation, adaptive capacity and vulnerability. Glob Environ Change 16:282–292

Sutton WR, Srivastava JP, Neumann JE, Boehlert B (2013) A world bank study: reducing the vulnerability of Moldova's agricultural systems to climate change. The World Bank, Washington, D.C. Available Online http://www.worldbank.org/en/news/feature/2013/01/29/sustaining-forests-livelihoods-changing-world

TEC UNFCCC (2014) Technologies for adaptation in the agriculture sector (No. 4). UNFCCC, Bonn. Available Online http://unfccc.int/ttclear/misc_/StaticFiles/gnwoerk_static/TEC_column_L/544babb207e344b88bdd9fec11e6337f/bcc4dc66c35340a08fce34f057e0a1ed.pdf

United Nations Framework Convention on Climate Change (UNFCCC) (2014) Fact sheet: The need for adaptation. Available Online: http://unfccc.int/press/fact_sheets/items/4985.php. Accessed 10 July 2016

UNFCCC (2007) Investment and financial flows to address climate change. UNFCCC Secretariat, Bonn

UNFCCC (2016) FOCUS: adaptation. Available via UNFCCC. http://unfccc.int/focus/adaptation/items/6999.php. Accessed 10 July 2016

Williams C, Fenton A, Huq S (Eds) (2015) Knowledge and adaptive capacity. Available via Nat. ClimChange 5. http://www.icccad.net/wp-content/uploads/2015/12/Knowledge-and-adaptive-capacity.pdf

WorldBank (2013) Sustaining forests and livelihoods in a changing world. World Bank. Available Online: http://www.worldbank.org/en/news/feature/2013/01/29/sustaining-forests-livelihoods-changing-world. Accessed 21 Dec 2017

World Vision Mongolia (2015) Evaluation report: forest protection and enhanced rural livelihood project. World vision international Mongolia

World Vision Timor Leste (2016) Evaluation report: building resilience to climate change and environment (BRACCE). World Vision Timor Leste

Cynthia Juwita Ismail Cynthia completed a four-year undergraduate in Chemistry at Bandung Institute of Technology (Institut Teknologi Bandung). In order to go deeply about energy and environment, she took Erasmus Mundus Management and Engineering of Energy and Environment (ME3) 2013–2015. Following that, she involved in climate change mitigation activities for about 2 years working at an electricity generation company and was responsible for mitigation projects and evaluation of CSR activities related to carbon trading. Subsequently, she joined PT Sustainability and Resilience (su-re.co) and this gave an opportunity to perform a vulnerability assessment on commodities in Indonesia such as coffee and cocoa. Also, she have been involving some adaptation activities such as increasing the resilience of coffee farmers through biogas utilization, for example.

Takeshi Takama Takeshi works as an international expert on climate change, environment and energy for international and bilateral agencies for more than 10 years, including ADB, JICA, GIZ, and UN agencies as well as an associate for Stockholm Environment Institute and a professor at Udayana Univesity. His expertise includes Adaptation/Vulnerability for Climate Change (10 years), Renewable energy especially policy, development, and demand assessment of bio-energy (10 years), Green economy including micro-finance, trainings, and agro-business (3 years), Food security in African and Asian countries including rice, cacao, maize, and teff (7 years), Transportation especially on demands and mode choices (3 years), Education supervising master students at Oxford (5 years). His works were published in prestigious international journals and reported to policy makers, including the Ethiopian President in person. The latest achievements can be found in Takeshi's LinkedIn and his popular publications in Google Scholar.

Ibnu Budiman Ibnu has been working as researcher since 2011. He has done some research projects from University of Toronto, National University of Singapore, etc., with various themes from sustainability development, social issues, politics, and culture. Ibnu is also writer, speaker, facilitator and moderator in some media outlets, workshops, trainings and seminars about climate finance, environment, scientific and popular writing, public speaking, education, life skill, and management of organization in some local and national events. He writes also for several media such as Jakarta Post. His degree is Bachelor of Science from Department of Geography Universitas Indonesia. His thesis was about political geography, which was published as a book.

Michele Knight The specialization of Michele include food security and natural resource management (NRM) with particular experience implementing, evaluating, and researching development projects under the lens of climate change adaptation. She has worked in projects promoting ecological agriculture, community-based NRM (predominantly sustainable forest regeneration and management), livelihood diversification, environmental markets, and nutrition-sensitive value chain development. She has been working across Asia, Pacific and Africa, particularly in Australia, Mongolia, Canada, Fiji, Vanuatu, and Myanmar.

Perceiving, Raising Awareness and Policy Action to Address Pollinator Decline in Nigeria

Thomas Aneni, Charles Aisagbonhi, Victor Adaigbe and Cosmas Aghayedo

Abstract Insect pollinators contribute to agricultural crop yield and beekeeping provides a major source of livelihoods for farmers in Nigeria. This study developed two survey questionnaires and collected data from beekeepers, researchers and government officials to generate quantitative indicators for the purpose of description as a guide to action. Evaluation and characterization of colony bee loses by beekeepers were assessed. The surveys conducted between October 2015 and March 2016 consisted of questions related to: the importance of pollinators, including managed honeybees (*Apis mellifera*), in agriculture and observations on factors associated with pollinator declines; and management of bee mortality. Evaluation and characterization of colony bee loses by beekeepers in Osun State was conducted. Responses were received from 31 beekeepers and 20 policy makers and researchers. 81% of beekeepers reported a reduction in number of colonies. The results inform policy action on pollinator benefits for increasing crop yield and helping smallholder farmers adapt to a decline in insect pollinators. This study emphasizes pollination and insect pollinators as drivers of agricultural crop production with a view to providing guidance for sustainable management of pollinators and achievement of green growth objectives.

Keywords Insect pollinators · Colony bee loses · Bee keepers
Policy makers · Crop yield

1 Introduction

Animal pollination, mainly performed by bees, is an important ecosystem service with almost 90% of flowering plants and 75% of the world's most common crops benefiting from animal flower visitation (Klein et al. 2007; Ollerton et al. 2011).

T. Aneni (✉) · C. Aisagbonhi · V. Adaigbe · C. Aghayedo
Entomology Division, Nigerian Institute for Oil Palm Research, P.M.B. 1030,
Benin City, Nigeria
e-mail: tomaneni1@yahoo.com

P. Castro et al. (eds.), *Climate Change-Resilient Agriculture and Agroforestry*,
Climate Change Management, https://doi.org/10.1007/978-3-319-75004-0_25

As the majority of the world's staple foods are wind- or passively self-pollinated (wheat, corn, rice), or are vegetatively propagated (potatoes), their production does not depend on and increase with animal pollinators (insects, birds, and bats). These crops account for 65% of global food production, leaving as much as 35% depending on pollinating animals (Klein et al. 2007). In Nigeria, crops that require pollination by bees and the crop's natural pollinators include: Mango, Runner beans, Guava, Pear, Cowpea, Tomato, Grape, Onion, Okra, Oil palm and Cashew. Currently, the main conservation strategy at present is to promote pollinators through establishment of protected areas. Pollination is the transfer of pollen from the stamen, or the male component of a flower, to the pistil, or the female part. The pollen grain reaches the ovary via the stigma to fertilize the ovules which produce the seeds and fruit. Several types of vectors may ensure fertilization of a flower: wind, water, and animals, especially insects. There is increasing evidence of a global decline in insect pollinators that threatens the reproductive cycle of many plants and may reduce the quality and quantity of fruit and seeds, many of which are of nutritional and medicinal importance to humans. Identification of appropriate actions is needed, especially given the uncertainty posed by gaps in both scientific knowledge and effective policy interventions. Insect pollinators, comprising both managed (e.g. honeybee *Apis mellifera*) and wild populations (species that exist as non-managed wild populations including wild *Apis* spp.), have become a focus of global scientific, political and media attention because of their apparent decline and the perceived impact of such declines on crop production (Cameron et al. 2011; Kerr et al. 2015). Pollinator declines are a consequence of multiple environmental pressures, e.g. habitat transformation and fragmentation, loss of floral resources, pesticides, pests and diseases, and climate change (Potts et al. 2010; Vanbergen 2013). Similar environmental pressures are faced in Nigeria where there is a high demand for pollination services. The fact that almost half the data on pollinator decline from recent studies comes from only five countries, with only 4% of the data from the continent of Africa (Archer et al. 2014), highlight the lack of information.

Despite the perceptions of global honeybee decline, long-term global data indicate an increase in managed honeybees (Aizen et al. 2008; Aizen and Harder 2009), except in the USA. However, agricultural demand could outstrip supply of managed honeybees (Aizen and Harder 2009) and greater demand for high value fruit and nut crops may further increase demand for pollination services (Gallai et al. 2009; Breeze et al. 2014). This demand implies that pollination services may experience constraints even without a dramatic decline in honeybees and highlights the need for effective strategies to safeguard reliable pollination services for agriculture. Such strategies could include: improved health of managed honeybees; identifying possible substitutes for managed honeybees (Corbet et al. 1991; Potts et al. 2011); increasing and diversifying the suite of wild pollinators where possible (Corbet et al. 1991); and increasing the effectiveness of wild pollinators (Brittain et al. 2013). The latter includes conserving suitable food sources and nesting habitat for wild pollinators within the agricultural matrix and raises the question: 'Is management to secure biodiversity benefits more rewarding for crop production than management less favourable to biodiversity?' If so, then strategies to improve pollination services need to be aligned with strategies to conserve biodiversity in agricultural landscapes (Ghazoul 2013). Another way to examine the likelihood or proximity of a pollination crisis is to examine delivery of pollination services. Although global honey bee stocks have increased by ~45%, demand has risen more than supply, for the fraction of global crops that require animal pollination has tripled over the same time period (Smith et al. 2013), making food production more dependent on pollinators than before. It has also emerged that the majority of crop pollination, at a global scale, is delivered by wild pollinators rather than honey bees. Yields correlate better with wild pollinator's abundance than with abundance of honey bees (Breeze et al. 2011; Garibaldi et al. 2013; Mallinger and Gratton 2014); hence increasing honey bee numbers alone is unlikely to provide a complete solution to the increasing demand for pollination. Reliance on a single species is also a risky strategy (Kearns et al. 1998). While Aizen et al. (2008) concluded from a global analysis of changing crop yields over time that there was not yet any clear evidence that a shortage of pollinators was reducing yield, a subsequent analysis of the same data set by Garibaldi et al. (2011) shows that yields of pollinator-dependent crops are more variable, and have increased less than crops that do not benefit from pollinators, to the extent that a shortage of pollinators is reducing the stability of agricultural food production. In a meta-analysis of 29 studies on diverse crops and contrasting biomes, Garibaldi et al. (2011) found that wild pollinator visitation and yields generally drop with increasing distance from natural areas, suggesting that yields on some farms are already impacted by inadequate pollination.

1.1 Nigeria Agricultural Transformation Agenda

The agricultural transformation agenda of the Federal Ministry of Agriculture and Rural Development is the pillar of Nigeria's current agricultural policy. The Agricultural sector is an important segment of the economy with high potentials for employment generation, food security and poverty reduction. Low productivity in Nigeria over the years has been largely attributed to low fertilizer and improved seed utilization and inadequate government expenditure and the inability to compete with others. The vision in the transformation strategy is to achieve a hunger-free Nigeria through an agricultural sector that drives income growth, accelerates achievement of food and nutritional security, generates employment and transforms Nigeria into a leading player in global food markets to grow wealth for millions of farmers. Transformation action plan for some priority agricultural commodities is focused in the six geopolitical zones of the country. The commodities are rice, cassava, sorghum, cocoa cotton, maize, dairy, beef, leather, poultry, oil palm, fisheries as well as agricultural extension. These are carried out through the value chains of each of the commodities while recognizing roles of the actors/stakeholders along the nodes of the chain, inputs requirements in achieving production targets, constraints faced and expected output. The main target is to grow the agricultural sector through the various commodities and also to generate employment opportunities. Pollination and pollinators as drivers of agricultural crop production yield was not emphasized in the agricultural transformation agenda policy with a view to providing guidance for sustainable management of pollinators.

Objectives:

(a) Assess honey bee colony population abundance in study area.
(b) Identify targeted activities and methods to manage and mitigate changes in pollinator abundance.
(c) Development of best pollinator management strategy.

2 Methods

2.1 *Study Area*

Osun State is an inland State in South-Western Nigeria with capital is Osogbo. Its situated in the tropical rainforest zone. It covers an area of approximately 14,875 km^2 and lies between latitude 7° 30′N and longitude 4° 30′E. Its boundaries are: Ogun State to the South; Kwara State to the North; Oyo State to the West; and Ekiti and Ondo State to the East. The State was selected due to the beekeeper's being well organized under the Federation of Beekeepers Association in Nigeria (FEBKAN), Osun State branch.

2.2 *Data Collection*

Evaluation and characterization of colony bee loses by beekeepers in Osun State, Nigeria, using a detailed questionnaire. A survey on insect pollination management was conducted among researchers and policy makers. Information was collected using interviews and two survey questionnaires for beekeepers (Annex 1), policy makers and researchers (Annex 2). The questions used for the beekeeper survey is adapted from the colony loss monitoring questionnaire (Van der Zee et al. 2013) for beekeepers in Osun State. Beekeeper assessments were based on production colony information between October 2014 and September 2015. The survey consisted of 22 major questions with some questions further divided into subparts. Although the majority of questions were intended to generate yes/no responses, several questions were multiple-choice or were open-ended to provide respondents with an opportunity to enter their own responses and supporting references. Thirty one (Kevan 2001) participating beekeepers returned their completed surveys to the author out of a total of thirty five beekeepers indicating 89% response rate. Data were excluded from the loss rate analysis if the essential questions about colony losses were not answered. Where necessary, translation was required in the indigenous language. Participant knowledge, expectations, experience through spoken or written forms were obtained. Transcripts were analyzed to provide salient information, including potential trends in responses. Participants for the pollination management survey included government officials (Corbet et al. 1991), researchers and agricultural scientists (Aizen et al. 2008). After survey results were collected, they were entered into a spread sheet and frequency of response tables calculated.

3 Results and Discussion

Overall, the Nigerian honeybee populations in the study area have not exhibited significant losses (number of dead bees in production colonies), probably because of the relatively unmanaged state of African honeybees and the fact that they are indigenous. However, the fairly recent advent of environmental change (Climate change) globally and in Nigeria suggests that our bees are now more vulnerable and stressed than was previously the case. There is need to ensure that we are tackling all the issues that place pressure on honeybees, because in so doing we will hopefully also ensure the survival of some of the other lesser-known pollinators.

3.1 Beekeeper's Survey

In this study, 31 beekeepers (89% of the total number of beekeepers in Osun State) participated out of a total number of 35 beekeepers. The summary beekeepers response received from respondents is presented in Annex 3.

3.2 Policy Maker's Survey

This study developed a questionnaire and collected information from researchers and government officials to generate policy indicators related to pollination and pollinators for the purpose of description as a guide to action. The summary policy survey response received from respondents is presented in Annex 4 (researchers) and Annex 5 (government officials).

3.3 Beekeeper's Survey Response

In relation to production colonies during a one year period (October 2014–September 2015), beekeepers reported a loss of production colonies without dead bees in the hive. However, only 3% of beekeepers reported loss of production colonies due to queen challenges (queen less or drone-laying queen). 9% of beekeepers reported a reduction in total production colony numbers due to uniting/merging. A majority of beekeepers (48%) that responded were ignorant of the cause of the death of their colonies while others attributed the cause to starvation (3%), poor queens (6%) and disease (6%) and other unknown factors (12%). The beekeepers unanimously agreed that origin of queens were through rearing by their colonies. Most beekeepers did not have to provide a new queen (94%) nor were their colonies treated with a product for disease condition (97%). Most beekeepers (97%) reported that their colonies were neither contracted for pollination services nor moved for honey production. Due to a large proportion of small holder farmers in Osun State, bee movement for crop pollination is not practiced presently. This is in contrast with large-scale agricultural production systems such as almonds, apples, melons and other cucurbits where large fields provide limited edges where wild pollinators may nest (Chagnon 2008). Beekeepers replaced on average, 47% of combs in the majority of production colonies and majority (81%) did not use any supplemental sugar feed while others used honey (19%), Beet sugar (6%) and inverted beet sugar (3%). Colony disturbance reported by beekeepers were mainly by ants, humans (theft), rats and squirrels.

3.4 Honey Bee Colony Population Abundance in Study Area

All beekeepers reported a reduction in production colonies without dead bees in the hive. All local identified stressors—climate change, habitat loss, disease, diets, and pesticides- which were observed not to act in isolation. Inadvertently making bees more susceptible to population inhibiting pressures, and driving honey bee colony losses and declines of managed bees in the study area.

3.5 Researchers and Government Officials Survey Response

Information from researchers and Government officials indicates that population abundance trends in honey bee and other pollinator populations have largely not been documented in Nigeria. A majority of respondents, researchers (75%) and Government officials (83%) were not aware or uncertain of active in-country pollination research on various native and non-native pollinators. This implies that there is an urgent need for special funding for pollination research. A majority of respondents for both researchers and Government official's (75%) were not aware or uncertain on the role of managed bees in pollinating major crops. This implies that there is need to create incentives and increase awareness for farmers to increase crop productivity with managed bees. All respondents (100%) were either not aware or uncertain if honey bee population declines have been documented in Nigeria. This is largely because there have not been large scale studies on honey bee abundance and distribution. There is need for a country wide bee abundance assessment. All respondents (100%) were either not aware or uncertain if other non-honey bee pollinator populations have been documented. Abundance in other non-honey bee pollinator populations has been documented. However, information on non-honey bee pollinator population variations over time is limited. There is need for further studies on non-honey bee pollinators of major crops. When asked if any Ministry has a formal insect pollination policy, a large percentage of the respondents, researchers (87%) and Government officials (92%) were not aware of any policy. An outcome from the survey indicates that there is no formal insect pollination policy by the Ministry of Agriculture. When asked if the Federal Ministry of Agriculture has conducted cross-ministerial work, with any other Ministry, incorporating insect pollination into national policies and programs, a large percentage of respondents, researchers (87%) and Government officials (100%) were not aware or expressed uncertainty. However, a general insect pest control policy is available for crop protection in Nigeria. There is need for incorporating insect pollination and pollinators into national policies and programs.

3.6 Perception of the Importance of Honey Bee Pollination Service by Policy Makers

The perception of policy makers surveyed were very low (75%) as most were not aware or uncertain on the role of managed bees in pollinating major crops. This is attributable to fragmentary information on the benefits of pollinators and pollination services and the need for broad policies to aid their conservation in other to boost crop production.

3.7 Risks to Pollination Decline that Deserve Future Study

Pollination is the transfer of pollen from the stamen, or the male component of a flower, to the pistil, or the female part. The pollen grain reaches the ovary via the stigma to fertilize the ovules which produce the seeds and fruit. A pollinating species is termed "wild" when its habitat is located in a natural environment or an environment with no human interference. A "native pollinator" refers to a species originating in, associated with, and established in a given habitat over a long period. Introduced (managed) pollinators refer to species in which reproduction and survival are controlled by man (Chagnon 2008). Over the past two decades, there has been considerable concern globally over the apparent reduction in populations of pollinators of all kinds. Several research projects, publications and public awareness campaigns have focused on determining the possible causes of decline in introduced pollinator numbers, particularly among honey bees. Some of the causes include.

3.8 Pesticides

Pesticides constitute a major threat to pollinators. It has been known for some time that the use of pesticides to control agricultural pests can have a negative impact on honey bee colonies (Johansen and Mayer 1990). For decades, there have been massive losses in bee colonies wherever agriculture and beekeeping have co-existed. Losses in bee numbers are often the result of poor handling and application procedures for pesticides or else failure to follow the recommendations printed on the label. Even when the instructions are closely followed, the pesticide will inevitably constitute a serious risk for all the pollinators, regardless of whether they are wild or introduced. Pesticides are potentially able to harm a large number of pollinating species and even to eliminate a certain number of populations of species occurring in an ecosystem (Nabhan and Buchmann 1997). The presence and abundance of suitable floral resources in an environment are therefore extremely important factors. A relatively new class of widely used systemic insecticides, the neonicotinoids, is highly toxic to insects, including bees, at very low concentrations. The group includes imidacloprid, thiamethoxam, clothianidine and several other compounds which are widely used to coat seeds. These compounds can be taken up via the roots and then carried by the sap to all parts of the plant as it grows. This ensures protection against root pests but also against insects attacking the aerial portions of the plant. Since they are active until the flowering stage, they can be picked up by pollinators in the pollen and nectar. Pesticide use in Nigeria has been on the increase over the decades (Asogwa and Dongo 2009). It has been estimated that about 125,000–130,000 metric tons of pesticides are applied every year in Nigeria (Ikemefuna 1998). There is currently a Pesticides Registration Regulations arising from the Drugs and Related Products (Act 19 of 1993). It is well established that the improper use of agricultural pesticides negatively affects development of honey bee colonies. Pesticides should be reduced or completely eliminated. Standard guidelines and label instructions should be correctly applied.

3.9 Transgenic Crops (GMOs)

Transgenic plants were developed specifically to reduce some of the undesirable and involuntary effects of pesticides. There are concerns, however, about potential impacts the direct effects of insecticide proteins in the pollen may have on non-targeted species, including some pollinators (Losey et al. 1999). These concerns focus on the lack of information on the lethal threshold of transgenic insecticide proteins and the sublethal effects of the proteins on the physiological and reproductive behavior of the insects feeding on them. Published results suggest that the impacts of transgenic plants on bees should be examined case by case and depend on the portion of the plant that is ingested (Malone and Pham-Delègue 2001). In Nigeria, no method has been developed to assess the impact of genetically modified organisms on pollinators under natural conditions.

3.10 Fragmentation and Habitat Loss

Fragmentation and habitat loss are two types of disruption that have been recognized as important factors in loss of biodiversity on a local as well as global scale. Habitat loss refers to the loss of a natural environment arising from a primary succession, i.e., a natural landscape. Fragmentation of a habitat refers to the break-up of a habitat into fragments that are often too small to ensure the viability of populations of all species. Pollinators and pollination-dependent plants are not protected from this type of disruption (Kevan 2001). In Nigeria, the traditional land tenure system in Nigeria coupled with increasing population encourages land fragmentation with attendant consequences for agricultural productivity and pollinator loss. Beekeepers in Osun State, observed that major challenges include pressures due to reduction in native vegetation area and indiscriminate pesticide use. Land fragmentation has severe consequences for agricultural development; it leads to scattering of plots, little incentive for improvements, lack of security of tenure, restricted scale of operations (Idowu and Oladebo 1999). In spite of these associated costs, land fragmentation is still persistent and wide spread in Nigerian agricultural practice. Land fragmentation practices not only reduce natural and semi-natural habitats, they also cause loss of diversity among cultivated plants, further impoverishing the range of floral resources available to the natural pollinators in the area. Habitat fragmentation and loss affect pollinators in two ways. First, they reduce the availability of the range of plants capable of meeting all food needs throughout a season (Kearns and Inouye 1997). Loss of access to resources could increase competition among local species for the limited resources. Secondly, habitat loss could also disrupt nesting among a number of bee species that dig their nests in burrows.

3.11 Climate Change

According to some specialists, behavioral changes linked to the species' physiology have already been observed in some pollinators. Over the past two decades, British butterflies have made their first appearance of the season earlier and earlier and the peak period has also been brought forward. Similar changes have also been observed in California's butterflies (Forister and Shapiro 2003). The average period for the first flight of 16 species studied tended to occur earlier. An average difference of 24 days for four of them represented a statistically significant trend. On the other hand, seven species tended to appear later in the season. Different species of pollinators are consequently going to react differently to climate change, which will affect the diversity and abundance of their populations in varying degrees. In terms of physiology, some factors like the photoperiod and temperature exert a control on endocrine activity and can modify fertility, the mode and rate of reproduction as well as the rate of development. These physiological reactions may differ from one species to the next. The underlying causes for changes within a pollinator community are therefore highly variable. Climate change and variability from 1961 to 2010 and projections up to 2050 and its impacts on the oil palm leaf miner—*Coelaenomenodera elaeidis* in Edo State, Nigeria has been evaluated (Aneni et al. 2015). Currently, honey bee farmers in Nigeria, have observed low yield and the crystallized honey combs in their hives (Centre for Bee Research and Development (CEBRAD) 2016) which has been attributed to increased rainfall intensity (scarcity period for honey bee activity) (Oyerinde et al. 2014). Information gathered from this study indicates that there are limited published studies on pollinators and climate change interaction in Nigeria. However, it can largely be deduced from other insect studies that climate change would have an impact on insect pollinators in Nigeria.

4 Conclusion

4.1 Recommendations and Proposed Actions

The clear message of this study is that pollination is a key factor in agricultural productivity and pollinators are essential in providing this service. Fears over pollinators and pollination services continue to build up in the scientific and public space. Therefore, there is the need to enhance local data for understanding the status and trends of pollinators to sustainably manage pollination services. All stakeholders need to ensure that pollination is well understood as a key limiting factor in agricultural productivity and that steps are taken to manage it in sustainable ways that maintain populations of pollinators and their habitats.

4.2 Pollinator Gap Analysis

There is mismatch between government and local understanding of the problem of pollination service loss and governance priorities. This points out that while larger institutions can form the pillar for wider activities, practical measures need to be adapted to facilitate rather than hinder local farmers. The insect pollination gap analysis highlighting strengths and challenges in Nigeria is presented Table 1.

4.3 Pollinator Management

Pollinator management practices have been identified, to conserve and manage pollinator populations. These practices not only benefit pollination ecosystem services, but contribute to crop diversity (biodiversity), soil health and reduced pesticide use. They include.

4.3.1 Reduced Pesticide Usage

Pest control practices such as Integrated Pest Management that enhances natural pest controls reduce or eliminate the use of pesticides. At the same time, this greatly benefits pollinators which may be heavily impacted by pesticides.

Table 1 Strengths and challenges in the management of insect pollination in Nigeria

Factors	Strengths	Challenges
Insect pollinators of major crops	Insect pollinators of major crops identified	Inadequate capacity for appropriate management of the pollinators
Potential drivers of pollinator decline	Potential drivers of pollinator decline defined	Limited knowledge on drivers of pollinator decline
Policy on pollinator management	Progress in the development of national policies on agriculture	Pollinators not taken into account in existing national policies
Coordination, collaboration and partnership	Formal and informal structures for collaboration of relevant sectors exist	Lack of mechanisms for coordination and collaboration among relevant sectors
Human resource capacity on pollinator management	National training institutions available	Inadequate human resource for pollinator management
Surveillance capacity for pollinators	Relevant government institutions available	Surveillance systems for pollinator monitoring generally absent
Laboratory capacity for testing pollinator pesticide lethal and sub-lethal levels	Reference laboratories that deal with most chemicals identified as being of major public concern to pollinators available	Inadequate laboratory equipment and essential reagents

4.3.2 Maintaining Hedgerows and Floral Diversity

Hedgerows provide habitat and forage resources for bees, and by diversifying the floral resources, insect pollinators are encouraged to remain on-site even in the following year. This also contributes to biodiversity conservation.

4.4 Proposed Actions

(a) Pollinators of major crops in Nigeria

  – Develop regulations, guidelines and tools for the safe management of insect pollinators

(b) Legislation and policy

  – Develop comprehensive policies for an integrated approach to insect pollinators management using a life-cycle approach

(c) Coordination, collaboration and partnership

  – Implement inter-sectoral coordination mechanisms for the safe management of insect pollinators
  – National multi-sectoral task forces that deal with issues related to crops and the environment to include insect pollinators on their agenda

(d) Human resource capacity

  – Develop training packages on pollinators that can be used to upgrade the capacity and capability of farmers

(e) Surveillance capacity

  – Enhance surveillance capacity for monitoring insect pollinators that could have impact on agricultural production
  – Foster inter-sectoral collaboration in the sharing of information and surveillance data

(f) Laboratory capacity

  – Develop at the minimum capability for laboratory analysis of lethal and sub-lethal pesticide levels in insect pollinators.

Further scientific research is needed to inform policy decisions that are underpinned by sound scientific basis. These include:

1. Quantifying the abundance of pollinators in Nigeria and the risks associated with the loss of pollination services.
2. Determine the economic value of pollinators for key crops.

3. Establish the conservation status of insect pollinators.
4. Investigate the drivers of pollinator loss.
5. Investigate honeybee forage resources under global change scenarios (land use change or climate change).
6. Detailed research on threats to honeybees in Nigeria.
7. Research alternative species of pollinators (other than the honeybee) for potential managed pollination.
8. More standardized monitoring and documentation of the occurrence and abundance of pollinators are needed to enable comprehensive assessment of pollinator trends.
9. Developing excellence in pollinator taxonomy.
12. Identification of native pollinators for agricultural production.
11. Studying plant-pollinator relationships.
12. Protecting foraging sites and restoring degraded habitats.
13. Studies of pesticide impact and pathogens of wild insect pollinators.
14. Using the honey bee as a bio-indicator.

4.5 Priority Actions

The Sustainable Development Goals recognizes that biodiversity and ecosystem services can play a role in poverty alleviation, and the need to integrate ecosystem services such as pollination into food production. Priority actions include:

- Dissemination of this report to all relevant stakeholders.
- In-depth on-site evaluation of pollinator numbers and diversity in selected states based on the findings of this report.
- Elaboration of a country 2017–2020 strategy for management of pollinators to address the issues and challenges identified in this report.
- Development where and as necessary on the capacities required for pollinator management.
- Development of a comprehensive training package for public agricultural professionals on pollinator management, working in close collaboration with relevant stakeholders.
- Provision of technical support to research institutions for the implementation, monitoring and evaluation of the 2017–2020 country strategy after it is developed.

Acknowledgements I acknowledge the funding provided through the African Climate Change Fellowship Program (ACCFP). The ACCFP is supported by a grant from the International Development Research Centre (IDRC), Canada. The International START Secretariat is the implementing agency in collaboration with the Institute of Resource Assessment (IRA) of the University of Dar Es Salaam. I appreciate the facilitation of the beekeeper's questionnaire by Mr. Kayode Ogundiran and Mr. Bidemi Ojeleye, Centre for Bee research and Development (CEBRAD).

Annex 1: Essential Beekeeper Information and Mortality Quantification

Respondent Information: *Please fill out this Sect.*

1. How many production colonies did you have on October 1st 2014?
   In the next questions you are asked for numbers of colonies lost. Please consider a colony as lost if it is dead, or reduced to a few hundred bees, or alive but with unsolvable queen problems
2. How many of your production colonies were lost between October 1st 2014 and September 30th 2015
3. How many of your production colonies were lost between October 1st 2014 and September 30th 2015 without dead bees in the hive or in the apiary (bee yard)?
4. How many of the production colonies were lost between October 1st 2014 and September 30th 2015 because of queen problems (queenless or drone-laying queen)?
   Please answer the next 2 questions only if you bought, sold, united or split colonies between October 1st 2014 and September 30th 2015
5. Between October 1st 2014 and September 30th 2015;
   What was the reduction in total production colony numbers due to uniting/merging?*
   *eg two colonies united/merged together = loss of one colony
6. How many production colonies did you have on September 30th 2015?
   Identification of Possible Risk Factors (If any)
   In the next question, please TICK the alternatives that best answers your situation (you may tick more than one)
7. To what do you attribute the major cause of the death colonies in your operation (If observed)?
   Don't know.....
   Starvation........
   oor queens.....
   Disease............
   Others..............
   The next question is about the origin of your queens. Please choose the items which describe your situation (you may tick more than one box).
8. What is the origin of your queens?

   (a) Reared by the colony it self
   (b) Reared from one of your own selected queens
   (c) Acquired from a queen breeder
   (d) Acquired for a queen breeder outside Nigeria
   The next question is about queen problems, please don't include normal requeening (e.g. when the queen is old) in your answer.

9. In how many of your colonies did you have to provide a new queen because of queen problems last year?
10. In what months and year have you treated your colonies with a product for disease condition during the period October 2014–September 2015?
11. How many of your colonies were contracted for pollination services last year?
12. How many of your colonies were moved for honey production last year? Please choose the honey flow sources in the next question, which best describe your situation
13. What percentage of combs did you replace in the majority of your production colonies last year?
14. If you gave your colonies a supplemental sugar feed last year, what product was used

    (a) Honey
    (b) Beet Sugar
    (c) Inverted Beet Sugar Syrup
    (d) High Fructose Corn Syrup (HFCS)
    (e) Other product, namely

15. Have your colonies suffered any disturbance by

    (a) Mice/rats
    (b) Ants
    (c) Squirrels
    (d) Humans (vandalism, robbery)
    (e) Other

Annex 2: Insect Pollination Management Survey

Respondent Information: *Please fill out this section.*
Name:
Phone number:
Email address:
Age: <21 21–30 31–40 41–50 >50
Gender:

1. Are you aware of research that has been conducted on the relative proportions of crops
   pollinated by various native and non-native pollinators?
   ___Yes ___No ___Uncertain
   If yes, please provide a reference.
2. Do managed bees pollinate major crops in Nigeria?
   ___Yes ___No ___Uncertain
   If yes, please list if known

3. Have declines in honey bee populations been documented in Nigeria?
   ___Yes ___No ___Uncertain
   If yes, please provide a reference.
4. Have declines in other pollinator (non-honey bee) populations been documented in Nigeria?
   _ __Yes ___No ___Uncertain
   If yes, please provide a reference for the study or survey.
5. Please describe your expertise
   ___Agricultural policy
   ___scientist
   ___research
   ___Other (*please specify*)____
   ____

   Insect pollination in National-Level Policies and Programmes
6. To the extent of your knowledge, which, if any, Ministries have a formal insect pollination policy, or include insect pollination considerations within their national-level policies and/or programmes?

Ministry	Yes	No	Unsure
Agriculture			
Water Resources			
Environment			
Science and Technology			
Lands and Housing			
Education			
Other			
Other			

   Please provide any further information: ________________________________
   ____
7. To the extent of your knowledge, Has the Federal Ministry of Agriculture conducted cross-ministerial work, with any other Ministry, incorporating insect pollination into national policies and programs?
   Yes_______No_______Unsure_______
   If yes, with which Ministry or Ministries? __________________________
   ____
   Please provide any further information: ______________________________
   ____
   ____
   ____
   ____
   ____

Do you have a colleague whom the bearer should contact for this survey? Please provide their information below.

Name: ____

Position: ____

Contact Information: ___________________________________

Thank you for taking time to complete this questionnaire. The information will help prioritize research efforts on topics of benefit for insect pollination, beekeepers and agricultural productivity.

Annex 3: Summary Bee Keeper's Response B

A	B														
	1	2	3	4	5	6	7	8	9	10	11	12	13	14	15
1	2670	0	0	0	0	–	–	A	75	0	0	0	25%	ABE	ABD
2	20	0	0	0	15	20	–	A	0	0	0	0	–	A	AB
3	12	4	4	–	–	8	–	A	0	0	0	0	–	0	ABD
4	10	3	3	–	–	7	A	A	0	0	0	0	–	0	AD
5	14	4	4	–	–	10	A	A	0	0	0	0	–	0	B
6	44	4	4	–	–	40	A	A	0	0	0	0	–	0	B
7	5	1	0	–	–	–	E	A	0	0	0	0	–	0	AB
8	12	4	4	–	–	8	E	A	0	0	0	0	–	0	D
9	15	3	3	–	–	12	A	A	0	0	0	0	60%	0	AD
10	24	5	2	0	0	20	BCDE	A	0	0	0	0	2%	AC	AD
11	40	10	5	–	–	30	A	A	0	0	0	0	–	A	D
12	4	1	1	–	–	3	A	A	0	0	0	0	–	0	D
13	27	8	4	–	–	27	–	A	0	0	0	0	–	0	D
14	82	6	2	–	46	82	E	A	0	0	0	0	–	0	BD
15	14	6	4	–	–	8	A	A	0	0	0	0	–	0	D
16	9	2	2	–	–	7	A	–	0	0	0	0	55%	0	B
17	15	6	6	–	–	9	A	A	0	0	0	0	55%	0	AB
18	18	18	17	–	–	1	–	A	0	0	0	0	–	0	D
19	20	2	2	–	–	18	A	A	0	0	0	0	60%	0	D
20	300	5	10	–	55	210	D	A	5	June & July	8	150	–	A	ABD
21	40	–	–	–	–	–	A	A	0	0	0	0	–	0	AB
22	10	–	–	–	–	10	–	A	0	0	0	0	60%	0	B
23	22	–	–	–	–	20	–	A	0	0	0	0	–	A	AE
24	100	–	–	–	–	100	A	A & B	0	0	0	0	–	B	BC
25	20	3	–	1	0	–	C	A	0	0	0	0	–	0	BC
26	43	15	15	–	–	28	A	A	0	0	0	0	–	0	B

(continued)

(continued)

A	B														
	1	2	3	4	5	6	7	8	9	10	11	12	13	14	15
27	5	1	1	–	–	–	A	A	0	0	0	0	55%	0	B
28	35	10	10	–	–	25	A	A	0	0	0	0	–	0	B
29	30	10	10	–	–	20	–	A	0	0	0	0	–	0	E
30	17	2	2	–	–	15	A	A	0	0	0	0	–	0	B
31	53	9	9	–	–	44	–	A	0	0	0	0	–	0	ABC

A - Respondents
B - Response to questions (Annex 1)

Annex 4: Summary Response for Researchers

Questions	Yes	No	Uncertain
Are you aware of research that has been conducted on the relative proportions of crops pollinated by various native and non-native pollinators?	2	2	4
Do managed bees pollinate major crops in Nigeria?	2	2	4
Have declines in honey bee populations been documented in Nigeria?	0	1	7
Have declines in other pollinator (non-honey bee) populations been documented in Nigeria?	0	1	7
To the extent of your knowledge, which, if any, Ministries have a formal insect pollination policy, or include insect pollination considerations within their national-level policies and/or programmes?	1	3	4
To the extent of your knowledge, Has the Federal Ministry of Agriculture conducted cross-ministerial work, with any other Ministry, incorporating insect pollination into national policies and programs?	0	2	5

Annex 5: Summary Response for Government Officials

Questions	Yes	No	Uncertain
Are you aware of research that has been conducted on the relative proportions of crops pollinated by various native and non-native pollinators?	2	4	6
Do managed bees pollinate major crops in Nigeria?	3	2	7
Have declines in honey bee populations been documented in Nigeria?	0	2	10

(continued)

(continued)

Questions	Yes	No	Uncertain
Have declines in other pollinator (non-honey bee) populations been documented in Nigeria?	0	3	9
To the extent of your knowledge, which, if any, Ministries have a formal insect pollination policy, or include insect pollination considerations within their national-level policies and/or programmes?	1	7	4
To the extent of your knowledge, Has the Federal Ministry of Agriculture conducted cross-ministerial work, with any other Ministry, incorporating insect pollination into national policies and programs?	0	5	7

Glossary

Beekeeping

The husbandry of bees, especially honeybees (the genus Apis) but can be applied to other bees.

Biodiversity

Short for "Biological diversity" which is the variety of life on Earth. The variability among living organisms from all sources including terrestrial, marine and other aquatic ecosystems and the ecological complexes of which they are a part; this includes diversity within species, between species and of ecosystems.

Diversity

The condition of having or comprising differing elements or qualities (peoples, organisms, methodologies, organizations, viewpoints, etc.).

Drivers, direct

Drivers (both natural and anthropogenic) that operate directly on nature (sometimes also called pressures).

Drivers, indirect

Drivers, that operates by altering the level or rate of change of one or more direct drivers.

Drivers, institutions and governance and other indirect

The way in which societies organize themselves. They are the underlying causes of environmental change that are external (exogenous) to the ecosystem in question.

Drivers, natural direct

Direct drivers that are not the result of human activities and are beyond human control.

Economic value

A measure of the benefit provided by a good or service to an economic agent (e.g. buyer or seller). It is not necessarily the same as market value. It is generally measured by units of currency, and can be interpreted to mean the maximum amount of money a specific actor is willing and able to accept or pay for the good or service.

Ecosystem

A community of living organisms (plants, animals, fungi and microbes) in conjunction with the nonliving components of their environment (such as energy, air, water and mineral soil), all interact as a system.

Ecosystem services

A service that is provided by an ecosystem as an intrinsic property of its functionality (e.g. pollination, nutrient cycling, Nitrogen fixation, fruit and seed dispersal). The benefits (and occasionally disbenefits) that people obtain from ecosystems. These include provisioning services such as food and water; regulating services such as flood and disease control; and cultural services such as recreation and sense of place. In the original definition of the Millennium Ecosystem Assessment the concept of "ecosystem goods and services" is synonymous with ecosystem services. Other approaches distinguish "final ecosystem services" that directly deliver welfare gains and/or losses to people through goods from this general term that includes the whole pathway from ecological processes through to final ecosystem services, goods and values to humans.

Farm

An area of land, a holding of any size from a small plot or garden (fractions of a hectare) to several thousand hectares, that is devoted primarily to agriculture or an area of water that is devoted primarily to aquaculture, to produce food, fibre, or fuel. A farm may be owned and operated by an individual, family, community, corporation or a company, may produce one to many types of produce.

Field

In agriculture, it is a defined area of cleared enclosed land used for cultivation or pasture.

Flowering plant

Plants that are characterized by producing flowers, even if inconspicuous. They are collectively called angiosperms and include most plants grown for food and fibre.

Food Security

The World Food Summit of 1996 defined food security as existing "when all people at all times have access to sufficient, safe, nutritious food to maintain a healthy and active life".

Global

Pertaining to the whole world.

Governance

All processes of governing, whether undertaken by a government, market or network, whether over a family, tribe, formal or informal organization or territory and whether through laws, norms, power or language. It relates to the processes of interaction and decision-making among the actors involved in a collective problem that lead to the creation, reinforcement, or reproduction of social norms and institutions.

Habitat fragmentation

A general term describing the set of processes by which habitat loss results in the division of continuous habitats into a greater number of smaller patches of lesser total and isolated from each other by a matrix of dissimilar habitats. Habitat fragmentation may occur through natural processes (e.g. forest and grassland fires, flooding) and through human activities (forestry, agriculture, urbanization).

Insecticide

A substance that kills insects. Insecticides may be synthetic chemicals, natural chemicals, or biological agents.

Introduced pollinator

A pollinator species living outside its native distributional range.

Invasive species

A species, that once it has been introduced outside its native distributional range, has a tendency to spread over space without direct human assistance.

IPM (integrated pest management)

It is a broadly based approach that integrates various practices for economic control of pests (q.v.). IPM aims to suppress pest populations below the economic injury level (EIL) (i.e. to below the level that the costs of further control outweigh the benefits derived). It involves careful consideration of all available pest control techniques and then integration of appropriate measures to discourage development of pest populations while keeping pesticides and other interventions to economically justifiable levels with minimal risks to human health and the environment. IPM emphasizes the growth of a healthy crop with the least possible disruption to agro-ecosystems and encourages natural pest control mechanisms.

Mitigation

Lessening the force or intensity of something that can result in disbenefits.

National

Pertaining to a nation state or people who define themselves as a nation. A nation can be thought of as a large number of people associated with a particular territory and who are sufficiently conscious of their unity to seek or to possess a government peculiarly its own.

Native pollinator

A pollinator species living in an area where it evolved, or dispersed without human intervention.

Parasite

An organism that lives on or within another organism of a different species (the host) from which it obtains nourishment and to which it causes harm.

Pest

An animal, plant, fungus, or other organism that thrives in places where it is not wanted by people, e.g. in fields, with livestock, in forests, gardens, etc.

Pollination

The transfer of pollen from an anther to a stigma. Pollination may occur within flowers of the same plant, between flowers of the same plant, or between flowers of different plants (or combinations thereof). Although pollination is a precursor to plant sexual reproduction, it does not assure same.

Pollinator
An agent that transports pollen. Such agents may be animals of many kinds or physical (wind or water), or both.
Pollinator decline
Decrease in abundance or diversity, or both, of pollinators.
Uncertainty
Any situation in which the current state of knowledge is such that (1) the order or nature of things is unknown, (2) the consequences, extent, or magnitude of circumstances, conditions, or events is unpredictable, and (3) credible probabilities to possible outcomes cannot be assigned.

References

Aizen MA, Harder LD (2009) The global stock of domesticated honey bees is growing slower than agricultural demand for pollination. Curr Biol 19:1–4

Aizen MA et al (2008) Long-term global trends in crop yield and production reveal no current pollination shortage but increasing pollinator dependency. Curr Biol 18:1–4

Aneni T, Aisagbonhi C, Adaigbe V, Iloba B (2015) Evaluation of climate variability impacts on the population of the oil palm leaf miner in Nigeria. Annu Res Rev Biol 8(1):1–15. https://doi.org/10.9734/arrb/2015/19895

Archer CR, Pirk CW, Wright GA, Nicolson SW (2014) Nutrition affects survival in African honeybees exposed to interacting stressors. Funct Ecol 28(4):913–923

Asogwa EU, Dongo LN (2009) Problems associated with pesticide usage and application in Nigerian cocoa production: a review. Afr J Agric Res 4(8):675–683

Breeze TD, Bailey P, Balcombe KG, Potts SG (2011) Pollination services in the UK: how important are honeybees? Agric Ecosyst Environ 142:137–143. https://doi.org/10.1016/j.agee.2011.03.020

Breeze TD, Vaissière BE, Bommarco R, Petanidou T, Seraphides N, Kozák L, Potts SG (2014) Agricultural policies exacerbate honeybee pollination service supply-demand mismatches across Europe. PLoS ONE 9(1):e82996

Brittain C, Williams N, Kremen C, Klein AM (2013) Synergistic effects of non-Apis bees and honey bees for pollination services. Proc R Soc B: Biol Sci 280(1754)

Cameron SA, Lozier JD, Strange JP, Koch JB, Cordes N, Solter LF, Griswold TL (2011) Patterns of widespread decline in North American bumble bees. Proc Natl Acad Sci USA 108:662–667

Centre for Bee Research and Development (CEBRAD) (2016) Bee farmers bemoan climate change on honey yield. CEBRAD publications, Ibadan, pp 1–6

Chagnon M (2008) Causes and effects of the worldwide decline in pollinators and corrective measures. Canadian Wildlife Federation, Quebec Regional Office

Corbet SA, Williams IH, Osborne JL (1991) Bees and the pollination of crops and flowers in the European Community. Bee World 72:47–59

Forister ML, Shapiro AM (2003) Climatic trends and advancing spring flight of butterflies in lowland California. Glob Change Biol 9(7):1130–1135

Gallai N et al (2009) Economic valuation of the vulnerability of world agriculture confronted with pollinator decline. Ecol Econ 68:810–821

Garibaldi LA, Aizen MA, Klein AM, Cunningham SA, Harder LD (2011a) Global growth and stability of agricultural yield decrease with pollinator dependence. Proc Natl Acad Sci U S A 108:5909–5914. https://doi.org/10.1073/pnas.1012431108

Garibaldi LA, Steffan-Dewenter I, Kremen C, Morales JM, Bommarco R, Cunningham SA, Carvalheiro LG, Chacoff NP, Dudenhöffer JH, Greenleaf SS, Holzschuh A, Isaacs R, Krewenka K, Mandelik Y, Mayfield MM, Morandin LA, Potts SG, Ricketts TH, Szentgyörgyi H, Viana BF, Westphal C, Winfree R, Klein AM (2011b) Stability of pollination services decreases with isolation from natural areas despite honey bee visits. Ecol Lett 14:1062–1072. https://doi.org/10.1111/j.1461-0248.2011.01669.x

Garibaldi LA, Steffan-Dewenter I, Winfree R, Aizen MA, Bommarco R, Cunningham SA, Kremen C, Carvalheiro LG, Harder LD, Afik O, Bartomeus I, Benjamin F, Boreux V, Cariveau D, Chacoff NP, Dudenhöffer JH, Freitas BM, Ghazoul J, Greenleaf S, Hipólito J, Holzschuh A, Howlett B, Isaacs R, Javorek K, Kennedy CM, Krewenka KM, Krishnan S, Mandelik Y, Mayfield MM, Motzke I, Munyuli T, Nault BA, Otieno M, Petersen J, Pisanty G, Potts SG, Rader R, Ricketts TH, Rundlöf M, Seymour CL, Schüepp C, Szentgyörgyi H, Taki H, Tscharntke T, Vergara CH, Viana BF, Wanger TC, Westphal C, Williams N, Klein AM (2013) Wild pollinators enhance fruit set of crops regardless of honey bee abundance. Science 339:1608–1611. https://doi.org/10.1126/science.1230200

Ghazoul J (2013) Pollination decline in context. Science 340:923–924. https://doi.org/10.1126/science.340.6135.923-b

Idowu FO, Oladebo JO (1999) The effects of scattered farm plots on agricultural production in the Guinea Savannah zone of Oyo state. J Rural Econ Dev 13:21

Ikemefuna PN (1998) Agrochemicals and the environment. NOVARTIS Newsletter 4:1–2

Johansen CA, Mayer DF (1990) Pollinator protection. A bee and pesticide handbook. Wicwas Press, Cheshire

Kearns CA, Inouye DW (1997) Pollinators, flowering plants, and conservation biology. Bioscience 47(5):297–307

Kearns CA, Inouye DW, Waser NM (1998) Endangered mutualisms: the conservation of plant-pollinator interactions. Annu Rev Ecol Evol Syst 29:83–112. https://doi.org/10.1146/annurev.ecolsys.29.1.83

Kerr JT, Pindar A, Galpern P, Packer L, Potts SG, Roberts SM, Rasmont P, Schweiger O, Colla SR, Richardson LL, Wagner DL, Gall LF, Sikes DS, Pantoja A (2015) Climate change impacts on bumblebees converge across continents. Science 349:177–180

Kevan PG (2001) Pollination: Plinth, pedestal, and pillar for terrestrial productivity. the why, how, and where of pollination protection, conservation, and promotion. In: Stubbs CS, Drummond FA (eds) Bees and crop pollination—crisis, crossroads, conservation. Thomas Say Publications in Entomology Entomological Society of America, Lanham, pp 7–68

Klein AM, Vaissière BE, Cane JH, Steffan-Dewenter I, Cunningham SA, Kremen C (2007) Importance of pollinators in changing landscapes for world crops. Proc R Soc Lond B 274:303–313

Losey JE, Rayor LS, Carter ME (1999) Transgenic pollen harms monarch larvae. Nature 399 (6744):214

Mallinger RE, Gratton C (2014) Species richness of wild bees, but not the use of managed honeybees, increases fruit set of a pollinator-dependent crop. J Appl Ecol https://doi.org/10.1111/1365-2664.12377

Malone LA, Pham-Delègue MH (2001) Effects of transgene products on honey bees (*Apis mellifera*) and bumblebees (*Bombus* sp.). Apidologie 32(4):287–304

Nabhan GP, Buchmann SL (1997) Services provided by pollinators. In: Daily G (ed) Nature's services. Island Press, Washington, DC, pp 133–150

Ollerton J, Winfree R, Tarrant S (2011) How many flowering plants are pollina ted by animals? Oikos 120:321–326

Oyerinde AA, Chuwang PZ, Oyerinde GT, Adeyemi SA (2014) Assessment of the impact of climate change on honey and propolis production in Nigeria. Acad J Environ Sci 2(3):037–042

Potts SG, Biesmeijer JC, Kremen C, Neumann P, Schweiger O, Kunin WE (2010) Global pollinator declines: trends, impacts and drivers. Trends Ecol Evol 25:345–353. https://doi.org/10.1016/j.tree.2010.01.007

Potts SG et al (2011) Developing European conservation and mitigation tools for pollination services: approaches of the STEP (Status and Trends of European Pollinators) project. J Res Apic Res 50:152–164

Smith KM, Loh EH, Rostal MK, Zambrana-Torrelio CM, Mendiola L, Daszak P (2013) Pathogens, pests, and economics: drivers of honey bee colony declines and losses. EcoHealth 10:434–445. https://doi.org/10.1007/s10393-013-0870-2

Van der Zee R, Gray A, Holzmann C, Pisa L, Brodschneider R, Chlebo R, Coffey MF, Kence A, Kristiansen P, Mutinelli F, Nguyen BK, Adlane N, Peterson M, Soroker V, Toposka G, Vejsnaes F, Wilkins S (2013) Standard survey methods for estimating colony losses and explanatory risk factors in *Apis mellifera*. J Apic Res 52(4):1. https://doi.org/10.3896/IBRA.1.52.4.18

Vanbergen AJ (2013) The insect pollinators initiative. Threats to an ecosystem service: pressures on pollinators. Front Ecol Environ 11:251–259

Promoting Circular Economy Through Sustainable Agriculture in Hidalgo: Recycling of Agro-Industrial Waste for Production of High Nutritional Native Mushrooms

María Virginia Ozcariz-Fermoselle, Gabriela de Vega-Luttmann, Fernando de Jesús Lugo-Monter, Cristina Galhano and Oscar Arce-Cervantes

Abstract The effect of climate change on agriculture and its implications on food security are demanding topics. It is crucial to convert the existing methods of food production into a more sustainable, resilient and productive agriculture. Reduction of food loss and waste will improve the efficiency of the food system, and simultaneously it will reduce the pressure put on natural resources and diminish the greenhouse gases emissions. This project aims to assess the potential use of agribusiness waste as a substrate for cultivation of *Pleurotus* spp., in order to contribute to the development of a more sustainable agriculture practices while promoting local development. The principles of Circular Economy are being applied to the most representative lignocellulosic waste of the Hidalgo State: pecan nutshell (PS), agave bagasse (AB), pine needles (PN), tamal leaves (TL) and coffee pulp (CP). It was studied the potential of these wasted plant material as substrates to grow mushrooms, of ten native *Pleurotus* spp. strains, with high nutritional value (seven *Pleurotus ostreatus* strains, one *Pleurotus eryngii* strain, one *Pleurotus*

M. V. Ozcariz-Fermoselle
Laboratorio de Tecnología de La Madera, Departamento de Ingeniería Agrícola Y Forestal, Universidad de Valladolid, Campus Palencia, 34071 Palencia, Spain

M. V. Ozcariz-Fermoselle · G. de Vega-Luttmann · Fernando de JesúsLugo-Monter
O. Arce-Cervantes
Instituto de Ciencias Agropecuarias, Universidad Autónoma Del Estado de Hidalgo, Rancho Universitario, 43600 Tulancingo, Mexico

C. Galhano (✉)
Department of Environmental Sciences, Coimbra College of Agriculture Polytechnic Institute of Coimbra, Bencanta 3045-601 Coimbra, Portugal
e-mail: cicgalhano@esac.pt

C. Galhano
Department of Life Sciences, Centre for Functional Ecology, University of Coimbra, 3000-456 Coimbra, Portugal

P. Castro et al. (eds.), *Climate Change-Resilient Agriculture and Agroforestry*, Climate Change Management, https://doi.org/10.1007/978-3-319-75004-0_26

djamor strain and one *Pleurotus opuntiae* strain). In order to evaluate the different waste usefulness in mushroom's growth, radial growth rate of the ten strains was assessed. Generally, the agave bagasse substrate promoted the highest growth rate. All *Pleurotus* spp. strains had slowest growth in control medium, PDA, due to the lack of lignocellulosic compounds.

Keywords Agro-industrial waste · Circular economy · *Pleurotus* spp
Sustainable agriculture

1 Introduction

Food security, sustainable development and poverty eradication are goals largely threatened by climate change, a global situation chiefly caused by the greenhouse gas emissions produced by human activities (FAO 2017). On the other hand, climate change is responsible for changes in function and composition of ecosystems, including agroecosystems. Consequently, agricultural practices have also been adapted to conditions resulted from climate change, seeking for alternatives for sufficient food production (FAO 2017).

It is expected that by 2050, the global population will be 50% larger than at present. Hence, to achieve the desired global political and social stability, further increases in agricultural yields are essential. Nevertheless, those yield increases are often based on the simplification of agroecosystems caused by the intensification of agricultural practices, which may affect important ecosystem functions via the loss of biodiversity. For example plant growth, pest control, pollination and decomposition processes. Together with global climate change and over-population, new challenging targets and refreshing prospects are expected from industrial and environmental biotechnology, in terms of impact, mitigation and adaptation strategies. The impacts of such environmental and societal pressures reflect on agriculture, land-use and water supply and, consequently, on the availability of food, energy and fresh water. Adaptation strategies may rely on the exploitation of alternative food products. Contributions to mitigation of the stressors can arise in the form of new or improved biomass conversion and renewable energies, carbon and greenhouse gas sequestration measures, and more effective waste management options (Orgiazzi et al. 2016).

Agribusiness activities produce tons of waste that can be valued by assigning them to other purposes; those byproducts can be used to produce other products with social, economic, or environmental value (Saval 2012). In fact, the use of natural resources should cause the least possible environmental impacts. To minimize those impacts, waste must not be immediately considered as waste, but as a potential source of essential environmental services (Smil 1999; Santana-Méridas et al. 2012).

Sustainability as a concept integrates social development, economic growth and environmental care (Roseland 2000). Considering this point of view, environmental care must provide knowledge regarding species interaction.

The United Nations Environment Programme (UNEP 1995) estimated that low-income countries burn approximately 25% of all waste. Considerable amounts of unused lignocellulosic by–products are available in tropical and subtropical areas. These by–products are left to rot in the field or incinerated. Nevertheless, their use as nutrient essential sources for fungal (e.g. *Pleurotus* genus) biomass growing, a bioconversion process, offers an opportunity to contribute to circular economy (Akyuz and Kirbag 2009; Madigan et al 1997). In fact, mushroom cultivation using locally available technologies may be a solution to convert these inedible wastes into a highly valued edible biomass (Tesfaw et al. 2015).

Annually, Mexico produces nearly 3,901,146.63 ton of lignocellulosic matter. Every kilogram of burned lignocellulosic matter releases 2 kg of carbon dioxide (CO_2). This gas, among many other greenhouse gases, is responsible for most of the impact on Earth's climate (Chauhan et al. 2005). It is known that carbon dioxide is the most potent switch controlling the greenhouse effect, and that CO_2 abundance determines the amount of water vapor in the atmosphere, making it responsible for more than 60% of the greenhouse effect.

Diverse Mexican industries (e.g. production of coffee, tequila, pecan nut, and woodcutting) produce ton of lignocellulosic by-products. Most of them are discarded or burned, generating carbon dioxide and nitrous oxide emissions, both substances contributing to greenhouse effect.

In order to avoid those source of greenhouse gases emissions, some strategies were proposed to break the chain of linear economy and start implementing circular economy systems, based on optimization of stocks and the flow of materials, energy and waste, in order to use resources more efficiently (MacArthur 2013; Bicket et al. 2014).

2 Pecan Nutshell (*Carya Illinoinensis*)

Mexico is the second world producer of pecan nut, generating over 79,000 ton (Camacho 2012). In 2012, the State of Hidalgo produced 2,751.05 ton of nuts, of which 97.9% was pecan nut (Fig. 1). In 2014, production increased to 2,929.92 ton (SIAP, 2014). Current practices exploit only $\sim$50% of the product's weight, the other $\sim$50% is nutshell (Frusso 2007), reaching 1,346.63 ton in 2012. This kind of by–product is usually wasted, harming the environment. For example, in the municipality of Atotonilco el Grande, nutshell is discarded with trash, disposed over public grounds, hindering the growth of local vegetation (Fig. 2), or burned, releasing two kilograms of CO_2 for each kilogram of lignocellulosic matter. Nevertheless, this waste has bioactive products such as poly and monomeric phenols and its potential as biofungicide was already mentioned (Santana-Méridas et al. 2012).

Fig. 1 A vending stall for pecan nut in Tula de Allende, Hidalgo, Mexico

3 Agave Bagasse (*Agave Tequilana* Weber)

Blue agave, *Agave tequilana*, is widely cultivated through Mexico for tequila production, a traditional a spirituous drink, made from the plant's "heart". Yearly, 100,000 ton of agave bagasse (Fig. 3), a type of lignocellulosic waste, was generated during tequila production, which was mostly used as compost (Abreu-Sherrer 2013; Saucedo-Luna et al. 2010). This type of bagasse contains cellulose, hemicellulose and lignin at 65, 5.5 and 17%, respectively (Quintana-Vega, 2014). Approximately 232,000 ton of agave was processed in 2016, providing a total of 93,000 ton of bagasse (El informador, 2016).

*3.1 Pine Needles (*Pinus Pseudostrobus*)*

Pine needles are scarcely used for craftsmanship (Fig. 4). In the forest, pine needles take a long time to decompose, prevent the sun to reach ground plants, increase soil acidity, and are highly inflammable, representing an important fire threat (Córdova-Ordóñez 2009). There are big volumes of pine needles in the forests in the State of Hidalgo, without any law or rule currently controlling its management.

Fig. 2 Pecan nutshell discarded in Atotonilco el Grande, Hidalgo, Mexico

3.2 Tamale Leaves

For Mexicans, maize (or corn) is an important part of daily life, ubiquitous in traditions and cultural roots. One of the most important crops in the country, maize is also integrated into Mexican identity. Even nowadays, despite one third of the production is being exported due to internal economy, maize harvesting is still a cultural symbol in rural communities, since food security relies on it. In 2014, Mexico reached the sixth place in global production of corn, with 23,273,256.54 ton (SIAP 2014), from which 70,000 ton were corn husk, also known as "totomoxtle" (Fig. 5). A minimal amount of this waste is used for construction, ornamental, or gastronomic purposes. Regarding cuisine, corn husks are used for

Fig. 3 Residual fiber, commonly known as bagasse, after agave juice extraction, Rudio Distiller, Jalisco, Mexico

Fig. 4 Pine needle residues accumulated near a residential hill in Tulancingo, Hidalgo, Mexico

wrapping tamales, a traditional dish. However, once consumed, the tamale leaf, is discarded once again. The tamale leaves contains 78.86% cellulose (Prado–Martínez 2012).

Fig. 5 Tamale leaves (corn husks) discarded in a dumpster at Tulancingo, Hidalgo, Mexico

3.3 Coffee Pulp

Coffee pulp is a by-product of the production of coffee beans (Fig. 6). Pulp represents approximately 40% of the weight in coffee cherries (Olivares et al. 1990). During 2016–2017 México produced 9,342,000 ton of coffee, also generating 3,736,800 ton of pulp. Coffee pulp contains 63% cellulose, 2.3% hemicellulose, and

Fig. 6 Coffee pulp piled at the countryside, Hidalgo, Mexico

17.5% lignin (Murthy and Naidu, 2012). Furthermore, coffee pulp has bioactive compounds as caffeine and chlorogenic acid, having also been pointed out its allelopathic characteristics (Santana-Méridas et al. 2012).

The main goal of this work was to evaluate the potential use of the most representative lignocellulosic waste of the Hidalgo State (pecan nutshell, agave bagasse, pine needles, tamal leaves, and coffee pulp) as substrates for cultivation of ten native strains of *Pleurotus* spp. with high nutritional value, in order to contribute to a more sustainable agriculture and promote local development applying the Circular Economy principles.

4 Materials and Methods

4.1 Pleurotus *Strains Propagation and Storage*

The ten native *Pleurotus* strains (seven strains of *Pleurotus ostreatus*, one of *Pleurotus eryngii,* one of *Pleurotus djamor* and one of *Pleurotus opuntiae*) used in this study were obtained from the Instituto de Ciencias Básicas e Ingeniería (ICBI) of Universidad Autónoma del Estado de Hidalgo (UAEH) collection. Mycelium from each strain was inoculated in Petri dishes with potato dextrose agar (PDA) culture medium for propagation and storage. Firstly, culture media was prepared and sterilized in Erlenmeyer flasks for 15 min, at 15 lb/in^2 and 121 °C, in an autoclave. After being distributed through Petri dishes, PDA was inoculated with fungi strain samples, and Petri dishes were incubated at 21 $\pm$ 2 °C, for 21 days before being used. Each used strain was also stored in Petri dishes, at 4 °C.

4.2 Lignocellulosic Waste Selection and Preparation to Be Used as Pleurotus *Strains Growth Substrate*

The waste selection criteria to be used as substrate for mushroom growth were based on those previously pointed out by Saval (2012). The main waste component: (a) should be usable as solid state fermentation (SSF) substrate for saprophytic mushrooms grow; (b) should have local and sufficient availability; (c) should not have other applications or uses competing with the intended process; (d) should not require extra treatment to be use or, if necessary, should be simple and cheap.

Based on these criteria the selected waste studied as potential mushroom growth substrate were: pecan nutshell (PS), *Agave tequilana* bagasse (AB), *Pinus pseudostrobus* needle (PN), tamale leaf (TL), and coffee pulp (CP).

Pecan nutshells were kindly provided by a nut processing company from the region of the Biosphere Reservation "Barranca de metztitlán"; *Agave tequilana* bagasse was provided by a company from Tequila, Jalisco; *Pinus pseudostrobus* needles were collected from the ICAp garden in Tulancingo; tamale leaves were offered by the ICAp cafeteria in Tulancingo. Coffee pulp was provided by coffee producers from Huejutla de Reyes, Hidalgo.

Before being used in growth mushroom experiments, these waste were firstly pulverized, using a manual mill and then sieved through a 35 mesh.

4.3 *Evaluation of Lignocellulosic Waste* as Pleurotus *Spp. Strains Growth Substrates*

To study the potential cultivation use of the five previously referred most Hidalgo State, representative lignocellulosic waste as substrates for ten native *Pleurotus* spp. strains, it was used the following described methodology.

For each individual waste (treatment), firstly, 25 g of waste prepared as substrate (as described above) were added for each liter of prepared PDA growth medium. PDA medium without any added waste was used as control.

Ten groups of Petri dishes for each of the five waste prepared culture growth media and control were separated and labeled for each of the 10 *Pleurotus* strains: six *P. ostreatus* strains, labeled as PL10, Leb, Gris, B2, BPR2 and P07; one *Pleurotus* sp. strain native of the Tlanchinol region, labeled as PT; one *P. eryngii* strain, labeled as PE; one *P. opuntiae* strain, labeled as PM1; and *P. djamor* labeled as SMR.

The center of each Petri dish was inoculated with a micelial area of 1 mm^2. Then, three straight lines (A, B, C) were drawn, at 60°, intersecting at the inoculation center (Imtiaj et al. 2009). Petri dishes were incubated at 25 °C, in absolute darkness. Radial mycelium growth rate on solid media was measured daily, for 21 days. Growth was measured as the mean of mycelium extension over lines A, B, C (Fig. 7). Growth rate was measured over a distance of 80 mm covered by

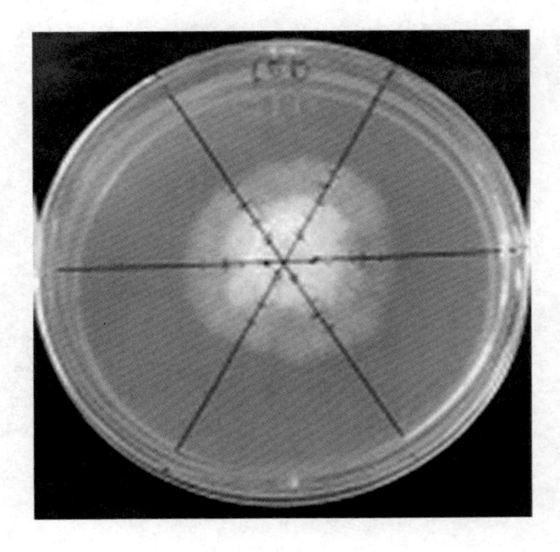

Fig. 7 Petri dish prepared for mycelium growth measurement

mycelia, divided by mean days of invasion. Five replicates were made for each treatment and control.

4.4 Data Analysis

First, the assumptions of independence, normality and homoscedasticity were tested for the studied variables. Secondly, given that the data structure comply with normality criteria, ANOVA was used to evaluate a mycelial growth rate, which were represented in terms of mean ± standard error of the realized tests. Significant differences between formulations were determined by paired student-t test (LSD) with $\alpha = 0.05$.

5 Results and Discussion

5.1 *Evaluation of Lignocellulosic Waste* as Pleurotus *Spp. Strains Growth Substrates*

A general view of the experiment can be seen in Fig. 8.

The growth rate for every combination of strains in each waste substrate is analyzed. It is evident that the best result was obtained in the medium with AB waste as substrate, and being the control medium, PDA, where the lowest growth was registered (Fig. 9).

This study showed that all mushroom strains grew slower in control, PDA media, than in any of the waste treatment media (Fig. 10). Strains PL10, Leb, Gris, B2, BPR3, P07, PE, and SMR grew faster in *Agave tequilana* bagasse (AB) substrate medium, while strains PT and PM1 grew faster in *Pinus pseudostrobus* needle (PN) substrate medium (Fig. 10).

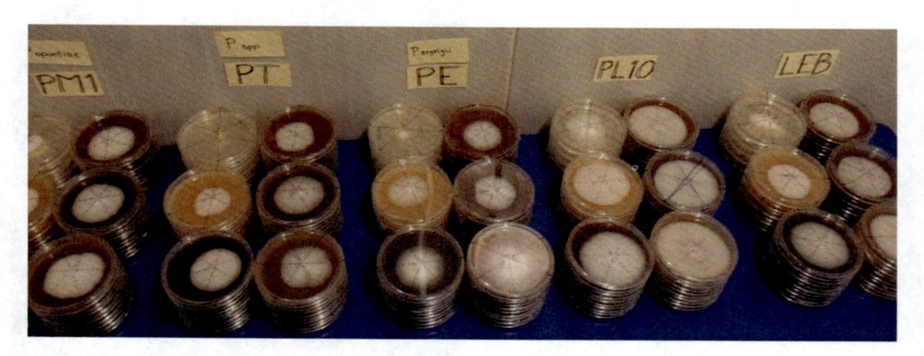

Fig. 8 Petri dishes grouped by mushroom strains, for mycelium growth measurement in six different waste media

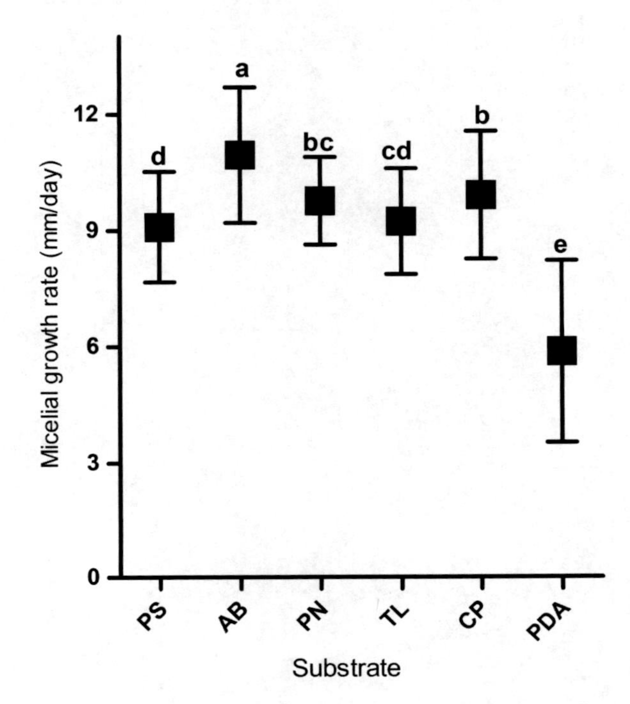

Fig. 9 Mycelium growth rate of the 10 *Pleurotus* strains studied using different waste substrates. PS—pecan nutshell; AB - *Agave tequilana* bagasse; PN—*Pinus pseudostrobus* needle; TL—tamale leaf; CP—coffee pulp; PDA—Control. Letters above boxes group data by statistically different means ($P < 0.05$)

A significant difference of growth was found between PDA and the other five substrates in strains Leb (7–9 days in AB, CP, PS, PN and TL; <21 days in PDA), P07 (8–11 days in AB, PN, CP, PS and TL; <21 days in PDA), PT (10–12 days in PN, PS, AB, TL y CP; <21 days in PDA), and PE (8–11 days in AB, PS, PN, CP and TL; <21 days in PDA).

The results suggest that growth can be stimulated by the addition of lignocellulosic waste as SSF substrates to growth medium. Gris strain showed similar growth rates in all treatments (7 days in AB and CP; 8 days in PS, PN and TL; 9 days in PDA). This could be explained by a complete domestication of the Gris strain, adapting it to commercial growth media, or because the strain is not affected by these studied lignocellulosic substrates.

Agave bagasse can be considered the best substrate, in combination with PDA growth media, for mycelium production in Petri dishes. Nevertheless, since bagasse must be acquired outside the Hidalgo State, a better option could be the use of pine needles, as this waste is easily obtained and it is a low cost option. Tamale leaf can be considered a second option, as it requires collecting and separating trash, with an intermediate washing step. Moreover, pecan nutshell and coffee pulp can be considered as good options in regions where they are easily acquired. Besides that, the SSF is able to produce bioactive compounds from agricultural waste (Santana-Méridas et al. 2012).

The results showed that it is possible to use the studied lignocellulosic waste as substrates to add to PDA, for mycelium producing in Petri dishes, at a quick rate. It

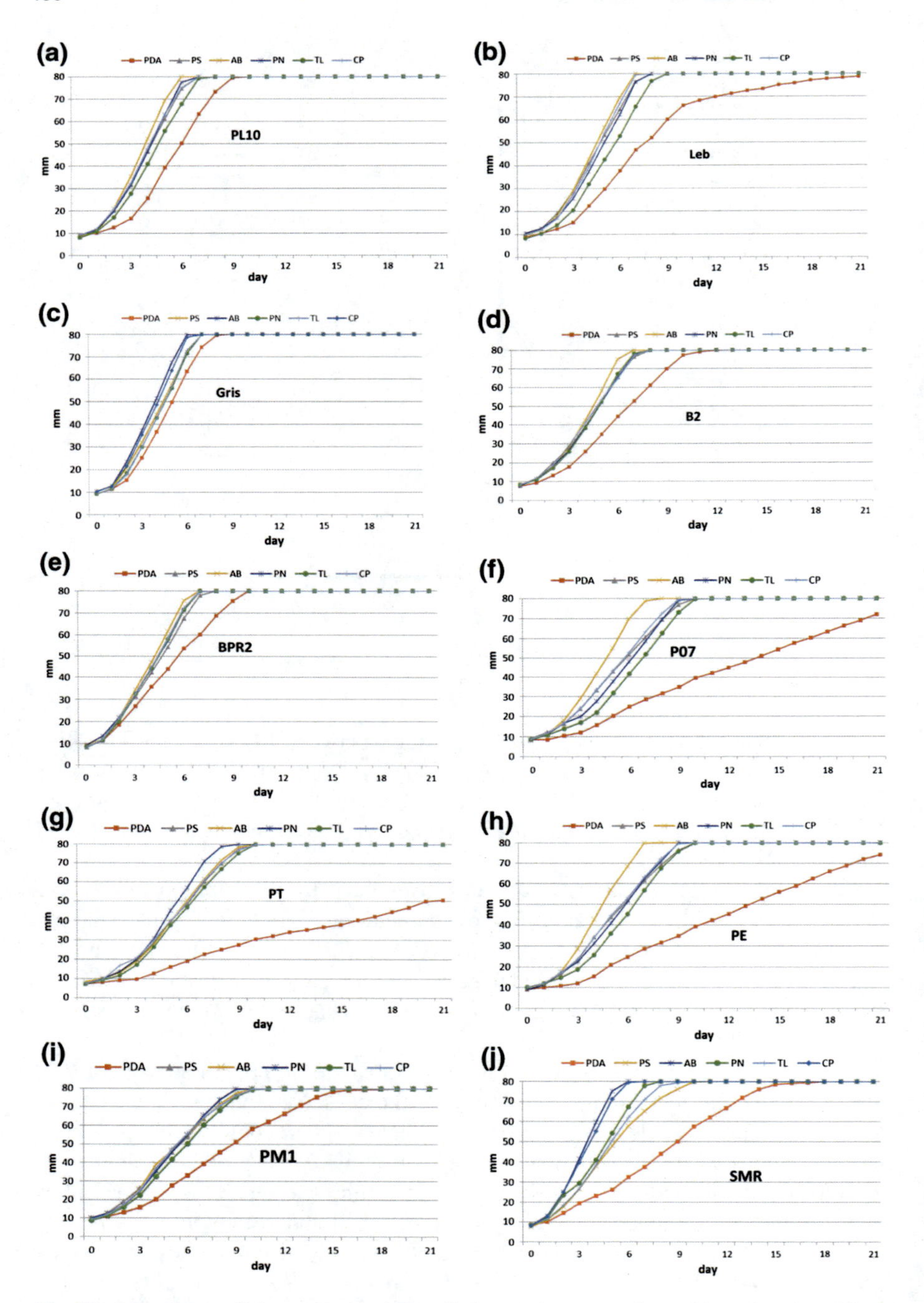

Fig. 10 Mycelium radial growth (mm) for all the mushroom strains cultivated under different waste substrates. **a** PL10 strain; **b** Leb strain; **c** Gris strain; **d** B2 strain; **e** BPR2 strain; **f** P07 strain; **g** PT strain; **h** PE strain; **i** PM1 strain; and **j** SMR strain

seems that for Gris strain growing, there is no need to add lignocellulosic substrates, PDA media is enough. On the other hand, the growth of PT and PM1 strains, both recently isolated from natural sources, is better when lignocellulosic compounds are added to PDA culture medium. Thus, it is possible to grow recently isolated strains under a laboratory setting using growth media complemented with lignocellulosic matter.

Thus, in this study it is pointed out a sustainable method for production of a high economic and healthy food product, contributing to boost Circular Economy at local level, enhancing simultaneously plant growth which will sequester carbon dioxide.

There is already a case of success in Mexico, at Las Vigas de Ramírez, in Veracruz State. A group of farmer women are producing edible mushrooms using agro-industrial waste (Mata et al. 2013).

Moreover, the microorganisms as SSF play a very important role in climate change, as there is a bioconversion of lignocellulosic waste, contributing to food production and simultaneously to CO_2, CH_4 and nitrogen, consumption, stabilizing climate change impacts (Chatzipavlidis et al. 2013).

In the future, it is very likely that new microbial strains will be developed offering a potential solution to problems related to food shortage. Besides that, taking into account the controversies that exist around the application of biotechnology in agriculture, there is a strong pressure and incentive to use natural biodiversity to meet the ever-growing consumer demands for such products in our increasingly environmentally focused society (Orgiazzi et al. 2016).

6 Conclusion

The five studied waste can be considered as sustainable alternatives to be used as growth media additives for *Pleurotus* spp. production, as this high nutritional value mushroom genus has high adaptability to grow on lignocellulosic materials.

Therefore, the use of lignocellulosic waste in this type of mycelium production can prevent the greenhouse gas emissions which will be produced by incineration of this kind of waste. Besides that, the removal of waste from the natural environment allows other plant species to grow, increasing CO_2 fixation and increase microorganism's biodiversity.

References

Abreu Sherrer JS (2013) Aprovechamiento de bagazo de Agave tequilana Weber para la produccion de bio-hidrógeno. http://hdl.handle.net/11627/86

Akyuz M, Kirbag S (2009) Antimicrobial activity of *Pleurotus eryngii* var. ferulae grown on various agro-wastes. EurAsian J BioSciences 3:58–63

Bicket M, Guilcher S, Hestin M, Hudson C, Razzini P, Tan A, Ten Brink P. Van Dijl E, Vanner R, Watkins E (2014) Scoping study to identify potential circular economy actions, priority sectors, material flows and value chains. Project Report. European Commission. Official URL: http://bookshop.europa.eu/en/scoping-study-to-identify-potential-circular-economy-actions-priority-sectors-material-flows-and-value-chains-pbKH0114775/

Camacho JA (2012) La Nuez Pecanera Mexicana "La reina de las frutas secas" Importancia nutricional y usos de la nuez pecanera. Obtenido de http://2006-2012.sagarpa.gob.mx/agricultura/productodetemporada/_layouts/mobile/dispform.aspx?List=75320ba8-c685-403d-a5aa-b32646bacf02&View=5050ddad-bb6c-4d20-8e46-97b7179e8410&ID=68

Chauhan S, Khandelwal RS, Prabhu KV, Sinha SK, Khanna-Chopra R (2005) Evaluation of usefulness of daily mean temperature studies on impact of climate change. J Agron Crop Sci 191(2):88–94

Chatzipavlidis I, Kefalogianni I, Venieraki A, Holzapfel W (2013) Commission on Genetic Resources for Food and Agriculture. FAO

Córdova-Ordóñez MM (2009) Estudio comparativo del crecimiento miceliar del hongo (*Pleurotus ostreatus*) en acícula de pino, bagazo de caña y bagazo de maíz. Bachelor's thesis, University of Azuay. http://dspace.uazuay.edu.ec/handle/datos/580

El Informador (2016) Registran aumento en la producción de tequila en 2016. http://www.informador.com.mx/economia/2016/662113/6/registran-aumento-en-la-%20produccion-de-tequila-%20en-2016.htm. Accessed 19 May 2016

FAO (2017) Conference on sustainable and climate-smart agriculture. Mumbai, India. Accessed 21 Apr 2017

Frusso EA (2007) Características morfológicas y fenológicas del pecán. En: RS Lavado y EA Frusso. Producción de Pecán en Argentina. Capítulo III

Imtiaj A, Jayasinghe C, Lee GW, Lee TS (2009) Comparative study of environmental and nutritional factors on the mycelial growth of edible mushrooms. J Cult Collect 6(1):97–105

MacArthur E (2013) Towards the circular economy. J Ind Ecol 23–44

Madigan MT, Martinko JM, Parker J (1997) Brock biology of microorganisms, vol 11. Prentice hall, Upper Saddle River, NJ

Mata G, Hernández RG, Salmones D (2013) Biotechnology for edible mushroom culture: a tool for sustainable development in Mexico. WIT Transactions on State-of-the-art in Science and Engineering 64

Murthy PS, Naidu MM (2012) Sustainable management of coffee industry by-products and value addition—A review. Resour Conserv Recycl 45–58

Olivares GN, González GV, Rojas GG, Favela M, Huerta E, Roussos S,…, Gutiérrez Rojas M (1990) Producción de enzimas a partir de pulpa de café y su aplicación en el beneficio húmedo. Seminario Internacional sobre Biotecnología en la Agroindustria Cafetalera. Universidad Autónoma Metropolitana, México (México) ORSTOM, París (Francia), 12–15 Abr 1989

Orgiazzi A, Bardgett RD, Barrios E, Behan-Pelletier V, Briones MJI, Chotte JL, De Deyn GB, Eggleton P, Fierer N, Fraser T, Hedlund K, Jeffery S, Johnson NC, Jones A, Andeler E, Kaneko N, Lavelle P, Lemanceau P, Miko L, Montanarella L, Moreira FMS, Ramirez KS, Scheu S, Singh BK, Six J, van der Putten WH, Wall DH (2016) Global soil biodiversity atlas. european commission, publications office of the european union, Luxembourg. 176 pp. (Eds.) European Union, 2016. L-2995 Luxembourg, Luxembourg

Prado-Martínez M, Anzaldo-Hernández J, Becerra-Aguilar B, Palacios-Juárez H, Vargas-Radillo JDJ, Rentería-Urquiza M (2012) Caracterización de hojas de mazorca de maíz y de bagazo de caña para la elaboración de una pulpa celulósica mixta. Madera y bosques 18(3):37–51

Quintana-Vega AM (2014) Aprovechamiento integral del bagazo de la piña de *Agave tequilana* Weber: Caracterización de fracciones lignocelulósicas obtenidas por un proceso organosolv. URI: http://tesis.ipn.mx:8080/xmlui/handle/123456789/12958 . Accesed 04 Jun 2014

Roseland M (2000) Sustainable community development: integrating environmental, economic, and social objectives. Prog Plann 54(2):73–132

Santana-Méridas O, González-Coloma A, Sánchez-Vioque R (2012) Agricultural residues as a source of bioactive natural products. Phytochem Rev 11(4):447–466

Saval S (2012) Aprovechamiento de residuos agroindustriales: Pasado, presente y futuro. BioTecnología 16(2):14–46

Saucedo-Luna J, Castro-Montoya AJ, Martínez-Pacheco M, Campos-García J (2010) Diseño de un bioproceso para la obtención de etanol anhidro a partir de bagazo del *Agave tequilana* Weber. Ciencia Nicolaita No, Especial

SIAP (2014) Anuario Estadístico de la Producción Agrícola. Obtenido de Servicio de Información Agroalimentaria y Pesquera. http://infosiap.siap.gob.mx/aagricola_siap/icultivo/index.jsp

Smil V (1999) Crop Residues: Agriculture's Largest Harvest: Crop residues incorporate more than half of the world's agricultural phytomass. Bioscience 49(4):299–308

Tesfaw A, Tadesse A, Kiros G (2015) Optimization of Oyster (*Pleurotus ostreatus*) mushroom cultivation using locally available substrates and materials in Debre Berhan, Ethiopia. J Appl Biol Biotechnol 3(1):15–20

UNEP (1995) United Nations Environmental Program. www.unep.org/publications/

People and Parks: On the Relationship Between Community Development and Nature Conservation Amid Climate Change in South-Eastern Zimbabwe

Wedzerai Chiedza Mandudzo

Abstract Wildlife conservation is a topic that has captured public imagination in both developed and developing nations. This is evident by the creation and establishment of protected areas such as national parks and trans-boundary protected areas. In addition to their fundamental role of protecting natural resources, protected areas largely have the vital task of supporting tourism and socio-economic development of local communities. However, with the establishment of protected areas, the concept of communities' dependence on natural resources has been ignored and protection of biodiversity taken precedence. Consequently, the prioritization of conservation over livelihoods has led to the widespread notion that conservation is a threat to development. Conservationists, on the other hand, assert that the onslaught of development is dependent on the same resources it threatens. This study evaluates the relationship between community development and nature conservation efforts among the Chitsa community and Gonarezhou National Park (GNP) in South-Eastern Zimbabwe amid climate change. In order to achieve the aim of the study, critical ethnography was employed, and utilized semi-structured interviews, focus group discussions and life histories as data collection methods. Findings of the study reveal that nature conservation and community development have long represented contrasts in both research and practice. Of significance are imbalances that favour analyses and prioritization of nature conservation over community development outcomes supported by natural resources in resource dependent communities. It appears that nature conservation focuses on the strict protection of natural resources and ignores aspects of social and political processes involved in it hence it limits the people's ability too adapt to climate change.

W. C. Mandudzo (✉)
Department of Anthropology and Archaeology, Faculty of Humanities,
University of Pretoria, Pretoria, South Africa
e-mail: wedzemandudzo@gmail.com

P. Castro et al. (eds.), *Climate Change-Resilient Agriculture and Agroforestry*,
Climate Change Management, https://doi.org/10.1007/978-3-319-75004-0_27

1 Introduction

Throughout the world, biodiversity conservation has attracted much attention in many countries. This is evident by the creation and continued establishment of protected areas such as national parks and trans-boundary protected areas; Zimbabwe is no exception to this. Since the dawn of mankind, a number of communities have lived adjacent to or within protected areas and rely on natural resources for various purposes such as food, trade, medicine and building materials. However, with the establishment of protected areas, the theory of communities' dependence on natural resources has been disregarded and the protection of biodiversity has become dominant. Consequently, the prioritization of conservation over livelihoods has led to the widespread notion that conservation is a threat to development (Colchester 2004; Boonzaaier and Wilson 2011; Van der Duim 2011). Conservationists, on the other hand, assert that the onslaught of development is dependent on the same resources it threatens (Miller et al. 2011; Dowie 2009). Accordingly, community development and nature conservation ideologies have been presented as distant cousins.

With the establishment of protected areas, human occupancy in parks was seen as a hindrance to conservation (Nelson 2008; Sanderson and Redford 2003). Nevertheless, considering the interconnection between human population and environment, separating conservation and community development was and is not feasible (Van der Duim 2011; Colchester 2004). These poor results can be attributed to top down approaches have been adopted by conservationists and the exclusion of indigenous communities in decision-making, management and utilization of natural resources within their locale. The question of whether community development and nature conservation are compatible has been a focus of debates on sustainable development and biodiversity conservation in the South (Duffy 2000; Neumann 1998). Environmentalists assume that community development is not compatible with conservation; this stems from the idea that objectives of community development and nature conservation are different. According to the United Nations Conference on Environment and Development UNCED held in Rio de Janeiro in 1992 there is a shift on how natural resources are viewed and valued between developed (Northern Hemisphere) and less developed (Southern hemisphere) nations. Developed nations' leaders assumed that development in its forms is not compatible with conservation and less developed nations' leaders argued that conservation is a luxury unaffordable to them since it prevents development. These differences also help explain the 'blame game' in development-conservation nexus. Besides increasing interest to link conservation with the lives of communities living in or adjacent to protected areas, little attempt has been made to assess the link between these two dichotomies (Coria and Calfucura 2012; Van der Duim 2011).

The herein presented study sought to:

2 Objectives

1. To examine and describe the history of both Chitsa community and Gonarezhou National Park in order to identify the key factors that shaped the interaction between "people and parks" in Gonarezhou national park.
2. To identify the power relations that shapes the relationship and alleged conflict between the Chitsa community, the Gonarezhou Park's management, the Zimbabwean Government and other stakeholders

3 Background to the Study

Compatibility of community development and nature conservation in Zimbabwe is an area that has not been prioritized. Yet, the interrelationship between wildlife and human population is so intricate. The socio-economic well-being of humans depends on good management of wildlife, their ability to adapt to climatic changes and other natural resources in national parks of the world and vice versa (Burnham 2000; Colchester 2004; Gandiwa et al. 2011; Romero et al. 2012; Diam et al. 2012). This means that protected areas have long been realized as the most important means of conserving wildlife and biodiversity but they are characterized by tension between wilderness preservation, tourism, climate change adaptation and community development. Balancing community development, rural livelihoods and nature conservation is one of the greatest challenges being faced in Africa in the 21st century (Van der Duim 2011; Boonzaaier and Wilson 2011).

In as much as it is acknowledged that poverty and biodiversity loss are linked and should be treated together, there is a huge debate about the success of community based approaches to conservation (Adams et al. 2004). This debate stems from the fact that realization of strategies and objectives of community development, conservation and tourism leads to interference between wildlife and human populations living within or adjacent to such parks. Mombeshora and Le Bel (2009) asserts that history of most national parks are characterized by conflicts between local communities and institutions that are trying to bring the concepts of community development and nature conservation to fruition. The causes of these conflicts can be attributed to the establishment of these parks based on American romanticism whereby indigenous people were not consulted. This has caused problems in the implementation of conservation strategies, utilization and management of natural resources.

The other problem that surrounds the relationship of conservation and development is that both concepts are used in different ways and they depict different meanings (Campbell et al. 2010; Miller et al. 2011). The conflict between these two emanates from the "general conservationist view" that large conservation areas such as transfrontier parks should embody the notion of "wilderness", that is, an area in which, except for employees and tourists, humans should be absent (Twyman

2001). However, while conservationists try to preserve and protect natural resources, development practitioners on the other hand vye for the sustainable use of these resources by indigenous communities. This diverge can also be attributed to the lack of harmony between community development and conservation (Miller et al. 2011; Brandon 2002; Schwartzman et al. 2000).

In Zimbabwe one of the strategies that have been adopted as a bridge between conservation and development is Communal Areas Management Program for Indigenous Resources (CAMPFIRE). CAMPFIRE was established in 1989 under the direction and sponsorship of Zimbabwe s Department of National Parks And Wildlife Management (Logan and Moseley 2002; Frost and Bond 2008). However, it was noted to have failed due to political situation in Zimbabwe, withdrawal by donors and the imbalance of CAMPFIRE objectives between ecology and poverty eradication (Wolmer 2003; Logan and Moseley 2002). This study is important in the body of knowledge because it provides an insight into the realities of the interaction between wildlife and people in Gonarezhou National Park given that there is a community residing inside the park (Chitsa community). The study sought to understand if community development can be a panacea for conservation and conservation a panacea for community development. An assessment was done on whether conservation-community development initiatives are not money making initiatives under the guise of nature conservation, community development and tourism. The quest for an understanding of the nature of interconnectedness of everything in conservation and development attracted attention. The investigation on the gaps between community development and conservation and possible solutions to marry them into practice lies at the bottom of this research.

3.1 Setting the Research in Context: Origins of GNP and Chitsa Community Interactions

Pre-colonial Africa has often been depicted as a land of "*uncivilized warring savages in loin cloths and spears with no knowledge of the environment or insight into its sustainable use*" (DeGeorges and Kevin 2008; Van der Duim 2011). During this period, it was believed that Europeans were required with their 'modern' wildlife practices and administrative structures to bring about conservation and sustainable use of natural resources. It can, however, be argued that colonialism and independence have hindered the evolution of traditional resource management systems, ignored them and in some instances, killed them (DeGeorges and Kevin 2008). Understanding the history of the establishment of GNP and its interaction with indigenous people in South-Eastern Zimbabwe helps in sparking debate and revolutionizing the community development and nature conservation thinking relationship in the 21st century.

GNP has beautiful scenery that is interspaced with numerous rivers, the largest of which are Mwenezi, Runde and Save. There are also several kopjes that cap the

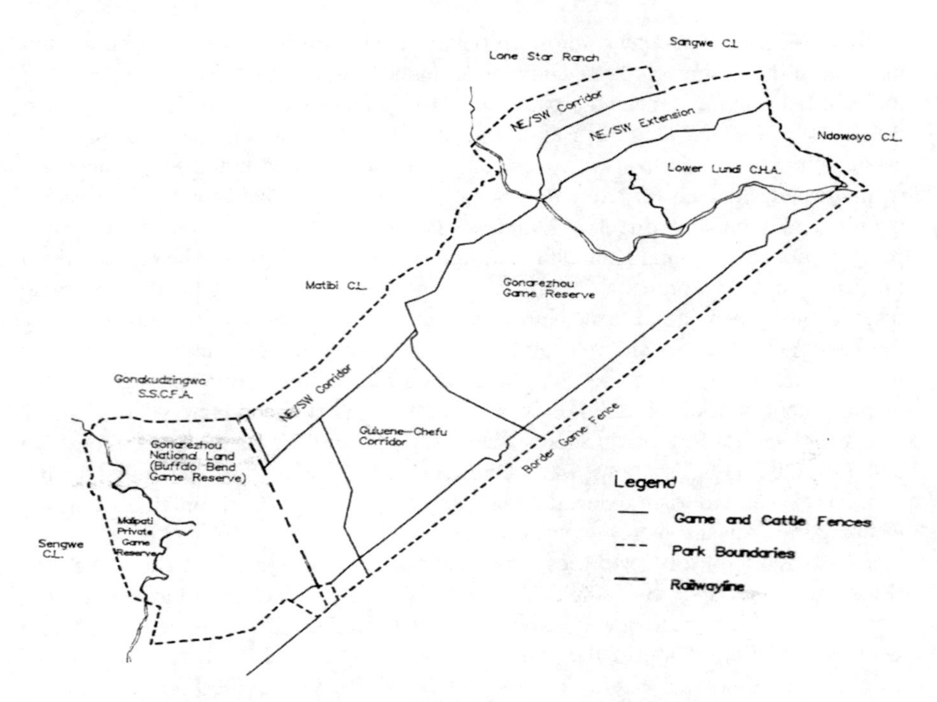

Fig. 1 GNP former land demarcations (Mugadza and Mandizodza 2006)

beauty of the park. Wolmer (2007) stated that 'humans and their changing environments are reciprocally inscribed in cosmological ideas and cultural indulgence: the forest is the people, in the same way that ancestors can be, in a sense, extensions of the living. Therefore, like the rest of Zimbabwean people, the Chitsa people regard nature and themselves as inseparable. Figure 1 shows the locality of Gonarezhou Park and adjacent Shangaan communal lands.

GNP was created from the land that historically belonged to the Shangaan people. As noted above, the Shangaan people in Chipinge and Chiredzi districts are segments of a larger group who were the major occupants of the South-Eastern Lowlands of Zimbabwe in pre-colonial times. The advent of colonialism brought forced removal and concentration of settlements in native reserves as land was taken by the government for conservation purposes. As GNP was being created, the Chitsa people were also affected by these land expropriations (Murombedzi 2003). Between 1890 and 1933, the area belonging to the Chitsa people at the confluence of Runde and Save Rivers was a fragment of the controlled hunting area. In these areas, the hunting of wild animals for food or sport was prohibited unless permission was sought and granted by the responsible authorities (Mombeshora and Le Bel 2009). Victoria Publicity Association strongly campaigned for the game reserve on the basis that it would stimulate tourism in the province (Wolmer 2007). They were strongly supported by the Umtali Publicity Association, who figured that the game reserve would attract tourists to the town of Mutare.

In 1934, a parliamentary verdict proclaimed Gonarezhou Game Reserve amid opposition from white cattle ranchers, who feared that it would lead to the spread of the tsetse fly and endanger the cattle industry (Scoones et al. 2012; Murombedzi 2003; Mombeshora and Le Bel 2009). Smallholder farmers also resisted the establishment of game reserves, especially since this would entail their relocation from areas designated for game reserves. Frontier disputes between the indigenous people and the Department of National Parks still transpire to this day, with pre-colonial and colonial land claims designated for national parks being reasserted in many areas (Murombedzi 2003). Furthermore, the establishment of these game reserves was not without cost. Numerous numbers of Africans were displaced by the new game reserves and relocated into the already crowded reserves. Thus, for instance, the proclamation of the Gonarezhou Game Reserve in 1934 led to the displacement of 1500 households who were relocated in the overcrowded Matibi 2 reserve in Natural Region 5 (Scoones et al. 2012; Murombedzi 2003; Mombeshora and Le Bel 2009; Tarutira 1988). One can deduce from the above that nature conservation and tourism reduced community access to natural resources; according to Dzingirai, "disenfranchisement at large".

In a bid to control tsetse flies, Gonarezhou Game Reserve was de-proclaimed (Mombeshora and Le Bel 2009; Wolmer 2007). The Chitsa community land was originally not part of the game reserve, but by 1956 the settlement was included in Gonarezhou. Redrawing of the game reserve boundaries also led to redrawing of chieftaincies (Scoones et al. 2012; Murombedzi 2003; Mombeshora and Le Bel 2009). Chief Chitsa, who had previously tried to fight inclusion of his area in the reserve, was relegated to a mere headman under Chief Tshovani (Mombeshora and Le Bel 2009). Following this, a game fence was erected along Chiwonja hills which separated Seven Jack (the area where the Chitsa community now lives) and Gonarezhou Game Reserve. In 1962, yet again, the Chitsa people were forcibly removed to the Seven Jack area to pave the way for another tsetse control procedure (Mombeshora and Le Bel 2009; Wolmer 2007). Consequently, Sangwe Communal Land was identified as a resettlement area for the Chitsa. The resettlement in the Ndali area was agreed upon on the basis that once the tsetse control program was completed, they would return to their land (Lan 1985). In the interim, boreholes were drilled by the government in the area, but the Seven Jack area was rented to Lone Star Ranch as a cattle grazing area (Wolmer 2003). This led to a veterinary cattle fence with iron poles being erected in 1974 around the disputed land. With the intensification of the liberation war struggle in the 1970s, Sangwe communal land people were moved into keeps (protected villages). Consequently in 1975 Gonarezhou was officially proclaimed a national park and the Seven Jack area which the Chitsa people claim as theirs was integrated into the park (Lan 1985; Mombeshora and Le Bel 2009).

After independence, the state wanted to transform the population living in chiefdoms into citizens who would participate in election of their leaders. Notwithstanding these determinations, Tshovani and Chitsa continued to operate as guardians of customary law (Mombeshora and Le Bel 2009; Wolmer 2007). The Zimbabwean government reverted to praising the role of chiefs in rural areas, mainly because of their vote broking role for the ruling party. Chiefs were given cars, electrification of their rural homes and paid high salaries by the government. Headman Chitsa invaded GNP in 2000 as a means of escalating his claim to the lost chieftainship, status and privileges (Murombedzi 2003; Tawuyanago and Makwara 2011).

Headman Chitsa was able to stake his land claim through a local alliance with a local councillor, war veterans and a sympathetic provincial governor who directed farm invasions (Tavuyanago and Makwara 2011; Mombeshora and Le Bel 2009). The invasion was then formalised during 2000 as the Department of Agricultural, Technical and Extension Services (Agritex) pegged the area and ten villages were laid out. With the approval of the provincial governor, the district administrator issued Chitsa people with endorsed documents to reside, cultivate and keep livestock in the park (Wolmer 2003; Mombeshora and Le Bel 2009; Wolmer 2007). Meanwhile the Ministry of Environment and Tourism and the Department of National Parks and Wildlife Management under which GNP falls did not have knowledge and consent of what was happening.

The South-Eastern Lowveld has been struck by a number of droughts in recent decades, most notably in 1982–1984, 1991–1992 and since 2001 (Wolmer 2007). Its communal areas are very remote as evidenced by poor access and limited infrastructure. The area is accessible through dust roads that go through big, uninhabited ranches. Cyclone Eline of January 2000 brought severe flooding and the washing away of numerous bridges, which worsened the inaccessibility of the area (Tavuyanago and Makwara 2011; Mombeshora and Le Bel 2009). The events between GNP and the Chitsa community show how inharmonious their history is. The question that comes to one's mind after being presented with such history is: can a land reform programme alone provide a solution to the strained relationship between community development and nature conservation in the South-Eastern marginal community of Zimbabwe?

Land and natural resources are not the only important factors to livelihoods nor are they always the most important ones (Wolmer 2003). As Scoones et al. (2012) notes, political, media, and academic commentary has been attracted by the publicity surrounding Zimbabwean land reform. Despite official declarations and guarantees that such areas will remain untouched, people have invaded them. From the history of GNP and the Chitsa community presented above, it can be concluded that the application of exotic conservation approaches in cultural areas ignores community knowledge (Pikirayi 2011). However, the question that remains is: did Chitsa people invade GNP solely as a way of acclaiming lost chieftaincy, or there are more reasons to the invasion?

4 Overview of the Research Methods

A qualitative research approach was adopted for the study, since the evaluation of development and conservation was to be studied in the natural context and could not be quantified ordinarily. The researcher was attracted to critical ethnography because this approach is apprehensible to inequalities within societies and focuses on advocacy for positive social change (Madison 2012: 5; Asher and Miller 2011; Bernard 2006: 342). It looks at how basic issues of social structure, power relationships and cultural practices influence human behaviour (Manias and Street 2001: 235) and it is influential in revealing embedded values and practices that are not obvious to people (Thomas 1993). De Vos (2011) acknowledges that in an effort to gain a better understanding of the phenomenon being studied, qualitative research is essential. Critical ethnography has been deduced as a modernization of orthodox ethnography which seeks empowerment and freedom. It gained momentum after mainstream ethnography was criticised for ignoring issues of empowerment and freedom of the researched. This research was not merely a study of socially marginal communities, but it also sought to help in the achievement of emancipatory goals, negate exploitive influences that lead to unnecessary social domination of groups. Consequently, critical ethnography was the most appropriate method of inquiry suitable for understanding and interpreting how the various participants construct realities of the world around them. In an effort to remain 'scientific' while simultaneously practising critique, the following data collection tools were employed: participant observation, life histories, semi-structured interviews and FGD (Fig. 2).

5 Reflections and Realities of Development-Conservation Nexus

5.1 The Historical Background of the Chitsa Community and the Establishment of Gonarezhou National Park

To obtain first hand historical facts about the community and establishment of GNP (Gonarezhou National Park), the researcher visited the local chief, Chitsa, who narrated the clan's life history recounting from the arrival of Munhumutapa in the ninth century to the arrival of the Chitsa people in the 17th century (1695) to the arrival of Europeans in the 19th century (1838). According to his narration, Chief Chitsa pointed out that original occupants of the area they are occupying were the Bushmen. Chief Chitsa indicated that people within the South-Eastern Lowveld are clustered around chiefs and headmen. In this particular area, there are two chiefs,

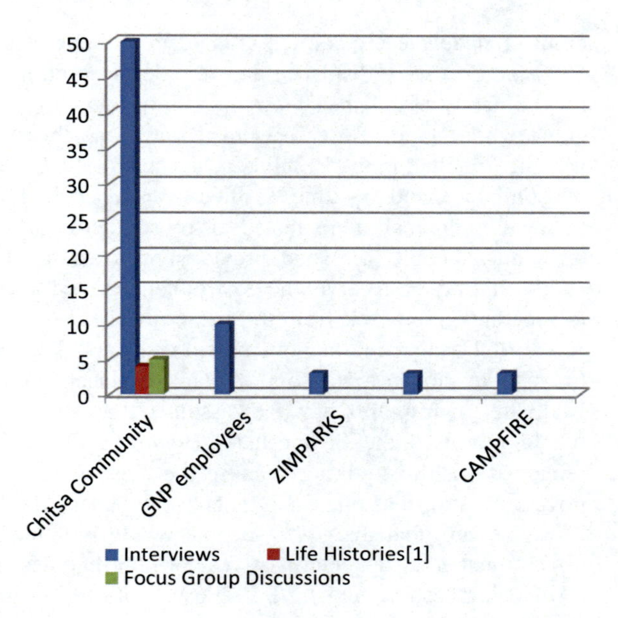

Fig. 2 Participant demographics, 2012

namely Chitsa (officially recognised as a headman) and Tshovani. Chitsa people came into Zimbabwe from Mozambique,[1] settling at the confluence of Runde and Save Rivers during the late 19th century (Wolmer 2007). Ever since they settled in this area they have lived side by side with wildlife. The aspect of living side by side with wildlife concurs with DeGeorges and Reilly's (2008) assertion that in Africa, human beings and wildlife have co-existed since the advent of humankind.

> We the Machangana (Shangaan) people of the South-Eastern Lowveld of Zimbabwe are the proud owners of a unique and vibrant social Xichangana culture.

Xichangana is the correct term if referring to the language. Machangana is a plural form of muchangana, referring to the people. Chief Chitsa mentioned that in the pre-colonial period, his community consisted of a hierarchy of land communities that nested one within another, and membership of these communities depended on acceptance by traditional governance authorities. Governance of natural resources like land and its related resources (wildlife included) were regulated by this structure and hence it functioned as a social control device.

However, the advent of colonialism in the 20th century had a serious negative impact on traditional governance of resources. Chief Chitsa acknowledged that communities were re-defined and they cut natural and cultural systems. Accordingly, between 1890 and 1933, their area was declared by the then Rhodesian government as a controlled hunting area. People lost access and control over natural resources within the area. In an attempt to conserve wildlife, in 1934 a

[1]Chitsa, J. Ndali Business Community (Life History, 10 December 2012).

ministerial decree proclaimed Gonarezhou Game Reserve out of Shangaan land. 1500 families were displaced and resettled in Matibi Nature Reserve to pave way for the newly established reserve.[2] During the early 1950s, the boundaries of the Gonarezhou Nature Reserve were redrawn and this time the Chitsa people's land was incorporated in the Reserve.[3] Once again, the Chitsa people were evicted from the confluence and this time resettled in an area called Seven Jack, which is near the Sangwe communal area situated in the southern part of GNP.[4] To make sure that the Chitsa people who had previously shown resistance would not attempt to encroach onto the reserve, the government erected a fence along the Chivonja hills to separate Seven Jack from the game reserve.[5]

In 1962, yet again, the government displaced the Chitsa people from Seven Jack to facilitate another tsetse control.[6] This displacement led to their resettlement in the Ndali area. During this stay, they assumed that once the tsetse control was over they would return to the Seven Jack. However, Chief Chitsa complained that this remained wishful thinking because the area was leased as a grazing area by the government to the Lone Star Ranch, now called Malilangwe Trust Conservancy.[7] To make sure that the Chitsa people would not 'illegally' settle in this area, the government erected a veterinary fence with iron poles around the Seven Jack area in 1974 (Mombeshora and Le Bel 2009). With the intensification of the war of liberation, the Chitsa people in the Sangwe area were moved into 'keeps'[8] just like any other people. It is apparent from the above account that in the eyes of government and colonial conservationists, human population was a hindrance and disturbance to 'natural' or nature conservation. These narrations, while preliminary, suggest an aspect of deep-rooted bitterness towards displacement and disempowerment of communities. It is justifiable then to note that during this time development and conservation had an inverse relationship, because for conservation objectives to be achieved, human presence in the reserve was prohibited. What further complicates the relationship between development and conservation here is the fact that people regarded land, together with other natural resources, as a source of ontological

[2]Chitsa, J., Kuruveli, J. & Musainge, R. Chitsa Community (Life Histories, 29 January 2013).

[3]Redrawing of reserve boundaries was accompanied with redrawing of chieftaincies and people were scattered. Those who resisted where either abolished or demoted. Chief Chitsa acknowledged that he was demoted because he resisted the inclusion of his land into park (Mombeshora and Le Bel 2009).

[4]Chitsa, J. Ndali Business Community (Life History, 10 December 2012).

[5]This fence is the one Chief Chitsa acknowledges as the official boundary to this day.

[6]Chitsa, J. Ndali Business Community (Life History, 10 December 2012).

[7]Chitsa, J. Ndali Business Community (Life History, 10 December 2012).

[8]Keeps are protected villages where ordinary citizens were put in during the liberation war struggle in Zimbabwe. Villagers were forcibly removed from their homes to these protected areas which in actual sense were nothing more than concentration camps in inhumane conditions. They were approximately 100 acres and surrounded by a high chain- link fence with barbed wire at the top (Zimbabwe Bulletin 1977).

origin, so displacement for them was unacceptable. This is justified by the continued occupation by Chitsa people in the GNP to this day.

The Reserve, in 1975, was subsequently gazetted as GNP by the government and this encompassed the Seven Jack area that belonged to the Chitsa people. After independence, Chief Chitsa and Comrade (Cde) Phikilele, a local Shangaan war veteran leader, teamed up and led an invasion on the portion of the park[9] (160 km^2) which they had always believed to be theirs, despite proclamations by the government (Scoones et al. 2012: 17). Consequently, 750 households were established in the park and they cleared some land for the purpose of farming. Occupants started farming cotton as well as sorghum and they managed to secure contracts with Delta Breweries, a Zimbabwean beverage manufacturer, for supplying sorghum. However, as the local people settled in the park, non-indigenous people, especially war veterans from outside the area, came and settled there as well.[10] Currently, there are ten villages with approximately 1500 households inside the park.

In a bid to legalise the invasion, the government, through the Department of Agricultural, Technical and Extension Services (Agritex), pegged the area and ten villages were formally laid out. As presented above, this pegging was supported by the provincial governor (Wolmer 2003; Mombeshora and Le Bel 2009: 2610; Wolmer 2007). Respondents from the ZIMParks said the pegging of plots in the park took place without permission of the Ministry of Environment and Tourism and the Zimbabwe Parks Wildlife Management Authority under which GNP falls.[11] The government drilled boreholes and deployed teachers to schools that were built by local people with mud and poles. While all this took place, the Ministry of Environment, conservationists and park authorities were not happy with this development and continued to vie for the eviction of the Chitsa people. As narrated above, it is significant that the reoccupation of the park, resistance by conservation actors and legalisation by the government goes beyond nature conservation and local economic development. This reveals how development and conservation are political processes in Zimbabwe.

The present historical findings are significant in at least two respects, namely, contested land and competing land uses. As shown in the above account, the land use contest emerges from forced displacements and land repossession complexity. While the land the Chitsa people claim to be theirs is officially a segment of GNP, this debate heightens an already complex relationship between development and conservation. As mentioned previously, spiritual associations between people and land through certain activities and practices connects them in a sensitive and complex web of meanings, responsibilities and reciprocities. This finding, though preliminary, suggests that when knowledge is expressed in a spiritual or social manner, scientists often find it challenging to acknowledge its relevance to NRM.

[9]Chitsa, J. Ndali Business Community (Life History, 10 December 2012).

[10]Chadenga, R. Harare (Interview, 10 December 2012).

[11]Chadenga, R., Sibanda, T. & Mapfumo Harare (Personal communication, 7 February 2013).

It can be concluded from the above that colonial and post-colonial NRM laws dispossessed local people of their customary rights over natural resources. There is a long-standing denial of local economic development through free access and use of natural resources, coupled with the enactment of official law transferring property to states, notably through imported legal concepts such as *terra nullius* (nobody's land). This highly politicised nature of community development and nature conservation increases the complexity of local economic development, nature protection, conflicts and resistance between development and conservation camps. As evidenced above, local history contributes to the understanding of development and conservation relationship in South-Eastern Lowveld.

5.2 Power Relations

Having examined the historical background of the interaction between the Chitsa community and GNP, the researcher saw it as imperative to examine how power relations shape the association between community development and nature conservation. The nature of power relations between the Zimbabwean Government, Chitsa community and other actors involved in the development, nature conservation and use of natural resources have proved to have a major bearing on the nature of the relationship between community development and nature conservation in the South-Eastern Lowveld of Zimbabwe. The effect, hostility and interference in the implementation of community development and nature conservation initiatives greatly influence the nature of the relationship and lead to unintended outcomes and possibly even to conflicts. Findings on how power relations affect the relationship between community development and nature conservation have been subcategorized into local traditional leadership, power relations between local traditional leadership, local government departments and central government.

5.3 Local Traditional Leadership

As noted earlier (Chief Chitsa Pers.comm, December 2012), people within the Chitsa community are under the traditional institutions of governance, which are chiefs and headmen. The two chiefs in this area, Tshovani and Chitsa, are embroiled in a dispute over local leadership. These two are from the same descendant named Zari, who had several sons. Amongst his sons were Mihingo and Tshovani.[12] The current Chief Chitsa is a descendant of Mihingo and Chief Tshovani is descendant of Tshovani (Mombeshora and Le Bel 2009: 2608). The genealogy is presented in Fig. 3 (Mombeshora and Le Bel 2009: 2608) below.

[12]Chitsa, J. Ndali Business Community (Life History, 10 December 2012).

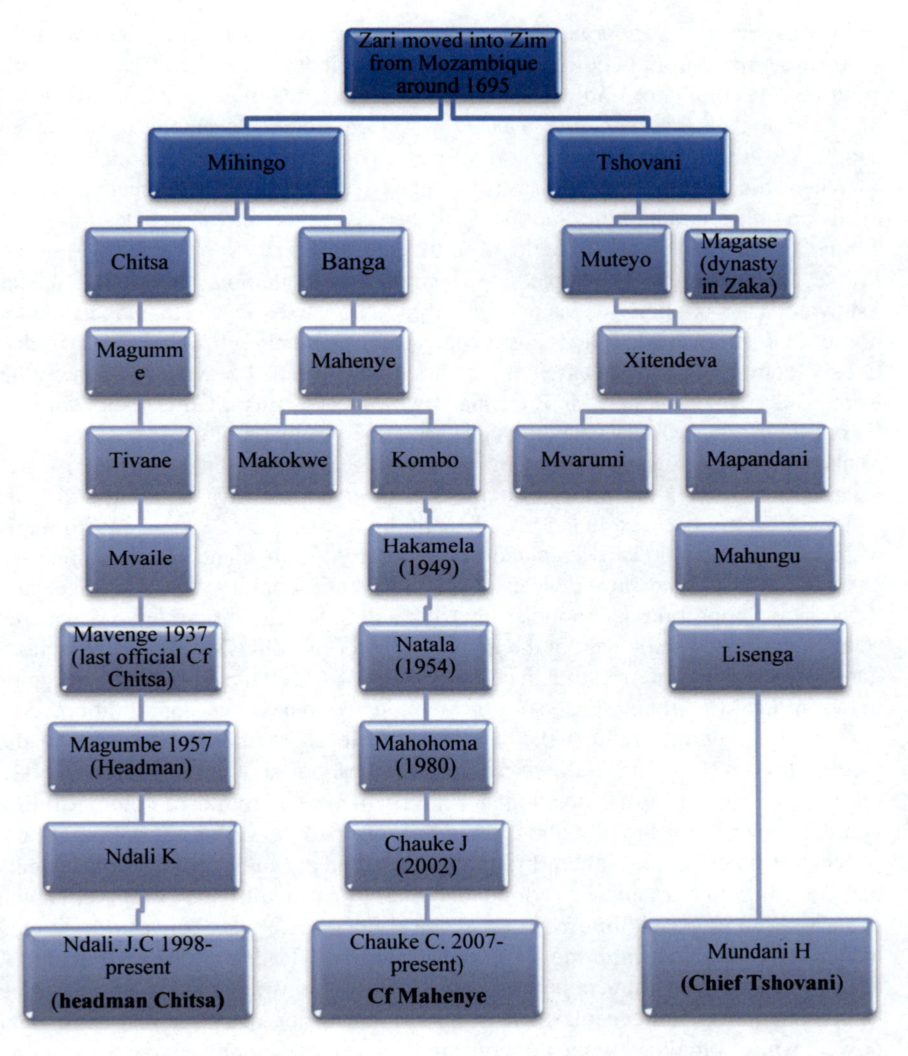

Fig. 3 Shangaan royal hierarchy in South-Eastern Zimbabwe (Chief Chitsa Pers. comm, December 2012; Mombeshora and Le Bel 2009: 2608)

Chief Chitsa, in justifying his chieftaincy, said Mihingo, from whom he is a descendant, was more senior than Tshovani, his brother, and therefore, he is the rightful chief of the area. He considers himself as the senior in the current royal hierarchy. He argued that being considered as the headman is mockery and a farce of uprightness that calls for correction. On the other hand, his rival, Chief Tshovani, acknowledged coming from the same descendant (Zari), but argued that the current royal status should be maintained as it is. Chief Chitsa traced the injustice of royal hierarchy to colonial times. He said when their ancestors came to reside in this place from Mozambique in approximately 1695 the area was total wilderness with no occupants, apart from wildlife. They then lived in the area harmoniously with wildlife. The coming of Europeans and proclamation of Gonarezhou Game Reserve impacted on their lives negatively. There were recurrent displacements of the Chitsa people, all in the name of tsetse control and paving way for the National Park.[13]

When the then Chief Chitsa (Mavenge) resisted these displacements and inclusion of his land into the National Park, he was demoted to a mere headman and Tshovani, who is believed to be loyal to the government, was rewarded with more people, since the Chitsa people now were placed under his chiefdom.[14] Chief Tshovani, who is allied to the provincial governor, lives in an upgraded modern house that is electrified and has been given partnership in the Save Valley Conservancy (SVC).[15] Seven chiefs in the South-Eastern Lowveld are benefitting from SVC: Chief Nhema, Mr Ranganai Bwawanda of Zaka, Chief Tshovani (Mr Felix Mundau of Chiredzi), Chief Gudo (Mr Mavivi Karukai of Chiredzi) and Chief Msikavanhu (Mr Vusani Mutumebvi of Chipinge), Chief Budzi of Bikita, Chief Chamutsa of Buhera and Chief Mutema of Chipinge.[16] As this last point indicates, Chief Chitsa has been excluded from benefitting from the SVC. Apparent from this account is how historical events and power struggles still shape and determine the nature of the relationship between development and conservation. Colonial land division and apportionment policies have left a deep imprint on present day patterns of land tenure and settlement in the South-Eastern Lowveld. Despite its exploratory nature, the study offers some insights into how chieftainship power struggles between local traditional leadership threatens development and conservation.

The above finding reflects the sentiments of local traditional leadership in the face of power struggles. As asserted by Mombeshora and Le Bel (2009: 2608), these arguments are to be positioned within wider background of colonial ruling and the deviations it brought about on indigenous people. Colonialism brought the concept of nature conservation through protected areas which led to displacements and concentration of people in communal land. Consequently, due to mixing and a concentration of people in certain areas, local power struggles between traditional leaders erupted there. Ignoring these local traditional leadership struggle realities, achievement of development and conservation objectives remains wishful thinking as chiefs continue influencing their people to embark on activities that show their powers while complicating the relationship between development and conservation.

---

[13]Chitsa, J. Ndali Business Community (Life History, 10 December 2012).

[14]Chitsa, J. & Kuruveli, J. Chitsa community (Personal communication, 26 May 2013).

[15]The Save Valley Conservancy is situated in the South eastern Lowveld of Zimbabwe. It is one of the largest conservancies in the world and covers approximately 845,044 acres (342,123 km^2). Historically, this area was predominantly used for cattle ranching. A massive drought served as a catalyst to change overall land use from cattle ranching to conservation in 1991 (Wolmer 2003: 3).

[16]Chitsa, J. Ndali Business Community (Life History, 10 December 2012).

5.4 The Nature of Power Relations Between Traditional Leadership, Central Government and Local Government Departments

As mentioned earlier, the particular study area is embedded in a land conflict with the government. The major communities demanding their land back are the Chitsa, Ndali and Chibemenene. Respondents of villages 1 to 10 in the GNP blamed the government for initially allowing them to stay in the park and subsequently, reneging on this promise.[17] The chief and other community members agreed that the government allowed them to stay in the park mainly for political reasons during the 2008 parliamentary and presidential elections. The chief said the then ZANU PF (Zimbabwe African National Union Patriotic Front) national chairman visited them in the park in 2007 and made assurances that the government would not remove them from their ancestral lands, but once the elections were over, they no longer cared about the Chitsa people.

> …..They assured us that we will stay in the park for good now that elections are over they want to chase us just like that……[18]

> …if Smith's administration achieved robbing our land an Act (legislation) why can't the government revise the same Act so that we can repossess our ancestral land? We don't want the entire of Gonarezhou but just a strip….[19]

Whilst the government tried to facilitate conservation by relocating the Chitsa people, the villagers resisted, claiming that they are the rightful owners of the land. In 2010 the government deferred, resettling the Chitsa people citing unavailability of funds. Conservationists who participated in this study said that after resettling in the park, the community started to cut down trees rampantly and pull down the fence, implying that they were likely to move further into the park.[20] Community members were given an opportunity to justify why they were pulling down the fence in a Focus Group Discussion. They indicated that the erection of the GNP fence was done by the government without consulting indigenous people. Therefore, they do not recognize the boundary. Very strong opinions pointed out that the fence has separated them from their ancestors and therefore, they will not respect this boundary. Musainge of Village 9 had this to say:

> Since they erected this new fence we have observed drastic changes in our lives, our land have become unproductive, we did not receive any rain last year and our children have all left. These are revealing signs that all is not well in the spiritual world.

---

[17]Hahlani, C., Kunene, J., Kunene, F., Masivamele, F. & Musainge, R. Ndali Business Centre (FGD, 6 May 2013).

[18]Phikilele, Y. Village 1, (Personal communication, 17 February January 2013).

[19]Chitsa, J. Chitsa community (Personal communication, 20 January 2013).

[20]Chadenga, R. Jongoni, S. & Ncube D Harare (Personal communication, 27 January 2013).

One can read from the above account that national parks have remained to protect the wilderness form of conservation which view indigenous people as obstacles to effective protection of biodiversity. However, anthropologically, humans and their environment have proved to be inseparable; thus, nature and environment are socially produced. This finding concurs with that of West and Brockington (2006: 613) that different power plays between development and conservation brings out the social effects of protected areas on people. An implication of this finding is that social structures, cultural values, power plays and individual behaviours are embedded within environmental services, and therefore cannot be ignored if successful conservation of development is to be achieved.

The provincial governor emphasized that it is impossible to move the Chitsa people because they are just claiming their land. To him, occupation in the park does not interfere with any NRM effort within the park. He indicated that their resettlement is history as the government could not secure funds to relocate and compensate them.[21] However, a representative within the Ministry of Environment and NRM said the government explained to the Chitsa people the importance of the park locally, nationally and regionally. Of cognitive importance to the responsible ministry is the contribution of the park to economic growth through tourism. The representative reiterated that the South-Eastern Lowveld is seen and should be maintained as a wilderness landscape.[22] It can be deduced from the above that in the eyes of conservationists, wildlife preservation has priority over people. Government, through representation by conservation actors and political leaders, views conservation as having top priority over resource access and use by indigenous people.[23] Be that as it may, development actors and community members agreed that the economic contribution of the park is realized more at national level than locally. Therefore, development actors deemed it difficult to support conservation efforts while the 'first people' are not directly benefitting from it.

While the community members blamed the erection of the fence for separating them from wildlife and adding to the misfortunes in their lives, high profile participants within the Ministry of Environment and NRM said the ministry is opposed to the invasion. Dr Chadenga voiced his opinion that villagers should be resettled elsewhere because they are invading the Gonarezhou sanctuary and it is not good for tourism.[24] These sentiments expressed by respondents support the second research objective, namely that of 'identifying key issues in the GNP's engagement with the Chitsa people who are living in the park'. Linked to the current relationship of development and conservation in the South-Eastern Lowveld is the epistemological differences between the two: origins, objectives, theory and practice. It can be deduced from the above that indigenous people claim that their lives are shaped and revolve around their history, culture, traditions and land tenure. It appears as

[21]Local Government Representative, Masvingo (Personal communication, 10 May 2013).

[22]Chadenga, R. Harare (Interview, 10 December 2012).

[23]Chadenga, R., Sibanda, T. & Mapfumo Harare (Personal communication, 7 February 2013).

[24]Chadenga, R. Harare (Personal Communication, 7 February 2013).

though development and conservation have different origins and thus, it is difficult to harmonise them. Due to their different origins, they are rooted in different operational frameworks in which development on one hand seeks to exercise equity, sustainability and self-reliance, and conservation on the other stresses maximising revenue generation whilst preserving wildlife. Caught in between these two discourses are the indigenous people.

One of the government officials interviewed said that in Zimbabwe, land for wildlife conservation should be solely used for preservation of wildlife rather than crop production; therefore, farmers allocated land under land reform program should practice wildlife conservancy.[25] This implies that conservationists are of the opinion that the practice of crop production, which is the Chitsa people's source of food security, is incompatible with stipulated land use in the area. On the other hand, community development actors object to the above mentioned view, saying that all land was declared as farming land. Consequently, people given land in the respective areas should practice crop farming regardless of "land use" options available in the areas.[26] In concurrence with the findings of Romero et al. (2012), the difference in the origins, actors and power play between conservation and development makes the two incompatible. These reflective opinion differences depict how land use has important implications for the livelihoods of resource users. Just as in much of rural Africa, land is of crucial importance to economies and societies, constituting the main livelihood basis for a large portion of the Chitsa population (Cotula 2007: 6). In this context, changes in land tenure systems bring about winners and losers. As land competition increases between wildlife conservation and crop production, resource access and use become more complicated.

The research also revealed a constricted relationship between local leadership, government departments and the central government. The local traditional leadership accused the government of not attending to their needs. The locals said they had to beg the government to build schools and clinics for them.[27] Chief Chitsa complained that the government viewed the community as a homogeneous entity, ignoring the diversity within the community, and this has led to problems of differential access to resources and benefits. He emphasized that the government assumed his rival, Chief Tshovani, has the ordinary people's interests at heart, but the reality is that people from outside have been allocated land in the park and other indigenous people continue staying in the unproductive communal area. This finding provides evidence with respect to Wassermann and Kriel's (1997: 67) erroneous assumptions of community development. This view concurs with Deville

[25]Sibanda, T. Harare (Personal communication, 8 March 2013).

[26]Mutumbi, A., Hove, C., Kubvakacha, P. & Muzari, T. Harare (Personal communication, 8 March 2013).

[27]Chitsa, J., Phikelele, H., Kuruveli, J., Hahlani, C., Kunene, J., Kunene, F., Masivamele, K. & Musainge, R. Chitsa community (Personal communication, February 2013).

in Cotula's (2007: 36) assertion that in customary systems, access to land and resources are an integral part of social relationships. Customary systems are alliance-based and rely on a number of principles. These usually include predominance of first occupants and access to resources linked to lineage membership.

6 Conclusions, Recommendations of the Study

Community development and nature conservation efforts which in turn influence adaptation capacity are not sufficiently recognized and managed in GNP. In the first instance, IKS, and roles and responsibilities for all stakeholders should be incorporated in strategies for both development and conservation. A knowledge base for nature conservation, rural livelihoods and community development seems disjointed, uncoordinated and rigid, and thus community development and nature conservation have a complex, but inseparable relationship. The pressure on natural resources is intense because of the inharmonious relationship that exists between community development and nature conservation. Consequently, this threatens the natural attractiveness of the park, sustainable use and management of resources, security of livelihoods by indigenous people, and facilitation of community development and nature conservation. The reality of minimal, if any, dividends returning to the community from tourism and hunting have been caused by lack of knowledge and ability to manage how these can simultaneously contribute to community development and nature conservation.

Although the study revealed the existence of a positive relationship between the Chitsa community and GNP, to conclude that community development and nature conservation can act as a panacea (remedies) for each other is rather unjustified, given the complex relationship revealed in the study. Evidence from the study indicates that pre-existing historical configurations influence long- term evolution of conservation and development. History has been shown to be influential in shaping up the current relationship between development and conservation in Zimbabwe's South-Eastern Lowveld. In particular, history is important because the line of movement, direction and flow of previous resource dependency cycles shape the current interaction of people and the park in this community. As asserted by Brennan (2009: 3), local culture has proved to be the basis for development and it serves to promote the use of local identity and management of resources. It mobilizes local population's way of behaviour. At the heart of the strand of the debate surrounding Africa's development and conservation relationship is history and culture. This study argues that history and culture embrace every attempt and activity of man in his life mentally, physically or spiritually, and therefore nature conservation needs to acknowledge and accommodate indigenous cultures.

The evaluation of development and conservation in the South-Eastern Lowveld points out two positions. Firstly, there is the claim that people live in harmony with natural resources and do not threaten nature conservation. As asserted by Orlove and Brush (1996: 335), this claim emanates from a long history of the presence of people in protected areas, rich knowledge in IKS, specific management practices based on IKS, and religious beliefs and ritual practices. These claims prove that indigenous people are committed to conservation. However, this is easily attacked by conservationists, who lack recognition that indigenous knowledge has a place in nature conservation.

Secondly, as illustrated by Sherman et al. (2011: 98), bureaucratic arrangements of government structures make the involvement of indigenous people in nature conservation difficult. As presented in above, the decentralized nature of NRM retained powers in the hands of the government. Communities have no governance and decision-making powers. Therefore, CAMPFIRE has failed to harmonise development and conservation in this resource dependent community. The study has shown discrepancies between theory and the practice of the devolution of powers in NRM to the lowest level of structures (community). The theory of CAMPFIRE has the potential to harmonise conservation and development efforts. However, in practice, these principles are rarely adoptable in resource-dependent communities (Dzingirai 2004; Frost and Bond 2008).

The study also recommends effective stakeholder analysis and engagement on the relationship between community development and nature conservation. It proposes promotion of effective dialogue between key stakeholders which have been identified as ZIMParks, Chitsa community, GNP employees, development actors, traditional leadership, the Department of Community Development and the government. The involvement of stakeholders in both development and conservation efforts/projects will instil ownership of the projects and thereby facilitate sustainability of those efforts through empowerment and capacity building. It is presumed that stakeholder analysis, consultation and engagement create harmonization of the relationship between development and conservation. Subsequently, through engagement, acknowledgement of GNP as containing multiple cultural activities of different groups who use and protect those natural resources will be achieved. In turn, interests and priorities of different stakeholders will be shaped in a participatory and inclusive model of the park management that will provide the underpinning for sharing different cultural approaches regarding resource values and benefits. In particular, as mentioned in the previous sections, current forms of participation do not support the needs, benefits and rights of indigenous people. The proposed effective stakeholder analysis, consultation and engagement on NRM has the potential of creating space in which local, national and global values relating to both community development and nature conservation can be expressed. It involves rethinking how natural resources within GNP can provide both benefits to the Chitsa community's development and support national conservation purposes at the same time.

References

Adams W, Aveling R, Brockington D, Dickson B, Elliot J, Hutton J, Roe D, Vira B, Wolmer W (2004) Biodiversity conservation and the eradication of poverty. Science 306(5699):1146–1149

Asher A, Miller S (2011) So you want to do anthropology in your library? Or a practical guide to ethnographic research in academic libraries. ERIAL Project

Bernard HR (2006) Research methods in anthropology: qualitative and quantitative approaches (4th edn). AltaMira Press.

Boonzaaier CC, Wilson GDH (2011) Institutionalisation of community involvement: the case of Masebe Nature Reserve, South Africa. In: Van der Duim R, Meyer D, Saarinen J, Zellmer K (eds) New alliances for tourism, conservation and development in eastern and Southern Africa. Delft, Eburon

Brandon K (2002) Putting the right parks in the right places. In: van Terborgh J, Schaik C, Davenport L, Rao M (eds) Making parks work: strategies for preserving tropical nature. Island Press, Washington, DC

Burnham P (2000) Native American country, god's country: native Americans and the national parks. Island Press, Washington, DC

Colchester M (2004) Conservation policy and indigenous peoples. Environ Sci Policy 7(2): 145–153

Coria J, Calfucura E (2012) Ecotourism and the development of indigenous communities: the good, the bad, and the ugly. Ecol Econ 73:47–55

Cotula L (ed) (2007) Changes in "customary" land tenure systems in Africa. International Institute for Environment and Development, Stevenage, Hertfordshire: SMI (Distribution Services) Ltd

DeGeorges PA, Reilly BK (2008) A critical evaluation of conservation and development in Sub-Saharan Africa. The Edwin Mellen Press, Lewiston, New York, NY, USA

De Vos AS (ed) (2011) Research at grassroots: a primer for the caring professions. Pretoria, Van Schaik publishers

Diam MS, Bakri AF, Kamarudin H, Zakaria SA (2012) Being a neighbor to a national park: are we ready for community participation. Prod Soc Behavioral Sci (36):211–220

Dowie M (2009) Conservation refugees: the hundred-year conflict between global conservation and native peoples. Cambridge: MIT Press

Duffy R (2000) Killing for conservation. The politics of wildlife in Zimbabwe. Oxford, James Currey

Dzingirai V (2004) Disenfranchisement at large: transfrontier zones, conservation, and local livelihoods. IUCN Regional Office for Southern Africa, Harare

Frost PGI, Bond I (2008) The CAMPFIRE programme in Zimbabwe: payments for wildlife services. Ecol Econ 65(4):776–787

Gandiwa P, Matsvayi W, Ngwenya MM, Gandiwa E (2011) Assessment of livestock and human settlement encroachment into the Northern Gonarezhou National Park. Zimb J Sustain Dev Afr 13(5):19–33

Lan D (1985) Guns and rain: Guerrillas and spirit mediums in Zimbabwe. James Currey, London

Logan BI, Moseley WG (2002) The political ecology of poverty alleviation in Zimbabwe's Communal Areas Management Program For Indigenous Resources (CAMPFIRE). Geo-forum 33:1–14

Madison DS (2012) Critical ethnography: methods, ethics and performance. London, SAGE Publications

Manias E, Street A (2001) Nurse–doctor interactions during critical care ward rounds. J Clin Nurs 10:442–450

Miller TR, Minteer BA, Malan LC (2011) The new conservation debate: the view from practical ethics. Biol Cons 144:943–951

Mombeshora S, Le Bel S (2009) Parks-people conflicts: the case of Gonarezhou National Park and the Chitsa Community in South-East Zimbabwe. Biodivers Conserv 18:2601–2623

Mugadza F, Mandizadza S (2006) The historical and social background to the chitsa community and Gonarezhou National Park Land Dispute. Consultant Report submitted to the Chitsa Dispute Taskforce

Murombedzi J (2003) Devolving the expropriation of nature: the devolution of wildlife management in Southern Africa. In: Adams WM, Mulligan M (eds) Decolonizing nature: strategies for conservation in a post-colonial era. Earthscan, London

Nelson F (2008) Livelihoods: conservation and community based tourism in Tanzania: potential and performance. In: Spenceley A (ed) Responsible tourism: critical issues for conservation and development. Earthscan Publishers, London

Neumann R (1997) Primitive ideas: protected areas buffer zones and the politics of land in Africa. Dev Change 28:559–582

Orlove B, Brush S (1996) Anthropology and the conservation of biodiversity. Annu Rev Anthropol 25:329–352

Pikirayi I (2011) Tradition, archaeological heritage protection and communities in the Limpopo Province of South Africa. OSSREA, Adis Ababa

Romero C, Athayde S, Collomb JD, Di-Giano M, Schmink M, Schamski S, Seales L (2012) Conservation and development in Latin America and Southern Africa: setting the stage. Ecol Soc 17(2):1–13

Sanderson S, Redford KH (2003) Contested relationships between biodiversity conservation and poverty alleviation. Oryx 37:389–390

Schwartzman S, Moreira A, Nepstad D (2000) Rethinking tropical forest conservation: perils in parks. Conserv Biol 14:1351–1357

Scoones I, Chaumba J, Mavedzenge B, Wolmer W (2012) The new politics of Zimbabwe's Lowveld: struggles over land at the margins. Afr Aff 00(00):1–24

Tarutira MT (1988) A review of tsetse and trypanosomiasis in Southern Rhodesia: economic significance up to 1955. M. A. Thesis, Department of History, University of Zimbabwe

Tavuyanago B, Makwara EC (2011) Contested landscape: the struggle for the control of Gonarezhou since the inception of colonial rule in Zimbabwe. J Sustain Dev Afr 13(7):46

Thomas J (1993) Doing critical ethnography. London, Sage Publications

Twyman C (2001) Natural resource use and livelihoods in Botswana's wildlife management areas. Appl Geogr 21(1):45–68

Van der Duim R (2011) New institutional arrangements for tourism, conservation and development in Sub-Saharan Africa. In: Van der Duim R, Meyer D, Saarinen J, Zellmer K (eds) New alliances for tourism, conservation and development in eastern and southern Africa. Eburon Academic Publishers, Delft

Wasserman I, Kriel JD (1997) Facts and fallacies. Perspectives on community development. Wasserman & Kriel, Pretoria

West P, Brockington D (2006) An anthropological perspective on some unexpected consequences of protected areas. Conserv Biol 20(3):609–616

Wolmer W (2003) Transboundary conservation: the politics of ecological integrity in the Great Limpopo Transfrontier Park. J South Afr Stud 29(10):261–278

Wolmer W (2007) From wilderness vision to farm invasions; conservation and development in Zimbabwe"S Southeast Lowveld. Weaver Press, Harare

Environmental Assets and Carbon Markets: Could It Be Amazônia's New *Belle Époque*?

Thiago Lima Klautau de Araújo, Amadeu M. V. M. Soares and Ulisses M. Azeiteiro

Abstract The Carbon and the Environmental Assets Markets are not regulated in Brazil. They are pointed out by experts and activists as sustainable alternatives of wealth generation and forest valuation. But will they be enough to make Amazônia able to experience once again a time of economic prosperity related to environmental preservation, just like in the *Belle Époque* (the golden time of Amazônia, supported by the rubber extraction)? This paper intends to discuss several issues—currently ignored—about the subject, considering historical, legal, social, environmental, economic and political backgrounds. Besides those contexts, there is an assessment of the public policies, current studies on regulation, and legislative trends about the environmental issues, the Carbon and the Environmental Assets Markets. Several inconsistencies/weaknesses were found in the legal system and if they are not properly considered, they might threaten the success of those markets and/or even preclude the social and economic return for the local populations, especially from Amazônia, in a possible future regulation. Without due care, instead of becoming an alternative of environmental preservation with economic development and decrease in social and regional inequalities, they may become another example of financial conglomerates income concentration, at the expense of the region.

Keywords Amazônia · Law and economics · Public policies · Environmental assets · Carbon markets

T. L. Klautau de Araújo (✉) · A. M. V. M. Soares · U. M. Azeiteiro
Department of Biology & CESAM—Centre for Environmental and Marine Studies, University of Aveiro, 3810-193 Aveiro, Portugal
e-mail: thiagoklautau@gmail.com

P. Castro et al. (eds.), *Climate Change-Resilient Agriculture and Agroforestry*, Climate Change Management, https://doi.org/10.1007/978-3-319-75004-0_28

1 Introduction

The last century (particularly in its second half) has presented a major dilemma to Brazil: economic growth or environmental conservation? Up until recent years, undoubtedly, the unsustainable and irresponsible economic growth prevailed over biodiversity issues, environmental and social affairs (Pinho et al. 2014), or even over justice and fraternity among the states of Federation, existing values in the successive Brazilian Constitutions. Amazônia, a long and uncharted territory, has been characterized by the most perverse government interventions. Not properly planned and truly ignoring the reality of the region, it were based on the eagerness for economic "progress" and began a process of social and environmental degradation (Klautau de Araújo 1995), land conflicts and political disputes.[1]

Initially fueled by Juscelino Kubistchek's developmental policies in the 1950s, followed by national integration initiatives of the military dictatorship[2] (Paulino 2014), and, finally, leading to the disregard from the governments after the redemocratization, Amazônia faces, nowadays, a real environmental, social and economic disruption. However, it was one of the richest and most important regions in Brazil between the second half of the 19th century and the first half of the 20th century. More recently, increased attention was therefore paid to the process of environmental, social and economic degradation of the region, which shelters the most relevant and significative storehouses of the planet's biodiversity

Despite several isolated initiatives, not only governmental ones, but also, and, mainly, from the civil society and international organizations, the social and environmental situation[3] from Amazônia continues to deteriorate. The inefficient work by the successive governments and the environmental agencies can be pointed out, as well as the incoherence and weaknesses of the Brazilian Legal System—from the Constitution to other assorted regulatory provisions—the failure of the environmental education policies, illegal practices, forest exploration with predatory economic activities (for instance: mining, logging, agriculture and cattle) and the extreme poverty of the populations from the region are considered to be some of the major determinants that contribute to aggravate the situation.

---

[1]Which triggered, for instance, several break-up states projects and a popular referendum on the division of Pará State in three other ones (Pará, Carajás e Tapajós), in 2011, that was rejected, above all, by the dissenting vote of the people who live in the metropolitan zone of the capital of the state, Belém. In the affected areas by the creation of the new states, the votes in favor of the division were more than 90%.

[2]Both have proved disastrous because not only the proposals were not achieved (creation of transport infrastructures, road links to urban centers of other regions, creation of agrarian settlements, among other initial intentions), but it also created several economic and social problems that had never existed before.

[3]An increase is observed not only in the number of conflicts, but also in the violence rate. Despite the fact that the North Region has a small population if compared to other Brazilian areas, it sees a serious increase of the violence rate, much faster than the more populated regions (FBSP and IPEA 2016).

In addition to the aforementioned elements of that disrupted context, the existing political resistance regarding the elaboration of appropriate public policies to environmental preservation[4] is noticeable.

The political forces from Amazônia are considered to be tiny if compared to the ones from South and Southeast states.[5] Therefore, there is no political pressure or bargaining power[6] since the representation is clearly small and fragmented among the recognizable opponents, including those who are against the environmental actions because, according to their point of view, they would hinder the economic growth.

To make things worse, Amazônia has the largest mineral province of the planet, located in the state of Pará, with significant production of iron ore, gold, bauxite (and its by-products alumina and aluminum), nickel, copper, kaolin, among others. That exploration results in environmental and/or social damage, without financial compensation. This is because, according to Article 155, §2°, X, of the Constitution of the Federative Republic of Brazil from 1988 (CF/88), the products intended for export cannot be taxed by the states. In other words, the ICMS (tax over operations related to the circulation of goods and interstate and intercity transport and communication services), main state tax in Brazil, cannot be charged by the states on the mineral extraction activity to export, which corresponds almost entirely to the total of the mineral exploitation. There is, also, oil and natural gas production in Amazonas state and it is likely that there are oil stocks on the Pará's Coast.

For those reasons, several political agents are still convinced that the environmental preservation, especially in Amazônia, is a strong obstacle to the Brazilian

---

[4]The already low level of funding for fighting against deforestation had been reduced in 72% between 2010 and 2014 (Leite 2015) and in 2017, there was a reduction of more than 50% in the Ministry of the Environment's budget—from 911 million (Ministério da Fazenda 2017), to R$ 446 million (Gesisky 2017; Moutinho and Guerra 2017). Meanwhile, the Federal Government has provided R$ 190.25 billion to the Agricultural and livestock plan 2017/2018, more than the R$ 185 billion available for the 2016/2017 Plan (Peduzzi 2017). The amounts invested in the agricultural plan represent more than 426 times the current budget of the Ministry of the Environment, which is the responsible for the federal environmental monitoring of the whole country, not only of the Amazônia.

[5]The North Region, that corresponds to almost the total of the Brazilian Legal Amazônia, has seven states, representing 45.25% of the national territory (IBGE 2016); however, it relies only on 65 of the 513 parliamentarians from the House of Representatives, since the population is proportionately represented, but this number is outdated and does not match the actual population—Amazônia would have more deputies if the division of places was updated. Nowadays, only the state of São Paulo has more than 70 representatives, 5 more than all states from the North Region together.

[6]Weinstein (1993), describing the negative fall in the prices of rubber latex in the international market, during the first decades from the 20th century, demonstrates that the lack of political power from the North Region is an old problem in Brazil: "(…) It is essential to consider the political component of the economic obstacles from the region. Since it needs political support at the national level, the elite of Amazônia failed many times in offering support to programs that intended to combat the devastating effects of price fluctuations. Moreover, their appeals for emergency assistance right after the collapse were largely ignored".

economic growth. Nevertheless, it is necessary to highlight that the environmental preservation may be an important complement of the economic growth and development of the region—and vice versa—like the one we have seen in the past, in the most relevant moments of Amazônia history.

This paper intends, therefore, to discuss about alternatives for Amazônia's development, the context of the current laws, the terms for establishing markets and the challenges for doing so. Despite the fact that the Carbon Market is the fashionable one, we comprehend that it is, in the context of Amazônia, just one of many renewable and non-predatory possibilities of rational use from the existing natural resources in that region. Due to those reasons, as we shall debate it later, we suggest the regulation, also, of the Environmental Assets Market (and the Carbon capture, obviously, is one of them), so that there can be established the maximum use of Amazônia potential, with the environmental preservation, climate context improvement and development for the region and its local populations.

2 Amazônia, the *Belle Époque* and the Environmental Assets

The Environmental Assets, some of the greatest Amazon riches, have fundamental importance to the sustainable development[7] of the region. As a matter of fact, the history of Amazônia shows two examples in which the economic growth was directly associated to the use of those environmental assets without forest damage: at the beginning of the colonization in that region, with the so-called backland drugs; and with the latex exploration, also called "rubber boom", divided into two phases, one during the 19th century and the other one during the 20th century.

The backland drugs were the main reason of Amazônia colonization, since Portugal could not control the Indian spices route, it then attempted to use Amazônia products as substitutes, among them, the cocoa, *pau-cravo* and achiote; that led to the cities and villages foundation in order to control the territory and the transit of foreigners in Amazônia. Belém, the capital of Pará, was founded and expanded in that context of monitoring, due to its geographic location and strategic importance, seen as one of the main entry doors for Amazônia (Cardoso 2015).

The rubber cycle, in turn, consisted in the exploitation of latex from Amazônia rubber tree *Hevea brasiliensis*. Its widespread use in the industry for a variety of purposes and its exclusiveness of production in the states in Brazil's North Region would promptly bring prosperity to the region and then enrich it, fact that made Belém one of the main (if not the main) financial center of the country in the second

[7]Based upon the economic growth, environmental preservation and improvement of the social conditions.

half of 19th century, not only due to the wealth brought by the rubber, but also due to its strategic proximity to Europe and the United States of America.[8]

All that abundance culminated in the building of imposing palaces, houses, parks, avenues, hotels, theaters, cities planning, development of significative infrastructure works like ports, railroads, public illumination, electricity, trams, among other various ventures that took the latest technologies to Amazônia. Clearly, all things worked and were built with European influence,[9] especially from Paris, hence the designation used for that period: *Belle Époque*, being Belém affectionately nicknamed as "Paris n'America".[10] The sumptuousness of the buildings and the ingenuity of the urban solutions were really impressive, especially if we consider the logistical and geographical challenges still found up to the present. Undoubtedly, it was Amazônia's Golden Era.

The first rubber cycle ended at the beginning of the 20th century because British Government Officials smuggled rubber trees seeds and established huge plantations in Malaysia (Garfield 2009), where did not exist the plagues that blocked the development of large fields of rubber trees in Amazônia, so they could reduce the costs of production and it led the Amazonian rubber plantations to the decline, followed up by their abandonment, since they were more expensive, with tricky logistic lines and more time-consuming collection and distribution, regarding the dispersed trees throughout the forest.[11]

However, the Malaysia rubber plantations were under the Nazis' control during the World War II, which led to three situations: the shortage of material, the consequent rise in latex price in the international market and the weakening of the Allies' armies (and their economies) due to the lack of this essential raw material for the industry. Hence, the rubber exploitation in Amazônia restarted, with shorter duration at that time, but brought fresh air to the economy of the region.[12]

There was, over those periods of time, an unmistaken direct relation between the produced wealth and the forest preservation.[13] Naturally, on account of the indigenous labor (during the backland drugs period) and the use of slave labor or

---

[8]The geographic importance of Belém was demonstrated, for instance, by the air links to Europe and to the USA, which were set up from Belém and not from São Paulo.

[9]With particular situations like, for example, rich families from Belém used to have the laundry done in Paris (Weinstein 1993) and the use of large and warm clothing in a region where the temperatures remain constantly above 30 °C during the whole year.

[10]Curiously, that is the name of a traditional store, which still works at the same place since 1900 in a building in the downtown of Belém.

[11]Regarding the existing problems in Amazônia for the wealth coming from the latex transition to a more industrialized economy—as it did not turn out to be, see Weinstein (1993).

[12]For more details on the relationship between Brazil and the USA, due to the importance of Amazônia for producing rubber latex during the World War II and the immigrating and economic impact to the region, see Garfield (2009, 2010).

[13]Different from other events in the Brazilian history, as in Pau Brasil exploitation (*Caesalpinia echinata*), that almost made that specie disappear, or as in the economic situation of Amazônia itself, which has been destroyed for soy-plantation, cattle breeding and timber extraction.

slave-like terms or even low wages (in latex production),[14] it is not possible to talk about sustainable development, since the social component was not covered, especially because those practices were common and legal at that time. However, despite the reprehensible and inadmissible practices related to social and labor terms, the mentioned times demonstrate that the maximum use of the Amazon potentials can occur not only with activities that do not deforest the region, but also can quite possibly be a complement to combat environmental degradation.

3 The Negative Change in the Amazônia Path

Despite the clear propensity of the region in considering a non-predatory use of the environmental assets,[15] the economic, political and infrastructural models applied in Amazônia since the 1950s did not consider basic cultural, natural and social characteristics of the place. That caused serious damage to the environment and to the society (Klautau de Araújo 1995, 2014; Paulino 2014). The Brazilian governments imposed solutions previously used in other states which were successful, but that, obviously, did not work in Amazônia because its environmental features demanded—and still demand—carefully conceived and personalized choices.[16]

Therefore, it was ignored the necessity to treat unequals unequally to achieve justice—Aristotle's thought plentifully discussed over the years.[17] Thus, there was no participation of the local populations in the definition of laws and public policies

---

[14]Weinstein (1993) states that it was the main reason behind the economic decline in Amazônia with the rubber price collapse, if compared to São Paulo, which also suffered from the coffee crisis effects, but it became industrialized: with no internal market of products, based upon wage labor, Amazônia did not have other options besides the latex extraction. That excessive dependency on a single product, without establishing a productive market, led it to its economic ruin.

[15]The proof lies in the fact that the richest period of time of the region history was based in a renewable and non-predatory product, with almost no deforestation, and without serious environmental impacts. On the other hand, currently the environmental degradation is noticeable, either because of mining, livestock or logging activities, while the population remains miserable.

[16]Amazônia is unique and, therefore, different from other Brazilian realities. The soil type, the rivers dimensions, the features of the weather are, sometimes, impossible challenges to solve traditionally. Examples of failed initiatives that did not take into account those elements are several: Madeira-Mamoré railway, which was not concluded due to project problems that brought a huge distress initially and due to the death of several workers who developed tropical diseases; Balbina hydroelectric, in the state of Amazonas, which has a reservoir similar to the one of Tucuruí hydroelectric power plant, in Pará, but that has about 3,3% of the energy produced in Tucuruí; the roads BR-163 (Cuiabá-Santarém) and BR-230 (Transamazônica) with no completion until nowadays, besides many other examples.

[17]The debate on the topic has begun with Aristóteles, who considered it in the "*Nicomachaen Ethics*", especially in Chaps. 4 and 5 of its 5th book. Still on that, there are works like the "*Discours sur l'origine et les fondements de l'inégalité parmi les hommes*" by Jean-Jacques Rousseau, e "*A Theory of Justice*", by John Rawls. In the Brazilian context, it is important to mention "Oração aos Moços" (Addressing the Young), by Rui Barbosa.

to Amazônia, or in their corresponding implementation and monitoring.[18] The region was treated merely as a new border for changing problems from the main urban centers of the country and also seen as a solution for housing the farmers who did not have lands in other regions.[19]

Under the slogan of national integration, with a development-oriented policy, the government of Juscelino Kubitschek promoted roads construction without proper planning, intending to connect the North Region with the whole country and with the new federal capital. In continuity, the military governments built more roads[20] and started a disordered landholding colonization process, under the motto "land without men for men without lands", and arguing that the settlement of the area could protect the national territory, safeguarding the national sovereignty. However, the attempt of landholding colonization without any support by the public agencies, lacking adequate infrastructure and even no researches on the soil types of the region[21] demonstrated that it was not an attempt to solve the Brazilian agrarian problem (taking into account that, until the present moment, the Land Reform in Brazil has not been enacted), but it was a political way to keep the popular pressure

---

[18]Until these days, the effective popular participation is a huge democratic challenge, even in developed regions. In Amazônia, it is particularly difficult, and there are studies and proposals with possible replication, as the one carried out by Folhes et al. (2015). However, it is necessary to emphasize that Amazônia is diversified and that not all designed models for it will have the same effects in different zones. Moreover, the participation model has to be defined properly and planned along with the community, considering each case specifically (Klautau de Araújo and Lima 1997a, b).

[19]Wolford (2016) points out some notes about the used strategies by the military governments for Colonization in Amazônia: "INCRA was created on July 9, 1970, as an autonomous agency tied to the Ministry of Agriculture (Decree-Law 1110, Article 4, July 9, 1970). The military government in power at the time created the agency to oversee the colonization and settlement of Brazil's vast and "underpopulated" northwestern frontier. The march westward was expected to fulfill Brazil's promise as a developed, modern nation, which meant extinguishing peasant protests in the Northeast and dealing with the presumed threat of communist guerrillas known to be hiding out in the Amazon rain forest (Martins 1984, 41; Bunker 1985). Colonization was also a means of combating external influence; the slogan "*integrar para não entregar*" (integrate to avoid delivering [the Amazon to foreigners]) was part of the substantial publicity campaign that accompanied frontier development (Reel 2010, 36). In the early 1970s, Brazilian theaters showed films weekly documenting the bulldozers and trucks cutting through the jungle to build new highways (Drosdoff 1986, 60–74). (…) Buttressed by a sense of manifest destiny, INCRA employees moved west to settle "men without land in a land without men," carving out thousands of 100-hectare plots, building houses and towns, and leading markets into relatively untapped regions of the Amazon rain forest (Hecht and Cockburn 1989, 108)".

[20]Including two that have left a trail of unprecedented environmental destruction: the Transamazônica and the Cuiabá-Santarém roads, already mentioned in the footnote 16.

[21]"The endless soil fertility of Amazônia, announced by natives and visitors of the region, has proven to be one of the several legends fostered by the luxurious and thick forests from the region. As a matter of fact, the soil of Amazônia highland, like the tropical soil in general, is thin and with nutrients easily exhausted as soon as the forest cover exposes it to the rain. After that, only the intense fertilizer application will facilitate the growing/cultivation. Regarding the small part of Amazônia surface classified as alluvial plain, that is less susceptible to leaching and hardening than the 'dry land', but can also burn out under intensive cultivation" (Weinstein 1993).

away. At no time, any of the plans or government interventions had considered the social-environmental aspects or eventual externalities that ended up occurring.[22] Concerning the national security, the argument has also fallen down because the Amazônia frontiers are still opened for the trafficking of arms, drugs, animals, human organs and people. The increase in population did not bring any benefits, neither to the region, nor to the local populations or even to the immigrants. Moreover, there was a worsening of the social conditions and deforestation.

The roads construction is also linked to the migratory flows towards the surrounding regions, where people would be attracted by the promise of economic development and jobs, since the landholding colonization programs had failed. The population growth with no basic structures like health, education, security, sanitation and other State services has set up a complex socioenvironmental context. The successful options have been marked by binding together both society and the governments, searching for a broader integrated understanding of the local situation,[23] instead of isolated mitigation measures, as it has commonly been done.

Another important aspect to be highlighted is that the built roads did not work as an alternative to the existing railroads, but rather as substitutes. Nowadays, the road transport corresponds to 61.1% of the transport of goods and to 95% of the transport of people (CNT; SEST; SENAT 2016). The concentration of cargo and people transportation in the road model, apart from being more expensive and polluting, is also slower.[24] It partly explains the difficulties in creating a local productive chain for Amazônia goods: the distance from major consumer centers of the country in addition to the lack of faster, safe and cheaper transport alternatives have reduced the interest of investments in that region. The establishment of companies could certainly contribute for generating jobs, income, local technology and for promoting a closer interaction between the populations and the production of non-predatory items derived from the forest, which arouses an increasing interest to conserving the environment. That would be an interesting option for adding value and regional development around its main potential: the forest; nowadays, nearly all Amazônia's products are exploited to be manufactured in other states or

[22]Even though the described facts have occurred during the military dictatorship time, lack of concern about Amazônia features still remains, and the governments, successively, keep on ignoring the basic aspects of the region context. Recently, for example, school buses were donated to Afuá city, in Marajó archipelago, Pará; but in Afuá, known as "Marajoara Venice", there are no streets, since all its buildings are on stilts and there are lowland soils and floodplains in which the use of motor vehicles is forbidden (Meirelles Filho 2015).

[23]As in Paragominas, State of Pará, where with strong participation from the mayor, managers and society, the context of social and environmental degradation of the city was changed, and Paragominas is an interesting example to be examined. See more in Klautau de Araújo (2014).

[24]And, if we consider the terrible road construction, especially in Amazônia, the numbers are more negative: due to the bad quality of the pavement, more than 77,488 million liters of diesel fuel are spent unnecessarily every year, which represents 2.07 $MtCO_2$ of additional emissions (CNT; SEST; SENAT 2016).

countries, leaving so little or nothing to the local people, leading to an understandable declining interest in environmental issues in the face of the current poverty.

4 Environmental Assets: A "New" Hope for Sustainable Development in Amazônia

Considering that context, the environmental assets can be decisive elements for changing the socioenvironmental situation. As we have already mentioned in this paper, the creation of productive chains for Amazônia products could lead to the economic and social development of the region. Fostering the Environmental Assets and considering the possibility of using them as an enhancing instrument for the forest are widely discussed alternatives (Ministério do Meio Ambiente 2016; Birdsall et al. 2014; CGD 2015; Seymour and Busch 2016), even before the consolidation of the use of that terminology.[25]

Amazônia is rich in environmental assets of all kinds,[26] and if is made the most of that potential, with the local populations engagement, the possibilities for the sustainable wealth generation in that region are feasible. For instance, by addressing some infrastructure issues, it is possible to develop the ecotourism, an alternative that remains unexplored in the area, or expand the food, pharmaceutical and cosmetics industry[27] which use the regional products.[28] Everything depends on the way that the environmental assets and their possible markets are regulated, conceived and defined, as well as the underlying issues discussed up to this point.

---

[25]The sociologist José Mariano Klautau de Araújo started to debate about the need of the Environmental Assets sustainable use in Amazônia since the 1970s, when the projects for national integration and settlement were at the top. His discussions, subsequently published on books and papers, added to the Socioenvironmental Method, written by him, are interesting points of reference to understand the historical and institutional background of public policies for Amazônia since the 1950s until nowadays. The Socioenvironmental Method was the pedagogical foundation for Escola Bosque, an initiative awarded internationally and that can be read in much more details in Klautau de Araújo and Lima (1997a, b, c), Klautau de Araújo (2016) and Lima (2013).

[26]As it can be seen in more depth in Seymour and Busch (2016), according to the different types of environmental assets classified by the authors.

[27]A huge Brazilian cosmetics company was accused of using, without permission, traditional knowledge from herb medicine women from the Ver-o-Peso market, in Belém (Weis 2006; Soares 2016); after a great controversy, there was then a refund for that (Soares 2016), and a soap and essential oils factory from that company was opened in the metropolitan region, in 2014 (Kafruni 2014).

[28]Products like açaí, cupuaçu, cumaru, camapu, among other fruits, plants, herbs and seeds derived from Amazônia have awakened a great international interest due to their nutritional, aesthetic and therapeutic properties. However, the most part of those products, with rare exceptions, is collected in the region, and the end products are produced somewhere else, with no return to the local people.

5 Current Legal Brazilian Situation Concerning the Carbon Market, Environmental Assets, Mitigation and Adaptation to Climate Change

The Brazilian environmental laws are dispersed, confusing, bureaucratic and complicated to deal with, due to their gaps, shortcomings and lack of regulation (Klautau de Araújo 2014, 2016). The few existing legal devices regarding the Environmental Assets trading, including the carbon, or related to the climate change are in that same direction, as well as the current legislative proposals and public policies.

As the main legal basis for that issue, the Law 12187/2009 (National Policy on the Climate Change), in spite of being recognized as an important legislative framework (Milaré 2014; Lopes et al. 2015; Pinheiro Pedro et al. 2015) comes up a bit short of the expected. In its art. 4º, VII, it is established the fostering to the development of the Brazilian Emissions Reduction Market, which would open the national Carbon Market and would contribute to the clearer participation of Brazil in the international market, but it does not mention how that aim would be reached (nor even the Decree 7390/2010, which regulates parts of the above-mentioned law).

In the art. 9º, the Law 12187/2009 grants carbon credits (here mentioned in a broad sense) of legal nature of securities and requires them to be traded in the stock exchange,[29] being the Securities Exchange Commission (Comissão de Valores Mobiiários—CVM) the responsible for monitoring it; however, the CVM disagrees with the classification established by the National Policy on Climate Change.[30] Since the articles 4º, VIII and 9º are too superficial and were not regulated by the mentioned decree, the Carbon Market has not yet been satisfactorily implemented

[29]In this context, the BM&F Bovespa.

[30]The problem in this case was the legal system inconsistency. That was because the Law 6385/76, about the securities, has a strict list in its art. 2º. Therefore, the Securities and Exchange Commission expressed its opinion, asserting that "Carbon Credits are securities issued by an organization associated with the United Nations which represent no-emission of certain amount of gas that cause global warming. The CVM discusses matters related to the carbon credits and why they must not be considered derivatives or collective investment securities—**thus, they are not securities, but assets and they are marketed to reach the targets of carbon emission reduction or aiming the investment. In addition, The CVM understands that it would be inconvenient to classify the carbon credits as securities through the edition of the law regarding the arrangement of those instruments**. The Securities and Exchange Commission also discusses the features of some financial products derived from the carbon credits, that, depending on their characteristics, might be defined as securities. The assessment of each financial product will be done by the Securities and Exchange Commission" *[highlighted by us]* (CVM 2009). Although that issue may appear to be a simple detail, it may be decisive in the future concerning the tax matters, fundraising from the Carbon Market and the legal certainty/validity of the contracts. More than that, by failing to amend the Law 6385/76 or to consult the Securities and Exchange Commission about the legal nature of an obligation modality that would be ruled by that institution shows the regulator's negligence and weakens the applicability of the law.

in Brazil[31] because it is not as significant as it could be and it does not offer sufficient legal guarantees to the investors. Last but not least, the art. 12 establishes voluntary reduction from 36.1% to 38.9% based on the projected emissions until 2020,[32] but it does not say how it should be done either,[33] not even mentioning the metrics to be used to verify the results.

Therefore, if there is no commitment to the targets, no solid legal basis to create the Carbon Market, to reduce the emissions, to implement concrete measures to be taken by the public authorities, the Law 12187/2009 and the Decree 7390/2010 seem to be only intentions that the federal government may not be interested in putting in practice.[34]

Regarding other aspects of combat, mitigation or adaptation to climate change, there are: Law 12114/2009, which creates the National Fund on Climate Change, regulated by the Decree 7343/2010; Law 11284/2006, which is "about the public forest management for sustainable production; it establishes, on the basis of Ministry for the Environment, the Brazilian Forestry Service; it creates the National Forest Development Fund", among other measures, as well as its regulatory Decrees 6063/2007 and 7167/2010; the Decree 6527/2008, which is about Fundo Amazônia; the Decree 8576/2015, which "establishes the National Commission to Reduce the Greenhouse Gas Emissions Resulting from Deforestation and Environmental Degradation, Forest Carbon Stocks Conservation, Sustainable Forest Management and Enhancement of Forest Carbon Stocks—REDD+"; Some of the Forest Code provisions (Law 12651/2012),[35] especially the Article 41, which

---

[31]Although Brazil holds 5% of the world's Carbon Market, while 20% was expected initially (Brasil 2012).

[32]That, according to the art. 6° of the Decree 7390/2010, means to reduce between 1168 million of tonCO_2eq and 1259 million of tonCO_2eq of emissions.

[33]The art. 6°, §1 of Decree 7390/2010 lists ten actions to be taken to make feasible the accomplishment of the goal, being foreseen and implemented by sectorial plans.

[34]Lopes et al. (2015) and Pinheiro Pedro et al. (2015) emphasize the importance of the Law 12187/2009 to the establishment, at first, of a voluntary market and of a mandatory one later, after adopting the compulsory measures of emissions reduction. Nevertheless, we understand that the enactment of the above mentioned law was only an attempt of political response to the national and international pressures/expectations and not a practical step indeed. The Brazilian legal experience in the last decades demonstrates that law enactments without regulation (or their ineffective regulation) has been a strategy of the governments in order to avoid the responsibility of polemic or complex issues (Klautau de Araújo 2014, 2016). Due to that, we believe that a norm with inconsistencies, omissions and inaccuracies is not able to substantiate a whole system of combating measures, mitigation and adaptation to climate change.

[35]One of the legislative provisions of the Forestry Code is the Environmental Reserve Quota (CRA), that consists in a portion of non-deforested land which exceeds the law requirements. It may be considered as an Environmental Asset because it may be negotiated with other landowners who have not fulfilled the minimal legal size of vegetation in their areas. It is provisioned by the articles 44 to 50 of the law 12651/2012. There is a recent study developed by Brito (2017) about more efficient regulations of that legal institute so that there is not an excess supply, pulling down the prices and inhibiting the restoration of the areas, taking Pará state as a reference for additional regulation to the recommended by the Forestry Code.

is about the Environmental Assets. Also, there are two National Congress Bills, in progress: the 212/2011 draft legislation, which intends to set up the "national system of reducing emissions from deforestation and forest degradation, conservation, sustainable forest management, maintenance and enhancement of forest carbon stocks (REDD+)",[36] and the 95/2012 draft legislation, which intends "to determine the securities trading in the Brazilian Emissions Reduction Market related to the avoided greenhouse gas emissions verified in indigenous lands must be previously authorized by FUNAI".

5.1 Legislative Trends

As previously observed, there were several laws and regulations concerning environmental and climate change issues and there are new ones already being planned. In other words, there is a tendency for legal provisions and sparse regulations that makes the legal system tricky. That tends to decrease the efficiency of implementation, the legal compliance and laws inspection, as well as to difficult the citizens' understanding on the Law field, if compared to the issues that are arranged in codes or in few and concise laws (Klautau de Araújo 2016). Milaré (2014) claims that the codification of the environmental issues can give legal security to the area, and we may infer that the same would happen if applied to the Environmental Assets Market. The same way that the environmental issues have been regulated, it became very difficult for citizens to participate in the elaboration, enforcement and, mainly, monitoring of the laws, since they do not know the technical terms, do not have resources or even because there is such a complex bureaucracy involved (Klautau de Araújo 2014).

Besides the regulation issues and the scattered laws, there is a worrying trend to weaken the federal environmental norms, despite the increase in deforestation. The text of the new Forest Code ended up confirming irregular deforestation before 2008.[37] In May 2017, the House of Representatives and the Senate ratified two provisional measures that reduced in 597 thousand hectares Pará and Santa Catarina states preserved areas. 587 thousand hectares of this reduction were only in the

---

[36]Although there is not any law project about it, we should consider the ENREDD+, the "REDD strategy development" planned by the Ministry of the Environment (Ministério do Meio Ambiente 2016).

[37]Legal devices like the articles 61-A, §§ 5° e 6°, 67, 68, among others from the new Forestry Code, reduce or leave behind the demand for reforestation of the Legal Reserve (minimum of native vegetation to be kept on the estate) or Permanent Protection Area (vegetation surrounding rivers and sources, for example) for several contexts, which means relaxation of the environmental rules and legalization of the irregular situations. The message to be understood is that the environmental rules may be not followed, since, at any time, there will be tolerance of validations of the illegalities in the name of the economic issues.

Jamanxim National Forest and in the Jamanxim National Park, located in Pará (Maisonnave 2017a, b).[38]

A data entry carried out by WWF indicates that bills in course in Congress reduce the protection of approximately 80,000 km^2[39] (Maisonnave 2017a).There is, also, another Law Draft (3729/2004), with 18 included projects, in course, that aims to relax the environmental license regulations, which has been pushed through to be approved (Miranda 2017).

The most recent environmental debate was derived from the edition of Decree 9142, from August 22, 2017, which extinguished The National Reserve of Copper and Associates (Renca)—an area between Pará and Amapá states with a forest area over 4 million hectares—and opened part of the reserve to mining. It had a great impact and the Government edited a new Decree (9147, from August, 28,2017) revoking the previous one, remaining the extinction, but addressing few points in detail which had been contested in Decree 9142; The Government and some experts say that the reserve has already been occupied by illegal loggers and miners, who extract the Brazilian resources avoiding taxes and polluting the rivers, also exposing fish and watercourses to mercury contamination, promoting deforestation, and the mining regulation would control the impact of those activities (Schreiber 2017). However, the polemic persisted throughout, the public opinion was overwhelmingly against the reserve extinction and, shortly after, the Federal Justice decided to suspend the Decree. There was another suspension order by the Supreme Court that determined the government to clarify the issue in 10 days. For those reasons, the Ministry of Mines and Energy (MME) published a Directive on September 5th, 2017, suspending the legal effects of the Decree 9147 during 120 days in order to broaden the discussion on the subject (G1 2017). On September 25th, 2017, the President decided to revoke the extinction of Renca, with the Decree 9159. The Ministry of Mines and Energy published a note affirming that: "(…) The country needs to grow and generate jobs, attract investments to the mineral sector, including to exploit the economic potential of the region. The MME reaffirms its commitment and of the entire government to the preservation of the environment, with the safeguards provided for in the environmental protection and preservation legislation, and that the debate on the subject must be taken up at a later time and must be extended to more people in the most democratic way possible" (Ministério de Minas e Energia 2017). It seems that the results of this story are far away, especially if we consider that the environmental question is, apparently, being used as an exchange for political power and support.

[38]However, after extreme national and international pressure, the President Michel Temer has gone back on his own proposal and has vetoed provisional measures 756 e 758, that established the reduction of the area. In a new twist, the Government sent a proposed law to the National Congress, reducing once again the area of Jamanxim National Forest, not in 486,000 hectares, but now in 349,000 hectares, that will be protected areas, with a few restrictions, if the project is approved (G1 PA 2017).

[39]Near Portugal's territory dimension, which has a little bit more than 90,000 km^2, including its islands.

Nowadays, as it can be seen, there is an overall weakening trend—rather than a strengthening one—of the Brazilian environmental legislation. Added to that, there are international uncertainties related to the success or to the failure of the countries cooperation against climate change,[40] endangering the Carbon and the Environmental Assets Markets in Brazil. Without a mandatory setting of a limit of emissions and a legal system efficient for punishing irregularities,[41] there will scarcely be any interest or economic feasibility of the markets. Moreover, regarding the Environmental Assets Market, directly affects the very existence of the assets to be traded.

6 Technical Challenges and Possible Distribution Models Originated from the Carbon and Environmental Assets Markets

6.1 Challenges—Not So New as They Seem

The economic development of Amazônia through activities linked to the regional vocation and to the sustainable management of the existing environmental resources is fundamental to equalize the social and environment preservation issues (Klautau de Araújo 1995, 2016). That is because Amazônia, currently, does not have the least necessary structure of transports, health, basic sanitation and education, with a few alternatives of decent livelihood remaining to the populations of its remote areas, who end up preying on the environment in exchange for some living, specifically (but not only limited to) cutting down the forest illegally for selling timber; with no guarantees for human dignity to the local populations, it is impossible to demand their support to the government against environmental degradation, or even expect that they worry about preserving Amazônia (Klautau de Araújo 2016). That situation fits within Paul A. Samuelson's argument (1976) about Economics of Forestry and the extinction of certain animal species.[42]

[40]Especially after the withdrawal of the United States from Paris Treaty. It is early to analyze the impacts and consequences of that decision, even with the response of leaders in the European Union and China reaffirming their commitment to the international agreement (Lusa 2017; Gomes 2017).

[41]The bureaucracy, the low values charged by the fines, added to the almost endless possibility of lodging an appeal create a sense of impunity and, at times, is more profitable to do something illegal and then pay or contest the possible fines (Klautau de Araújo 2016).

[42]"When people in a poor society are given a choice between staying alive in lessened misery or increasing the probability that certain species of flora and fauna will not go extinct, it is understandable that they may reveal a preference for the former choice. Once a society achieves certain average levels of well-being and affluence, it is reasonable to suppose that citizens will democratically decide to forego some calories and marginal private consumption enjoyments in favor of helping to preserve certain forms of life threatened by extinction." (Samuelson 1976).

The current policies of forest preservation are not only based upon a confusing and flawed system, but also transform the forests in a source of expenses, not of possible income, since the environmental protection is not well regarded by the governments (Klautau de Araújo 2016). Some environmental assets, like the Carbon Market—which generates wealth through the simple preservation of environmental areas—can change this context, adding value to the forest as heritage sites to be preserved not only due to its ecological, climate and biodiversity aspects, but also due to its inherent social and economic potential.

The environmental issue is not—and it has never been—into Brazilian political focus. In a particularly troubled moment for the Federal Public Management, with discussions and debates on electoral, political, labor and social security reforms and so many corruption scandals, the concern with the environment is far from being a priority. Another relevant aspect is the continuous disinvestment in the environmental monitoring sector, which has worsened over recent years (Klautau de Araújo 2016). Arousing the economic interests in maintaining the green areas is a way to bridge the gap between the environmental issues and legislators and governments.

Therefore, despite the existing limitations to the Environmental Assets, they are considered to be, in Amazônia context, the most feasible alternative at the moment, since they offer the possibility to generate wealth, income, employment, life conditions with dignity to the local populations and, simultaneously, the forest preservation. However, in the Brazilian case, despite all the problems pointed out by some scholars, the Environmental Assets, particularly the Carbon Market, do not face great difficulties in the international scenery or even in terms of long-term economic viability, but in the legal system. Maybe this is the most significant barrier for the implementation of a comprehensive policy focused on the sustained use of those resources.

The Brazilian legal system is extremely confusing and disconnected, making the procedures too bureaucratic, long and difficult. For instance, the first practical difficulty in the implementation of a Carbon Market in Brazil would be a more precise definition of land ownership. Odd as that may seem, there is no consensus on the exact public land delimitation and dimension for each one of the federal entities, setting judicial disputes among the states[43] and between The State and private individuals. One of the reasons for that is the lack of clarity (and/or regulation) of the art. 20, I to XI and §2º, and art. 26 of the 1988 Federal Constitution, as well as the art. 16 of the Transitory Constitutional Disposition Act.[44] Added to it, the Land Reform in Brazil has not been enacted, fact that brings legal uncertainty,

---

[43]Acre and Amazonas are disputing a 12,000 km^2, while Mato Grosso contests for 22,000 km^2 with Pará, and Ceará complains about 2821 km^2 with Piauí (Mariani et al. 2016).

[44]Referring to the public land ownership.

which gives ground to frauds against individual and state patrimonies.[45] Lastly, the lack of data links among registries makes the system even less reliable.

Defining public and individual land ownership in Brazil is essential to make the most out of the Environmental Assets potential, but in particular to make the Carbon Market feasible for the national territory. Without knowing which areas belong to whom, it is not possible to estimate neither how much land each market player (especially the public bodies) has for carbon absorption, nor how many tons might be absorbed. Moreover, without solving this problem several judicial disputes may occur focusing the money raised in market, leading purchasers' migration to other markets.

Although it is a complicating factor and that certainly will bring many practical difficulties and legal disputes over the years, that issue does not block the Brazilian Carbon Market from being implemented.[46] The purchasers, naturally, will seek for agents appropriately settled and they will select those who will remain or will leave the market. Something to worry about is that Brazil, until this moment, was not able to carry out structural adjustment reforms and any regulation arranged will only be a slight repair over a confusing and inconsistent legal system.

It is necessary to remember that the 1988 Federal Constitution—as has dealt with the public land ownership—established the environmental protection in a general way, regarding as a joint competence the environmental protection, air pollution combating, forest, flora and fauna preservation to the Member States, to Federal District and to the municipalities,[47,48] Added to that, there is a constitutional determination on the preservation of the environment contained in Art. 225[49] from CF/88, and its paragraphs, in which the expression "Public Power" is used in the broadest sense of the term. It means that all power spheres are responsible, conformity with their respective competences, for the environmental protection.

---

[45]Without the Agrarian Reform and a standardization of the securities, all proofs of ownership modalities are accepted, if the necessary requirements at the time of the alleged purchase are met. It happens that, if there is no control of the registers, it is possible, for example, to have a land title document from the sixteenth century validated. That is one of the reasons why a very common fraudulent practice in Brazil used to be the document forgery, putting the papers inside drawers with crickets, because their secretion would make them appear older than they actually are. That practice was called "grilagem", a nickname given, nowadays, to all kinds of fraudulent attempts of land regulation.

[46]Especially because it already exists, but it is not based on laws, as we will confirm later in this paper.

[47]Art. 23, VI and VII of the 1988 Federal Constitution.

[48]There is also legislative concurrent competence/authority to the Union, states and Federal District, regarding "forest, hunting, fishing, fauna, conservation of nature, soil and natural resources preservation, protection of the environment and pollution control", according to the art. 24, VI from 1988 Federal Constitution.

[49]"Art. 225. Everyone has the right to an ecologically balanced environment, which is an asset of common use and essential to a healthy quality of life, and both Governments and community shall have the duty to defend and preserve it for present and future generations."

The legislative intent was good: by expanding the number of the responsible ones for monitoring, the efficiency of surveillance would increase. But the results did not meet the expectations and the attempt to generically share the responsibility for the environmental monitoring caused the opposite effect.[50]

As we have already mentioned, in theory, if the Carbon Market, generates revenues, instead of being a burden for the State budget, the forest would be a source of income which would increase the interest for its conservation.[51] What happens is that the collection of taxes and revenues is strongly controlled by the Federal Government, but the charges of environmental monitoring belong to everyone.

If there is not any system restructuring of the environmental agencies and the clear liabilities establishment, the problem will remain unsolved, despite the influx in revenue of the Carbon Market.[52]

Another reason why the regulation must be a careful process is that the sale of carbon credits is, all in all, an offset, since it keeps a reduced economic activity in order to preserve the forests. Nowadays, the Brazilian cities that present the lowest human development indicators are located in Amazônia. Brazil shows extremely serious regional disparities and the Environmental Assets sustainable exploitation is one of the few available alternatives to reduce the disparities without further aggravating environmental degradation.[53]

Therefore, we point out four possibilities of collecting and distributing revenues from Carbon Market—or, why not, from the Environmental Assets market?—if the Brazilian regulation is possibly established.

---

[50]In another study (Klautau de Araújo 2016), by the use of the game theory and an adaptation of one game, it was pointed out that when all the constitutional powers are generically shared it is likely that none of the agencies act. That is because all tasks might be shared, but not the budgets. Without a clear division of competences and tasks, it is almost impossible to demand an agency responsibility in case of default since it is stated that you cannot blame a single absent agency for the unsuccessful public policies when other agencies did not act either.

[51]Not only for public entities, but also for individuals. The Carbon Market is an alternative for obtaining economic income from preserved areas required by law inside the states, as the Legal Reserve, which currently are unpaid. Concerning that, the then-Senator and current foreign minister Aloysio Nunes stated that: “the Carbon Market would be a stimulus for the grower to maintain his legal reserve, so that he can generate income from that. It would fulfill our goals not only internationally, but also inside Brazil, implementing an important law, welcome by the whole world, that is the Forestry Code law” (Altafin 2016).

[52]In the discussions on this topic, this point was also raised by Senator Jorge Viana: “who can trade the carbon absorption in the forests? Is that the country and in a centralized manner? Or are the states, which are guardians of the forests? There is almost no regulation of it. It is a new market, not widely-known, which is part of climate change” (Altafin 2016).

[53]The areas with higher biodiversity in Brazil are those in which the people have lower incomes and little provision of public services (Kageyama 2009; Klautau de Araújo 2016).

6.2 *Possible Methods for Collecting and Distributing Revenues*

6.2.1 Federal Government Revenue Collection and Distribution of the Total Value or a Portion of It Proportionally

That hypothesis would consist of the Carbon Market values centralized in the hands of Federal Government and further distribution proportionally among the States and cities.

In this model, the Federal Government would be the only entity of public power able to make trades—as it is already seen, for example, in the capital market. There are strong chances to adopt that option, since it is about repeating the existing practices in which the power is centered at the Federal level and the sharings are distributed in a standard way, taking into account the region of the federative entity and the income *per capita.*[54] However, with that regulation form neither the regional inequalities in Brazil will be reduced, nor the forest conservation will be encouraged. This is because in the distribution according to the previous criteria, the states with the highest capacity to absorb Carbon[55] will have to divide the main compensation for maintaining their forest area. However, all the conservation burden will remain individualized. Therefore, a scenario of strong regional differences is eternalized, enhancing those differences, due to the fact that the forest conservation duty is imposed (involving costs and the forgoing of tax revenues) in exchange for specified economic benefits that will be shared.

We could consider a possible comparison between that model and the regulation of royalty payments related to oil and natural gas exploration, ruled by the Laws 9478/97, 12351/2010 and 12858/2013. There is a share of the royalties aimed at the producing states and cities and a shared aimed at the Federal Government and other non-producers, what may seem unfair at first. But the oil exploration creates a lot of jobs and attracts investments, which is the opposite of the Carbon Market: you pay for the forest maintenance by using credits derived from the option of not making economic activities. Moreover, the oil and the places where the exploration is more common belong to the Federal Government, according to the Article 20 V, VI e IX of the 1988 Federal Constitution. The states, the Federal District and the municipalities are allowed to participate in the results of those resources exploration or

[54]As it is used, for instance, in the division of the State Participation Fund (FPE) and in the Municipalities Participation Fund (FPM). For further details, read: art. 159 of the 1988 Federal Constitution; art. 34, §2°, I, II and III of the Transitory Constitutional Disposition Act; the arts. 90, 91 e 92 of the Law 5172/66 (National Tax Code); and the Complementary Laws 62/89 and 91/2001.

[55]Consequently, with less economic activity, lower GDP (gross domestic product) growth, and higher monitoring costs and forest conservation.

they can be compensated by the exploration itself if it takes place in their area, according to the §2° of the forementioned article, but the property is still of the Federal Government. That is why that analogy is not valid, considering that in the Carbon Market the land ownership to be conserved for CO_2 capture is assured, and it may belong to the Federal Government, to the states, to the Federal District, to the cities or even to individuals.

Thus, that model is unwelcome because it would disregard the individual capacity of Carbon absorption and the negotiation of federative entities and individuals in the Market. It also would increase the regional inequalities by not compensating economic activity decrease (or little maintenance) of the states where there are preservation areas.

6.2.2 Federal Government Revenue Collection and Distribution Among the Entities in Their Proper Quotas

In this system there is a similarity between what happens in the capital market and the previous model, the only one entity of public power able to operate in the Carbon Market would be the Federal Government, but the difference is the way of sharing the funds after raising them. Sharing without regarding the proportional criteria of population, location of the states and per capita income, but considering the actual results of environmental preservation and carbon absorption, there would be a more effective offset for the carbon absorption and a greater economic interest of the states and the Federal Government in the environment preservation. Therefore, this option could be more fitting and fairer that the first one.

In order to distribute the amount collected by the Federal Government in the Carbon Market to the states and the cities, there are two options: carbon trading assignment of rights of the federal entities to the federal government and payment for the proper quotas[56] or the royalties payment of the Union to the states and cities regarding their preserved areas, in a similar system of the Financial Compensation for the Exploration of Mineral Resources (CFEM).[57]

[56]What, in practice, would work as though the Union had bought the carbon credits from the states and cities and had resold them in the market with its own credits.

[57]This system is regulated by the law 7990/89 and 8001/90 and distribute the royalties obtained with mining as it follows: 65% to the cities where the production occurs (and to the Federal District, if applicable), 15% to the states (or the Federal District) where the production occurs, 15% to the cities and to the Federal District affected by the mining activity, and 10% divided between Federal Agencies. Observe that, different from the oil royalties, only the producer states and cities (and those which are not producers, but are directly affected by mining infrastructures) receive a financial compensation.

The bureaucracy involved, however, is the problem of the last modality because it gives grounds for corruption[58] due to the excessive centralization of powers of some agencies and political players.[59]

Besides those difficulties, there are also complications in budget planning of the states and the cities in the context of this model. That is because depending on the Union transfer of resources, it is not entirely clear when and how much will be available in the states and cities coffers.[60] Due to all these uncertainties and to the insistence on imposing a model of powers centralization, we believe that this alternative would not fit properly within the Brazilian context, especially regarding Amazônia sustainable development.

6.2.3 Linking of Earnings for Sustainable Development Projects in the Region

There are strong indications that this hypothesis could be confirmed, maybe, as a complementary manner to others, especially because it has already been used in the regulamentation of the oil royalties, but to a fund for Education and Health. It would be interesting to use part of the revenue obtained by the Union through Carbon Market as an investment for environmental preservation, for sustainable development of regions with rich biodiversity and natural resources as well as for restoring degraded areas. Here, that possible regulation is attractive because it qualifies the purpose of those obtained values, otherwise, there would be common revenues.

Thus, the repetitive discourse which affirms that the environmental preservation sucks resources dry and that it blocks the economic growth is not valid, and with this alternative, this narrative loses its meaning, because providing funds to the country may boost the economy and safeguarding part of revenue to invest in environmental recovery and protection makes the system self-contained. By promoting sustainable development, apart from social conditions improvements, there is a chance for Brazil to develop economically, heading towards a more

[58]In 2016, a serious corruption scheme was uncovered involving the National Department of Mineral Production (DNPM), dismantled by the Federal Police, through Operação Timóteo, in 11 states and the Federal District (G1 DF 2016; Affonso et al. 2016). This department is responsible for collecting and distributing the mining royalties in Brazil and handled approximately R$ 1.6 billion in 2015. In this specific case, the corruption scheme was related to the amounts which would be allocated directly to municipalities.

[59]Klautau de Araújo (2016) points out that one of the biggest problems of the public policies implementation process in Brazil is the centralization in the planning and decision-making processes and the dispersion of the implementation. In that context, the situation is similar: the concentration of power in few agents attracts strong political interest, makes the proper use of the money difficult and increases the possible deviations from ethical conduct.

[60]The event at the National Department of Mineral Production had to do with it: there were late payments and the agents from the department received bribes (concealed by law and consultancy offices) in order to speed up the money deposits and to increase the values.

environmentally friendly production with a strong focus on the rational and conscious use of the natural resources. The path towards an environmentally friendly economy is neither about loss of competitiveness nor about impoverishment,[61] but it needs more investments and also a paradigm shift in the relationship between State and society, as well between those and the nature.

However, for the same reasons presented in the previous alternative, there are strong evidences that it would be a failure as the main model. That would happen because the resource centralization in a single fund, besides leading to the greed of malicious public managers,[62] will also lead to the problem of over-centralization pointed out by Klautau de Araújo (2016). More than that, it is important that the environmental assets can be seen as attractive business opportunities, as well as income generation and job creation opportunities. The advantage of using them in a coordinated manner is that they can complement each other. That means it is possible to generate wealth, for instance, through ecotourism or adventure tourism, extraction and cultivation of medicinal herbs, flowers and fruits from Amazônia, to achieve the carbon absorption and to sell securities on the Market, all those things by using the same area. The allocation of revenue to a single fund (even though if it would be only the public one) obtained from a specific environmental asset would reduce the flexible usage of sustainable environmental assets and mutually complementary. Consequently, their competitiveness and attractiveness for the private sector would also be reduced.

Specifically in the case of Amazônia, the community and individuals deep involvement will be necessary since the area to be monitored is very large, the governments do not have sufficient means to be present throughout the territory (even if they had, the costs would be extremely high), and the current methods are more repressive than preventive.[63] Moreover, the failures in implementing public policies in Amazônia in previous experiences, as we have already discussed in this paper have demonstrated the need for understanding, agreement, contribution and

[61]If we analyze the examples of the developed countries, the technology, the productive processes, the entrepreneurial dynamics, the environmental awareness and planning are more connected to the success of the economy than to the plenty of natural resources. Japan, for instance, relies on few natural resources, has a rough ground, hard to be used, besides being a group of islands, what would make its development difficult. However, despite all those facts, it is one of the most developed economies, a balanced society and with low level of poverty.

[62]Even though the study of Public Administration in Law courses, most of the times, recommends to consider corruption in the spheres of government as an exception, not as a rule, the Brazilian experience shows us that the reality is different and that the current political system has many examples of a corrupt governance. There are denouncements against all the political spectra and against almost all political parties. Therefore, the planning of alternatives to the country must be designed in a way that the flaws may be minimized and the public funds may be protected.

[63]The satellite surveillance, inspection activities, among others, intend to punish the responsible for the environment damage, what is valid, but if the damage has already been done, the environmental restoration is hard, even with the modern procedures of reforestation, due to the maintenance of ecological balance and biodiversity, it is best to keep the original wood instead of reforesting a cleared area.

participation of local populations in the decision-making and executive process. Without that, it is almost impossible to think about monitoring and managing such a large and difficult access area like Amazônia.

6.2.4 License for Individuals and Public Entities to Operate Freely in the Market

For the given reasons in the previous topic, the participation of the local people and of all the federative entities is necessary for finding solutions to the environmental issue. There are many doubts concerning the possible ways of arousing the interest. By allowing a broad participation of public and private entities in those markets, increasing their profitability with the natural resources conservation and biodiversity, there is a clear valorization of the forest and growing interest of the individuals in taking part in the initiatives for conservation.

Thus, the model we believe to be the most interesting for implementing the Carbon Market in Brazil is a hybrid model of collection, distribution and allocation of the obtained revenue, in which all the federative entities and individuals could freely participate, making their rights, duties and obligations clear, (topic that will be discussed further in this paper), also considering some part of the revenue allocated to a supervisory, monitoring, conservation and reforestation fund.

7 Practical Challenges

Besides all the technical challenges, in order to design a management model of the Environmental Assets that work efficiently and to make the Carbon and the Environmental Assets Markets reach the goals of economic and social sustainability, it is necessary to regard some practical challenges, namely: fostering an economic interest for conservation and maintenance of forests; designing a regulation system appropriate for the reality and accessible to the interested people, in which there is an understanding of the protected legal assets importance, the duties, obligations and rights of the parties involved in this market, whether they are small rural landowners or major corporations; and the strengthening of the educational system, especially the environmental education.

The first challenge is to find ways that make the Environmental Assets viable sustenance alternatives so that the local people cease predatory practices and become interested in maintaining and preserving the forests. In addition, engage them in the sense of not allowing third parties to make predatory activities. It means to make the local populations and the State partners in the monitoring of forests and ecosystems. It is necessary to note that the Environmental Assets may be economically attractive, as stated above in this paper, because it is possible to combine the exploitation of several environmental assets simultaneously, increasing their profitability.

The second challenge is to communicate effectively with those populations, making them access, in a real and not fictious way, the legal documents and legal regulations about the issue.[64]

It is not appropriate to use a very technical language (as it has been used) with terminologies which people do not know and can barely understand. Most of the times, it is necessary to have appropriate knowledge of the field in order to communicate with the people responsible for the studies and the regulation projects. For this reason, on behalf of honesty, transparency and good faith of the decision-making processes and enforcement proceedings, the communication must be improved. A context in which the regulation of a sector is not understood by almost any of the interested parties is an enabling environment for the spread of corrupt practices and vested interests that do not fit the exercise of citizenship and democracy.

The third one is the implementation of a qualified education system for the populations. That is the key component to make children, teenagers, young people and future generations environmentally aware, even for those who are not inserted into the forest reality, so that they will be able to help in the environment preservation.

If some of those things (especially the communication with the population) fail, the effectiveness and efficiency of that action as a sustainable way to Amazônia and other Brazilian biomes that are endangered will possibly be reduced, since the three presented components complement each other. Nevertheless, the tackling of those issues depend on a regulation system suited for the specificities of Brazil and its regions in order to arrange the Carbon and Environmental Assets Markets as economic development alternatives, not only as simple chances of some groups to make money out of the natural resources that are of interest to the society.

Especially because the Carbon and Environmental Assets Markets are not going to save Amazônia—or the world, but they are sustainable alternatives for economic diversification and natural resource use. Overly concentrating investments and to pay attention exclusively to that possibility is risky and reckless. Doing that is like

[64]A brief comment on the popular participation and interaction effectiveness is welcome: they are possible only if there is an open, transparent and accessible language. The overuse of technical terms, jargons, acronyms and abbreviations makes the documents reading unpleasant and sets up a bar for any person who is not in the limited circle of "specialists". Even the Law or Environmental field professionals who are unfamiliar with the Carbon Market and the Environmental Assets experience serious difficulties when dealing with this topic due to the language unnecessarily tight, the challenge is greater to, for instance, a person from Amazônia countryside with poor education and with no access to public services, a very common profile found in the local community (brought about by the indifference of the governments). In order to make those people—who have so much to contribute to the understanding and preservation of their areas—participate in the debate, clarity, respect and transparency from the specialists are demanded and they may learn a lot about Amazônia with its natives. Unfortunately, until the present moment, frequently, the decision-making processes and the explanations of open initiatives to the participation of local people are not arranged in that way, resembling an accession model in which the "knowledge" is taken to the people; there is no dialogue, there is no participation.

getting back to *Belle Époque* but for bad reasons: that excessive economic dependency on the rubber led Amazônia to chaos with a decrease in that product prices in the international market.[65] If the Carbon and the Environmental Assets markets are not used as an instrument for the region to find two ways of sustainable development, the same past negative outcome may happen once again.[66] But the consequences may be even worse: without appropriate rules, a steep rise of the values may lead to a land purchase race,[67] fact that would evict or marginalize the local populations,[68] and that would increase risks to fraud for obtaining land.

All those issues cannot be dropped off the market regulation proposal, but the studies have been mostly carried out focusing the market functioning. That, over the years, can become a problem, since the external or secondary issues and the interaction of market and its surroundings is essential for its development.

The manner the legal regulation will be arranged may mean the success or the failure of this alternative and the real participation, or not, of the local people. According to what present proposals and studies establish, there are few chances for this effective participation. Most of the suggested forms are based upon existing models in other countries, what can lead to the great financial conglomerates, once more, profiting from Amazônia. That is, obviously, a failed path, designed to failure in few years. Since they do not have local connections, after taking profits on the

[65]Weinstein (1993), regarding the *Belle Époque* period, highlights that: "The fast growth of the main port of the area established a market for some few local industries; it promoted, also, the development of public works and of municipal improvements that transformed Belém in one of the most impressive capitals of state of Brazil. However, it seems to have created an interest larger in non-productive activities, such as the real estate speculation and the import of luxury goods". That concentration of activities, which made the collapse of the rubber cycle even worse, with no alternative economic sectors was due to the close monitoring of the rubber elite on the poorest population that was involved in the productive chain and that acted against any development that could transform the extractive economy (Weinstein 1993).

[66]That is the reason why we also understand that international donations for preservation funds, such as the donation of US$ 1 Billion from Norway to the Amazônia Fund—assessed by some scholars (Birdsall et al. 2014)—is not as beneficial as it sounds. This is because the donation of values from other countries is a one-off aid, which causes dependence on the outside, without generating value for the forest, and without causing the necessary cultural change on the value of living forest for the local populations. Governments may need money to keep the forests preserved. However, people living in these places need sustainable alternatives to avoid predating the forest, living with dignity and helping to preserve the environment. An Environmental Assets Market can provide this; international donations, no. Donations without other supports (scientific, medical or technological) may cause dependence of the poor countries, and it may hinder their development (Deaton 2017).

[67]There are, also, bills that seek to allow the purchase of lands for foreigners, almost without limits (Estadão Conteúdo 2017). The content of the proposals does not include that permission when the biomes demand Legal Reserve of 80%, that it is the case of Amazônia, but for being just of a project, that provision can be dropped easily.

[68]As it was seen in the areas where there was mining, for instance, or huge agricultural projects. The population was no longer the owner of the lands and experienced living on activities related to the flow of people attracted by the area. However, that flow of people brought violence, overload of the facilities and of the public services, marginalizing the local populations.

situation or, maybe, if there is not any interest, those huge companies will leave the region, creating a big social and economic empty hole.[69] Thus, what was conceived for this opportunity will not be designed and required: a model of sustainable development, with the decrease of social disparities in Amazônia and in the rest of the country, and at the same time, the forest preservation, an issue of international interest.

8 Conclusions

In this paper, we have tried to discuss several issues related to the Environmental Assets and Carbon Markets as sustainable alternatives for Amazônia (if regulated), in a multi and interdisciplinary perspective. Considering all the gathered data and data analysis, as well as the authors' professional and personal experience, we believe that the environmental issue, in Brazil, is not properly regulated and ruled. Moreover, the success of those markets establishment and their potential of social and economic transformation in the region will depend upon the way the regulamentation will be arranged. It is a valuable and unique opportunity and perhaps the last chance to preserve and value Amazônia Forest. However, if its regulamentation is not well arranged, there is a risk of concentrating, even more, the riches of the region in the hands of large corporations, making, once more, the local people helpless.

Also, we have raised the issue of the Carbon Market as a relevant alternative but not the only one available to Amazônia. We have suggested the Environmental Assets Market regulation, so that there will possibly be a coordinated, intelligent and systematic valorization towards the social, economic and environmental development of the region, free from the dependency of a single activity. That all hinges upon the participation of local people and on the manner the subject will be regulated.

References

Affonso J, Macedo F, Fabrini F, Serapião F (2016) Operação Timóteo investiga esquema de corrupção em cobrança de royalties. Retrieved from: http://bit.ly/2lCQAe9

Altafin I (2016) Participação brasileira no mercado de carbono será analisada na CMA. Retrieved from: http://bit.ly/2yGlF9L

Aristóteles (2009) Ética a Nicômaco. Tradução de Antônio de Castro Caeiro. Atlas, São Paulo

Barbosa R (1999) Oração aos Moços – Edição popular anotada por Adriano da Gama Kury. 5ª edição. Rio de Janeiro: Fundação Casa de Rui Barbosa

---

[69]Such as that seen in the second rubber cycle, in Fordlândia project (Pará), in the exploration of manganese at Serra do Navio (Amapá), amongst other examples.

Birdsall N, Savedoff W, Seymour F (2014) The Brazil-Norway agreement with performance-based payments for forest conservation: successes, challenges, and lessons. Retrieved from: http://bit.ly/2y34KKr

Brasil (1966) Lei nº 5172, de 25 de outubro de 1966. Retrieved from: http://bit.ly/2vZZveQ

Brasil (1976) Lei nº 6385, de 7 de dezembro de 1976. Retrieved from: http://bit.ly/2yGE3PY

Brasil (1988) Constituição da República Federativa do Brasil de 1988, promulgada em 05 de outubro de 1988. Retrieved from: http://www.planalto.gov.br/ccivil_03/Constituicao/Constituicao.htm

Brasil (1989) Lei Complementar nº 62, de 28 de dezembro de 1989. Retrieved from: http://bit.ly/2xk8o3h

Brasil (1989) Lei nº 7990, de 28 de dezembro de 1989. Retrieved from: http://bit.ly/2fL0awx

Brasil (1990) Lei nº 8001, de 13 de março de 1990. Retrieved from: http://bit.ly/2velUIb

Brasil (1997) Lei Complementar nº 91, de 22 de dezembro de 1997. Retrieved from: http://bit.ly/2xA8Bi0

Brasil (1997) Lei nº 9478, de 6 de agosto de 1997. Retrieved from: http://bit.ly/1tfLEPg

Brasil (2004) Projeto de Lei nº 3729/2004. Retrieved from: http://bit.ly/2iCpMwQ

Brasil (2006) Lei nº 11284, de 02 de março de 2006. Retrieved from: http://www.planalto.gov.br/ccivil_03/_ato2004-2006/2006/lei/l11284.htm

Brasil (2007) Decreto nº 6063, de 20 de março de 2007. Retrieved from: http://bit.ly/2yLjZts

Brasil (2008) Decreto nº 6527, de 1º de agosto de 2008. Retrieved from: http://bit.ly/2y2M2Hr

Brasil (2009) Lei nº 12114, de 9 de dezembro de 2009. Retrieved from: http://bit.ly/2zMO5N9

Brasil (2009) Lei nº 12187, de 29 de dezembro de 2009. Retrieved from: http://www.planalto.gov.br/ccivil_03/_ato2007-2010/2009/lei/l12187.htm

Brasil (2010) Decreto nº 7167, de 5 de maio de 2010. Retrieved from: http://bit.ly/2gHbsDh

Brasil (2010) Decreto nº 7343, de 26 de outubro de 2010. Retrieved from: http://bit.ly/2leoold

Brasil (2010) Decreto nº 7390, de 9 de dezembro de 2010. Retrieved from: http://bit.ly/2yNja39

Brasil (2010) Lei nº 12351, de 22 de dezembro de 2010. Retrieved from: http://bit.ly/1rcaWMi

Brasil (2011) Projeto de Lei do Senado n° 212, de 2011. Retrieved from: http://bit.ly/2gBnWZ7

Brasil (2012) Entenda como funciona o mercado de crédito de carbono. Retrieved from: http://bit.ly/2htE1lb

Brasil (2012) Lei nº 12651, de 25 de maio de 2012. Retrieved from: http://bit.ly/1zecCID

Brasil (2012) Projeto de Lei do Senado n° 95, de 2012. Retrieved from: http://bit.ly/2gzIFN1

Brasil (2013) Lei nº 12858, de 9 de setembro de 2013. Retrieved from: http://bit.ly/2xloGbO

Brasil (2015) Decreto nº 8576, de 26 de novembro de 2015. Retrieved from: http://bit.ly/2iz1t39

Brasil (2016) Medida Provisória nº 756, de 19 de dezembro de 2016. Retrieved from: http://bit.ly/2z68aRF

Brasil (2016) Medida Provisória nº 758, de 19 de dezembro de 2016. Retrieved from: http://bit.ly/2y3AGDa

Brasil (2017) Decreto nº 9142, de 22 de agosto de 2017. Retrieved from: http://bit.ly/2g5fkgv

Brasil (2017) Decreto nº 9147, de 28 de agosto de 2017. Retrieved from: http://bit.ly/2vHFaKr

Brasil (2017) Decreto nº 9159, de 25 de setembro de 2017. Retrieved from: http://bit.ly/2lgebor

Brito B (2017) Potential trajectories of the upcoming forest trading mechanism in Pará State, Brazilian Amazon. PLoS ONE 12(4):e0174154. https://doi.org/10.1371/journal.pone.0174154

Cardoso A (2015) Especiarias na Amazônia portuguesa: circulação vegetal e comércio atlântico no final da monarquia hispânica. Revista Tempo 21(37):116–133

CGD—Center for Global Development (2015) Look to the Forests: how performance payments can slow climate change. CGD, Washington, DC. Retrieved from: http://bit.ly/2w6pVyQ

CNT; SEST; SENAT (2016) Pesquisa CNT de Rodovias 2016: relatório gerencial. 20ª ed. Brasília: CNT.

CVM (2009) CVM comunica seu entendimento sobre créditos de carbono e produtos que deles derivam. Retrieved from: http://bit.ly/2i18xlc

Deaton A (2017) A grande saída: saúde, riqueza e a origem das desigualdades. Intrínseca, Rio de Janeiro

Estadão Conteúdo (2017) Projeto para permitir venda de terra a estrangeiro vai ao Congresso. Retrieved from: https://glo.bo/2zMHIcF

FBSP; IPEA (2016) Atlas da violência 2016; nota técnica. Brasília: Instituto de Pesquisa Econômica Aplicada – IPEA. Retrieved from: http://bit.ly/1R16zsQ

Folhes R et al (2015) Multi-scale participatory scenario methods and territorial planning in the Brazilian Amazon. Futures 73:86–99

G1 DF (2016) PF desarticula esquema de corrupção na cobrança de royalties de mineração. Retrieved from: https://glo.bo/2y3ans0

G1 PA (2017) Governo envia ao Congresso projeto de lei que reduz floresta nacional no Pará. Retrieved from: https://glo.bo/2yMS3qH

Garfield S (2009) A Amazônia no imaginário norte-americano em tempo de guerra. In: Revista Brasileira de História, v. 29, nº 57, pp. 19–65. São Paulo

Garfield S (2010) The environment of wartime migration: labor transfers from the Brazilian Northeast to the Amazon during World War II. J Soc Hist Summer 2010 (George Mason University Press, Fairfax)

Gesisky J (2017) Meio Ambiente perde metade dos recursos para 2017. Retrieved from: http://bit.ly/2zA0wuP

Gomes M (2017) China reafirma compromisso com o Acordo de Paris. Retrieved from: http://bit.ly/2ldCWlk

IBGE—Instituto Brasileiro de Geografia e Estatística (2016) Área Territorial Brasileira. Retrieved from: http://bit.ly/2xxyleQ

Kafruni S (2014) Natura inaugura complexo industrial na Amazônia e gera 500 empregos. Retrieved from: http://bit.ly/2h4B49Y

Kageyama PY (2009) Biodiversidade e Biopirataria: contradição entre a biodiversidade e a pobreza no mundo. In: Aleixo A, Azevedo-Ramos C, Camargo E, Kageyama PY, Maio MC, Nascimento DM, Oliveira NS (eds) Amazônia e Desenvolvimento Sustentável. Fundação Konrad Adenauer, Rio de Janeiro

Klautau de Araújo JM (1995) Caligrafias de Belém – vol I: a dimensão insular. Imprensa Oficial do Estado do Pará, Belém

Klautau de Araújo JM, Lima DMB (1997a) Projeto Escola Bosque do Amapá – Centro de Referência em Educação Ambiental da Ilha de Santana. Governo do Estado do Amapá, Macapá

Klautau de Araújo JM, Lima DMB (1997b) Projeto Escola Bosque do Amapá – O método sócio-ambiental. Macapá: Governo do Estado do Amapá

Klautau de Araújo JM, Lima DMB (1997c) Projeto Escola Bosque do Amapá – Projeto de Socialização. Macapá: Governo do Estado do Amapá

Klautau de Araújo TL (2014) Environmental law, public policies, and climate change: a social-legal analysis in the Brazilian context. In: Leal Filho W (eds) Handbook of climate change adaptation. Springer, Berlin, pp 973–982. https://doi.org/10.1007/978-3-642-40455-9_115-1; ISBN: 978-3-642-40455-9

Klautau de Araújo TL (2016) Public policies and education for biodiversity: Brazilian challenges in a new global context. In: Castro P, Azeiteiro UM, Bacelar Nicolau P, Leal Filho W, Azul AM (eds) Biodiversity and education for sustainable development. Springer, Berlin, pp 219–235

Leite M (2015) Dilma Corta 72% da verba contra desmatamento na Amazônia. Retrieved from: http://bit.ly/2yOPbd9

Lima DMB (2013) The Escola Bosque project: building ways to an ecological society. In Proceedings 7th world environmental Education Congress, Marrakech, Morocco, June 9–14, 2013

Lopes L et al (2015) Estudos sobre Mercado de Carbono no Mercado de Carbono no Brasil: Análise Legal de Possíveis Modelos Regulatórios. Banco Interamericano de Desenvolvimento, Monografia No. 307. BID, Washington, DC

Lusa (2017) Juncker alerta Trump: "Os norte-americanos não podem sair sem mais nem menos do acordo de Paris". Retrieved from: http://bit.ly/2yOzfrB

Maisonnave F (2017a) Câmara aprova reduzir proteção de áreas de conservação no PA e em SC. Retrieved from: http://bit.ly/2zzHUen
Maisonnave F (2017b) Senado ratifica redução na proteção de áreas de conservação na Amazônia. Retrieved from: http://bit.ly/2xkIOea
Mariani D, Demasi B, Almeida R (2016) Três disputas de território entre os Estados brasileiros. Retrieved from: http://bit.ly/2zzX3wi
Meirelles Filho J (2015) Quem sabe do lugar é quem vive nele, *in* Revista Página 22, nº 98, set/out 2015. São Paulo: Fundação Getúlio Vargas
Milaré É (2014) Direito Ambiental, 12th edn. Revista dos Tribunais, São Paulo
Ministério da Fazenda—Brasil (2017) Despesas Contingenciáveis na LOA 2017. Retrieved from: http://bit.ly/2zM5cyy
Ministério de Minas e Energia – Brasil (2017) Governo revoga decreto que extingue a Renca. Retrieved from: http://bit.ly/2h7dE3O
Ministério do Meio Ambiente – Brasil (2016) ENREDD+ - Estratégia Nacional para Redução das Emissões Provenientes do Desmatamento e da Degradação Florestal, Conservação dos Estoques de Carbono Florestal, Manejo Sustentável de Florestas e Aumento de Estoques de Carbono Florestal. Retrieved from: http://bit.ly/2yMm607
Miranda G (2017) Projeto de lei quer afrouxar licenciamento ambiental no Brasil. Retrieved from: http://bit.ly/2itPiVj
Moutinho P, Guerra R (2017) O vexame de cortar pela metade a ínfima verba para o Meio Ambiente. Retrieved from: http://bit.ly/2y2grWz
Paulino E (2014) The agricultural, environmental and socio-political repercussions of Brazil's land governance system. Land Use Policy 36:134–144
Peduzzi P (2017) Governo anuncia R$ 190,25 bilhões para Plano Agrícola e Pecuário 2017/2018. Retrieved from: http://bit.ly/2sg2l00
Pinheiro Pedro A et al (2015) Organização do Mercado Local de Carbono: Sistema Brasileiro de Controle de Carbono e Instrumentos Financeiros relacionados. Retrieved from: http://bit.ly/2yK49ix
Pinho P et al (2014) Ecosystem protection and poverty alleviation in the tropics: perspective from a historical evolution of policy-making in the Brazilian Amazon. Ecosyst Serv 8:97–109
Rawls J (2008) Uma teoria da Justiça, 3ª edn. Martins Editora, São Paulo
Rousseau J-J (2008) Discurso Sobre a Origem e os Fundamentos da Desigualdade Entre os Homens. LP&M Pocket, São Paulo
Samuelson P (1976) Economics of forestry in an evolving society. Econ Inq XIV
Schreiber M (2017) Reação ao fim da Renca foi 'histeria', 'infantilidade' e 'desinformação', dizem geólogos. Retrieved from: http://bbc.in/2yJPI0B
Seymour F, Busch J (2016) Why Forests? Why Now? The science, economics, and politics of tropical forests and climate change. CGD, Washington, DC
Soares G (2016) Proteção dos conhecimentos tradicionais e repartição de benefícios: uma reflexão sobre o caso da empresa Natura do Brasil e dos erveiros e erveiras do mercado Ver-o-Peso. Retrieved from: http://bit.ly/2z4OIo5
Weinstein B (1993) A borracha na Amazônia: expansão e decadência (1850-1920). Editora da Universidade de São Paulo, São Paulo
Weis B (2006) Polêmica entre Natura e Ver-o-peso expõe dilemas na proteção de conhecimentos tradicionais no Brasil. Retrieved from: http://bit.ly/2yKb4bE
Wolford W (2016) The casa and the causa: institutional histories and cultural politics in Brazilian Land Reform. In: Latin American research review, vol 51, No 4. Latin American Studies Association, Austin

Printed in the United States
By Bookmasters